New Spin on Metal-Insulator Transitions

New Spin on Metal-Insulator Transitions

Editor

Andrej Pustogow

MDPI • Basel • Beijing • Wuhan • Barcelona • Belgrade • Manchester • Tokyo • Cluj • Tianjin

Editor
Andrej Pustogow
Institute of Solid State Physics,
TU Wien
Vienna, Austria

Editorial Office
MDPI
St. Alban-Anlage 66
4052 Basel, Switzerland

This is a reprint of articles from the Special Issue published online in the open access journal *Crystals* (ISSN 2073-4352) (available at: https://www.mdpi.com/journal/crystals/special_issues/MITs).

For citation purposes, cite each article independently as indicated on the article page online and as indicated below:

LastName, A.A.; LastName, B.B.; LastName, C.C. Article Title. *Journal Name* **Year**, *Volume Number*, Page Range.

ISBN 978-3-0365-7058-7 (Hbk)
ISBN 978-3-0365-7059-4 (PDF)

Cover image courtesy of Andrej Pustogow

© 2023 by the authors. Articles in this book are Open Access and distributed under the Creative Commons Attribution (CC BY) license, which allows users to download, copy and build upon published articles, as long as the author and publisher are properly credited, which ensures maximum dissemination and a wider impact of our publications.

The book as a whole is distributed by MDPI under the terms and conditions of the Creative Commons license CC BY-NC-ND.

Contents

About the Editor vii

Preface to "New Spin on Metal-Insulator Transitions" ix

Andrej Pustogow
New Spin on Metal-Insulator Transitions
Reprinted from: *Crystals* **2023**, *13*, 64, doi:10.3390/cryst13010064 1

Olga Iakutkina, Roland Rösslhuber, Atsushi Kawamoto and Martin Dressel
Dielectric Anomaly and Charge Fluctuations in the Non-Magnetic Dimer Mott Insulator λ-(BEDT-STF)$_2$GaCl$_4$
Reprinted from: *Crystals* **2021**, *11*, 1031, doi:10.3390/cryst11091031 3

Ka-Ming Tam, Hanna Terletska, Tom Berlijn, Liviu Chioncel and Juana Moreno
Real Space Quantum Cluster Formulation for the Typical Medium Theory of Anderson Localization
Reprinted from: *Crystals* **2021**, *11*, 1282, doi:10.3390/cryst11111282 17

Shusaku Imajo and Koichi Kindo
The FFLO State in the Dimer Mott Organic Superconductor κ-(BEDT-TTF)$_2$Cu[N(CN)$_2$]Br
Reprinted from: *Crystals* **2021**, *11*, 1358, doi:10.3390/cryst11111358 33

Shiori Sugiura, Hiroki Akutsu, Yasuhiro Nakazawa, Taichi Terashima, Syuma Yasuzuka, John A. Schlueter and Shinya Uji
Fermi Surface Structure and Isotropic Stability of Fulde-Ferrell-Larkin-Ovchinnikov Phase in Layered Organic Superconductor β''-(BEDT-TTF)$_2$SF$_5$CH$_2$CF$_2$SO$_3$
Reprinted from: *Crystals* **2021**, *11*, 1525, doi:10.3390/cryst11121525 45

Andrej Pustogow, Daniel Dizdarevic, Sebastian Erfort, Olga Iakutkina, Valentino Merkl, Gabriele Untereiner and Martin Dressel
Tuning Charge Order in (TMTTF)$_2$X by Partial Anion Substitution
Reprinted from: *Crystals* **2021**, *11*, 1545, doi:10.3390/cryst11121545 55

Yuki Matsumura, Shusaku Imajo, Satoshi Yamashita, Hiroki Akutsu and Yasuhiro Nakazawa
Electronic Heat Capacity and Lattice Softening of Partially Deuterated Compounds of κ-(BEDT-TTF)$_2$Cu[N(CN)$_2$]Br
Reprinted from: *Crystals* **2022**, *12*, 2, doi:10.3390/cryst12010002 65

Hiroshi Ito, Motoki Matsuno, Seiu Katagiri, Shinji K. Yoshina, Taishi Takenobu, Manabu Ishikawa, Akihiro Otsuka, et al.
Metallic Conduction and Carrier Localization in Two-Dimensional BEDO-TTF Charge-Transfer Solid Crystals
Reprinted from: *Crystals* **2022**, *12*, 23, doi:10.3390/cryst12010023 77

Yoshitaka Kawasugi and Hiroshi M. Yamamoto
Simultaneous Control of Bandfilling and Bandwidth in Electric Double-Layer Transistor Based on Organic Mott Insulator κ-(BEDT-TTF)$_2$Cu[N(CN)$_2$]Cl
Reprinted from: *Crystals* **2022**, *12*, 42, doi:10.3390/cryst12010042 85

Reizo Kato, Masashi Uebe, Shigeki Fujiyama and Hengbo Cui
A Discrepancy in Thermal Conductivity Measurement Data of Quantum Spin Liquid β'-EtMe$_3$Sb[Pd(dmit)$_2$]$_2$ (dmit = 1,3-Dithiol-2-thione-4,5-dithiolate)
Reprinted from: *Crystals* **2022**, *12*, 102, doi:10.3390/cryst12010102 101

Mathieu Taupin and Silke Paschen
Are Heavy Fermion Strange Metals Planckian?
Reprinted from: *Crystals* **2022**, *12*, 251, doi:10.3390/cryst12020251 . . . . **109**

Liang Si, Paul Worm and Karsten Held
Fingerprints of Topotactic Hydrogen in Nickelate Superconductors
Reprinted from: *Crystals* **2022**, *12*, 656, doi:10.3390/cryst12050656 . . . . **127**

Yoshihiko Ihara and Shusaku Imajo
Superconductivity and Charge Ordering in BEDT-TTF Based Organic Conductors with β''-Type Molecular Arrangement
Reprinted from: *Crystals* **2022**, *12*, 711, doi:10.3390/cryst12050711 . . . . **141**

Nikolina Novosel, David Rivas Góngora, Zvonko Jagličić, Emil Tafra, Mario Basletić, Amir Hamzić, Teodoro Klaser, et al.
Grain-Size-Induced Collapse of Variable Range Hopping and Promotion of Ferromagnetism in Manganite $La_{0.5}Ca_{0.5}MnO_3$
Reprinted from: *Crystals* **2022**, *12*, 724, doi:10.3390/cryst12050724 . . . . **157**

Kenichiro Hashimoto, Ryota Kobayashi, Satoshi Ohkura, Satoru Sasaki, Naoki Yoneyama, Masayuki Suda, Hiroshi M. Yamamoto, et al.
Optical Conductivity Spectra of Charge-Crystal and Charge-Glass States in a Series of θ-Type BEDT-TTF Compounds
Reprinted from: *Crystals* **2022**, *12*, 831, doi:10.3390/cryst12060831 . . . . **179**

Owen Ganter, Kevin Feeny, Morgan Brooke-DeBock, Stephen M. Winter and Charles C Agosta
A Database for Crystalline Organic Conductors and Superconductors
Reprinted from: *Crystals* **2022**, *12*, 919, doi:10.3390/cryst12070919 . . . . **189**

Yuting Tan, Vladimir Dobrosavljević and Louk Rademaker
How to Recognize the Universal Aspects of Mott Criticality?
Reprinted from: *Crystals* **2022**, *12*, 932, doi:10.3390/cryst12070932 . . . . **201**

Nicholas Walker, Samuel Kellar, Yi Zhang, Ka-Ming Tam and Juana Moreno
Neural Network Solver for Small Quantum Clusters
Reprinted from: *Crystals* **2022**, *12*, 1269, doi:10.3390/cryst12091269 . . . . **225**

Kira Riedl, Elena Gati and Roser Valentí
Ingredients for Generalized Models of κ-Phase Organic Charge-Transfer Salts: A Review
Reprinted from: *Crystals* **2022**, *12*, 1689, doi:10.3390/cryst12121689 . . . . **239**

About the Editor

Andrej Pustogow

Prof. Andrej Pustogow obtained his PhD in 2017 at Universität Stuttgart, where he studied organic charge-transfer salts and some inorganic correlated electron systems using optical spectroscopy, transport experiments, and other methods. Following that, he performed a Postdoc at UCLA where he tackled longstanding issues of unconventional superconductivity and frustrated magnetism using nuclear magnetic resonance. In 2020, he received a tenure track position at TU Wien, where he is currently building up his research group at the Institute of Solid State Physics. Pustogow Spectroscopy Laboratory investigates metal–insulator transitions, unconventional superconductivity, quantum spin liquids, and other interesting phenomena in correlated electron systems throughout the electromagnetic spectrum.

Preface to "New Spin on Metal-Insulator Transitions"

Electrons in solids behave differently compared to free electrons in vacuum, which is incorporated in the effective mass m* of electronic quasiparticles. In extreme cases, their mutual interactions can even result in qualitatively different behavior such as the localization of conduction electrons upon a metal–insulator transition (MIT). Near such electronic instabilities, unconventional superconductivity, non-Fermi-liquid transport, exotic magnetism, and other interesting electronic phases are stabilized in correlated electron systems. This Special Issue provides a view into the ongoing research endeavors investigating emergent phenomena around MITs.

Andrej Pustogow
Editor

Editorial

New Spin on Metal-Insulator Transitions

Andrej Pustogow

Institute of Solid State Physics, TU Wien, 1040 Vienna, Austria; pustogow@ifp.tuwien.ac.at

Metal-insulator transitions (MITs) constitute a core subject of fundamental condensed-matter research. The localization of conduction electrons has been observed in a large variety of materials and gives rise to intriguing quantum phenomena such as unconventional superconductivity and exotic magnetism. Nearby an MIT, minuscule changes of interaction strength via chemical substitution, doping, physical pressure or even disorder can trigger spectacular resistivity changes from zero in a superconductor to infinity in an insulator near T = 0. While approaching an insulating state from the conducting side, deviations from Fermi-liquid transport in bad and strange metals are the rule rather than the exception, discussed in terms of spatial inhomogeneity and quantum criticality. Moreover, charge localization upon MITs has a crucial impact on the magnetic degrees of freedom that are studied for the possible realization of a quantum spin liquid.

Solving the challenges of correlated electron systems and the emergent phenomena around MITs has attracted much interest. As the drosophila of electron–electron interactions, the Mott MIT receives particular attention as it can be studied using the Hubbard model. On the experimental side, the topic has been recently promoted by the advent of twisted Moiré bilayer systems; however, true bulk materials, such as organic charge-transfer salts, fullerides and transition-metal oxides, remain indispensable for elucidating macroscopic quantum phases such as unconventional superconductivity and frustrated magnetism. Various novel methods have become available lately to tune and map the complex evolution of the metallic and insulating phases at cryogenic temperatures, including uniaxial strain and imaging techniques such as near-field microscopy. The controlled variation of disorder has also been utilized to study Griffiths phases and Anderson-type MITs.

Investigating MITs requires minute control of the relevant tuning parameters, such as the electronic bandwidth and band filling. While doping is the preferential tool in oxides, such as superconducting nickelates that are impacted by topotactic hydrogen [1], pressure tuning is the method of choice for organic charge-transfer salts. Kawasugi et al. achieved simultaneous control of band filling and bandwidth via in situ strain and gate tuning on κ-(BEDT-TTF)$_2$X crystals [2]. Another powerful tuning method is partial chemical substitution, as applied in κ-type systems [3] and quasi one-dimensional (TMTTF)$_2$X [4]. In the latter case, the Fabre salts with quarter-filled bands exhibit textbook-like charge-ordered states driven by inter-site Coulomb interactions, which also give rise to unconventional superconductivity in β''-(BEDT-TTF)$_2$X, as reviewed by Ihara and Imajo [5]. Such layered organic superconductors are well suited to study the Fulde–Ferrell–Larkin–Ovchinnikov (FFLO) state [6,7]. On the theoretical side, Riedl and coworkers review generalized models of κ-systems [8], while Tan et al. assessed universal aspects of Mott criticality [9]. Disorder and Anderson localization are studied using cluster methods [10,11] and experiments on BEDO-TTF crystals [12]. A useful tool to systematically investigate and compare all these phenomena is the database for crystalline organic conductors and superconductors provided by Ganter et al. [13]. The interplay of charge and spin degrees of freedom, prominently seen in manganites [14], is considered important in dimerized organic Mott systems, where dielectric anomalies have been controversially discussed [15]. In geometrically frustrated systems, the charge order can transform into a charge glass [16] and magnetic order can transform into a quantum spin liquid. Regarding the latter scenario, R. Kato et al. discuss discrepancies of thermal transport measurements on β'-EtMe$_3$Sb[Pd(dmit)$_2$]$_2$ [17].

Citation: Pustogow, A. New Spin on Metal-Insulator Transitions. *Crystals* 2023, 13, 64. https://doi.org/10.3390/cryst13010064

Received: 23 December 2022
Accepted: 26 December 2022
Published: 30 December 2022

Copyright: © 2022 by the authors. Licensee MDPI, Basel, Switzerland. This article is an open access article distributed under the terms and conditions of the Creative Commons Attribution (CC BY) license (https://creativecommons.org/licenses/by/4.0/).

Taupin and Paschen inspect the controversies of condensed-matter research, namely the topic of whether heavy Fermion systems exhibit 'Planckian dissipation' [18].

This Special Issue provides a glimpse into the latest progress in answering the existing questions around MITs, including various topics of solid-state physics.

Funding: This research received no external funding.

Data Availability Statement: No data were published in this Editorial.

Conflicts of Interest: The author declares no conflict of interest.

References

1. Si, L.; Worm, P.; Held, K. Fingerprints of Topotactic Hydrogen in Nickelate Superconductors. *Crystals* **2022**, *12*, 656. [CrossRef]
2. Kawasugi, Y.; Yamamoto, H.M. Simultaneous Control of Bandfilling and Bandwidth in Electric Double-Layer Transistor Based on Organic Mott Insulator κ-(BEDT-TTF)$_2$Cu[N(CN)$_2$]Cl. *Crystals* **2022**, *12*, 42. [CrossRef]
3. Matsumura, Y.; Imajo, S.; Yamashita, S.; Akutsu, H.; Nakazawa, Y. Electronic Heat Capacity and Lattice Softening of Partially Deuterated Compounds of κ-(BEDT-TTF)$_2$Cu[N(CN)$_2$]Br. *Crystals* **2022**, *12*, 2. [CrossRef]
4. Pustogow, A.; Dizdarevic, D.; Erfort, S.; Iakutkina, O.; Merkl, V.; Untereiner, G.; Dressel, M. Tuning Charge Order in (TMTTF)2X by Partial Anion Substitution. *Crystals* **2021**, *11*, 1545. [CrossRef]
5. Ihara, Y.; Imajo, S. Superconductivity and Charge Ordering in BEDT-TTF Based Organic Conductors with β''-Type Molecular Arrangement. *Crystals* **2022**, *12*, 711. [CrossRef]
6. Sugiura, S.; Akutsu, H.; Nakazawa, Y.; Terashima, T.; Yasuzuka, S.; Schlueter, J.A.; Uji, S. Fermi Surface Structure and Isotropic Stability of Fulde-Ferrell-Larkin-Ovchinnikov Phase in Layered Organic Superconductor β''-(BEDT-TTF)$_2$SF$_5$CH$_2$CF$_2$SO$_3$. *Crystals* **2021**, *11*, 1525. [CrossRef]
7. Imajo, S.; Kindo, K. The FFLO State in the Dimer Mott Organic Superconductor κ-(BEDT-TTF)$_2$Cu[N(CN)$_2$]Br. *Crystals* **2021**, *11*, 1358. [CrossRef]
8. Riedl, K.; Gati, E.; Valentí, R. Ingredients for Generalized Models of κ-Phase Organic Charge-Transfer Salts: A Review. *Crystals* **2022**, *12*, 1689. [CrossRef]
9. Tan, Y.; Dobrosavljević, V.; Rademaker, L. How to Recognize the Universal Aspects of Mott Criticality? *Crystals* **2022**, *12*, 932. [CrossRef]
10. Walker, N.; Kellar, S.; Zhang, Y.; Tam, K.M.; Moreno, J. Neural Network Solver for Small Quantum Clusters. *Crystals* **2022**, *12*, 1269. [CrossRef]
11. Tam, K.M.; Terletska, H.; Berlijn, T.; Chioncel, L.; Moreno, J. Real Space Quantum Cluster Formulation for the Typical Medium Theory of Anderson Localization. *Crystals* **2021**, *11*, 1282. [CrossRef]
12. Ito, H.; Matsuno, M.; Katagiri, S.; Yoshina, S.K.; Takenobu, T.; Ishikawa, M.; Otsuka, A.; Yamochi, H.; Yoshida, Y.; Saito, G.; et al. Metallic Conduction and Carrier Localization in Two-Dimensional BEDO-TTF Charge-Transfer Solid Crystals. *Crystals* **2022**, *12*, 22. [CrossRef]
13. Ganter, O.; Feeny, K.; Brooke-deBock, M.; Winter, S.M.; Agosta, C.C. A Database for Crystalline Organic Conductors and Superconductors. *Crystals* **2022**, *12*, 919. [CrossRef]
14. Novosel, N.; Rivas Góngora, D.; Jagličić, Z.; Tafra, E.; Basletić, M.; Hamzić, A.; Klaser, T.; Skoko, Ž.; Salamon, K.; Kavre Piltaver, I.; et al. Grain-Size-Induced Collapse of Variable Range Hopping and Promotion of Ferromagnetism in Manganite La$_{0.5}$Ca$_{0.5}$MnO$_3$. *Crystals* **2022**, *12*, 724. [CrossRef]
15. Iakutkina, O.; Rosslhuber, R.; Kawamoto, A.; Dressel, M. Dielectric Anomaly and Charge Fluctuations in the Non-Magnetic Dimer Mott Insulator λ-(BEDT-STF)$_2$GaCl$_4$. *Crystals* **2021**, *11*, 1031. [CrossRef]
16. Hashimoto, K.; Kobayashi, R.; Ohkura, S.; Sasaki, S.; Yoneyama, N.; Suda, M.; Yamamoto, H.M.; Sasaki, T. Optical Conductivity Spectra of Charge-Crystal and Charge-Glass States in a Series of θ-Type BEDT-TTF Compounds. *Crystals* **2022**, *12*, 831. [CrossRef]
17. Kato, R.; Uebe, M.; Fujiyama, S.; Cui, H. A Discrepancy in Thermal Conductivity Measurement Data of Quantum Spin Liquid β'-EtMe$_3$Sb[Pd(dmit)$_2$]$_2$ (dmit = 1,3-Dithiol-2-thione-4,5-dithiolate). *Crystals* **2022**, *12*, 102. [CrossRef]
18. Taupin, M.; Paschen, S. Are Heavy Fermion Strange Metals Planckian? *Crystals* **2022**, *12*, 251. [CrossRef]

Disclaimer/Publisher's Note: The statements, opinions and data contained in all publications are solely those of the individual author(s) and contributor(s) and not of MDPI and/or the editor(s). MDPI and/or the editor(s) disclaim responsibility for any injury to people or property resulting from any ideas, methods, instructions or products referred to in the content.

Article

Dielectric Anomaly and Charge Fluctuations in the Non-Magnetic Dimer Mott Insulator λ-(BEDT-STF)$_2$GaCl$_4$

Olga Iakutkina [1,*], Roland Rösslhuber [1], Atsushi Kawamoto [2] and Martin Dressel [1]

[1] 1. Physikalisches Institute, Universität Stuttgart, 70569 Stuttgart, Germany;
roland.roesslhuber@pi1.physik.uni-stuttgart.de (R.R.); dressel@pi1.physik.uni-stuttgart.de (M.D.)

[2] Department of Physics, Graduate School of Science, Hokkaido University, Sapporo 060-0810, Japan;
atkawa@phys.sci.hokudai.ac.jp

[*] Correspondence: olga.iakutkina@pi1.uni-stuttgart.de

Abstract: The dimer Mott insulator λ-(BEDT-STF)$_2$GaCl$_4$ undergoes no magnetic order down to the lowest temperatures, suggesting the formation of a novel quantum disordered state. Our frequency and temperature-dependent investigations of the dielectric response reveal a relaxor-like behavior below $T \approx 100$ K for all three axes, similar to other spin liquid candidates. Optical measurement of the charge-sensitive vibrational mode $\nu_{27}(b_{1u})$ identifies a charge disproportionation $\Delta \rho \approx 0.04e$ on the dimer that exists up to room temperature and originates from inequivalent molecules in the weakly coupled dimers. The linewidth of the charge sensitive mode is broader than that of typical organic conductors, supporting the existence of a disordered electronic state.

Keywords: strongly correlated systems; organic conductors; relaxor-ferroelectrics; dielectric spectroscopy; infrared spectroscopy; disordered systems

1. Introduction

Strongly correlated electron systems attract much interest due to some novel magnetic, dielectric, and superconducting properties; both electron–electron interactions and quantum fluctuations are considered crucial for understanding these phenomena. Among these systems, the quasi-two-dimensional organic charge-transfer salts are renowned for their versatility and enormously rich phase diagrams, comprising superconducting, spin-liquid, antiferromagnetic, or charge-ordered phases that arise from the interplay of spin, charge, and lattice degrees of freedom [1–5]. Besides the most popular examples κ-(BEDT-TTF)$_2X$, another dimerized family has drawn large attention, the λ-salts, where the lattice system consists of triangular and square tiling as depicted in Figure 1b. In addition to the well-studied unconventional superconducting properties, such as a Fulde–Ferrell–Larkin–Ovchinnikov state and field-induced superconductivity at strong magnetic fields [6–9], a spin-liquid-like state was discovered recently [10,11].

The electronic phase of the insulating λ-(BEDT-STF)$_2$GaCl$_4$ is situated between the antiferromagnet λ-(BEDT-TTF)$_2$GaCl$_4$ and the superconductor λ-(BETS)$_2$GaCl$_4$ (cf. Figure 1a for the molecular structure). While the most insulating compound λ-(BEDT-TTF)$_2$GaCl$_4$ undergoes an antiferromagnetic transition at $T_N = 13$ K, no magnetic order occurs in λ-(BEDT-STF)$_2$GaCl$_4$ down to 1.63 K regardless of the strong coupling $J = 194$ K [12]. The temperature dependence of the magnetic susceptibility can be described by a $S = \frac{1}{2}$ two-dimensional antiferromagnetic Heisenberg model on the triangular lattice, suggesting geometrically frustrated spin-liquid-like behavior. However, a nuclear magnetic resonance (NMR) study found an inhomogeneous electronic state; after an increase of $1/T_1$, the NMR relaxation rate saturates at a low temperature, which is in stark contrast to the magnetic properties of other spin liquid candidates. Hence, λ-(BEDT-STF)$_2$GaCl$_4$ is considered a realization of a novel quantum disordered state [11].

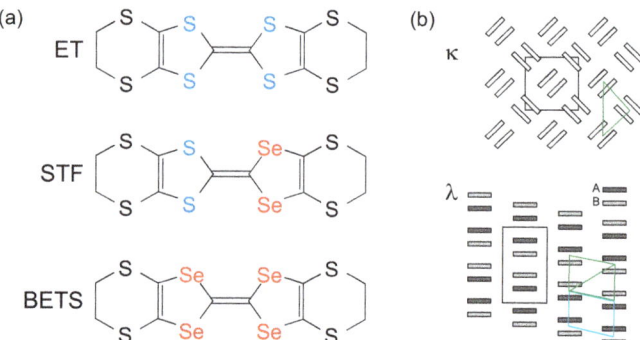

Figure 1. (**a**) The donor molecule ET = BEDT-TTF, i.e., bis-(ethylenedithio)-tetrathiafulvalene), is the most common building block, but sulfur can be replaced by selenium, leading to STF = BEDT-STF, i.e., bis-(ethylenedithio)-diseleniumdithiafulvalene, and BETS = BEDT-TSF, i.e., bis-(ethylenedithio)-tetraselenafulvalene. The more extended orbitals cause a larger bandwidth favoring better conductivity. (**b**) The two-dimensional charge transfer salts form different dimer patterns, where two crystallographically independent donors crystallize face-to-face. In the κ-phase the dimers are rotated with respect to each other, while the λ-pattern is organized in stacks with two dimers per unit cell; here, the constituent molecules A and B differ by symmetry. The unit cell given by black contains four molecules (A, B) and (B′, A′). The triangular arrangement of the dimers is indicated in green, where—depending on the particular transfer integrals—a high degree of frustration can be reached. Due to the weaker diagonal interaction, a square tiling occurs, shown by blue lines.

Having in mind that numerous organic conductors including antiferromagnets and spin-liquid candidates exhibit dielectric anomalies [13–24], with the charge-order driven ferroelectric state detected for some of them [25–27], and the growing numbers of possible applications of ferroelectric materials [28–32], understanding the mechanism of dielectric anomaly and investigating the charge dynamics of the electronic state in organic conductors are of great interest. In this study, we focused on the disordered quantum state of λ-type salts.

To this end, we employ dielectric and vibration spectroscopies to explore the charge state and the presence of the dielectric anomaly in λ-(BEDT-STF)$_2$GaCl$_4$; in addition, the compound is investigated by infrared spectroscopy as the standard and very powerful tool to elucidate the charge distribution on the molecules [33–38].

2. Experimental Details

Single crystals of λ-(BEDT-STF)$_2$GaCl$_4$ were synthesized at Hokkaido University by the standard electrochemical oxidation method [39]. In contrast to BEDT-TTF molecules, in BEDT-STF, two central sulfur atoms are substituted by Se atoms, leading to asymmetric BEDT-STF molecules as sketched in Figure 1a. The crystals have a needle-like shape parallel to the c-axis and typical dimensions of 1 mm $\times$ 0.2 mm $\times$ 0.05 mm. The donor molecules are dimerized with the pairs arranged in the ac-plane. As shown in Figure 2, the conducting layers of donor molecules alternate with insulating anion sheets along the b-axis, giving rise to a quasi-two-dimensional structure. The morphology of the crystals corresponds to the ($1\bar{1}0$), (110), and (001)-planes. For clarity ($1\bar{1}0$) and (110) planes are depicted in Figure 2c, while the (001)-plane coincides with the c-axis.

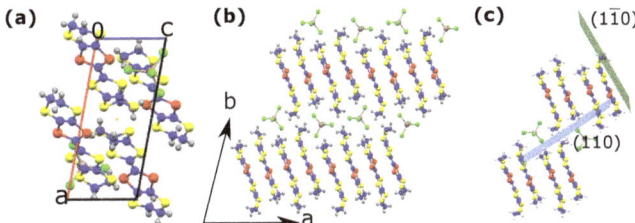

Figure 2. Crystal structure of λ-(BEDT-STF)$_2$GaCl$_4$. (**a**) The λ-type arrangement of donor molecules within the highly conducting ac-plane. (**b**) The layered structure becomes obvious when looking along the c-direction, where the alternation of donor and anion layers are seen. (**c**) Green and blue planes correspond to (110) and ($1\bar{1}0$), respectively.

Dielectric measurements between $T = 295$ and 7 K are carried out using an Agilent A4294 impedance analyzer that covers the frequency range 100 Hz–10 MHz. The spectra of the complex dielectric permittivity

$$\hat{\epsilon}(\omega) = \frac{\hat{\sigma}(\omega) - \sigma_0}{i\epsilon_0 \omega} = \epsilon'(\omega) + i\epsilon''(\omega) \qquad (1)$$

are obtained along all three directions, [110], [$1\bar{1}0$], and [001], using the two-contact method, where gold wires are attached on both sides of the crystal with carbon paste; the other ends of the wires are connected to the sample holder with silver paint. Here, σ_0 is the conductivity caused by free electrons in the material, and ϵ_0 is the vacuum permittivity; ϵ' and ϵ'' are the real and imaginary parts of the permittivity. To have reliable data, the sample holder open-loop contribution is subtracted [40]. Using a continuous helium-flow cryostat (KONTI by CryoVac, Troisdorf, Germany), the samples can be cooled down from room temperature to $T = 7$ K.

Optical reflectivity measurements off the (110)-plane of λ-(BEDT-STF)$_2$GaCl$_4$ single crystals are carried out with a Bruker Hyperion infrared microscope attached to a Bruker Vertex 80v Fourier-transform infrared spectrometer. The experiments are performed with the light polarized parallel to [$1\bar{1}0$], i.e., in the direction most sensitive to the charge-sensitive infrared-active intramolecular vibrational mode $\nu_{27}(b_{1u})$ [35,41,42]. The spectra are recorded in a frequency range from 500 to 8000 cm^{-1} between $T = 295$ and 12 K. The optical conductivity is calculated via the Kramers–Kronig transformation with constant extrapolation of reflectivity below 500 cm^{-1}, which is common for insulators, and using standard ω^{-4} decay as high-frequency extrapolation.

3. Results and Analysis

3.1. Dielectric Properties

Figure 3 displays the real part of the dielectric permittivity $\epsilon'(T)$ as a function of temperature measured at various frequencies $f = \omega/2\pi$ along the three directions [001], [110], and [$1\bar{1}0$] of a λ-(BEDT-STF)$_2$GaCl$_4$ crystal.

For the orientations $E \parallel$ [001] and $E \parallel$ [$1\bar{1}0$], broad maxima develop below $T = 100$ K, which are strongly frequency dependent. With decreasing frequency, the peak shifts toward lower temperatures and becomes sharper. For $E \parallel$ [110], a clear step can be seen in the real part of the dielectric constant around $T = 60$ K that shifts toward higher temperatures with increasing frequency. This behavior is typical for relaxor ferroelectrics, and similar dielectric anomalies are frequently observed in organic conductors [19–21,43,44].

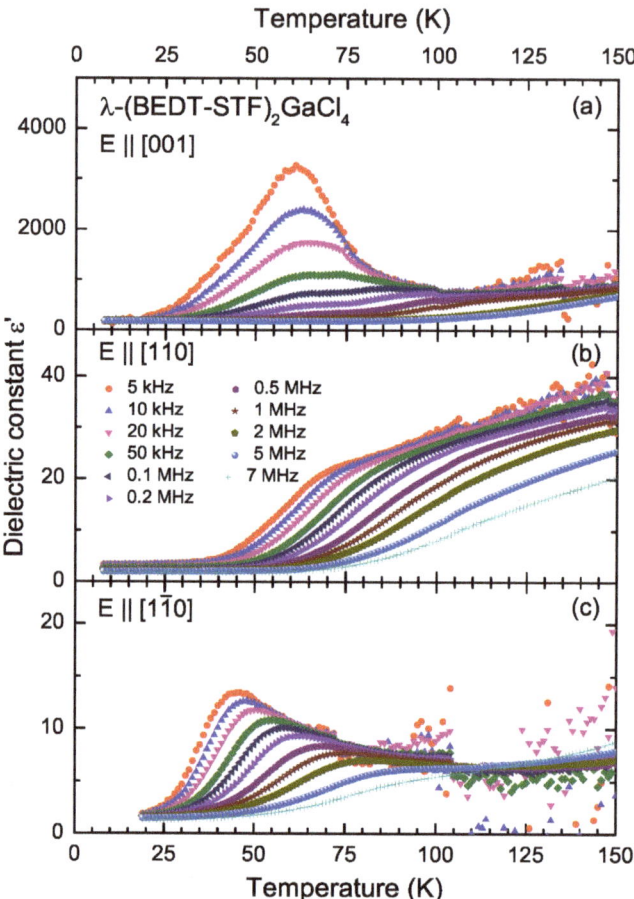

Figure 3. Temperature-dependent real part of the permittivity of λ-(BEDT-STF)$_2$GaCl$_4$ recorded at different frequencies for the electric field along the three different directions, i.e., (**a**) $E \parallel [001]$, (**b**) $E \parallel [110]$, and (**c**) $E \parallel [1\bar{1}0]$.

Dielectric relaxation appears in a rather broad temperature range; for intermediate temperatures, the frequency dependence of the real and imaginary parts of the dielectric permittivity is plotted in Figure 4. The overall behavior can be described by the generalized Debye model [45,46]:

$$\hat{\epsilon}(\omega) - \epsilon_\infty = \frac{\Delta\epsilon}{1 + (i\omega\tau_0)^{1-\alpha}}, \tag{2}$$

where $\Delta\epsilon = \epsilon'(\omega \to 0) - \epsilon'(\omega \to \infty)$ is a measure of the dielectric strength, and $\epsilon'(\omega \to 0)$ and $\epsilon'(\omega \to \infty)$ are the limiting static and the high-frequency values of the dielectric constant, respectively. τ_0 is the mean relaxation time, and $(1 - \alpha)$ is the symmetric broadening. The drop in $\epsilon'(\omega)$ with increasing frequency implies that the dipoles cannot follow the ac electric field at high frequencies [47]. Since $\epsilon'(\omega)$ and $\epsilon''(\omega)$ are linked via the Kramers–Kronig relation, the step in the real part results in a peak in the absorption ϵ''. The solid black lines in Figure 4 represent the fit of the data according to Equation (2), and the dashed green line is the dc contribution σ_0.

The temperature dependence of the parameters of λ-(BEDT-STF)$_2$GaCl$_4$ extracted from the fit of the data by Equation (2) is plotted in Figure 5 as a function of inverse temperature $1/T$. The dielectric strength $\Delta\epsilon$ exhibits a broad peak around $T = 67$ K for the electric field oriented within the $(1\bar{1}0)$-plane and around $T = 40$ K for the out of plane direction, $E \parallel [1\bar{1}0]$. This behavior resembles the temperature dependence of $\epsilon'(T)$ for low frequencies. At high temperatures, the mean relaxation time $\tau_0(T)$ shows thermally activated behavior for all three directions; at the same time, the symmetric broadening $(1 - \alpha)$ decreases with decreasing temperature. These features are signatures of cooperative behavior and glass-like freezing of molecular motion [48].

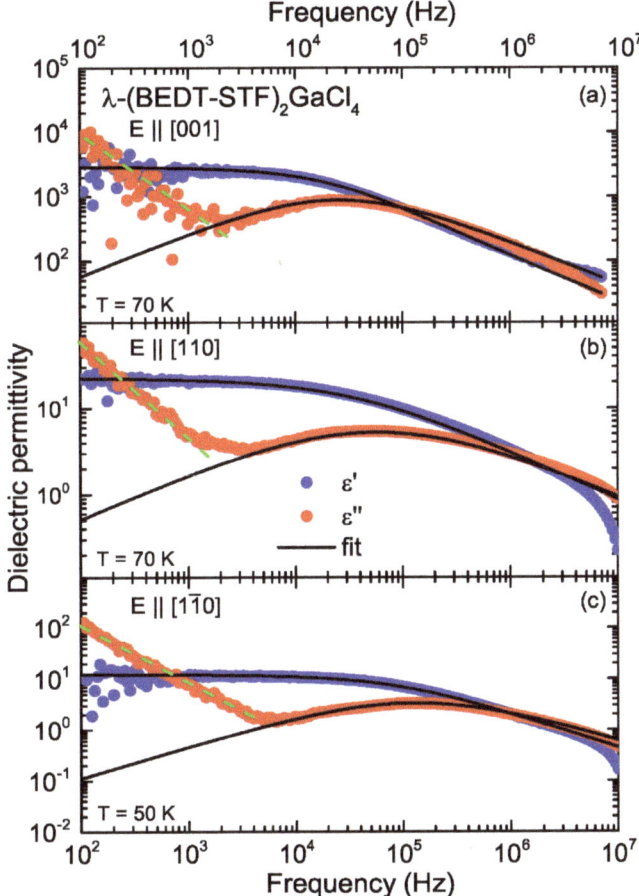

Figure 4. Double logarithmic presentation of the frequency-dependent real and imaginary parts of the dielectric permittivity of λ-(BEDT-STF)$_2$GaCl$_4$, ϵ' (blue symbols) and ϵ'' (red symbols). The data are recorded along (**a**) the [001]- and (**b**) [110]-directions at $T = 70$ K and (**c**) for $E \parallel [1\bar{1}0]$ at $T = 50$ K. The solid black lines represent fits by the generalized Debye model; the dashed green lines indicate the dc contribution to the imaginary part of the permittivity.

The temperature dependence of the mean relaxation time for disordered glassy systems with critical slowing down is commonly described by the Vogel–Fulcher–Tammann (VFT) expression [48,49]:

$$\tau_0(T) = \tau_{\text{VFT}} \exp\left\{\frac{\Delta_{\text{VFT}}}{T - T_{\text{VFT}}}\right\}, \tag{3}$$

where Δ_{VFT} is the activation energy for reorientational motion, T_{VFT} is the temperature where the mean relaxation time diverges, and τ_{VFT} is the time scale for the response in the high-temperature limit.

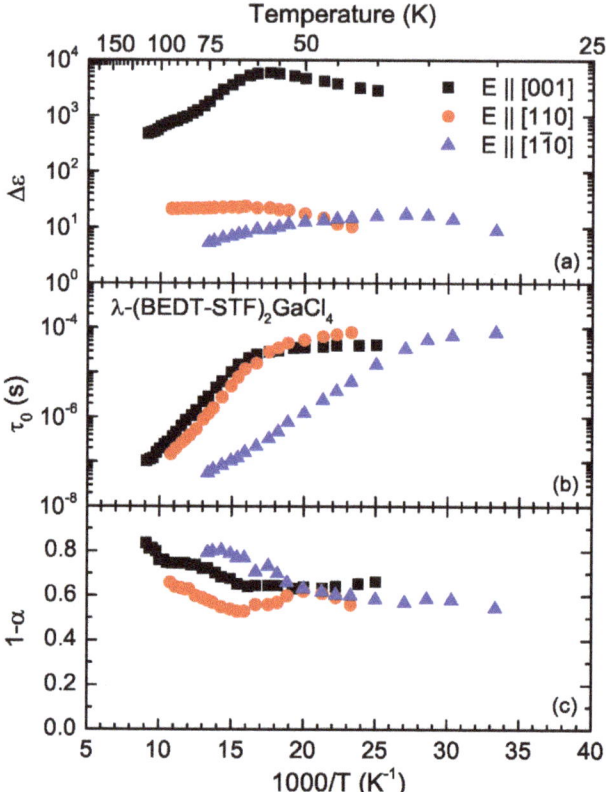

Figure 5. (a) Dielectric strength $\Delta\epsilon(T)$, (b) mean relaxation time $\tau_0(T)$, and (c) symmetric broadening $(1-\alpha)$ as a function of inverse temperature for all the directions of λ-(BEDT-STF)$_2$GaCl$_4$, as indicated.

In Figure 6, the mean relaxation time $\tau_0(T)$ of λ-(BEDT-STF)$_2$GaCl$_4$ is plotted for all three orientations as a function of $1/(T - T_{VFT})$ together with fits by Equation (3). The Vogel–Fulcher–Tammann law explains how the peak seen in the temperature-dependent plot $\epsilon'(T)$ of Figure 3 shifts with frequency. The parameters of the mean relaxation time obtained by the fits are listed in Table 1 for the different axes. We also see that T_{VFT} is equal to 15 K for all three directions; the value is slightly higher than $T_{VFT} \approx 6$ K, extracted for the spin liquid candidate κ-(BEDT-TTF)$_2$Cu$_2$(CN)$_3$, adjusted to the anomaly observed in numerous other quantities [44].

Table 1. The Vogel–Fulcher–Tammann parameters of λ-(BEDT-STF)$_2$GaCl$_4$: the mean relaxation time τ_0, activation energy Δ_{VFT}, and glass temperature T_{VFT} obtained for $E \parallel [001]$, $E \parallel [110]$, and $E \parallel [1\bar{1}0]$.

Directions of Measurement	τ_{VFT} (s)	Δ_{VFT} (K)	T_{VFT} (K)
[001]-axis	2.5×10^{-10}	560	15
[110]-axis	1.85×10^{-10}	525	15
[1$\bar{1}$0]-axis	8.2×10^{-10}	250	15

The anisotropy of the activation energy extracted in κ-(BEDT-TTF)$_2$Cu$_2$(CN)$_3$ also ranged up to a factor of 2 in the Vogel–Fulcher–Tammann fit with remarkable deviations between different single crystals, reaching up to 510 K and 330 K, respectively [19,20]. Hence, from the slowing down of the relaxation time according to an Arrhenius behavior, Pinterić et al. obtain values comparable to the ones given in Table 1 with a glass temperature between 10 and 15 K. Again, the sample-to-sample deviation indicates disorder being important for κ-(BEDT-TTF)$_2$Cu$_2$(CN)$_3$ and also for κ-(BEDT-TTF)$_2$Ag$_2$(CN)$_3$ [20–22]

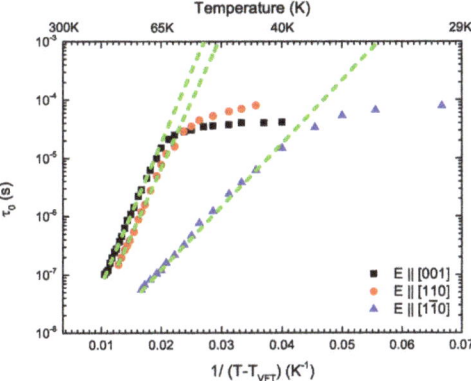

Figure 6. Arrhenius presentation of the mean relaxation time $\tau_0(T)$ of λ-(BEDT-STF)$_2$GaCl$_4$ measured along the [001] (black square), [110] (red circle), and [1$\bar{1}$0] (blue triangle) axes. The solid green line corresponds to fits by the Vogel–Fulcher–Tammann expression (3).

Alternatively, the temperature dependence of the real part of the dielectric permittivity displayed in Figure 3 might be described by a Curie–Weiss behavior. The Curie–Weiss law for the static dielectric constant as a function of temperature has the form

$$\epsilon'(T) = \frac{C}{T - T_C}, \qquad (4)$$

where C is the Curie constant, and T_C is the Curie temperature. In Figure 7, we plot $\epsilon'(T)$ along $E \parallel [001]$ and $E \parallel [1\bar{1}0]$ for several frequencies.

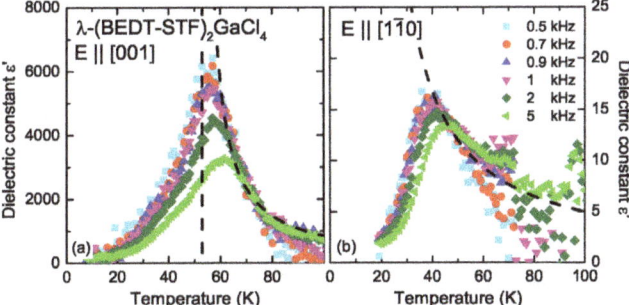

Figure 7. The real part of the dielectric constant $\epsilon'(T)$ plotted as a function of temperature at certain frequencies for (**a**) $E \parallel [001]$ and (**b**) $E \parallel [1\bar{1}0]$. The dashed black line corresponds to the Curie–Weiss fit according to Equation (4).

The best fit by Equation (4) for frequencies less than 5 kHz is given by the dashed line; no clear Curie–Weiss peak is visible for the direction $E \parallel [110]$. The obtained parameters are listed in Table 2. From the Curie constant C for $E \parallel [1\bar{1}0]$ (out of plane), we can estimate the dipole strength following the procedure described by Pinterić et al. [20]. Assuming that the dielectric behavior is a result of charge imbalance within the dimers, we can estimate the amount of charge disproportionation $\Delta\rho \approx 0.05e$.

Table 2. Parameters C and T_C of λ-(BEDT-STF)$_2$GaCl$_4$ obtained from the fit of $\epsilon'(T)$ by the Curie–Weiss Equation (4) for $E \parallel [001]$ and $E \parallel [1\bar{1}0]$.

Directions of Measurement	C (K)	T_C (K)
[001]-axis	4×10^4	53
[1$\bar{1}$0]-axis	420	15

3.2. Vibrational Spectroscopy

In order to learn about the possible charge disproportionation in λ-(BEDT-STF)$_2$GaCl$_4$ on a microscopic scale, we apply infrared spectroscopy frequently used for investigating organic charge-transfer salts [33–38,41,50]. Here, we focus on the asymmetric infrared-active vibrational mode $\nu_{27}(b_{1u})$ that involves the C=C bonds and is thus very sensitive to the electronic charge per molecule. It can be probed best perpendicular to the plane, i.e., for $E \parallel [1\bar{1}0]$. Figure 8a displays the optical conductivity $\sigma_1(\omega)$ of λ-(BEDT-STF)$_2$GaCl$_4$ in the corresponding spectral range for several temperatures. The broad band observed around 1465 cm^{-1} is assigned to the ν_{27} vibration for half a hole per BEDT-TTF molecule. The feature becomes more pronounced upon cooling, but soon it becomes obvious that it is composed of two modes; eventually, two peaks are well separated.

For the quantitative characterization, the conductivity spectra of the ν_{27} mode can be described satisfactorily with one Fano function above $T = 90$ K, while two Lorentzians are necessary below. For the optical conductivity, the Fano function and one Lorentzian give σ_1^{Fano} and $\sigma_1^{Lorentz}$, respectively, as described in, e.g., Equation (5a,b):

$$\sigma_1^{Fano}(\nu) = \sigma_0 \frac{\gamma\nu[\gamma\nu(q^2-1) + 2q(\nu^2 - \nu_0^2)]}{(\nu^2 - \nu_0^2)^2 + \gamma^2\nu^2} \quad (5a)$$

$$\sigma_1^{Lorentz}(\nu) = \frac{\sigma_0 \gamma \nu^2}{(\nu^2 - \nu_0^2)^2 + \gamma^2 \nu^2} \quad (5b)$$

where $\nu = f/c = \omega/(2\pi c)$, σ_0, and γ are frequency, amplitude, and damping, respectively, and q is a phenomenological coupling constant in the Fano function, which gives the Lorentzian shape in the case $q = \pm \infty$.

In Figure 8b,c, we present examples of the fits to $\sigma_1(\omega)$ at $T = 295$ and 12 K; in addition, a broad Lorentzian accounts for the electronic background. The peak frequencies and linewidths are plotted in panels (d) and (e) as a function of temperature. While some hardening is observed for $T > 150$ K, the mode frequency saturates when cooling further. This also holds when we fit the spectra by two modes at low temperatures. Besides some thermal narrowing, when cooling starts at room temperature, the overall linewidth remains constant at approximately 13 cm^{-1} below 150 K.

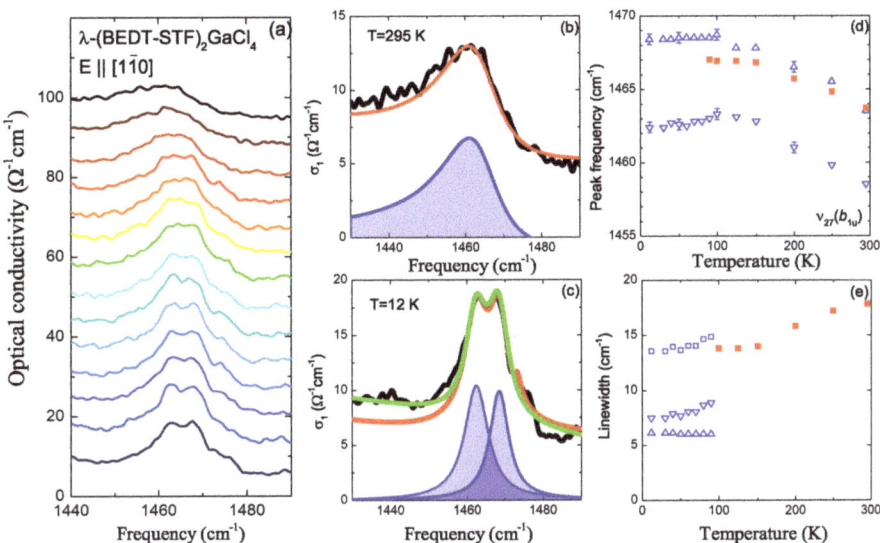

Figure 8. (a) Temperature evolution of the optical conductivity of λ-(BEDT-STF)$_2$GaCl$_4$ in the region of the molecular vibration $\nu_{27}(b_{1u})$. The data are shifted with respect to each other by a constant offset for clarity reasons. (b,c) Fits of the vibrational mode at T = 295 and 12 K. The experimental data are shown in black, the red lines correspond to the overall fit, and the blue lines are separate contributions to the mode. From the two-state jump model [Equation (7)], we obtain the green line. (d,e) Temperature dependence of the resonance frequency and linewidth of the charge-sensitive mode $\nu_{27}(b_{1u})$. While the red squares correspond to the fit by a single contribution for $T \geq 90$ K, the blue symbols represent the two-mode description, where triangles and rotated triangles are related to different Lorentzians; the open blue squares in panel (e) correspond to the sum of both, demonstrating that the overall width does not change.

In Figure 8d, the temperature-dependent results from fitting the $\nu_{27}(b_{1u})$ vibrational feature of λ-(BEDT-STF)$_2$GaCl$_4$ by two Lorentzian modes are displayed by blue symbols. For $T < 90$ K, the peaks are well separated, but we can also extend this approach to higher temperatures, as an alternative to the description by a single Fano-line (red squares). Obviously, there is a kink in the temperature evolution of the vibrational frequency around $T = 100$ K that is independent of the fit procedure. At elevated temperatures, the blue-shift upon cooling follows the typical thermal hardening. The kink in this behavior at around 100 K infers some modification in the physical properties. Even though the origin of this kink is unclear, it can be related to the realization of an inhomogeneous electronic state suggested from NMR measurements, where an increasing linewidth was observed in the same temperature range, and the temperature dependence of the dc resistivity follows variable-range hopping or soft Hubbard gap models below 100 K, characteristic for systems with disorder [51,52]. It is also interesting to note that this anomaly occurs exactly at the temperature where the dielectric dispersion starts to develop, as shown in Figure 3.

Let us first have a look at this separation. The splitting of the $\nu_{27}(b_{1u})$ molecular vibration is commonly taken as evidence that there are two distinct molecules containing unequal charge, and the charge imbalance $\Delta\rho$ can be simply determined as the differences between charges on these two molecules, $\Delta\rho = \rho(\text{charge rich}) - \rho(\text{charge poor})$. From the separation of the two peaks by $\Delta\nu = 6$ cm^{-1} extracted from Figure 8d, we can estimate the charge imbalance $\Delta\rho$ according to

$$\Delta\nu_{27} = -(140 \text{ cm}^{-1}/e)\Delta\rho \quad , \tag{6}$$

suggested for BEDT-TTF compounds [35,41]; here we would like to note that, despite this relation was established for the BEDT-TTF molecules, it will also hold for BEDT-STF, as the replacement of some S ions by Se in BEDT-STF or BETS leads to only a small shift in ν_{27} of less than 2 or 3 cm^{-1}, respectively, in absolute value [53,54]. From our data we obtain $\Delta\rho \approx 0.043e$ which is independent of temperature. Although the mode is thermally broadened at higher temperatures, the fit by two terms can be extended up to $T = 300$ K without change in the frequency separation. This implies that the charge disproportionation of $\Delta\rho$ is already present at an ambient condition and remains unaffected by temperature. In other words, there is no charge-order phase transition in λ-(BEDT-STF)$_2$GaCl$_4$, comparable to the one seen in one-dimensional charge transfer salts as a mean-field development of the charge disproportionation [44,55].

The behavior is also distinct from κ-(BEDT-TTF)$_2$Hg(SCN)$_2$Br, where a second peak develops around 18 cm^{-1} below the main peak, which, in fact, also exhibits a double structure with a sideband 5 cm^{-1} apart [24]. In the present case, we do not see a shoulder on one side gradually developing towards full peaks; instead, we observe a vibrational feature that is rather broad, and it becomes more pronounced upon cooling without strongly increasing or decreasing in width.

This observation is in line with the presence of two crystallographically inequivalent donor molecules, A and B, forming the dimer in the λ-type salts, as depicted in Figure 1b. We conclude that these molecules are not only distinct by symmetry but also carry different charges. The charge imbalance is rather small when compared to the non-dimerized α-(BEDT-TTF)$_2$I$_3$, for instance, where a charge disproportionation of more than $\Delta\rho \approx 0.1e$ is already present at room temperature well above the charge-order transition [56]. To our knowledge, there are no systematic vibrational studies of the family of λ-salts. In the related compound λ-(BETS)$_2$GaCl$_4$, a weak charge disproportionation was concluded from the broadening of the ^{77}Se NMR spectrum and its angular dependence [57].

Most dimerized charge-transfer systems, such as κ-(BEDT-TTF)$_2$Cu$_2$(CN)$_3$ or κ-(BEDT-TTF)$_2$Cu[N(CN)$_2$]Cl, do not develop any charge disproportionation beyond 1%, which is about the experimental resolution [58]. However, with approximately 6.5 cm^{-1}, the vibrational features of κ-(BEDT-TTF)$_2$Cu$_2$(CN)$_3$ are significantly broader than what is observed in typical charge-ordered compounds, such as α-(BEDT-TTF)$_2$I$_3$, where the linewidth is less than 3 cm^{-1} [59,60], or in κ-(BEDT-TTF)$_2$Hg(SCN)$_2$Cl where the individual width is around 4 to 5 cm^{-1} at $T = 10$ K [24]. This was explained by intradimer charge fluctuations, using a two-state jump model [36]. A similar conclusion can be drawn from investigations of the Raman-active fully symmetric vibrations, ν_2 and ν_3 [61].

As seen from Figure 8e, for λ-(BEDT-STF)$_2$GaCl$_4$ the width of the ν_{27} modes also decreases only slightly with a reduced temperature and remains at about 13 cm^{-1} in total. If we assume an electronic charge fluctuating within the dimer, depending on the fluctuation rate, the broadening or splitting of the mode can be described by the Kubo formula [36]:

$$\mathcal{L}(\omega) = \frac{\mathcal{F}[(\gamma + 2v_{ex}) - i(\omega - \omega_w)]}{\mathcal{R}^2 - (\omega - \omega_1)(\omega - \omega_2) - 2i\Gamma(\omega - \omega_{av})} \quad . \tag{7}$$

Here, $\mathcal{F} = f_1 + f_2$, with f_1, f_2 being the oscillator strengths of the bands at frequency ω_1 and ω_2 and halfwidth γ. The charge fluctuation velocity is v_{ex}, $\Gamma = \gamma + v_{ex}$ is the resulting width, and the abbreviation $\mathcal{R}^2 = 2\gamma v_{ex} + \gamma^2$. Finally, we define the average and weighted frequency, ω_{av} and ω_w, by

$$\omega_{av} = \frac{\omega_1 + \omega_2}{2} \quad \text{and} \quad \omega_w = \frac{f_2\omega_1 + f_1\omega_2}{f_1 + f_2} \quad . \tag{8}$$

When the charge oscillations are slow, $v_{ex} \ll |\omega_1 - \omega_2|/2$, Equation (7) yields two separated bands centered around ω_1 and ω_2, while for $v_{ex} \gg |\omega_1 - \omega_2|/2$, the motional narrowing will give one single band centered at the intermediate frequency ω_{av}. Finally, when $v_{ex} \approx |\omega_1 - \omega_2|/2$, we shall observe one broad band shifted towards the mode with larger oscillator strength.

The green line in Figure 8c represents the fit of the data by Equation (7), with a slitting of 6.2 cm^{-1} and a fluctuation rate $v_{ex} = 0.3$ cm^{-1} corresponding to 9×10^{10} s^{-1}. This exchange frequency is certainly slower than estimated for κ-(BEDT-TTF)$_2$Cu$_2$(CN)$_3$ but much faster than the $\nu_{ex} = 40$ cm^{-1} obtained from Raman measurements on κ-(BEDT-TTF)$_2$Hg(SCN)$_2$Cl [62].

Although the estimated charge disproportionation of $\Delta\rho \approx 0.043e$ is in good agreement with the value obtained from our dielectric measurements, we should keep in mind that this sort of charge fluctuation is much too fast to be the sole cause for the dielectric response observed in the kHz and MHz range of frequency. In addition, we do not observe any significant temperature dependence of the charge disproportionation among the molecules, which could be related to the significant temperature dependence of the dielectric behavior. The important facts in λ-(BEDT-STF)$_2$GaCl$_4$ are the intrinsic disorder due to the asymmetric BEDT-STF molecules and the domain wall formation due to charge order, as discussed previously [44,59,60]. The random orientation of the asymmetric BEDT-STF molecules introduces inhomogeneous charge localization, giving rise to enhanced linewidth. Hence, it is more plausible that the disordered donor molecule structure plays a role for the broad linewidth as it provides a different chemical environment, suggesting that charge fluctuations are not dominant in the insulating phase next to the SC phase in the λ-salts. This supports NMR studies claiming that magnetic fluctuation should contribute to the SC pairing mechanism [11,63]. To check the effect of the charge fluctuation in detail, ultrasonic measurements will be useful. Of course, Raman scattering experiments should eventually be performed to verify our findings.

4. Conclusions

Dielectric and vibrational spectroscopies were performed on λ-(BEDT-STF)$_2$GaCl$_4$ in order to elucidate the charge degrees of freedom. Our temperature and frequency-dependent investigations of the dielectric properties reveal relaxor-like ferroelectric behavior below $T \approx 100$ K. The vibration spectroscopy found two ν_{27} modes which can be related to inequivalent donor molecules. The amount of charge disproportionation is consistently estimated to be approximately $\Delta\rho = 0.04$–$0.05e$, which remains temperature independent, ruling out a charge-order transition. At this point, we cannot give a final answer as to what causes the kink in the vibrational properties around $T = 100$ K and the concomitant occurrence of the anomaly in the dielectric constant. The linewidth of the ν_{27} mode is broader than that of typical BEDT-TTF salts, indicating that the asymmetric BEDT-STF molecules constitute a different chemical environment. This supports that the electronic state in λ-(BEDT-STF)$_2$GaCl$_4$ is strongly influenced by disorder, leading to some novel quantum state, as previously suggested.

Author Contributions: Conceptualization, M.D. and O.I.; measurements and analysis, O.I. and R.R.; resources, M.D.; manuscript preparation, O.I. and M.D.; sample preparation: A.K. All authors have read and agreed to the published version of the manuscript.

Funding: This research was funded by by the DeutscheForschungsgemeinschaft (DFG) via DR228/39-3.

Institutional Review Board Statement: Not applicable.

Informed Consent Statement: Not applicable.

Data Availability Statement: Data are available from O.I. upon request.

Acknowledgments: We thank Gabriele Untereiner for the crucial technical support and Yohei Saito for fruitful discussions.

Conflicts of Interest: The authors declare no conflict of interest.

References

1. Lebed, A.G. *The Physics of Organic Superconductors and Conductors*; Springer Series in Materials Science; Springer: Berlin/Heidelberg, Germany, 2008; Volume 110.
2. Dressel, M. Quantum criticality in organic conductors? Fermi liquid versus non-Fermi-liquid behaviour. *J. Phys. Condens. Matter* **2011**, *23*, 293201. [CrossRef]

3. Dressel, M.; Tomić, S. Molecular quantum materials: Electronic phases and charge dynamics in two-dimensional organic solids. *Adv. Phys.* **2020**, *69*, 1–120. [CrossRef]
4. Powell, B.J.; McKenzie, R.H. Quantum Frustration in Organic Mott Insulators: From Spin Liquids to Unconventional Superconductors. *Rep. Progr. Phys.* **2011**, *74*, 056501. [CrossRef]
5. Kurosaki, Y.; Shimizu, Y.; Miyagawa, K.; Kanoda, K.; Saito, G. Mott Transition from a Spin Liquid to a Fermi Liquid in the Spin-Frustrated Organic Conductor κ-(ET)$_2$Cu$_2$(CN)$_3$. *Phys. Rev. Lett.* **2005**, *95*, 177001. [CrossRef] [PubMed]
6. Uji, S.; Terashima, T.; Nishimura, M.; Takahide, Y.; Konoike, T.; Enomoto, K.; Cui, H.; Kobayashi, H.; Kobayashi, A.; Tanaka, H.; et al. Vortex dynamics and the Fulde-Ferrell-Larkin-Ovchinnikov state in a magnetic-field-induced organic superconductor. *Phys. Rev. Lett.* **2006**, *97*, 157001. [CrossRef]
7. Uji, S.; Brooks, J.S. Magnetic-Field-Induced Superconductivity in Organic Conductors. *J. Phys. Soc. Jpn.* **2006**, *75*, 051014. [CrossRef]
8. Uji, S.; Shinagawa, H.; Terashima, T.; Yakabe, T.; Terai, Y.; Tokumoto, M.; Kobayashi, A.; Tanaka, H.; Kobayashi, H. Magnetic-field-induced superconductivity in a two-dimensional organic conductor. *Nature* **2001**, *410*, 908. [CrossRef]
9. Ardavan, A.; Brown, S.; Kagoshima, S.; Kanoda, K.; Kuroki, K.; Mori, H.; Ogata, M.; Uji, S.; Wosnitza, J. Recent Topics of Organic Superconductors. *J. Phys. Soc. Jpn.* **2012**, *81*, 011004. [CrossRef]
10. Minamidate, T.; Oka, Y.; Shindo, H.; Yamazaki, T.; Matsunaga, N.; Nomura, K.; Kawamoto, A. Superconducting Phase in λ-(BEDT-STF)$_2$GaCl$_4$ at High Pressures. *J. Phys. Soc. Jpn.* **2015**, *84*, 063704. [CrossRef]
11. Saito, Y.; Nakamura, H.; Sawada, M.; Yamazaki, T.; Fukuoka, S.; Matsunaga, N.; Nomura, K.; Dressel, M.; Kawamoto, A. Disordered quantum spin state in the stripe lattice system consisting of triangular and square tilings investigated by ^{13}C NMR. *arXiv* **2019**, arXiv:1910.09963.
12. Saito, Y.; Fukuoka, S.; Kobayashi, T.; Kawamoto, A.; Mori, H. Antiferromagnetic Ordering in Organic Conductor λ-(BEDT-TTF)$_2$GaCl$_4$ Probed by ^{13}C NMR. *J. Phys. Soc. Jpn.* **2018**, *87*, 013707. [CrossRef]
13. Hotta, C. Quantum electric dipoles in spin-liquid dimer Mott insulator κ-ET$_2$Cu$_2$(CN)$_3$. *Phys. Rev. B* **2010**, *82*, 241104. [CrossRef]
14. Hotta, C. Theories on Frustrated Electrons in Two-Dimensional Organic Solids. *Crystals* **2012**, *2*, 1155–1200. [CrossRef]
15. Dayal, S.; Clay, R.T.; Li, H.; Mazumdar, S. Paired electron crystal: Order from frustration in the quarter-filled band. *Phys. Rev. B* **2011**, *83*, 245106. [CrossRef]
16. Clay, R.; Mazumdar, S. From charge- and spin-ordering to superconductivity in the organic charge-transfer solids. *Phys. Rep.* **2019**, *788*, 1–89. [CrossRef]
17. Lunkenheimer, P.; Müller, J.; Krohns, S.; Schrettle, F.; Loidl, A.; Hartmann, B.; Rommel, R.; De Souza, M.; Hotta, C.; Schlueter, J.A.; et al. Multiferroicity in an organic charge-transfer salt that is suggestive of electric-dipole-driven magnetism. *Nat. Mater.* **2012**, *11*, 755–758. [CrossRef]
18. Tomić, S.; Pinterić, M.; Ivek, T.; Sedlmeier, K.; Beyer, R.; Wu, D.; Schlueter, J.A.; Schweitzer, D.; Dressel, M. Magnetic ordering and charge dynamics in κ-(BEDT-TTF)$_2$Cu[N(CN)$_2$]Cl. *J. Phys. Condens. Matter* **2013**, *25*, 436004. [CrossRef] [PubMed]
19. Abdel-Jawad, M.; Terasaki, I.; Sasaki, T.; Yoneyama, N.; Kobayashi, N.; Uesu, Y.; Hotta, C. Anomalous dielectric response in the dimer Mott insulator κ-(BEDT-TTF)$_2$Cu$_2$(CN)$_3$. *Phys. Rev. B* **2010**, *82*, 125119. [CrossRef]
20. Pinterić, M.; Čulo, M.; Milat, O.; Basletić, M.; Korin-Hamzić, B.; Tafra, E.; Hamzić, A.; Ivek, T.; Peterseim, T.; Miyagawa, K.; et al. Anisotropic charge dynamics in the quantum spin-liquid candidate κ-(BEDT-TTF)$_2$Cu$_2$(CN)$_3$. *Phys. Rev. B* **2014**, *90*, 195139. [CrossRef]
21. Pinterić, M.; Lazić, P.; Pustogow, A.; Ivek, T.; Kuveždić, M.; Milat, O.; Gumhalter, B.; Basletić, M.; Čulo, M.; Korin-Hamzić, B.; et al. Anion effects on electronic structure and electrodynamic properties of the Mott insulator κ-(BEDT-TTF)$_2$Ag$_2$(CN)$_3$. *Phys. Rev. B* **2016**, *94*, 161105. [CrossRef]
22. Pinterić, M.; Rivas Góngora, D.; Rapljenović, Ž.; Ivek, T.; Čulo, M.; Korin-Hamzić, B.; Milat, O.; Gumhalter, B.; Lazić, P.; Sanz Alonso, M.; et al. Electrodynamics in Organic Dimer Insulators Close to Mott Critical Point. *Crystals* **2018**, *8*, 190. [CrossRef]
23. Dressel, M.; Lazić, P.; Pustogow, A.; Zhukova, E.; Gorshunov, B.; Schlueter, J.A.; Milat, O.; Gumhalter, B.; Tomić, S. Lattice vibrations of the charge-transfer salt κ-(BEDT-TTF)$_2$Cu$_2$(CN)$_3$: Comprehensive explanation of the electrodynamic response in a spin-liquid compound. *Phys. Rev. B* **2016**, *93*, 081201. [CrossRef]
24. Ivek, T.; Beyer, R.; Badalov, S.; Čulo, M.; Tomić, S.; Schlueter, J.A.; Zhilyaeva, E.I.; Lyubovskaya, R.N.; Dressel, M. Metal-insulator transition in the dimerized organic conductor κ-(BEDT-TTF)$_2$Hg(SCN)$_2$Br. *Phys. Rev. B* **2017**, *96*, 085116. [CrossRef]
25. Nad, F.; Monceau, P Dielectric Response of the Charge Ordered State in Quasi-One-Dimensional Organic Conductors. *J. Phys. Soc. Jpn.* **2006**, *75*, 051005. [CrossRef]
26. Ishihara, S. Electronic ferroelectricity in molecular organic crystals. *J. Phys. Condens. Matter* **2014**, *26*, 493201. [CrossRef]
27. Monceau, P.; Nad, F.Y.; Brazovskii, S. Ferroelectric Mott-Hubbard Phase of Organic (TMTTF)$_2$X Conductors. *Phys. Rev. Lett.* **2001**, *86*, 4080–4083. [CrossRef]
28. Horiuchi, S.; Tokura, Y. Organic ferroelectrics. *Nat. Mater.* **2008**, *7*, 357–366. [CrossRef]
29. Mistewicz, K. Recent Advances in Ferroelectric Nanosensors: Toward Sensitive Detection of Gas, Mechanothermal Signals, and Radiation. *J. Nanomater.* **2018**, *2018*, 2651056. [CrossRef]
30. Zhang, S.; Malič, B.; Li, J.F.; Rödel, J. Lead-free ferroelectric materials: Prospective applications. *J. Mater. Res.* **2021**, *36*, 985–995. [CrossRef]
31. Asadi, K. (Ed.) *Organic Ferroelectric Materials and Applications*; Woodhead Publishing: Sawston, UK, 2021.

32. Kim, T.Y.; Kim, S.K.; Kim, S.W. Application of ferroelectric materials for improving output power of energy harvesters. *Nano Converg.* **2018**, *5*, 30. [CrossRef]
33. Dressel, M.; Drichko, N. Optical Properties of Two-Dimensional Organic Conductors: Signatures of Charge Ordering and Correlation Effects. *Chem. Rev.* **2004**, *104*, 5689–5715. [CrossRef] [PubMed]
34. Drichko, N.; Kaiser, S.; Sun, Y.; Clauss, C.; Dressel, M.; Mori, H.; Schlueter, J.; Zhyliaeva, E.; Torunova, S.; Lyubovskaya, R. Evidence for charge order in organic superconductors obtained by vibrational spectroscopy. *Phys. B* **2009**, *404*, 490–493. [CrossRef]
35. Girlando, A. Charge Sensitive vibrations and electron-molecular vibration coupling in bis (ethylenedithio)-tetrathiafulvalene (BEDT-TTF). *J. Phys. Chem. C* **2011**, *115*, 19371. [CrossRef]
36. Girlando, A.; Masino, M.; Schlueter, J.; Drichko, N.; Kaiser, S.; Dressel, M. Spectroscopic characterization of charge order fluctuations in BEDT-TTF metals and superconductors. *Phys. Stat. Sol.* **2012**, *249*, 953–956. [CrossRef]
37. Yakushi, K. Infrared and Raman Studies of Charge Ordering in Organic Conductors, BEDT-TTF Salts with Quarter-Filled Bands. *Crystals* **2012**, *2*, 1291–1346. [CrossRef]
38. Girlando, A.; Masino, M.; Kaiser, S.; Sun, Y.; Drichko, N.; Dressel, M.; Mori, H. Charge-order fluctuations and superconductivity in two-dimensional organic metals. *Phys. Rev. B* **2014**, *89*, 174503. [CrossRef]
39. Mori, H.; Suzuki, H.; Okano, T.; Moriyama, H.; Nishio, Y.; Kajita, K.; Kodani, M.; Takimiya, K.; Otsubo, T. Positional order and disorder of symmetric and unsymmetric BEDT-STF salts. *J. Solid State Chem.* **2002**, *168*, 626. [CrossRef]
40. Kremer, F.; Schönhals, A. *Broadband Dielectric Spectroscopy*; Springer Science & Business Media: Berlin/Heidelberg, Germany, 2002.
41. Yamamoto, T.; Uruichi, M.; Yamamoto, K.; Yakushi, K.; Kawamoto, A.; Taniguchi, H. Examination of the Charge-Sensitive Vibrational Modes in Bis(ethylenedithio)tetrathiafulvalene. *J. Phys. Chem. B* **2005**, *109*, 15226–15235. [CrossRef]
42. Painelli, A.; Girlando, A. Electron–molecular vibration (e–mv) coupling in charge-transfer compounds and its consequences on the optical spectra: A theoretical framework. *J. Chem. Phys.* **1986**, *84*, 5655. [CrossRef]
43. Iguchi, S.; Sasaki, S.; Yoneyama, N.; Taniguchi, H.; Nishizaki, T.; Sasaki, T. Relaxor ferroelectricity induced by electron correlations in a molecular dimer Mott insulator. *Phys. Rev. B* **2013**, *87*, 075107. [CrossRef]
44. Tomić, S.; Dressel, M. Ferroelectricity in molecular solids: A review of electrodynamic properties. *Rep. Prog. Phys.* **2015**, *78*, 096501. [CrossRef] [PubMed]
45. Jonscher, A.K. *Dielectric Relaxation in Solids*; Chelsea Dielectric Press: London, UK, 1983.
46. Jonscher, A.K. Dielectric relaxation in solids. *J. Phys. D* **1999**, *32*, R57. [CrossRef]
47. Lunkenheimer, P.; Loidl, A. Dielectric spectroscopy on organic charge-transfer salts. *J. Phys. Condens. Matter* **2015**, *27*, 373001. [CrossRef] [PubMed]
48. Cross, L.E. Relaxor ferroelectrics. *Ferroelectrics* **1987**, *76*, 241. [CrossRef]
49. Levstik, A.; Kutnjak, Z.; Filipič, C.; Pirc, R. Glassy freezing in relaxor ferroelectric lead magnesium niobate. *Phys. Rev. B* **1998**, *57*, 11204–11211. [CrossRef]
50. Hattori, Y.; Iguchi, S.; Sasaki, T.; Iwai, S.; Taniguchi, H.; Kishida, H. Electric-field-induced intradimer charge disproportionation in the dimer-Mott insulator $\beta'-(BEDT-TTF)_2ICl_2$. *Phys. Rev. B* **2017**, *95*, 085149. [CrossRef]
51. Saito, Y.; Rösslhuber, R.; Löhle, A.; Sanz Alonso, M.; Wenzel, M.; Kawamoto, A.; Pustogow, A.; Dressel, M. Bandwidth-tuning from insulating Mott quantum spin liquid to Fermi liquid via chemical substitution in κ-[(BEDT-TTF)$_{1-x}$(BEDT-STF)$_x$]$_2$Cu$_2$(CN)$_3$. *arXiv* **2021**, arXiv:1911.06766. doi:10.1039/D1TC00785H.
52. Shinaoka, H.; Imada, M. Theory of Electron Transport near Anderson–Mott Transitions. *J. Phys. Soc. Jpn.* **2010**, *79*, 113703. [CrossRef]
53. Iakutkina, O.; Uykur, E.; Kobayashi, T.; Kawamoto, A.; Dressel, M.; Saito, Y. Charge imbalance in λ-(BETS)$_2$GaCl$_4$ and their interplay with superconductivity. *Phys. Rev. B* **2021**, *104*, 045108. [CrossRef]
54. Olejniczak, I.; Graja, A.; Kushch, N.D.; Cassoux, P.; Kobayashi, H. Polarized IR Reflectance Studies of the Organic Conductor κ-(BETS)$_2$FeCl$_4$. *J. Phys. I Fr.* **1996**, *6*, 1631–1641. [CrossRef]
55. Dressel, M.; Dumm, M.; Knoblauch, T.; Masino, M. Comprehensive Optical Investigations of Charge Order in Organic Chain Compounds (TMTTF)$_2$X. *Crystals* **2012**, *2*, 528–578. [CrossRef]
56. Beyer, R.; Dengl, A.; Peterseim, T.; Wackerow, S.; Ivek, T.; Pronin, A.V.; Schweitzer, D.; Dressel, M. Pressure-dependent optical investigations of α-(BEDT-TTF)$_2$I$_3$: Tuning charge order and narrow gap towards a Dirac semimetal. *Phys. Rev. B* **2016**, *93*, 195116. [CrossRef]
57. Hiraki, K.; Kitahara, M.; Takahashi, T.; Mayaffre, H.; Horvatić, M.; Berthier, C.; Uji, S.; Tanaka, H.; Zhou, B.; Kobayashi, A.; et al. Evidence of Charge Disproportionation in λ Type BETS Based Organic Superconductors. *J. Phys. Soc. Jpn.* **2010**, *79*, 074711. [CrossRef]
58. Sedlmeier, K.; Elsässer, S.; Neubauer, D.; Beyer, R.; Wu, D.; Ivek, T.; Tomić, S.; Schlueter, J.A.; Dressel, M. Absence of charge order in the dimerized κ-phase BEDT-TTF salts. *Phys. Rev. B* **2012**, *86*, 245103. [CrossRef]
59. Ivek, T.; Korin-Hamzić, B.; Milat, O.; Tomić, S.; Clauss, C.; Drichko, N.; Schweitzer, D.; Dressel, M. Collective Excitations in the Charge-Ordered Phase of α-(BEDT-TTF)$_2$I$_3$. *Phys. Rev. Lett.* **2010**, *104*, 206406. [CrossRef] [PubMed]
60. Ivek, T.; Korin-Hamzić, B.; Milat, O.; Tomić, S.; Clauss, C.; Drichko, N.; Schweitzer, D.; Dressel, M. Electrodynamic response of the charge ordering phase: Dielectric and optical studies of α-(BEDT-TTF)$_2$I$_3$. *Phys. Rev. B* **2011**, *83*, 165128. [CrossRef]

61. Yakushi, K.; Yamamoto, K.; Yamamoto, T.; Saito, Y.; Kawamoto, A. Raman spectroscopy study of charge fluctuation in the spin-liquid candidate κ-(BEDT-TTF)$_2$Cu$_2$(CN)$_3$. *J. Phys. Soc. Jpn.* **2015**, *84*, 084711. [CrossRef]
62. Hassan, N.; Cunningham, S.; Mourigal, M.; Zhilyaeva, E.I.; Torunova, S.A.; Lyubovskaya, R.N.; Schlueter, J.A.; Drichko, N. Evidence for a quantum dipole liquid state in an organic quasi–two-dimensional material. *Science* **2018**, *360*, 1101–1104. [CrossRef]
63. Kobayashi, T.; Kawamoto, A. Evidence of antiferromagnetic fluctuation in the unconventional superconductor λ-(BETS)$_2$GaCl$_4$ by ^{13}C NMR. *Phys. Rev. B* **2017**, *96*, 125115. [CrossRef]

Article

Real Space Quantum Cluster Formulation for the Typical Medium Theory of Anderson Localization

Ka-Ming Tam [1,2,*], Hanna Terletska [3,*], Tom Berlijn [4], Liviu Chioncel [5,6] and Juana Moreno [1,2]

1. Department of Physics and Astronomy, Louisiana State University, Baton Rouge, LA 70803, USA; moreno@phys.lsu.edu
2. Center for Computation & Technology, Louisiana State University, Baton Rouge, LA 70803, USA
3. Department of Physics and Astronomy, Middle Tennessee State University, Murfreesboro, TN 37132, USA
4. Center for Nanophase Materials Sciences, Oak Ridge National Laboratory, Oak Ridge, TN 37831, USA; tberlijn@gmail.com
5. Theoretical Physics III, Center for Electronic Correlations and Magnetism, Institute of Physics, University of Augsburg, D-86135 Augsburg, Germany; liviu.chioncel@physik.uni-augsburg.de
6. Augsburg Center for Innovative Technologies, University of Augsburg, D-86135 Augsburg, Germany
* Correspondence: phy.kaming@gmail.com (K.-M.T.); Hanna.Terletska@mtsu.edu (H.T.)

Abstract: We develop a real space cluster extension of the typical medium theory (cluster-TMT) to study Anderson localization. By construction, the cluster-TMT approach is formally equivalent to the real space cluster extension of the dynamical mean field theory. Applying the developed method to the 3D Anderson model with a box disorder distribution, we demonstrate that cluster-TMT successfully captures the localization phenomena in all disorder regimes. As a function of the cluster size, our method obtains the correct critical disorder strength for the Anderson localization in 3D, and systematically recovers the re-entrance behavior of the mobility edge. From a general perspective, our developed methodology offers the potential to study Anderson localization at surfaces within quantum embedding theory. This opens the door to studying the interplay between topology and Anderson localization from first principles.

Keywords: metal insulator transition; Anderson localization; random disorder; typical medium theory; dynamical mean field theory; coherent potential approximation; dynamical cluster approximation; cellular dynamical mean field theory; cluster mean field theory

1. Introduction

The localization problem in disordered electronic systems was introduced in Anderson's seminal paper [1] in the late fifties, and it still remains in the forefront of research in materials science and condensed matter physics [2–5].

In disordered media, the scattering of charge carriers off random impurities may inhibit their propagation across the sample leading to a spatial confinement of carriers, a phenomenon known as Anderson localization [1]. Weak localization and strong Anderson localization have been conjectured and subsequently observed in experiments [6–10]. As a wave phenomenon, Anderson localization has been demonstrated for electrons [11–16], sound [17], photons [18–25], and ultra cold atoms [26].

To model disorder, Anderson proposed a simplified model of electrons hopping between lattice sites being subject to static scattering processes on locally disordered centers. The stochastic character of the problem is encoded into the on-site energies (disordered scattering centers) considered as random variables distributed according to a chosen probability distribution. The Green's function imaginary part, the local density of states (LDOS), turns out to be an important quantity which characterizes the disordered system. For example, the LDOS is finite for extended states, while the spectrum of localized states is discrete. A decade later, an approach based on the distribution of the site and energy dependent self-energies was formulated [27]. This approach leads to a self-consistent equation for the self-energy, which can be solved on a Cayley tree (Bethe lattice). However, for general lattices, only an approximate solution can be provided.

Computations for substitutionally disordered three-dimensional materials with ordinary lattice structures are therefore difficult to perform within the framework of tight-binding models [1,27]. Suitable modeling in such cases can be constructed based on effective medium theories. Among them, single site effective medium methods, such as the coherent potential approximation (CPA) [28–35] and the typical medium theory (TMT) [36], proved to be simple and transparent theories that are able to capture important features of the disorder effects in electron systems. Common to these two methods is the mapping of the lattice problem into the impurity placed in a self-consistently determined effective medium. In both methods, the measured quantity is the disorder averaged Green's function; however, in CPA, the Green's function is linearly (algebraically) averaged, while, in the TMT, the geometric average of the LDOS is used. This difference in disorder averaging defines the average and the typical effective media, respectively.

Unlike the algebraically averaged Green's function of the CPA effective medium, the geometric averaged LDOS, called the typical density of states (TDOS), drops to zero [36–47], at the Anderson transition. The geometrically averaged TDOS is an approximation to the most probable value in the distribution of the LDOS. At the Anderson transition, the system is not self-averaged, hence the distribution of the LDOS is highly skewed with long tails [37,48]. Therefore, the average and most probable values of the LDOS will be very different close to the transition [37,49–51]. Dobrosavljevic et al. [36] incorporated such statistical properties of the LDOS within an effective medium approach, called the TMT. They showed that the TDOS successfully captures the main signatures of the Anderson localization transition, with the TDOS being an order parameter to detect the localized states. In Refs. [52–54], the momentum-space cluster extension of the TMT [54], the typical medium dynamical cluster approximation (TMDCA) has been developed. The TMDCA is the typical medium extension of the Dynamical Cluster Approximation (DCA) [55,56], a momentum-space cluster extension of the CPA. The TMDCA overcomes the shortcomings of the local single site TMT and accurately predicts the critical disorder strength of the Anderson localization transition in a single-band Anderson model. For model Hamiltonian systems, the TMDCA has been applied to non-interacting and weakly interacting disordered three-dimensional systems [52,53,57,58], systems with off-diagonal disorder [59], phonon localization [60,61], and multi-orbital models [62]. Some of the methods inspired by the typical medium theories have been combined with first-principles calculations [63–65].

Complementary to the momentum space cluster methods, described above, techniques using embedding in real space provide an interesting alternative. This constitutes the aim of the present work. We have previously formulated the embedding into the effective typical medium which allows for addressing the Anderson localization transition in the framework of a locally self-consistent approach [66]. In addition, the locally self-consistent formulation opens up the possibility to formulate linear scaling methods. Unlike the previous typical medium cluster extensions of TMT, formulated in the momentum space (TMDCA) [54,67], or in a mixed representation (locally self-consistent approach) [66,68], here we propose an exclusively real space cluster extension of TMT (cluster-TMT). This construction is formally equivalent to the real space cluster extension of the dynamical mean field theory (DMFT) [69–73].

The key accomplishment of the present study is the development of a cluster mean field theory for the description of Anderson localization. The developed cluster version is based on a real-space approach, and presents an alternative to the existing momentum space version of TMDCA. To demonstrate the validity of the method, we apply it to the three-dimensional Anderson model with box disorder distribution, and reproduce the full phase diagram and the critical disorder strength, W_c, for the metal-insulator transition. The cluster mean field theory we designed is an extension of the local single site typical medium theory. The developed real space cluster extension method incorporates the spatial non-local effects systemically; therefore, the re-entrance behavior of the 3D Anderson model is recovered.

We find that cluster extensions of TMT are necessary to properly capture the non-local effects in the Anderson transition. Quantitatively, our results are in good agreement with the existing data in the literature. In particular, we find that the converged cluster value of $W_c \approx 17.05$ is superior to the value of 13.4 provided by single site TMT calculations. We demonstrate that non-local spatial correlations are significant in the 3D Anderson model, and hence going beyond a single site approximation is necessary to properly describe the metal-insulator transition. Unlike the single site TMT, the present real-space cluster computation captures the re-entrance behavior driven by non-local multiple scattering effects which are missing in local approximations [5,36,48,74]. In addition, just like the TMDCA, the real space cluster-TMT allows for a computationally efficient treatment of the non-local effects in Anderson localization. In addition, however, the real-space cluster TMT opens the door to treating problems with open boundary conditions, which offers the possibility to study the localization of surface states. One potential application of this capability would be the search for a material realization of the topological Anderson insulator [75] via first principles calculations. In addition, the presented formalism being a real space cluster opens venues for an easier embedding with ab initio Green's function electronic structure methods, which offer a more natural approach (in contrast to the momentum space TMDCA) to disordered real materials, including high entropy alloys and disordered metals [63–65].

It is worth noting that there is a long history of applications of the CPA both as a tool for model calculations and in computational studies of real materials. We refer the interested readers to the review of Yonezawa and Morigaki [76]. An extensive review of the early development of the DMFT method can be found in Georges et al. [77], and a review of more recent cluster extensions can be found in Maier et al. [78]. A review of current research on the Anderson localization using cluster methods can be found in Terletska et al. [54].

This paper is organized as follows: in Section 2, we present the Anderson model. In Section 3, we first briefly review the algorithm of the single site TMT and discuss the algorithm for the real-space cluster extension of the TMT. In Section 4, we present the results obtained with our cluster-TMT for the 3D Anderson model with box disorder distribution. We conclude in Section 5 and discuss possible future developments.

2. Model

Anderson proposed [1] that non-interacting electrons on site-disordered lattices may localize because of the destructive interference of wave functions. Subsequent theoretical and numerical studies [79] support the picture that, in three dimensions and for large enough disorder strength, single particle wave functions are localized and decay exponentially on the scale of the localization length.

The Anderson model Hamiltonian has the form:

$$H = -t \sum_{<i,j>,\sigma} (c_{i\sigma}^\dagger c_{j\sigma} + H.c.) + \sum_{i,\sigma} V_i n_{i\sigma}, \quad (1)$$

where $c_{i\sigma}^\dagger$ and $c_{i\sigma}$ are the creation and annihilation operators for electrons at site i with spin σ. $n_{i\sigma}$ is the number operator for site i of spin σ; t is the hopping energy between nearest neighbors. We consider a 3D simple cubic lattice. We set $t = 1$ to serve as the energy scale. The local random disorder is given by V_i. Here, we consider a so-called box disorder with $P(V_i) = \frac{1}{W}\Theta(W - V_i)$. This allows the disorder strength to be characterized by W. Other distributions are also considered in the literature; some common ones included bi-modal, Gaussian, and Lorentzian distributions [53,80].

The Anderson model has been the focus of numerous studies of the disorder-induced electron localization. Highly accurate numerical calculations based on the transfer matrix method and multifractal analysis have been used to study the model extensively, especially for zero energy [5,74,81–90].

Relatively few studies have been devoted to energy away from zero. A prominent feature at higher energy is the re-entrance from a metal to an insulator to a metal, as the disorder strength increases [74,82,91,92]. A heuristic argument for the nature of the re-entrance behavior is based on the tunneling mechanism for energies beyond the bandwidth of the hopping model. The width of the density of states increases as the disorder increases, though the states are localized. At sufficiently large disorder, the localized density of states is large enough to allow tunneling. The tunneling could become sufficiently long range that the localized states become extended, thus the insulator becomes a metal. This explains the lower transition in the re-entrance. Further increasing the disorder strength, the localized state will be more sparse in energy and tunneling becomes less likely to happen and insulating state resumes.

The above argument depends on the distribution of disorder, and the tunneling effect is maximised when the localized states are close in energy. A bounded random distribution is favored as compared to other distributions which are more widely spread over a range of energies, such as the Lorentzian distribution. The tunneling argument can only be supported in a system with multiple sites. For example, the TMT, which is a single site approximation, does not capture the re-entrance behavior. Thus, the capability of describing the re-entrance can serve as a good test for our real space cluster-TMT.

3. The Real Space Quantum Cluster Extension of TMT

3.1. Typical Medium Theory: TMT

To set the stage for the discussion of the real-space cluster extension of the TMT, here we briefly review the main steps of the TMT analysis. The TMT can be considered as a typical medium generalization of the CPA [28–34]. In a similar way to the CPA, the TMT employs the mapping of the original lattice problem into the impurity placed in a self-consistently determined effective medium. However, in the TMT, the typical (geometrically averaged over disorder) local density of states is used to construct the mean field bath for the effective impurity problem.

The numerical algorithm for the TMT procedure is shown in Figure 1. First, the guess for the effective medium self-energy $\Sigma(\omega)$ is made, usually zero. Then, the local (coarse-grained) lattice Green's function is calculated as $\bar{G}(\omega) = \frac{1}{N} \sum_k \frac{1}{\omega - \epsilon_k - \Sigma(\omega)}$. Using the Dyson's equation, we then obtain the impurity-excluded Green's function (bath Green's function) $\mathcal{G}^{-1}(\omega) = \bar{G}^{-1}(\omega) + \Sigma(\omega)$.

The next step is to solve the impurity problem. For each randomly chosen disorder configuration V, we calculate the impurity Green's function $G_{imp}(\omega, V) = (\mathcal{G}^{-1}(\omega) - V)^{-1}$. From this quantity, we obtain the typical (geometrically averaged density of states) $\rho_{typ}(\omega)$, which is constructed as $\rho_{typ}(\omega) = e^{\langle \ln(\rho(\omega,V)) \rangle}$. Here, $\rho(\omega, V) = -\frac{1}{\pi} \Im G_{imp}(\omega, V)$, and $\langle \ldots \rangle$ stands for the disorder averaging. In general, the geometrical average is not equivalent to the typical value. However, for log-normal distributions, the geometrical average is the same as the typical value and, since numerical studies have shown that near the localization transition the local density of states is log-normal distributed [37], this assumption is appropriate.

The output of the TMT impurity solver is the typical Green's function which is obtained using the Hilbert transform: $G_{typ}(w) = \frac{1}{\pi} \int d\omega' \frac{\rho_{typ}(\omega')}{\omega - \omega'}$. This step is the only difference between the CPA and the TMT self-consistency loop. For example, in the CPA, instead of the typical, the algebraically average DOS is calculated $\rho_{ave} = \langle \rho(\omega, V) \rangle$, with the average Green's function $G_{ave}(w) = \frac{1}{\pi} \int d\omega' \frac{\rho_{ave}(\omega')}{\omega - \omega'}$ being the output of the CPA impurity solver. Note that, for the CPA case, one can just do the disorder averaging over Green's function without the Hilbert transform of the average density.

Finally, the TMT self-consistency loop is closed by getting a new estimate of the self-energy $\Sigma(\omega) = \mathcal{G}^{-1}(\omega) - G_{typ}^{-1}(\omega)$, which is then used to calculate the coarse-grained local lattice Green's function. The whole procedure then repeats, until convergence is reached at which the impurity and the local lattice Green's function are equal within the desired accuracy.

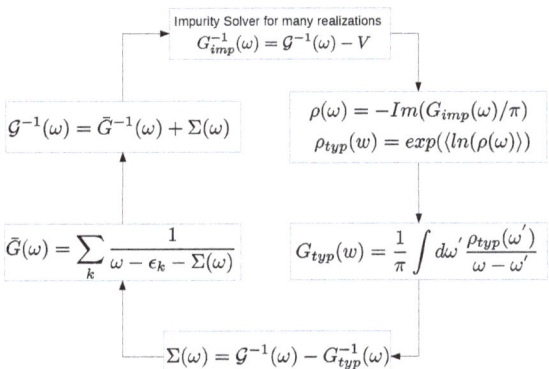

Figure 1. Numerical algorithm for the typical medium theory.

3.2. Real Space Cluster-TMT

To properly capture the effect of multiple impurities scattering in the disorder-driven Anderson localization, the cluster extension of the TMT is needed. Here, we present the real-space cluster extension of the TMT. Such real space variant of the cluster extension of the TMT is formally equivalent to the cluster DMFT solver, which has been extensively used in strongly-correlated electron systems to study non-local effects beyond DMFT. Here, we use the cluster DMFT approach as a tool to capture spacial non-local correlations beyond the TMT in disordered non interacting systems.

In the real space cluster-TMT, the infinite lattice in real space is tiled with identical clusters of size N_c [93]. In such construction, the scattering of electrons by impurities within a cluster is treated exactly, while the effects of impurities outside the cluster are replaced by the non-disordered effective medium (bath) that is determined self-consistently. There is no implicit assumption that the translational invariance is obeyed within the cluster. Therefore, the Green's function of the cluster is represented by an $N_c \times N_c$ matrix, which we denote as $\hat{G}_c(\omega)$. For the same reason, the self-energy and the bath Green's function are also represented in terms of matrices.

The self-consistency procedure for our real space cluster-TMT is shown in Figure 2. First, we start with the guess of the self-energy matrix $\hat{\Sigma}(\omega)$ (usually zero). Then, we calculate the lattice Green's function projected onto the cluster space $\mathring{\hat{G}}(\omega) = \frac{N_c}{N} \sum_{\mathbf{k} \in R.B.Z.} [\omega - \hat{t}(\mathbf{k}) - \hat{\Sigma}(\omega)]^{-1}$, where R.B.Z. stands for the reduced Brillouin Zone of the cluster with $-\frac{2\pi}{L_c} < k_x, k_y, k_z < \frac{2\pi}{L_c}$. In addition, $\hat{t}(\mathbf{k})$ is the dispersion of the lattice model expressed as a partial Fourier transform over the reduced Brillouin zone. Any element of this dispersion matrix is given as $t_{\mathbf{r},\mathbf{r}'}(\mathbf{k}) \equiv \sum_{\mathbf{R}} exp(i\mathbf{k} \cdot (\mathbf{R} + \mathbf{r} - \mathbf{r}')) t_{\mathbf{r},\mathbf{r}'+\mathbf{R}'}$, where $\mathbf{R}$ is the location vector of the super-cell, and $\mathbf{r}$ and $\mathbf{r}'$ are the vectors for the location of each site within a super-cell [93].

Next, using the Dyson's equation, we calculate the bath Green's function matrix, $\hat{\mathcal{G}}^{-1}(\omega) = \mathring{\hat{G}}^{-1}(\omega) + \hat{\Sigma}(\omega)$, which is used to construct the cluster problem. Then, for each disorder configuration V, we calculate the cluster Green's function by solving the matrix equation $G_c^{-1}(\omega, i, j) = \mathcal{G}^{-1}(\omega, i, j) - V(i, j)\delta_{ij}$.

The key to incorporate the typical medium into the analysis is to connect the Green's function matrix to the typical density of states. For this, we generalize the procedure we

21

used for the multi-orbital problem of the TMDCA [62], and define the typical density of states matrix in a similar way:

$$\hat{\rho}_{typ}(\omega) \equiv \begin{pmatrix} e^{\langle |\ln \rho_{11}(\omega)| \rangle} \frac{\langle \rho_{11} \rangle}{\langle |\rho_{11}| \rangle} & \cdots & e^{\langle \ln |\rho_{1N_c}(\omega)| \rangle} \frac{\langle \rho_{1N_c} \rangle}{\langle |\rho_{1N_c}| \rangle} \\ \vdots & \vdots & \vdots \\ e^{\langle \ln |\rho_{N_c 1}(\omega)| \rangle} \frac{\langle \rho_{N_c 1} \rangle}{\langle |\rho_{N_c 1}| \rangle} & \cdots & e^{\langle \ln |\rho_{N_c N_c}(\omega)| \rangle} \frac{\langle \rho_{N_c N_c} \rangle}{\langle |\rho_{N_c N_c}| \rangle} \end{pmatrix}, \quad (2)$$

Here, the diagonal entries will be just equal to $e^{\langle \rho_{ii}(\omega) \rangle}$ because $\rho_{ii} > 0$ is always positive definite; $\rho_{ii} = -\frac{1}{\pi} \Im[G_{ii}(\omega)]$; and for the off-diagonal terms $\rho_{ij} = \frac{i}{2\pi}[G_{ij}(\omega) - G_{ji}(\omega)]$ [94]. The role of the non-local off-diagonal components in the geometrically averaged cluster Green's function is explained in the Appendix A.

Notice that the real space cluster extension of the CPA, with the average effective medium, can be obtained by replacing the typical DOS with the linearly average DOS in the above Equation (2), i.e.,

$$\hat{\rho}_{ave}(\omega) \equiv \begin{pmatrix} \langle \rho_{11}(\omega) \rangle & \cdots & \langle \rho_{1N_c}(\omega) \rangle \\ \vdots & \vdots & \vdots \\ \langle \rho_{N_c 1}(\omega) \rangle & \cdots & \langle \rho_{N_c N_c}(\omega) \rangle \end{pmatrix}. \quad (3)$$

The $\hat{\rho}_{typ}(\omega)$ of Equation (2) possesses the following properties: (1) for $N_c = 1$, it reduces to the local TMT with $\rho_{typ}(\omega) = e^{\langle |\ln \rho(\omega)| \rangle}$; (2) At small disorder strength $W \ll W_c$, we observe numerically that $<\ln \rho(\omega)> \approx \ln <\rho(\omega)>$, i.e., the typical density of states (DOS) reduces to the average DOS calculated using algebraic averaging over disorder: $\rho_{typ} \to \rho_{ave}(\omega)$. Hence, in this regime, the typical DOS obtained with the cluster-TMT is expected to be close in magnitude to the one obtained with the real-space cluster-CPA with averaged effective medium. Such real space cluster extension of CPA is different from other existing cluster extensions, including the DCA [55,56] and non-local CPA [95]. The difference is that, in the real space cluster-CPA, all the quantities are matrices in the real space, and the coarse-graining step for $\hat{G}$ uses a projected lattice dispersion which is constrained to the real space cluster space.

In the next step of the cluster-TMT self-consistency loop, we must calculate the cluster typical Green's function $\hat{G}_{typ}$ ($\hat{G}_{ave}$ for the cluster-CPA) using the Hilbert transform. The Hilbert transform is performed for each matrix element individually, $G_{typ,ij}(w) = \frac{1}{\pi} \int d\omega' \frac{\rho_{typ,ij}(\omega')}{\omega - \omega'}$.

Next, using the Dyson's equation, we get the updated self-energy $\hat{\Sigma}(\omega) = \hat{\mathcal{G}}^{-1}(\omega) - \hat{G}_{typ}^{-1}(\omega)$, which is then used to calculate the coarse-grained lattice Green's functions matrix $\hat{G}$. The whole procedure then repeats, until convergence is reached with the desired accuracy.

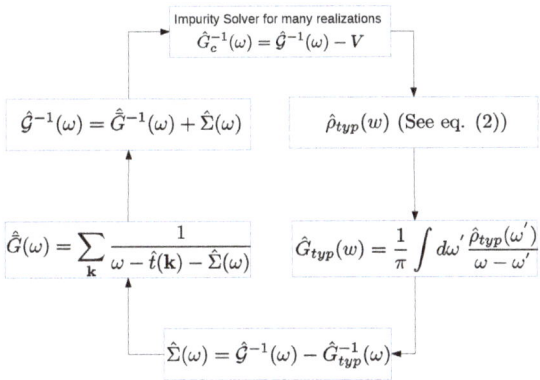

Figure 2. The self-consistency algorithm for the real space cluster-TMT formalism.

4. Results

We start the discussion of our results for the 3D Anderson model (for a box disorder distribution) by first focusing in panel a of Figure 3. This panel displays the $N_c = 3^3$ cluster average DOS (ADOS=$\frac{1}{N_c}\Sigma_i(\frac{-1}{\pi})\Im \hat{G}_{c,ii}(\omega)$) obtained using the average effective medium (constructed from Equation (3)) in the cluster self-consistent loop. These results correspond to the real-space cluster extension of the CPA. The data show that, as disorder strength W increases, the ADOS broadens and gets smaller, but does not go through significant qualitative changes when the metal-insulator transition is approached.

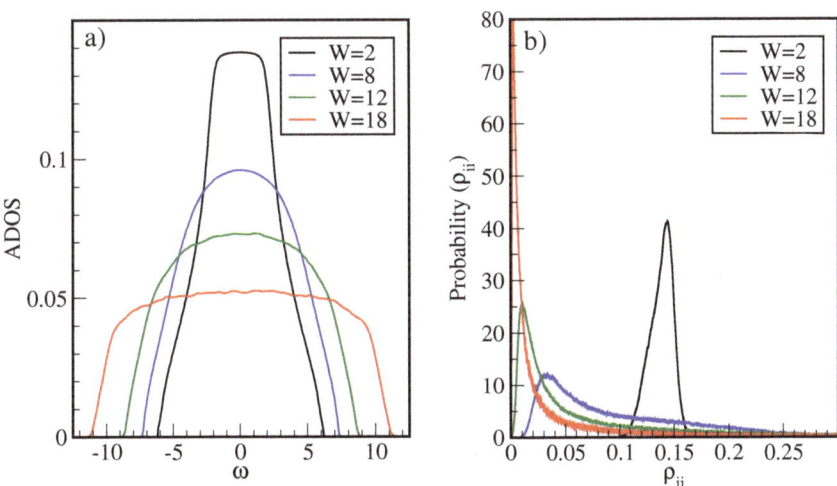

Figure 3. (a) The ADOS calculated for $N_c = 3^3$ at several disorder strengths $W = 2, 8, 12, 18$; (b) the probability distribution function of the local density of states ρ_{ii} for several values of disorder strengths, $W = 2, 8, 12, 18$.

To demonstrate why the ADOS fails to describe the Anderson transition, we display the probability distribution of the local density of states in panel b of Figure 3. At small disorder $W = 2$, the distribution of the LDOS is Gaussian-like. However, as disorder strength increases, the probability distribution becomes skewed with long tails (indicating that the system is not self-averaging). At even larger disorder strength ($W = 18$), the probability distribution peaks at values very close to zero. Such skewness of the distribution functions for large disorder strengths implies that the average and the most probable (typical) values of the DOS will differ significantly, and hence the numerical algorithms that employ the globally averaged Green's function in the self-consistency loop (e.g., the CPA, the DCA) will fail to describe the Anderson transition.

These results clearly demonstrate that the typical medium treatment is required to capture the non self-averaging behavior through the Anderson transition. To show this, in Figure 4, we compare the data for the energy resolved ADOS and the TDOS calculated for a cluster of $N_c = 3^3$ sites. The $\text{TDOS}(\omega) = exp(\frac{1}{N_c}\Sigma_i ln((\frac{-1}{\pi})\Im \bar{G}_{c,ii}(\omega)))$ is obtained from the present real-space cluster-TMT procedure which employs the geometric averaging in the self-consistency loop. At weak disorder strength ($W = 2.5$), as expected from our analytical arguments, both ADOS and TDOS are practically the same, indicating that, when $W \ll W_c$, the real space cluster-TMT reduces to the cluster-CPA scheme. As disorder strength increases, the ADOS and TDOS behave very differently. While the ADOS(ω) broadens and remains finite, the TDOS(ω) gets continuously suppressed ($W = 10$) and vanishes at even larger disorder strength ($W = 16$). Such vanishing of the TDOS at strong disorder values indicates that geometrically average DOS can be used as an order parameter for the Anderson localized states.

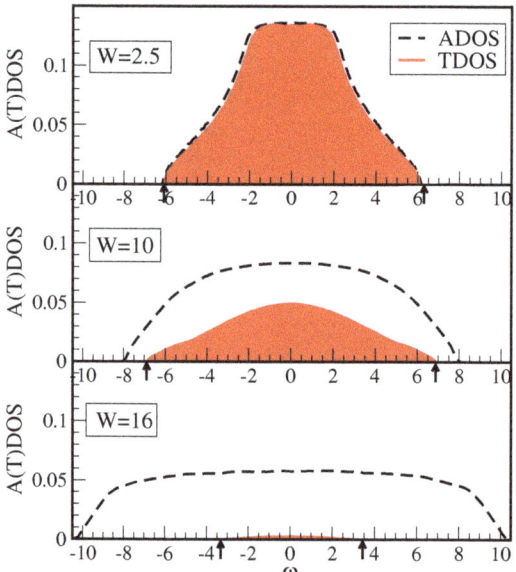

Figure 4. Evolution of the ADOS (dash lines) and the TDOS (shaded areas) as function of frequency ω at different disorder strengths $W = 2.5, 10, 16$, calculated using a cluster of $N_c = 3^3$. The approximate positions of the mobility edge boundaries are marked by vertical arrows.

Notice that, below the Anderson transition, for $W \ll W_c$, localization of states starts at the band tails. This is indicated by the vanishing TDOS(ω) and the finite ADOS(ω) at higher frequencies ω. The mobility edge (shown by arrows), i.e., the energy which separates the extended (with a finite TDOS) from the localized states (with zero TDOS) follows the expected re-entrance trajectory [52]: the mobility edge first expands beyond the zero disorder edge boundary, and then retracts at larger disorder strengths.

Next, we consider the evolution of the critical disorder strength W_c for the Anderson transition as a function of the cluster size N_c. The critical disorder W_c is extracted from the vanishing TDOS at the band center (TDOS($\omega = 0$)). In Figure 5, we plot TDOS($\omega = 0$) as a function of disorder strength W for several cluster sizes $N_c = 1, 2, 3, 2^3, 3^3, 4^3$. For $N_c = 1$ (the local TMT case), the critical disorder is $W_c \approx 13.4$. Since TMT is a mean field theory, it is expected that the critical disorder strength is underestimated and thus it is lower than the exact value. As the cluster size N_c increases, more spatial fluctuations are taken into account, which improves the value of W_c. With increasing N_c, the W_c converges quickly to $W_c \approx 17.05$ (see the inset of Figure 5), which is in good agreement with the values of W_c reported in the literature [90]. In addition, notice that, unlike the TDOS, the ADOS($w = 0$) (shown by the dashed line in Figure 5) remains finite as the disorder strength W increases, indicating that it can not be used as an order parameter for the Anderson transition, and hence the typical medium treatment is needed.

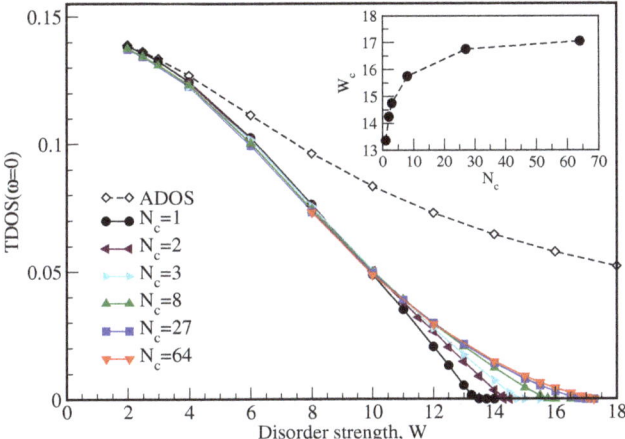

Figure 5. The typical density of states (solid lines) at the band center, $TDOS(\omega = 0)$, as a function of disorder strength W calculated for different cluster sizes $N_c = 1, 2, 3, 8, 27, 64$. The ADOS($\omega = 0$) as a function of disorder strength W is obtained for $N_c = 4^3$ (dashed line). Inset: the cluster size N_c dependence of the critical disorder strength W_c determined from the vanishing TDOS($\omega = 0$).

Finally, in Figure 6, we present the disorder strength W vs. frequency ω phase diagram. Here, we plot the cluster size N_c dependence of the mobility edge boundaries at different disorder strengths W obtained by our real space cluster-TMT formalism. In addition, we also show the band edges, which are defined by the frequencies at which ADOS(w) = 0. As we discussed above, a signature of the cluster mean field theory is the re-entrance at high energy. At $N_c > 1$, the mobility edge boundaries first expand and then retract back with increasing W. As Figure 6 displays, such re-entrance behavior is missing in the single site ($N_c = 1$) TMT case, and is recovered for $N_c > 1$ clusters. This indicates that non-local spacial correlations and multiple-scattering effects in the Anderson transition are important, and capturing such effects requires the usage of finite cluster methods. To benchmark our results even further, we also present the mobility edge trajectories obtained from the highly accurate transfer matrix method (TMM) [54]. For $N_c = 4^3$, the cluster-TMT results are already rather close to those of the TMM. These results demonstrate that our cluster-TMT method can be used to successfully describe the electron localization in the 3D Anderson model.

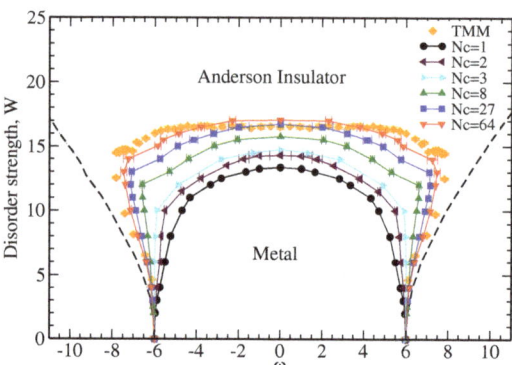

Figure 6. Disorder strength W vs. frequency ω phase diagram of 3D Anderson model obtained from cluster-TMT calculations. The mobility edge boundaries (solid lines) are obtained for $N_c = 1, 2, 3, 8, 27, 64$ cluster sizes. The dashed lines mark the band edges obtained from the ADOS(ω). The transfer matrix method (TMM) mobility edge boundaries are taken from Ref. [52].

5. Conclusions

We develop a real space quantum cluster theory based on the typical medium theory for random disorder systems. Unlike the coherent potential approximation with the algebraically average effective medium, the TMT captures the localization transition by considering the geometrically averaged local density of states to construct an effective medium. However, being a single site theory, the TMT underestimates the critical disorder strength of the transition, and misses the re-entrance behavior, which is due to the combined effects from multiple scattering sites. Recent studies based on the dynamical cluster approximation already confirmed that such non-local effects can be captured by considering momentum-space clusters extension of TMT [54].

In this paper, we construct the real space variant of the cluster-TMT. This method by construction is similar to the cellular dynamical mean field theory [72], which is a popular cluster method effectively used for strongly interacting electron systems. Here, we adopt such a real space cluster approach to disordered systems. Applying our real-space cluster-TMT approach to the 3D Anderson model with a box distribution, we demonstrate that the cluster-TMT is a successful self-consistent numerical approach to capture the Anderson localization transition. Performing N_c cluster-size analysis, we demonstrate the importance of including non-local spacial effects to properly describe the Anderson localization physics. Quantitatively, our results are in good agreement with existing data; in particular, we find that the converged cluster value of $W_c^{cluster-TMT} \approx 17.05$ is superior to the value predicted by single-site TMT, $W^{TMT} \approx 13.4$. Unlike the single site approach, the present real-space cluster-TMT captures the re-entrance behavior and correctly reproduces the phase diagram of the 3D Anderson model. The method, in principle, can also be used to calculate two particle quantities [96]. Furthermore, while the cluster TMT in this study has been restricted to periodic boundary conditions, the same methodology can be used to simulate Anderson localization in surfaces. This will be relevant, for example, to unraveling the role of disorder in topological materials [75,97]. Another interesting topic is to combine this approach with the multiple scattering theory [58], and the locally self-consistent multiple scattering method [66] for the study of materials with random disorder.

Author Contributions: Investigation, K.-M.T., H.T., T.B., L.C. and J.M.; Writing—original draft, K.-M.T. and H.T.; Writing—review & editing, K.-M.T., H.T., T.B., L.C. and J.M. All authors have read and agreed to the published version of the manuscript.

Funding: This manuscript is based upon work supported by the U.S. Department of Energy, Office of Science, Office of Basic Energy Sciences under Award Number DE-SC0017861. This work used the high performance computational resources provided by the Louisiana Optical Network Initiative http://www.loni.org, accessed on 16 September 2021, and HPC@LSU computing. This work also used the Extreme Science and Engineering Discovery Environment (XSEDE) through allocation DMR130036. K.-M.T. is partially supported by NSF DMR-1728457 and NSF OAC-1931445. H.T. has been supported by NSF OAC-1931367 and NSF DMR-1944974 grants. L.C. acknowledges the financial support by the Deutsche Forschungsgemeinschaft through TRR80 (project F6) project No. 107745057. A portion of this research was conducted at the Center for Nanophase Materials Sciences, which is a DOE Office of Science User Facility (TB). The manuscript has been authored by UT-Battelle, LLC under Contract No. DE-AC05-00OR22725 with the U.S. Department of Energy. The United States Government retains and the publisher, by accepting the article for publication, acknowledges that the United States Government retains a non-exclusive, paid-up, irrevocable, worldwide license to publish or reproduce the published form of this manuscript, or allow others to do so, for United States Government purposes. The Department of Energy will provide public access to these results of federally sponsored research in accordance with the DOE Public Access Plan (http://energy.gov/downloads/doe-public-access-plan, accessed on 16 September 2021).

Institutional Review Board Statement: Not applicable.

Informed Consent Statement: Not applicable.

Data Availability Statement: Correspondence and requests for data should be addressed to K.-M.T. or H.T.

Acknowledgments: The authors would like to thank V. Dobrosavljevic and S. Iskakov for useful comments and discussions.

Conflicts of Interest: The authors declare no conflict of interest.

Appendix A

In this section, we discuss the role of the non-local off-diagonal components in the ansatz for the geometrically averaged cluster Green's function of Equation (2). For this, we consider the "local" ansatz given by Equation (A1), where we set all off-diagonal terms in the typical DOS equal to zero:

$$\hat{\rho}_{typ}^{local}(\omega) \equiv \begin{pmatrix} e^{\langle \ln \rho_{11}(\omega) \rangle} & \cdots & 0 \\ \vdots & \ddots & \vdots \\ 0 & \cdots & e^{\langle \ln \rho_{N_c N_c}(\omega) \rangle} \end{pmatrix}. \tag{A1}$$

In Figure A1, we then compare the $N_c = 8$ results for the TDOS(ω) obtained with the "full" ansatz (Equation (2)) and the "local" ansatz (Equation (A1)) calculated at several values of the disorder strength: $W = 8.0, 10, 12, 14$. Our data indicate that the majority of the contribution to the TDOS(ω) is actually coming from the local terms in Equation (2). The critical behavior at the Fermi level ($\omega = 0$) is the same for both the "local" and the "full" ansatz. However, the non-local contribution seems to be important for properly capturing the mobility edge behavior (marked by vertical arrows in Figure A1). Here, at the mobility edges, we observe the most pronounced difference between the TDOS(ω) obtained using the "local" and the "full" ansatz. These results indicate that, while the critical behavior at the band center is captured properly by the "local" ansatz, the mobility edge trajectories of the "local" ansatz, however, will converge slower with the cluster size N_c. To demonstrate this explicitly, in Figure A2 (left panel), we plot the typical density of states as a function of disorder strength W at the band center (TDOS($\omega = 0$)). The critical value of disorder strength (W_c) at which the TDOS($\omega = 0$) = 0 vanishes at the band center is the same for both local and non-local ansatzes. However, as shown in the right panel of Figure A2, there is a substantial difference in the phase boundary near the band edges. Specifically, with the off-diagonal components, the re-entrance effect is much more pronounced even if the cluster size is relatively small. The off-diagonal components provide the contribution from the scattering among multiple sites, and hence generate more accurate results which are much closer to the results from the highly accurate transfer matrix method.

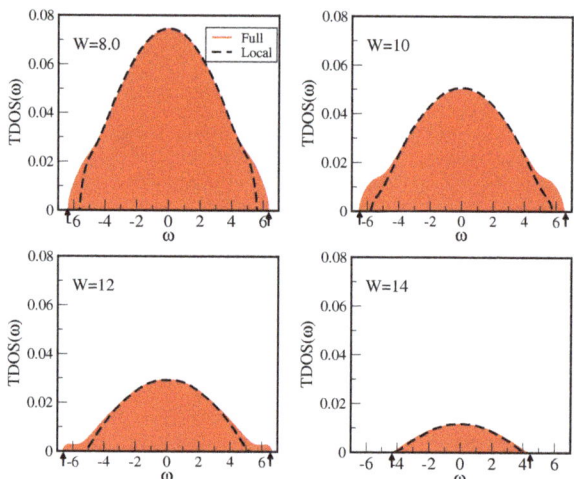

Figure A1. $N_c = 8$ results for the TDOS(ω) at increasing disorder strengths $W = 8, 10, 12, 14$. The data for the TDOS obtained using the "full" ansatz of Equation (2) (red shaded region), and the TDOS curves obtained using the simplified "local" ansatz of Equation (A1) (dashed lines), where the off-diagonal non-local contributions are set to zero. Vertical arrows mark the mobility edge boundaries.

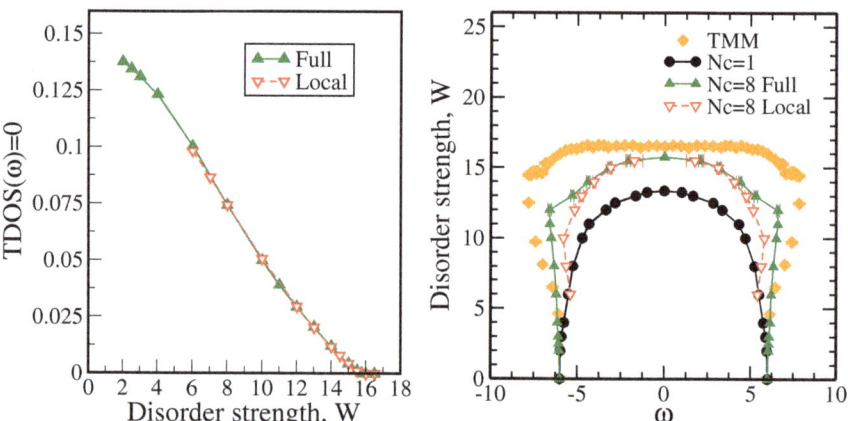

Figure A2. (**Left**): $N_c = 8$ cluster $TDOS(w = 0)$ vs disorder strength W calculated using the "full" (Equation (2)) and "local" (Equation (A1)) ansatzes. (**Right**): Disorder strength W vs. frequency ω phase diagram for the 3D Anderson model for $N_c = 1$ and $N_c = 8$ clusters. The mobility edge boundaries for $N_c = 8$ clusters are obtained using the "full" and "local" ansatzes. The transfer matrix method (TMM) mobility edge boundaries are taken from Ref. [52].

References

1. Anderson, P.W. Absence of Diffusion in Certain Random Lattices. *Phys. Rev.* **1958**, *109*, 1492–1505. [CrossRef]
2. Abrahams, E. (Ed.) *50 Years of Anderson Localization*; World Scientific: Singapore, 2010.
3. Vollhardt, D.; Wölfle, P. Anderson Localization in $d < \sim 2$ Dimensions: A Self-Consistent Diagrammatic Theory. *Phys. Rev. Lett.* **1980**, *45*, 842–846. [CrossRef]
4. Vollhardt, D.; Wölfle, P. Diagrammatic, self-consistent treatment of the Anderson localization problem in $d \leq 2$ dimensions. *Phys. Rev. B* **1980**, *22*, 4666–4679. [CrossRef]
5. Kramer, B.; MacKinnon, A. Localization: Theory and experiment. *Rep. Prog. Phys.* **1993**, *56*, 1469. [CrossRef]
6. John, S. Electromagnetic Absorption in a Disordered Medium near a Photon Mobility Edge. *Phys. Rev. Lett.* **1984**, *53*, 2169–2172. [CrossRef]
7. John, S. Strong localization of photons in certain disordered dielectric superlattices. *Phys. Rev. Lett.* **1987**, *58*, 2486–2489. [CrossRef]
8. Wolf, P.E.; Maret, G. Weak Localization and Coherent Backscattering of Photons in Disordered Media. *Phys. Rev. Lett.* **1985**, *55*, 2696–2699. [CrossRef] [PubMed]
9. Albada, M.P.V.; Lagendijk, A. Observation of Weak Localization of Light in a Random Medium. *Phys. Rev. Lett.* **1985**, *55*, 2692–2695. [CrossRef] [PubMed]
10. Tsang, L.; Ishimaru, A. Backscattering enhancement of random discrete scatterers. *J. Opt. Soc. Am. A* **1984**, *1*, 836–839. [CrossRef]
11. Wiersma, D.S.; Bartolini, P.; Lagendijk, A.; Righini, R. Localization of light in a disordered medium. *Nature* **1997**, *390*, 671–673. [CrossRef]
12. Störzer, M.; Gross, P.; Aegerter, C.M.; Maret, G. Observation of the critical regime near Anderson localization of light. *Phys. Rev. Lett.* **2006**, *96*, 063904. [CrossRef] [PubMed]
13. Sperling, T.; Buehrer, W.; Aegerter, C.M.; Maret, G. Direct determination of the transition to localization of light in three dimensions. *Nat. Photonics* **2013**, *7*, 48–52. [CrossRef]
14. Skipetrov, S.; Page, J.H. Red light for Anderson localization. *New J. Phys.* **2016**, *18*, 021001. [CrossRef]
15. Skipetrov, S.E.; Sokolov, I.M. Absence of Anderson Localization of Light in a Random Ensemble of Point Scatterers. *Phys. Rev. Lett.* **2014**, *112*, 023905. [CrossRef]
16. Sperling, T.; Schertel, L.; Ackermann, M.; Aubry, G.J.; Aegerter, C.M.; Maret, G. Can 3D light localization be reached in 'white paint'? *New J. Phys.* **2016**, *18*, 013039. [CrossRef]
17. Ángel, J.C.; Guzmán, J.T.; de Anda, A.D. Anderson localization of flexural waves in disordered elastic beams. *Sci. Rep.* **2019**, *9*, 3572. [CrossRef] [PubMed]
18. Frank, R.; Lubatsch, A.; Kroha, J. Theory of strong localization effects of light in disordered loss or gain media. *Phys. Rev. B* **2006**, *73*, 245107. [CrossRef]
19. Lubatsch, A.; Frank, R. Self-consistent quantum field theory for the characterization of complex random media by short laser pulses. *Phys. Rev. Res.* **2020**, *2*, 013324. [CrossRef]
20. Razo-López, L.; Fernández-Marín, A.; Méndez-Bermúdez, J.; Sánchez-Dehesa, J.; Gopar, V. Delay time of waves performing Lévy walks in 1D random media. *Sci. Rep.* **2020**, *10*, 20816. [CrossRef]
21. Kostadinova, E.G.; Padgett, J.L.; Liaw, C.D.; Matthews, L.S.; Hyde, T.W. Numerical study of anomalous diffusion of light in semicrystalline polymer structures. *Phys. Rev. Res.* **2020**, *2*, 043375. [CrossRef]
22. Ziegler, K. Ray Modes in Random Gap Systems. *Ann. Phys.* **2017**, *529*, 1600345. [CrossRef]
23. Leseur, O.; Pierrat, R.; Sáenz, J.J.; Carminati, R. Probing two-dimensional Anderson localization without statistics. *Phys. Rev. A* **2014**, *90*, 053827. [CrossRef]
24. Chabé, J.; Rouabah, M.T.; Bellando, L.; Bienaimé, T.; Piovella, N.; Bachelard, R.; Kaiser, R. Coherent and incoherent multiple scattering. *Phys. Rev. A* **2014**, *89*, 043833. [CrossRef]
25. Mafi, A.; Karbasi, S.; Koch, K.W.; Hawkins, T.; Ballato, J. Transverse Anderson Localization in Disordered Glass Optical Fibers: A Review. *Materials* **2014**, *7*, 5520–5527. [CrossRef] [PubMed]
26. White, D.H.; Haase, T.A.; Brown, D.J.; Hoogerland, M.D.; Najafabadi, M.S.; Helm, J.L.; Gies, C.; Schumayer, D.; Hutchinson, D.A.W. Observation of two-dimensional Anderson localisation of ultracold atoms. *Nat. Commun.* **2020**, *11*, 4942. [CrossRef] [PubMed]
27. Abou-Chacra, R.; Thouless, D.J.; Anderson, P.W. A Selfconsistent Theory of Localization. *J. Phys. C Solid State Phys.* **1973**, *6*, 1734–1752. [CrossRef]
28. Soven, P. Coherent-Potential Model of Substitutional Disordered Alloys. *Phys. Rev.* **1967**, *156*, 809–813. [CrossRef]
29. Shiba, H. A Reformulation of the Coherent Potential Approximation and Its Applications. *Prog. Theor. Phys.* **1971**, *46*, 77. [CrossRef]
30. Velický, B.; Kirkpatrick, S.; Ehrenreich, H. Single-Site Approximations in the Electronic Theory of Simple Binary Alloys. *Phys. Rev.* **1968**, *175*, 747–766. [CrossRef]
31. Kirkpatrick, S.; Velický, B.; Ehrenreich, H. Paramagnetic NiCu Alloys: Electronic Density of States in the Coherent-Potential Approximation. *Phys. Rev. B* **1970**, *1*, 3250–3263. [CrossRef]
32. Onodera, Y.; Toyozawa, Y. Persistence and Amalgamation Types in the Electronic Structure of Mixed Crystals. *J. Phys. Soc. Jpn.* **1968**, *24*, 341–355. [CrossRef]

33. Taylor, D. Vibrational Properties of Imperfect Crystals with Large Defect Concentrations. *Phys. Rev.* **1967**, *156*, 1017–1029. [CrossRef]
34. Yonezawa, F. A Systematic Approach to the Problems of Random Lattices. I: A Self-Contained First-Order Approximation Taking into Account the Exclusion Effect. *Prog. Theor. Phys.* **1968**, *40*, 734–757. [CrossRef]
35. Weh, A.; Zhang, Y.; Östlin, A.; Terletska, H.; Bauernfeind, D.; Tam, K.M.; Evertz, H.G.; Byczuk, K.; Vollhardt, D.; Chioncel, L. Dynamical mean-field theory of the Anderson-Hubbard model with local and nonlocal disorder in tensor formulation. *Phys. Rev. B* **2021**, *104*, 045127. [CrossRef]
36. Dobrosavljević, V.; Pastor, A.A.; Nikolić, B.K. Typical medium theory of Anderson localization: A local order parameter approach to strong-disorder effects. *EPL Europhys. Lett.* **2003**, *62*, 76. [CrossRef]
37. Schubert, G.; Schleede, J.; Byczuk, K.; Fehske, H.; Vollhardt, D. Distribution of the local density of states as a criterion for Anderson localization: Numerically exact results for various lattices in two and three dimensions. *Phys. Rev. B* **2010**, *81*, 155106. [CrossRef]
38. Byczuk, K.; Hofstetter, W.; Vollhardt, D. Mott-Hubbard Transition versus Anderson Localization in Correlated Electron Systems with Disorder. *Phys. Rev. Lett.* **2005**, *94*, 056404. [CrossRef] [PubMed]
39. Semmler, D.; Byczuk, K.; Hofstetter, W. Mott-Hubbard and Anderson metal-insulator transitions in correlated lattice fermions with binary disorder. *Phys. Rev. B* **2010**, *81*, 115111. [CrossRef]
40. Murphy, N.C.; Wortis, R.; Atkinson, W.A. Generalized inverse participation ratio as a possible measure of localization for interacting systems. *Phys. Rev. B* **2011**, *83*, 184206. [CrossRef]
41. Aguiar, M.C.O.; Dobrosavljević, V.; Abrahams, E.; Kotliar, G. Critical Behavior at the Mott-Anderson Transition: A Typical-Medium Theory Perspective. *Phys. Rev. Lett.* **2009**, *102*, 156402. [CrossRef]
42. Aguiar, M.C.O.; Dobrosavljević, V. Universal Quantum Criticality at the Mott-Anderson Transition. *Phys. Rev. Lett.* **2013**, *110*, 066401. [CrossRef]
43. Oliveira, W.S.; Aguiar, M.C.O.; Dobrosavljević, V. Mott-Anderson transition in disordered charge-transfer model: Insights from typical medium theory. *Phys. Rev. B* **2014**, *89*, 165138. [CrossRef]
44. Bragança, H.; Aguiar, M.C.O.; Vučičević, J.; Tanasković, D.; Dobrosavljević, V. Anderson localization effects near the Mott metal-insulator transition. *Phys. Rev. B* **2015**, *92*, 125143. [CrossRef]
45. Dobrosavljević, V. Typical-Medium Theory of Mott–Anderson Localization. *Int. J. Mod. Phys. B* **2010**, *24*, 1680–1726. [CrossRef]
46. Byczuk, K.; Hofsletter, W.; Yu, U.; Vollhardt, D. Correlated electrons in the presence of disoder. *Eur. Phys. J. Spec. Top.* **2009**, *180*, 135–151. [CrossRef]
47. Byczuk, K.; Hofstetter, W.; Vollhardt, D. Anderson Localization VS. Mott-Hubbard Metal-Insulator Transition in Disordered, Interacting Lattice Fermion Systems. *Int. J. Mod. Phys. B* **2010**, *24*, 1727–1755. [CrossRef]
48. Alvermann, A.; Schubert, G.; Weiße, A.; Bronold, F.; Fehske, H. Characterisation of Anderson localisation using distribution. *Phys. B Condens. Matter* **2005**, *359–361*, 789–791. [CrossRef]
49. Janssen, M. Mutifractal Analysis of Broadly Distributed Observables at Criticality. *Int. J. Mod. Phys. B* **1994**, *8*, 943. [CrossRef]
50. Janssen, M. Statistics and scaling in disordered mesoscopic electronic systems. *Phys. Rep.* **1998**, *295*, 1–91. [CrossRef]
51. Logan, D.E.; Wolynes, P.G. Dephasing and Anderson localization in topologically disordered systems. *Phys. Rev. B* **1987**, *36*, 4135–4147. [CrossRef]
52. Ekuma, C.E.; Terletska, H.; Tam, K.M.; Meng, Z.Y.; Moreno, J.; Jarrell, M. Typical medium dynamical cluster approximation for the study of Anderson localization in three dimensions. *Phys. Rev. B* **2014**, *89*, 081107. [CrossRef]
53. Ekuma, C.E.; Moore, C.; Terletska, H.; Tam, K.M.; Moreno, J.; Jarrell, M.; Vidhyadhiraja, N.S. Finite-cluster typical medium theory for disordered electronic systems. *Phys. Rev. B* **2015**, *92*, 014209. [CrossRef]
54. Terletska, H.; Zhang, Y.; Tam, K.M.; Berlijn, T.; Chioncel, L.; Vidhyadhiraja, N.; Jarrell, M. Systematic quantum cluster typical medium method for the study of localization in strongly disordered electronic systems. *Appl. Sci.* **2018**, *8*, 2401. [CrossRef]
55. Jarrell, M.; Krishnamurthy, H.R. Systematic and causal corrections to the coherent potential approximation. *Phys. Rev. B* **2001**, *63*, 125102. [CrossRef]
56. Jarrell, M.; Maier, T.; Huscroft, C.; Moukouri, S. Quantum Monte Carlo algorithm for nonlocal corrections to the dynamical mean-field approximation. *Phys. Rev. B* **2001**, *64*, 195130. [CrossRef]
57. Sen, S.; Terletska, H.; Moreno, J.; Vidhyadhiraja, N.S.; Jarrell, M. Local theory for Mott-Anderson localization. *Phys. Rev. B* **2016**, *94*, 235104. [CrossRef]
58. Terletska, H.; Zhang, Y.; Chioncel, L.; Vollhardt, D.; Jarrell, M. Typical-medium multiple-scattering theory for disordered systems with Anderson localization. *Phys. Rev. B* **2017**, *95*, 134204. [CrossRef]
59. Terletska, H.; Ekuma, C.E.; Moore, C.; Tam, K.M.; Moreno, J.; Jarrell, M. Study of off-diagonal disorder using the typical medium dynamical cluster approximation. *Phys. Rev. B* **2014**, *90*, 094208. [CrossRef]
60. Mondal, W.R.; Berlijn, T.; Jarrell, M.; Vidhyadhiraja, N.S. Phonon localization in binary alloys with diagonal and off-diagonal disorder: A cluster Green's function approach. *Phys. Rev. B* **2019**, *99*, 134203. [CrossRef]
61. Mondal, W.R.; Vidhyadhiraja, N.S. Effect of short-ranged spatial correlations on the Anderson localization of phonons in mass-disordered systems. *Bull. Mater. Sci.* **2020**, *43*, 314. [CrossRef]
62. Zhang, Y.; Terletska, H.; Moore, C.; Ekuma, C.; Tam, K.M.; Berlijn, T.; Ku, W.; Moreno, J.; Jarrell, M. Study of multiband disordered systems using the typical medium dynamical cluster approximation. *Phys. Rev. B* **2015**, *92*, 205111. [CrossRef]

63. Zhang, Y.; Nelson, R.; Siddiqui, E.; Tam, K.M.; Yu, U.; Berlijn, T.; Ku, W.; Vidhyadhiraja, N.S.; Moreno, J.; Jarrell, M. Generalized multiband typical medium dynamical cluster approximation: Application to (Ga,Mn)N. *Phys. Rev. B* **2016**, *94*, 224208. [CrossRef]
64. Zhang, Y.; Nelson, R.; Tam, K.M.; Ku, W.; Yu, U.; Vidhyadhiraja, N.S.; Terletska, H.; Moreno, J.; Jarrell, M.; Berlijn, T. Origin of localization in Ti-doped Si. *Phys. Rev. B* **2018**, *98*, 174204. [CrossRef]
65. Östlin, A.; Zhang, Y.; Terletska, H.; Beiuşeanu, F.; Popescu, V.; Byczuk, K.; Vitos, L.; Jarrell, M.; Vollhardt, D.; Chioncel, L. Ab initio typical medium theory of substitutional disorder. *Phys. Rev. B* **2020**, *101*, 014210. [CrossRef]
66. Zhang, Y.; Terletska, H.; Tam, K.M.; Wang, Y.; Eisenbach, M.; Chioncel, L.; Jarrell, M. Locally self-consistent embedding approach for disordered electronic systems. *Phys. Rev. B* **2019**, *100*, 054205. [CrossRef]
67. Terletska, H.; Moilanen, A.; Tam, K.M.; Zhang, Y.; Wang, Y.; Eisenbach, M.; Vidhyadhiraja, N.; Chioncel, L.; Moreno, J. Non-local corrections to the typical medium theory of Anderson localization. *Ann. Phys.* **2021**, 168454. [CrossRef]
68. Tam, K.M.; Zhang, Y.; Terletska, H.; Wang, Y.; Eisenbach, M.; Chioncel, L.; Moreno, J. Application of the locally self-consistent embedding approach to the Anderson model with non-uniform random distributions. *Ann. Phys.* **2021**, 168480. [CrossRef]
69. Georges, A.; Kotliar, G.; Krauth, W. Superconductivity in the Two-Band Hubbard Model in Infinite Dimensions. *Z. Phys. B Condens. Matter* **1993**, *92*, 313–321. [CrossRef]
70. Biroli, G.; Kotliar, G. Cluster methods for strongly correlated electron systems. *Phys. Rev. B* **2002**, *65*, 155112. [CrossRef]
71. Biroli, G.; Parcollet, O.; Kotliar, G. Cluster dynamical mean-field theories: Causality and classical limit. *Phys. Rev. B* **2004**, *69*, 205108. [CrossRef]
72. Kotliar, G.; Savrasov, S.; Palsson, G.; Biroli, G. Cellular Dynamical Mean Field Approach to Strongly Correlated Systems. *Phys. Rev. Lett.* **2001**, *87*, 186401. [CrossRef]
73. Lichtenstein, A.I.; Katsnelson, M.I. Antiferromagnetism and d-wave superconductivity in cuprates: A cluster dynamical mean-field theory. *Phys. Rev. B* **2000**, *62*, R9283–R9286. [CrossRef]
74. Bulka, B.; Kramer, B.; MacKinnon, A. Mobility edge in the three-dimensional Anderson model. *Z. Phys. B Condens. Matter* **1985**, *60*, 13–17. [CrossRef]
75. Li, J.; Chu, R.L.; Jain, J.K.; Shen, S.Q. Topological Anderson Insulator. *Phys. Rev. Lett.* **2009**, *102*, 136806. [CrossRef]
76. Yonezawa, F.; Morigaki, K. Coherent Potential Approximation. Basic concepts and applications. *Prog. Theor. Phys. Supp.* **1973**, *53*, 1–76. [CrossRef]
77. Georges, A.; Kotliar, G.; Krauth, W.; Rozenberg, M. Dynamical mean-field theory of strongly correlated fermion systems and the limit of infinite dimensions. *Rev. Mod. Phys.* **1996**, *68*, 13–125. [CrossRef]
78. Maier, T.; Jarrell, M.; Pruschke, T.; Hettler, M.H. Quantum cluster theories. *Rev. Mod. Phys.* **2005**, *77*, 1027–1080. [CrossRef]
79. Lee, P.A.; Ramakrishnan, T.V. Disordered electronic systems. *Rev. Mod. Phys.* **1985**, *57*, 287–337. [CrossRef]
80. Selvan, R.; Genish, I.; Perelshtein, I.; Moreno, J.; Gedanken, A. Single step, low temperature synthesis of submicron-sized rare earth hexaborides. *J. Phys. Chem. C* **2008**, *112*, 1795. [CrossRef]
81. Bulka, B.; Schreibe, M.; Kramer, B. Localization, Quantum Interference, and the Metal-Insulator Transition. *Z. Phys. B* **1987**, *66*, 21–30. [CrossRef]
82. Kramer, B.; Schreiber, M. Localization, quantum interference and transport in disordered solids. In *Fluctuations and Stochastic Phenomena in Condensed Matter*; Lecture Notes in Physics; Garrido, L., Ed.; Springer: Berlin/Heidelberg, Germany, 1987; Volume 268, pp. 351–375. [CrossRef]
83. Kramer, B.; MacKinnon, A.; Ohtsuki, T.; Slevin, K. Finite Size Scaling Analysis of the Anderson Transition. *Int. J. Mod. Phys. B* **2010**, *24*, 1841–1854. [CrossRef]
84. Rodriguez, A.; Vasquez, L.J.; Slevin, K.; Römer, R.A. Critical Parameters from a Generalized Multifractal Analysis at the Anderson Transition. *Phys. Rev. Lett.* **2010**, *105*, 046403. [CrossRef]
85. Rodriguez, A.; Vasquez, L.J.; Slevin, K.; Römer, R.A. Multifractal finite-size scaling and universality at the Anderson transition. *Phys. Rev. B* **2011**, *84*, 134209. [CrossRef]
86. Slevin, K.; Ohtsuki, T. Numerical verification of universality for the Anderson transition. *Phys. Rev. B* **2001**, *63*, 045108. [CrossRef]
87. Slevin, K.; Ohtsuki, T. Critical exponent for the Anderson transition in the three-dimensional orthogonal universality class. *New J. Phys.* **2014**, *16*, 015012. [CrossRef]
88. Slevin, K.; Ohtsuki, T. Corrections to Scaling at the Anderson Transition. *Phys. Rev. Lett.* **1999**, *82*, 382–385. [CrossRef]
89. Chang, T.; Bauer, J.D.; Skinner, J.L. Critical exponents for Anderson localization. *J. Chem. Phys.* **1990**, *93*, 8973–8982. [CrossRef]
90. MacKinnon, A.; Kramer, B. The scaling theory of electrons in disordered solids: Additional numerical results. *Z. Phys. B Condens. Matter* **1983**, *53*, 1–13. [CrossRef]
91. de Queiroz, S.L.A. Reentrant behavior and universality in the Anderson transition. *Phys. Rev. B* **2001**, *63*, 214202. [CrossRef]
92. Grussbach, H.; Schreiber, M. Determination of the mobility edge in the Anderson model of localization in three dimensions by multifractal analysis. *Phys. Rev. B* **1995**, *51*, 663–666. [CrossRef]
93. Sénéchal, D. An introduction to quantum cluster methods. *arXiv* **2010**, arXiv:0806.2690.
94. Kraberger, G.J.; Triebl, R.; Zingl, M.; Aichhorn, M. Maximum entropy formalism for the analytic continuation of matrix-valued Green's functions. *Phys. Rev. B* **2017**, *96*, 155128. [CrossRef]
95. Rowlands, D.A. Investigation of the nonlocal coherent-potential approximation. *J. Phys. Condens. Matter* **2006**, *18*, 3179–3195. [CrossRef]

96. Zhang, Y.; Zhang, Y.F.; Yang, S.X.; Tam, K.M.; Vidhyadhiraja, N.S.; Jarrell, M. Calculation of two-particle quantities in the typical medium dynamical cluster approximation. *Phys. Rev. B* **2017**, *95*, 144208. [CrossRef]
97. Roy, B.; Slager, R.J.; Juričić, V. Global Phase Diagram of a Dirty Weyl Liquid and Emergent Superuniversality. *Phys. Rev. X* **2018**, *8*, 031076. [CrossRef]

Article

The FFLO State in the Dimer Mott Organic Superconductor κ-(BEDT-TTF)$_2$Cu[N(CN)$_2$]Br

Shusaku Imajo * and Koichi Kindo

Institute for Solid State Physics, The University of Tokyo, Kashiwa 277-8581, Japan; kindo@issp.u-tokyo.ac.jp
* Correspondence: imajo@issp.u-tokyo.ac.jp

Abstract: The superconducting phase diagram for a quasi-two-dimensional organic superconductor, κ-(BEDT-TTF)$_2$Cu[N(CN)$_2$]Br, was studied using pulsed magnetic field penetration depth measurements under rotating magnetic fields. At low temperatures, H_{c2} was abruptly suppressed even by small tilts of the applied fields owing to the orbital pair-breaking effect. In magnetic fields parallel to the conducting plane, the temperature dependence of the upper critical field H_{c2} exhibited an upturn and exceeded the Pauli limit field H_P in the lower temperature region. Further analyses with the second derivative of the penetration depth showed an anomaly at 31–32 T, which roughly corresponded to H_P. The origin of the anomaly should not be related to the orbital effect, but the paramagnetic effect, which is almost isotropic in organic salts, because it barely depends on the field angle. Based on these results, the observed anomaly is most likely due to the transition between the Bardeen-Cooper-Schrieffer (BCS) and the Fulde-Ferrell-Larkin-Ovchinnikov (FFLO) states. Additionally, we discuss the phase diagram and physical parameters of the transition by comparing them with other FFLO candidates.

Keywords: FFLO; organic superconductor; penetration depth measurement

Citation: Imajo, S.; Kindo, K. The FFLO State in the Dimer Mott Organic Superconductor κ-(BEDT-TTF)$_2$Cu[N(CN)$_2$]Br. *Crystals* **2021**, *11*, 1358. https://doi.org/10.3390/cryst11111358

Academic Editor: Andrej Pustogow

Received: 25 October 2021
Accepted: 5 November 2021
Published: 8 November 2021

Publisher's Note: MDPI stays neutral with regard to jurisdictional claims in published maps and institutional affiliations.

Copyright: © 2021 by the authors. Licensee MDPI, Basel, Switzerland. This article is an open access article distributed under the terms and conditions of the Creative Commons Attribution (CC BY) license (https://creativecommons.org/licenses/by/4.0/).

1. Introduction

Superconductivity is one of the most intriguing topics in material science, both in terms of basic research and applications. Superconductivity appears when electron pairs are formed and condense in metals. The BCS theory explains the conventional superconductivity that appears in a variety of simple metals and alloys. However, the details of unconventional superconductivity are yet to be elucidated. Unconventional superconductivity is commonly realized nearby metal–insulator transitions, where electron correlations are enhanced. Even in unconventional superconductivity, the electrons are paired by attraction, as suggested by the BCS theory; thus, the details of the pairing mechanism are one of the main topics for unconventional superconductivity studies. The FFLO state, which is one of the unconventional pairings, was proposed by Fulde and Ferrell [1] as well as Larkin and Ovchinnikov [2] in 1964. In the FFLO state, the electrons in a pair have unbalanced momenta, and their total center-of-mass momentum q is finite. The finite q, which modifies the superconducting order parameter with the additional term, exp(iqr) for the FF state [1] and cos(qr) for the LO state [2], induces the spatial modulation of the superconductivity. The superconducting region and the normal-state region appear alternately in real space because the normal state appears at the node positions where the additional term becomes zero. The FFLO state is regarded as an inhomogeneous state, which breaks the rotational symmetry [1–5]. At zero field, the uniform BCS-type pairing with $q = 0$ is more stable than the inhomogeneous FFLO state; however, when applying magnetic fields, the Zeeman effect causes the Fermi surface to split depending on the spin directions. Above the field where the Zeeman splitting is comparable with the condensation energy of the superconductivity, known as the Pauli limit H_P [6], the BCS superconductivity is destroyed. This is known as the paramagnetic pair-breaking effect. However, the FFLO state can be favorable even above H_P by pairing on the split Fermi surfaces owing to the finite q. Thus, the FFLO state can appear only at high fields above the H_P. In higher magnetic fields, the FFLO state is also suppressed, and many theories [7–9] predict that the stability in magnetic fields is affected by various parameters. For example, in the case of isotropic three-dimensional superconductivity, the FFLO phase is very small in the H-T phase diagram [1,2]. Moreover, the difference between the FF state and LO state becomes larger in fields sufficiently higher than H_P. It is expected that the superconducting symmetry and the strength of orbital pair-breaking effects also play an important role in the stability of the FFLO state. The investigation of H_{c2} curve above H_P is be important to discuss the details of the FFLO state.

To realize an FFLO state, two conditions need to be met: first, the electronic system should be in a clean limit, and second, the orbital pair-breaking effect should be sufficiently suppressed. The FFLO state hosts the spatial modulation in real space owing to the additional vector q, and impurities smear this modulated pattern with scattering. Therefore, a clean electronic system, in which the mean-free path l is sufficiently larger than the coherence length $\xi_\parallel$, is typically required [4,10]. Some theories suggest that the FFLO state can survive even in some disordered systems [11,12]. For the orbital effect, the Maki parameter α_M, $2^{1/2} H_{orb}/H_P$, where H_{orb} is the orbital limit, must exceed 1.8 [13,14], because the superconductivity gets destroyed at lower fields before the FFLO state appears if the orbital effect is strong. Basically, for the most superconductivity, the orbital effect is so strong that superconductivity does not survive up to H_P. The orbital effect is suppressed when the vortices do not penetrate the superconductor, or the coherence length is sufficiently small, because it originates from the kinetic energy of the supercurrent around the vortices by the Lorentz force. Therefore, the FFLO state may be possible when a magnetic field is precisely applied parallel to the conduction plane of the low-dimensional superconductor, to prevent the magnetic flux from penetrating the superconducting plane, or in the case of heavy electron systems. [4,15,16].

The charge-transfer complex κ-(BEDT-TTF)$_2$Cu[N(CN)$_2$]Br (hereafter, abbreviated as κ-Br) is known as a high-T_c (~12 K) organic superconductor. This salt consists of the organic donor BEDT-TTF layer and the counter anion layer Cu[N(CN)$_2$]Br, as shown in Figure 1. κ-Br has intensively been investigated because of its unconventional superconductivity and proximity to the Mott metal–insulator transition [17,18]. The superconductivity is presumably classified in the d-wave symmetry originating from antiferromagnetic spin fluctuations [17–20], which grow near the antiferromagnetic Mott insulator phase. The superconductivity has a large superconducting energy gap, leading to a large upper critical field H_{c2}. Due to the experimental difficulty in performing high-field measurements up to H_{c2}, the superconducting phase diagram in magnetic fields has not been clarified completely until our recent report [21]. The field-temperature superconducting phase diagram exhibited an upturn of H_{c2} in a low-temperature and high-field region, which may exceed H_P. Moreover, this can be scaled with that of κ-(BEDT-TTF)$_2$Cu(NCS)$_2$ (κ-NCS), which is one of the prime FFLO candidates [22–25]. Basically, the large effective mass and the electronically quasi-two-dimensionality, suppressing the orbital effect and enhancing the nesting of Fermi surfaces, are advantageous for stabilizing the FFLO state. Although this implies that κ-Br also hosts the FFLO state above H_P, there have been no reports on the FFLO phase for κ-Br. Since the electronic structures in the other organic FFLO candidates found so far, such as λ-(BETS)$_2$GaCl$_4$ (λ-GaCl$_4$), β″-(BEDT-TTF)$_2$SF$_5$CH$_2$CF$_2$SO$_3$ (β″-SF$_5$), and β″-(BEDT-TTF)$_4$[(H$_3$O)Ga(C$_2$O$_4$)$_3$]PhNO$_2$ (β″-GaPhNO$_2$), are different in various aspects, it is necessary to consider several factors for discussing the parameters of the FFLO state. The comparison κ-Br with κ-NCS, which has a very similar electronic state, must be useful to discuss common points underlying the FFLO state. To detect the BCS-FFLO transition, a probe that can yield information even in the superconducting state is needed. As is found in a number of previous studies [23,26,27], the penetration depth is a very sensitive and high-resolution probe of the superconducting state even in short-time pulsed magnetic fields. Therefore, we performed penetration depth measurements to identify κ-Br as the FFLO candidate by detecting the phase boundary between the uniform superconductivity and the FFLO state. Additionally, compared with other FFLO candidates, universal features unique to the FFLO state are discussed.

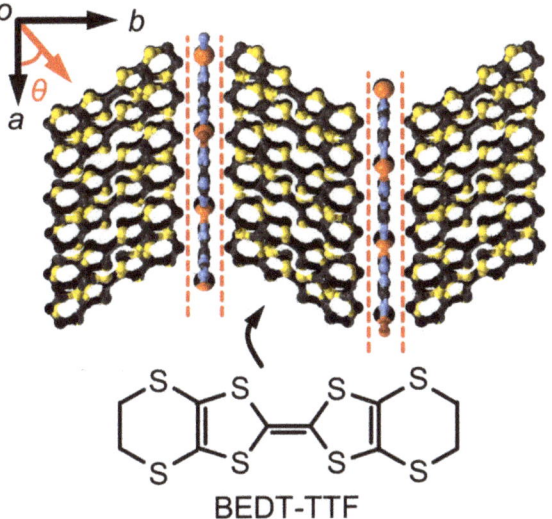

Figure 1. Crystal structure of κ-(BEDT-TTF)$_2$Cu[N(CN)$_2$]Br. As divided by the red dashed lines, this material has the two-dimensional layered structure. θ represents the angle from a-axis to b-axis, used for the magnetic-field direction applied in this study.

2. Radiofrequency (rf) Penetration Depth Measurements

The single crystal measured in the present study was synthesized electrochemically. The out-of-plane electrical resistance of the sample we measured in this study has been reported in [21]. For the rf penetration depth measurements with a tunnel diode oscillator (TDO), the sample, whose dimension was approximately 0.5 mm × 0.5 mm × 0.1 mm, was placed in one of two circles of a 0.7 mm-diameter 8-shaped coil, which could cancel out the voltage induced by the field change of pulsed magnetic fields. The direction of the magnetic field was changed by rotating the sample stage. The TDO circuit was operated at $F\sim$82 MHz with LC oscillators, similar to the reported design [26,28]. In this setup, the skin depth of the normal state significantly exceeded the sample thickness, and therefore, the change in the frequency ΔF originated only from the penetration depth of the superconductivity. These measurements were performed in a ^{4}He cryostat placed in a 60 T pulse magnet, installed at the International MegaGauss Science Laboratory, Institute for Solid State Physics, The University of Tokyo.

3. Results

3.1. Characterization of the Measured Sample

To evaluate whether the present sample was enough clean to host the FFLO state or not, we first estimated the mean-free path l from the quantum oscillations in resistivity, as shown in Figure 2a. The low-field behavior was related to the suppression of the superconductivity. The origin of the peak structure at approximately 10 T has been discussed in previous studies [29,30]. Above 40 T, the Shubnikov–de Haas oscillation was observed. The oscillation frequency was approximately 3900 T, which is consistent with the reported value [31]. Using the Lifshitz–Kosevich formula, the mean-free path l was obtained as ~130 nm, which was several times larger than the typical values 30–70 nm [31] and 20 times larger than the in-plane coherence length $\xi_{||}$=6–7 nm [21,32]. This implies that the present sample was sufficiently clean to form a spatially modulated pattern of the FFLO state.

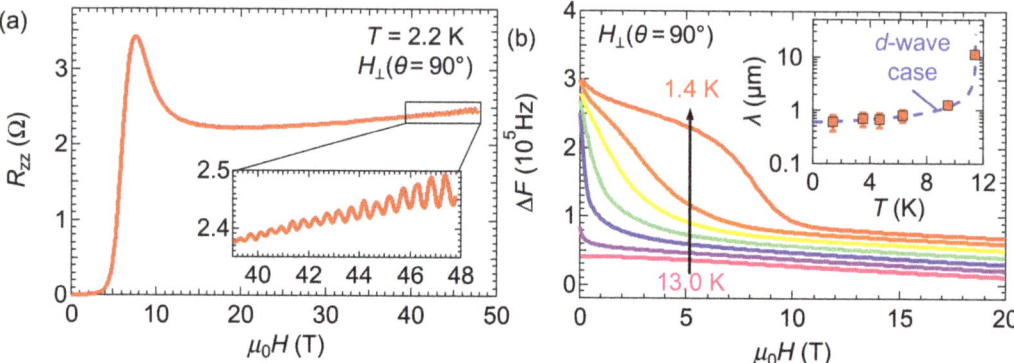

Figure 2. (a) Magnetoresistance at 2.2 K in a perpendicular field ($\theta = 90°$). The inset is the enlarged plot above 40 T to make the quantum oscillation clearer. (b) Shift of the penetration depth ΔF at various temperatures as a function of field when $\theta = 90°$. For clarity, the datasets include offsets. The inset is the temperature dependence of the penetration depth obtained by the Equations (1) and (2). The blue dashed curve indicates behavior for simple d-wave superconductivity.

In Figure 2b, we present the field dependence of ΔF in the perpendicular fields. At 13 K, which is higher than the critical temperature T_c, the field dependence originated from the magnetoresistance of the Cu wires composing the coil. Below T_c, a large response of ΔF was observed at low fields owing to the emergence of superconductivity. The data above 15 T indicated that the magnetoresistance of Cu did not have a large temperature dependence in this temperature region. The difference in ΔF between 0 T and 20 T was directly related to the shift of the penetration depth of the superconductivity $\Delta \lambda$, as shown in the following equation:

$$[\Delta F(0\ T) - \Delta F(20\ T)]/F = x\Delta\lambda/r, \quad (1)$$

where r and x represent the effective sample radius and filling factor of the sample in the coil, respectively. The absolute value of the penetration depth $\lambda(T)$ is given by the sum of the change and zero-temperature value, $\lambda(T) = \Delta\lambda + \lambda(0)$. Because the superfluid density $\rho(T)$ is determined by the relation $\rho(T) = [\lambda(0)/\lambda(T)]^2$, the Rutgers equation [33],

$$16\pi^2 \Delta C_p(T_c) \lambda(0) = \varphi_0 T_c d H_{c2'}(T_c) \rho'(T_c), \quad (2)$$

leads to $\lambda(0) = 0.6 \pm 0.2$ μm with the reported parameters, heat capacity jump $\Delta C_p(T_c) = 0.6$–0.7 J/Kmol [18–21] and the slope of H_{c2} at T_c, $\mu_0 H_{c2'}(T_c) \sim -15$ T/K [21,34]. Despite the large error, the value showed a good agreement with the reported values $\lambda(0) = 0.5$–0.7 μm [35–37]. The inset shows $\lambda(T)$ with a fit to the d-wave case (blue dashed curve) [38]. Although a large error made the precise determination of the pairing symmetry difficult, the d-wave model was acceptable for the present data. This result indicated that ΔF reflects the change in penetration depth in the superconducting state.

3.2. Magnetic-Field Dependence of ΔF and $d^2(\Delta F)/dH^2$ in Nearly Parallel Fields

In Figure 3a, we present the ΔF data at 1.4 K in fields almost parallel to the conducting plane ($\theta \sim 0°$), because the FFLO state occurs at lower temperatures when the orbital effect is sufficiently suppressed. At $\theta = 0°$, the onset of the change in ΔF from the normal state was approximately 37 T, which was almost consistent with the reported value of H_{c2} [21]. By tilting the angle from $\theta = 0°$, H_{c2} was suppressed. Figure 3b shows the second-field derivative of ΔF. The black lines represent the field-independent baseline of the normal state, and the black dotted curves are the eye guides. This plot further indicates that H_{c2} was approximately 37 T at $\theta = 0°$. Notably, these curves had some anomalies (green box and blue triangle) below H_{c2}, which were not clear in the ΔF data in Figure 3a. This behavior was similar to the results reported in earlier rf penetration depth studies for other organic FFLO candidates [23,26,27]. The anomaly at 31–32 T indicated by the blue triangles appeared to have a bare angle dependence, while the anomaly indicated by the green boxes showed the angle dependence. The angle dependence of the transition should not be significant in organics with weak spin-orbit coupling because the phase transition to the FFLO state is determined by the Zeeman effect; therefore, the anomaly indicated by the blue triangles at 31–32 T is considered to be the BCS-FFLO transition, namely $H_{\text{FFLO}} = 31$–32 T. In fact, the angle-independent behavior was observed in other FFLO candidates [23,25,26,39,40].

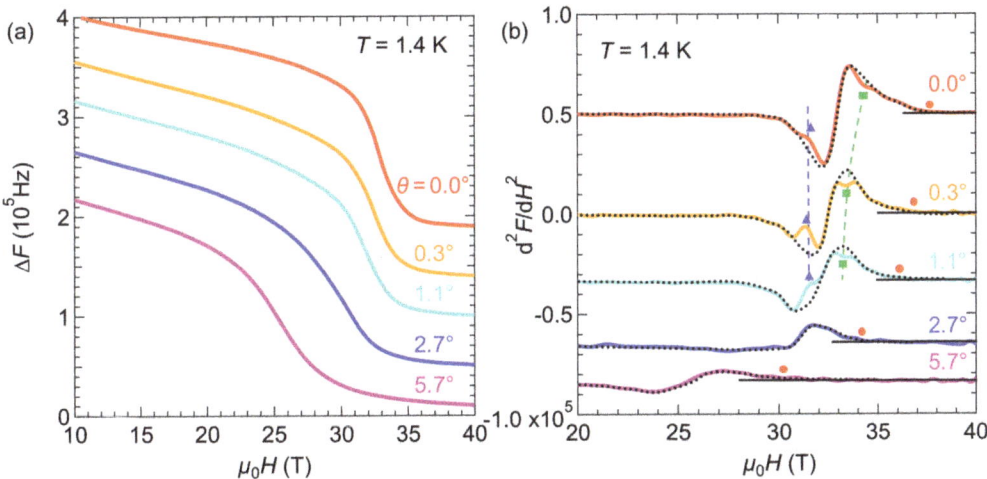

Figure 3. (a) Magnetic-field dependence of ΔF at 1.4 K with changing field angle θ. (b) Second derivative of (a) d^2F/dH^2 as a function of field. The black solid lines and the red dots show the background of the normal state and H_{c2}. The blue triangles and green squares indicate the anomalous fields of d^2F/dH^2. The black dotted, blue dashed, and green dashed curves are eye guides.

3.3. Temperature Dependence of ΔF and $d^2(IF)/dH^2$ in Perfectly Parallel Fields ($\theta = 0°$)

To discuss the stability of the FFLO state against temperature, in Figure 4, we present the field-dependent ΔF (a) and its second derivative (b) at $\theta = 0°$. The symbols shown here are the same as those used in Figure 3b. The BCS–FFLO transition (blue triangle) was observed below 4.0 K and showed a slight temperature dependence, as shown by the blue dashed curve. The additional anomaly in the FFLO state (green box) was immediately smeared out with increasing temperature above 1.4 K. Considering the angle-dependent behavior and the observation at low temperatures, the anomaly indicated by the green box may be related to the vortex transitions in the FFLO state [40,41].

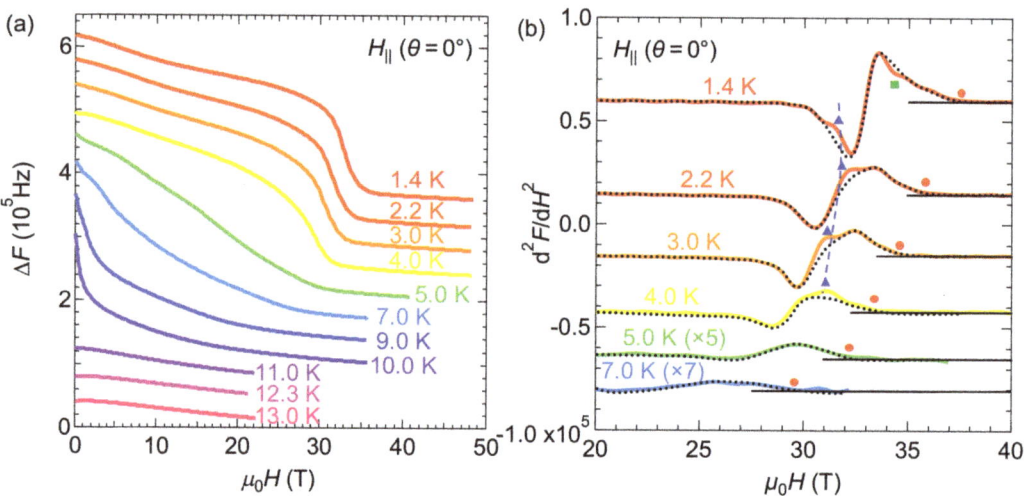

Figure 4. (a) ΔF ($\theta = 0°$) as a function of field at various temperatures. (b) Magnetic-field dependence of d^2F/dH^2. The symbols and curves are described using the same definitions as those used in Figure 3b.

4. Discussion

In Figure 5a, we organized the obtained H_{c2} and H_{FFLO} in parallel fields as an H-T phase diagram. The H_{c2} data reported in the previous study [21] were also plotted. The H_{c2} obtained in this study (red circle) was consistent with the reported data (gray box). The blue triangles denote the fields in which a kink is observed in Figure 4b. From the initial slope of the H_{c2} curve (solid line) near T_c, the orbital limit field H_{orb} and the perpendicular coherence length $\xi_\perp$ were estimated to be approximately 130 T, larger than H_{c2}, and 0.3 nm, five times smaller than the interlayer spacing 1.5 nm [31], respectively. These values indicate that the superconductivity was two-dimensional, and the orbital pair-breaking effect was quenched in parallel fields. To discuss the destruction of superconductivity, only the paramagnetic pair-breaking effect was considered. In a simple assumption based on the BCS theory, this effect gave the Pauli limit $\mu_0 H_P = 1.76 k_B T_c / (2^{1/2} \mu_B) \sim 1.84 (T_c[K])[T]$ from the balance between the superconducting energy gap $\Delta_0 = 1.76 k_B T_c$ and the Zeeman energy $g \mu_B H$. However, this assumption often does not work for organic superconductors, because the superconductivity in organics is usually strong-coupling and has unconventional pairing symmetry [19–21]. In Agosta's papers [16,23], to discuss the relation between H_P and H_{FFLO} more precisely, the following formula:

$$\mu_0 H_P = \alpha k_B T_c / \{2^{1/2} (g^*/g) \mu_B\}, \quad (3)$$

which includes the electron correlation and the coupling strength based on McKenzie's paper [42], was employed. Notably, g^* is the effective g-factor, including all many-body effects, and α is the coupling constant of the superconducting gap amplitude. The renormalized ratio g^*/g is experimentally determined using quantum oscillation measurements. In addition, g^*/g can be estimated by the ratio of the electronic heat capacity coefficient γ and the Pauli paramagnetism χ_P, because g^*/g can be equal to Wilson's ratio R_w [42]. For κ-Br, this relation led to $\mu_0 H_P = 31$–32 T, which corresponded to the present H_{FFLO}. This coincidence indicates that the anomaly observed in this study was caused by the transition to the FFLO state. Moreover, the Maki parameter $\alpha_M \sim 5.7$ was sufficiently larger than required. From the angle dependence of H_{c2} shown in Figure 3, the FFLO state should exist only in the limited region near the parallel direction $\theta \sim 0°$ because of the disappearance by the slight misalignment. This fragility to the orbital effect is also a characteristic of the FFLO state [14,43].

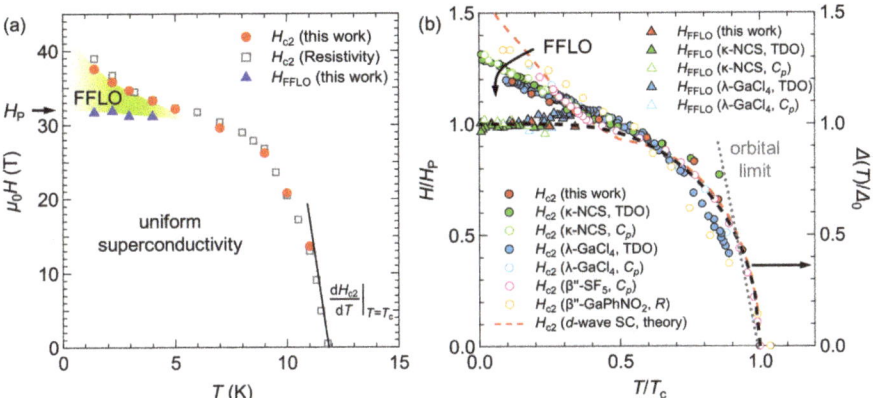

Figure 5. (a) H-T superconducting phase diagram of κ-Br in parallel fields. The gray boxes are the reported H_{c2} data [21]. The solid line indicates the slope of the H_{c2} curve near T_c at 0 T. The light green region above H_P is the FFLO state. (b) Superconducting phase diagram scaled by H_P and T_c, H/H_P vs. T/T_c. The circles and triangles represent H_{c2} and H_{FFLO}, respectively. The color of the symbols denotes the material. The filled symbols are taken from the TDO measurements [23,26,27], whereas the unfilled symbols are from other measurements [25,39,44,45]. The dotted gray line represents an example of the temperature dependence of H_{orb}, which is higher than H_{c2} at low temperatures. The red dotted curve is a simple theoretical calculation [9] of H_{c2} for the FFLO state with d-wave superconducting symmetry. The black dashed curve (right axis) is the temperature dependence of the reduced BCS-type superconducting gap $\Delta(T)/\Delta_0$.

For comparison with other organic FFLO candidates, the H-T phase diagram in a parallel field was reduced by H_P and T_c, as shown in Figure 5b. The H_{c2} and H_{FFLO} data for other salts are also shown [23,25–27,39,44,45]. The parameters, T_c, g^*/g, and α, which were used for the estimation of H_{FFLO}, are listed in Table 1. We used a typical value by referring to several references, because there is often some sample dependence in these parameters, and T_c depends on the measurement method and definition. Despite the large differences in their electronic states, such as the Fermi surface and dimensionality, these superconductors shared similar H/H_P-T/T_c phase diagrams. In the κ-type dimer-Mott electronic phase diagram, κ-Br and κ-NCS were located near the Mott metal–insulator boundary [17,18], indicating strong electron correlations originating from the large onsite Coulomb repulsion with the growth of the antiferromagnetic fluctuations. This characteristic resulted in a relatively large g^*/g and α of κ-Br and κ-NCS. Although λ-GaCl$_4$ and β″-SF$_5$ have significantly different Fermi surfaces [46,47], their parameters were almost identical and gave a similar H_{FFLO}. For β″-GaPhNO$_2$, the electronic state was expected to be near the charge-ordered phase on its electronic phase diagram and had a strong charge instability, which induced strong-coupling superconductivity α~2.5 [48]. Regardless of the variety in these electronic systems, the calculated H_P coincided with H_{FFLO}, as listed in Table 1. This fact demonstrated that the paramagnetic effect, which was the factor underlying H_P, mainly governed the transition between the BCS and FFLO states, as predicted by a number of theories. Indeed, the dashed curve shown in the right axis in Figure 5b, which is the temperature dependence of the BCS-type superconducting gap, roughly describes the BCS region (light orange). Importantly, the paramagnetic pair-breaking was determined by the competition between the superconducting gap and the Zeeman effect. However, the H_{c2} curves above H_P indicated that there were small differences in the stability of the FFLO state at high fields. The theoretical curve for simple d-wave superconductivity (red dotted curve) [9] was qualitatively similar to the obtained phase diagram. Nevertheless, the data in Figure 5b were not accurate enough to discuss small differences with the simple model, and therefore, it would be necessary to discuss with an appropriate theoretical model for the electronic system of each material rather than the simple model. The parameters related to the stability of the FFLO state are likely the dimensionality and the shape of the Fermi surface as well as the gap symmetry. For example, (TMTSF)$_2$ClO$_4$, which was expected to exhibit the FFLO state [49], had a quasi-one-dimensional system, and the difference was expected to be significant. Although its superconducting phase showed strange differences in electrical resistivity and specific heat measurements [50,51], it might be interesting to discuss it through H_P. As for the research method to discuss the FFLO state in more detail, for example, the in-plane angular dependence from Refs. [51,52] may be useful. Further research should be completed along with theoretical predictions.

Table 1. Reported H_{FFLO} and calculated H_P with parameters characterizing the FFLO state. The abbreviations of the material names are described in the main text. The shown H_P is estimated by the Equation (3) and the parameters shown here, which are typical values taking into account sample dependence, etc. For the estimation of the values of g^*/g, the Wilson' ratio R_W, calculated by γ and χ_P, is also used to compare with g^*/g determined by angle-dependent quantum oscillations. The values of α are taken from heat capacity measurements.

Material	T_c (K)	g^*/g	α	H_P (T)	H_{FFLO} (T)	Refs.
κ-Br	11.7	1.3	3.3	31	31–32	[17,19–21,42,53,54]
κ-NCS	9.0	1.3	2.9	21	21–22	[17,19,22–25,42,55]
λ-GaCl$_4$	4.7	1.0	2.1	10	9–10	[26,39,46,56–58]
β″-SF$_5$	4.3	1.0	2.1	9.5	9–10	[27,40,41,44,59–61]
β″-GaPhNO$_2$	4.8	0.8	2.5	16	16	[45,48,62]

5. Conclusions

We performed high-field rf-penetration depth measurements to determine whether the FFLO state manifested as a high-field superconducting state distinct from the BCS state in the organic superconductor κ-(BEDT-TTF)$_2$Cu[N(CN)$_2$]Br. From the quantum oscillation and the phase diagram, it was confirmed that the electronic system was sufficiently clean and two-dimensional to stably host the FFLO state. As has been discussed for the FFLO state in other candidates, the transition field between the BCS and FFLO states had no angle dependence, whereas the FFLO state was very sensitive to the field angle and was immediately smeared out by the slight misalignment. Compared to other organic FFLO candidates, their H/H_P-T/T_c superconducting phase diagrams suggest that H_P certainly corresponds to H_{FFLO}, regardless of the electronic states underlying the superconductivity. This verifies that the BCS-FFLO transition is determined by the competition between the Zeeman energy and the superconducting condensation energy. The FFLO state appears at very high fields above 31–32 T, because κ-Br can also be discussed in this framework, and its H_P is enhanced by the large superconducting gap originating from the strong electron correlations growing in proximity to the Mott metal-insulator boundary.

Author Contributions: Conceptualization, S.I.; methodology, S.I. and K.K.; investigation, S.I. writing, S.I.; funding acquisition, S.I. and K.K. All authors have read and agreed to the published version of the manuscript.

Funding: This work was partially supported by the Japan Society for the Promotion of Science KAKENHI Grant No. 20K14406.

Institutional Review Board Statement: Not applicable.

Informed Consent Statement: Not applicable.

Data Availability Statement: Data are available from the corresponding author upon reasonable request.

Acknowledgments: We thank Y. Kohama (ISSP, the University of Tokyo) for advice on the rf-TDO measurement.

Conflicts of Interest: The authors declare no conflict of interest.

References

1. Fulde, P.; Ferrell, R.A. Superconductivity in a strong spin exchange field. *Phys. Rev.* **1964**, *135*, A550. [CrossRef]
2. Larkin, A.I.; Ovchinnikov, Y.N. Inhomogeneous state of superconductors. *Sov. Phys. JETP* **1965**, *20*, 762.
3. Casalbuoni, R.; Nardulli, G. Inhomogeneous superconductivity in condensed matter and QCD. *Rev. Mod. Phys.* **2004**, *76*, 263. [CrossRef]
4. Matsuda, Y.; Shimahara, H. Fulde-Ferrell-Larkin-Ovchinnikov state in heavy fermion superconductors. *J. Phys. Soc. Jpn.* **2007**, *76*, 051005. [CrossRef]
5. Imajo, S.; Nomura, T.; Kohama, Y.; Kindo, K. Nematic response of the Fulde-Ferrell-Larkin-Ovchinnikov state. *arXiv* **2021**, arXiv:2110.12774.
6. Clogston, A.M. Upper limit for the critical field in hard superconductors. *Phys. Rev. Lett.* **1962**, *9*, 266–267. [CrossRef]
7. Shimahara, H.; Rainer, D. Crossover from Vortex States to the Fulde-Ferrell-Larkin-Ovchinnikov State in Two-Dimensional s- and d-Wave Superconductors. *J. Phys. Soc. Jpn.* **1997**, *66*, 3591. [CrossRef]
8. Shimahara, H. Transition from the vortex state to the Fulde–Ferrell–Larkin–Ovchinnikov state in quasi-two-dimensional superconductors. *Phys. Rev. B* **2009**, *80*, 214512. [CrossRef]
9. Croitoru, M.D.; Buzdin, A.I. In Search of Unambiguous Evidence of the Fulde–Ferrell–Larkin–Ovchinnikov State in Quasi-Low Dimensional Superconductors. *Condens. Matter* **2017**, *2*, 30. [CrossRef]
10. Aslamazov, L.G. Influence of impurities on the existence of an imhomogeneous state in a ferromagnetic superconductor. *Sov. Phys. JETP* **1969**, *28*, 773.
11. Cui, Q.; Yang, K. Fulde-Ferrell-Larkin-Ovchinnikov state in disordered s-wave superconductors. *Phys. Rev. B* **2008**, *78*, 054501. [CrossRef]
12. Vorontsov, A.B.; Vekhter, I.; Graf, M.J. Pauli-limited upper critical field in dirty d-wave superconductors. *Phys. Rev. B* **2008**, *78*, 180505. [CrossRef]
13. Maki, K.; Tsuneto, T. Pauli Paramagnetism and Superconducting State. *Prog. Theor. Phys.* **1964**, *31*, 945. [CrossRef]
14. Gruenberg, L.W.; Gunther, L. Fulde-Ferrell effect in type-II superconductors. *Phys. Rev. Lett.* **1966**, *16*, 996. [CrossRef]
15. Wosnitza, J. Spatially Nonuniform Superconductivity in Quasi-Two-Dimensional Organic Charge-Transfer Salts. *Crystals* **2018**, *8*, 183. [CrossRef]
16. Agosta, C.C. Inhomogeneous Superconductivity in Organic and Related Superconductors. *Crystals* **2018**, *8*, 285. [CrossRef]
17. Kanoda, K. Metal-Insulator Transition in κ-(ET)$_2$X and (DCNQI)$_2$M: Two Contrasting Manifestation of Electron Correlation. *J. Phys. Soc. Jpn.* **2006**, *75*, 051007. [CrossRef]
18. Nakazawa, Y.; Imajo, S.; Matsumura, Y.; Yamashita, S.; Akutsu, H. Thermodynamic Picture of Dimer-Mott Organic Superconductors Revealed by Heat Capacity Measurements with External and Chemical Pressure Control. *Crystals* **2018**, *8*, 143. [CrossRef]
19. Taylor, O.J.; Carrington, A.; Schlueter, J.A. Specific-Heat Measurements of the Gap Structure of the Organic Superconductors κ-(ET)$_2$Cu[N(CN)$_2$]Br and κ-(ET)$_2$Cu(NCS)$_2$. *Phys. Rev. Lett.* **2007**, *99*, 057001. [CrossRef]
20. Imajo, S.; Kindo, K.; Nakazawa, Y. Symmetry change of d-wave superconductivity in κ-type organic superconductors. *Phys. Rev. B* **2021**, *103*, L060508. [CrossRef]
21. Imajo, S.; Nakazawa, Y.; Kindo, K. Superconducting Phase Diagram of the Organic Superconductor κ-(BEDT-TTF)$_2$Cu[N(CN)$_2$]Br above 30 T. *J. Phys. Soc. Jpn.* **2018**, *87*, 123704. [CrossRef]
22. Lortz, R.; Wang, Y.; Demuer, A.; Bottger, P.H.M.; Bergk, B.; Zwicknagl, G.; Nakazawa, Y.; Wosnitza, J. Calorimetric Evidence for a Fulde-Ferrell-Larkin-Ovchinnikov Superconducting State in the Layered Organic Superconductor κ-(BEDT−TTF)$_2$Cu(NCS)$_2$. *Phys. Rev. Lett.* **2007**, *99*, 187002. [CrossRef]
23. Agosta, C.C.; Jin, J.; Coniglio, W.A.; Smith, B.E.; Cho, K.; Stroe, I.; Martin, C.; Tozer, S.W.; Murphy, T.P.; Palm, E.C.; et al. Experimental and semiempirical method to determine the Pauli-limiting field in quasi-two-dimensional superconductors as applied to κ-(BEDT-TTF)$_2$Cu(NCS)$_2$: Strong evidence of a FFLO state. *Phys. Rev. B* **2012**, *85*, 214514. [CrossRef]
24. Mayaffre, H.; Krämer, S.; Horvatić, M.; Berthier, C.; Miyagawa, K.; Kanoda, K.; Mitrović, V.F. Evidence of Andreev bound states as a hallmark of the FFLO phase in κ-(BEDT-TTF)$_2$Cu(NCS)$_2$. *Nat. Phys.* **2014**, *10*, 928. [CrossRef]
25. Agosta, C.C.; Fortune, N.A.; Hannahs, S.T.; Gu, S.; Liang, L.; Park, J.; Schlueter, J.A. Calorimetric Measurements of Magnetic-Field-Induced Inhomogeneous Superconductivity Above the Paramagnetic Limit. *Phys. Rev. Lett.* **2017**, *118*, 267001. [CrossRef]
26. Coniglio, W.A.; Winter, L.E.; Cho, K.; Agosta, C.C.; Fravel, B.; Montgomery, L.K. Superconducting phase diagram and FFLO signature in λ-(BETS)$_2$GaCl$_4$ from rf penetration depth measurements. *Phys. Rev. B* **2011**, *83*, 224507. [CrossRef]
27. Cho, K.; Smith, B.E.; Coniglio, W.A.; Winter, L.E.; Agosta, C.C.; Schlueter, J.A. Upper critical field in the organic superconductor β″-(ET)$_2$SF$_5$CH$_2$CF$_2$SO$_3$: Possibility of Fulde-Ferrell-Larkin-Ovchinnikov state. *Phys. Rev. B* **2009**, *79*, 220507. [CrossRef]
28. Imajo, S.; Sugiura, S.; Akutsu, H.; Kohama, Y.; Isono, T.; Terashima, T.; Kindo, K.; Uji, S.; Nakazawa, Y. Extraordinary π-electron superconductivity emerging from a quantum spin liquid. *Phys. Rev. Res.* **2021**, *3*, 033026. [CrossRef]
29. Zuo, F.; Schlueter, J.A.; Kelly, M.E.; Williams, J.M. Mixed-state magnetoresistance in organic superconductors κ-(BEDT-TTF)$_2$Cu(NCS)$_2$. *Phys. Rev. B* **1996**, *54*, 11973. [CrossRef]
30. Zuo, F.; Schlueter, J.A.; Williams, J.M. Interlayer magnetoresistance in the organic superconductor κ-(BEDT−TTF)$_2$Cu[N(CN)$_2$]Br near the superconducting. *Phys. Rev. B* **1999**, *60*, 574. [CrossRef]

31. Mielke, C.H.; Harrison, N.; Rickel, D.G.; Lacerda, A.H.; Vestal, R.M.; Montgomery, L.K. Fermi-surface topology of κ-(BEDT−TTF)$_2$Cu[N(CN)$_2$]Br at ambient pressure. *Phys. Rev. B* **1997**, *56*, R4309. [CrossRef]
32. Ito, H.; Watanabe, M.; Nogami, Y.; Ishiguro, T.; Komatsu, T.; Saito, G.; Hosoito, N. Magnetic Determination of Ginzburg-Landau Coherence Length for Organic Superconductor κ-(BEDT-TTF)$_2$X (X=Cu(NCS)$_2$,Cu[N(CN)$_2$]Br): Effect of Isotope Substitution. *J. Phys. Soc. Jpn.* **1991**, *60*, 3230. [CrossRef]
33. Kim, H.; Kogan, V.G.; Cho, K.; Tanatar, M.A.; Prozorov, R. Rutgers relation for the analysis of superfluid density in superconductors. *Phys. Rev. B* **2013**, *87*, 214518. [CrossRef]
34. Kovalev, A.E.; Ishiguro, T.; Kondo, T.; Saito, G. Specific heat of organic superconductor κ-(BEDT-TTF)$_2$Cu[N(CN)$_2$]Br under a magnetic field parallel to the conducting plane. *Phys. Rev. B* **2000**, *62*, 103. [CrossRef]
35. Yoneyama, N.; Higashihara, A.; Sasaki, T.; Nojima, T.; Kobayashi, N. Impurity Effect on the In-plane Penetration Depth of the Organic Superconductors κ-(BEDT-TTF)$_2$X (X = Cu(NCS)$_2$ and Cu[N(CN)$_2$]Br). *J. Phys. Soc. Jpn.* **2004**, *73*, 1290. [CrossRef]
36. Lang, M.; Toyota, N.; Sasaki, T.; Sato, H. Magnetic penetration depth of κ-(BEDT-TTF)$_2$Cu[N(CN)$_2$]Br, determined from the reversible magnetization. *Phys. Rev. B* **1992**, *46*, 5822. [CrossRef]
37. Wakamatsu, K.; Miyagawa, K.; Kanoda, K. Superfluid density versus transition temperature in a layered organic superconductor κ-(BEDT-TTF)$_2$Cu[N(CN)$_2$]Br under pressure. *Phys. Rev. Res.* **2020**, *2*, 043008. [CrossRef]
38. Prozorov, R.; Giannetta, R.W. Magnetic penetration depth in unconventional superconductors. *Supercond. Sci. Technol.* **2006**, *19*, R41. [CrossRef]
39. Imajo, S.; Kobayashi, T.; Kawamoto, A.; Kindo, K.; Nakazawa, Y. Thermodynamic evidence for the formation of a Fulde-Ferrell-Larkin-Ovchinnikov phase in the organic superconductor λ-(BETS)$_2$GaCl$_4$. *Phys. Rev. B* **2021**, *103*, L220501. [CrossRef]
40. Sugiura, S.; Isono, T.; Terashima, T.; Yasuzuka, S.; Schlueter, J.A.; Uji, S. Fulde–Ferrell–Larkin–Ovchinnikov and vortex phases in a layered organic superconductor. *NPJ Quantum Mater.* **2018**, *4*, 7. [CrossRef]
41. Sugiura, S.; Terashima, T.; Uji, S.; Yasuzuka, S.; Schlueter, J.A. Josephson vortex dynamics and Fulde-Ferrell-Larkin-Ovchinnikov superconductivity in the layered organic superconductor β"-(BEDT-TTF)$_2$SF$_5$CH$_2$CF$_2$SO$_3$. *Phys. Rev. B* **2019**, *100*, 014515. [CrossRef]
42. McKenzie, R.H. Wilson's ratio and the spin splitting of magnetic oscillations in quasi-two-dimensional metals. *arXiv* **1999**, arXiv:9905044.
43. Uji, S.; Terashima, T.; Nishimura, M.; Takahide, Y.; Konoike, T.; Enomoto, K.; Cui, H.; Kobayashi, H.; Kobayashi, A.; Tanaka, H.; et al. Vortex Dynamics and the Fulde- Ferrell-Larkin-Ovchinnikov State in a Magnetic-Field-Induced Organic Superconductor. *Phys. Rev. Lett.* **2006**, *97*, 157001. [CrossRef] [PubMed]
44. Beyer, R.; Bergk, B.; Yasin, S.; Schlueter, J.A.; Wosnitza, J. Angle-Dependent Evolution of the Fulde-Ferrell-Larkin-Ovchinnikov State in an Organic Superconductor. *Phys. Rev. Lett.* **2012**, *109*, 027003. [CrossRef]
45. Uji, S.; Iida, Y.; Sugiura, S.; Isono, T.; Sugii, K.; Kikugawa, N.; Terashima, T.; Yasuzuka, S.; Akutsu, H.; Nakazawa, Y.; et al. Fulde-Ferrell-Larkin-Ovchinnikov superconductivity in the layered organic superconductor β"-(BEDT-TTF)$_4$[(H$_3$O)Ga(C$_2$O$_4$)$_3$]C$_6$H$_5$NO$_2$. *Phys. Rev. B* **2018**, *97*, 144505. [CrossRef]
46. Mielke, C.; Singleton, J.; Nam, M.-S.; Harrison, N.; Agosta, C.C.; Fravel, B.; Montgomery, L.K. Superconducting properties and Fermi-surface topology of the quasi-two-dimensional organic superconductor λ-(BETS)$_2$GaCl$_4$ (BETS≡bis(ethylenedithio)tetraselenafulvalene). *J. Phys. Condens. Matter* **2001**, *13*, 8325. [CrossRef]
47. Yasuzuka, S.; Uji, S.; Terashima, T.; Sugii, K.; Isono, T.; Iida, Y.; Schlueter, J.A. In-Plane Anisotropy of Upper Critical Field and Flux-Flow Resistivity in Layered Organic Superconductor β"-(ET)$_2$SF$_5$CH$_2$CF$_2$SO$_3$. *J. Phys. Soc. Jpn.* **2015**, *84*, 094709. [CrossRef]
48. Imajo, S.; Akutsu, H.; Kurihara, R.; Yajima, T.; Kohama, Y.; Tokunaga, M.; Kindo, K.; Nakazawa, Y. Anisotropic Fully Gapped Superconductivity Possibly Mediated by Charge Fluctuations in a Nondimeric Organic Complex. *Phys. Rev. Lett.* **2020**, *125*, 177002. [CrossRef]
49. Aizawa, H.; Kuroki, K.; Tanaka, Y. Pairing Competition in a Quasi-One-Dimensional Model of Organic Superconductors (TMTSF)$_2$X in Magnetic Field. *J. Phys. Soc. Jpn.* **2009**, *78*, 124711. [CrossRef]
50. Yonezawa, S.; Maeno, Y.; Bechgaard, K.; Jérome, D. Nodal superconducting order parameter and thermodynamic phase diagram of (TMTSF)$_2$ClO$_4$. *Phys. Rev. B* **2012**, *85*, 140502. [CrossRef]
51. Yonezawa, S.; Kusaba, S.; Maeno, Y.; Auban-Senzier, P.; Pasquier, C.; Bechgaard, K.; Jérome, D. Anomalous In-Plane Anisotropy of the Onset of Superconductivity in (TMTSF)$_2$ClO$_4$. *Phys. Rev. Lett.* **2008**, *100*, 117002. [CrossRef] [PubMed]
52. Croitoru, M.D.; Buzdin, A.I. Resonance in-plane magnetic field effect as a means to reveal the Fulde-Ferrell-Larkin-Ovchinnikov state in layered superconductors. *Phys. Rev. B* **2012**, *86*, 064507. [CrossRef]
53. Weiss, H.; Kartsovnik, M.V.; Biberacher, W.; Balthes, E.; Jansen, A.G.M.; Kushch, N.D. Angle-dependent magnetoquantum oscillations in κ−(BEDT−TTF)$_2$Cu[N(CN)$_2$]Br. *Phys. Rev. B* **1999**, *60*, R16259. [CrossRef]
54. Imajo, S.; Dong, C.; Matsuo, A.; Kindo, K.; Kohama, Y. High-resolution calorimetry in pulsed magnetic fields. *Rev. Sci. Instrum.* **2021**, *92*, 043901. [CrossRef] [PubMed]
55. Wosnitza, J.; Crabtree, G.W.; Wang, H.H.; Geiser, U.; Williams, J.M.; Carlson, K.D. de Haas–van Alphen studies of the organic superconductors α-(ET)$_2$(NH$_4$)Hg(SCN)$_4$ and κ-(ET)$_2$Cu(NCS)$_2$ [with ET = ethelenedithio)-tetrathiafulvalene]. *Phys. Rev. B* **1992**, *45*, 3018. [CrossRef]
56. Tanatar, M.A.; Ishiguro, T.; Tanaka, H.; Kobayashi, H. Magnetic field–temperature phase diagram of the quasi-two-dimensional organic superconductor λ-(BETS)$_2$GaCl$_4$ studied via thermal conductivity. *Phys. Rev. B* **2002**, *66*, 134503. [CrossRef]

57. Imajo, S.; Kanda, N.; Yamashita, S.; Akutsu, H.; Nakazawa, Y.; Kumagai, H.; Kobayashi, T.; Kawamoto, A. Thermodynamic Evidence of d-Wave Superconductivity of the Organic Superconductor λ-(BETS)$_2$GaCl$_4$. *J. Phys. Soc. Jpn.* **2016**, *85*, 043705. [CrossRef]
58. Imajo, S.; Yamashita, S.; Akutsu, H.; Kumagai, H.; Kobayashi, T.; Kawamoto, A.; Nakazawa, Y. Gap Symmetry of the Organic Superconductor λ-(BETS)$_2$GaCl$_4$ Determined by Magnetic-Field-Angle-Resolved Heat Capacity. *J. Phys. Soc. Jpn.* **2019**, *88*, 023702. [CrossRef]
59. Koutroulakis, G.; Kühne, H.; Schlueter, J.A.; Wosnitza, J.; Brown, S.E. Microscopic Study of the Fulde-Ferrell-Larkin-Ovchinnikov State in an All-Organic Superconductor. *Phys. Rev. Lett.* **2016**, *116*, 067003. [CrossRef]
60. Beckmann, D.; Wanka, S.; Wosnitza, J.; Schlueter, J.A.; Williams, J.M.; Nixon, P.G.; Winter, R.W.; Gard, G.L.; Ren, J.; Whangbo, M.-H. Characterization of the Fermi surface of the organic superconductor β″-(ET)$_2$SF$_5$CH$_2$CF$_2$SO$_3$ by measurements of Shubnikov-de Haas and angle-dependent magnetoresistance oscillations and by electronic band-structure calculations. *Eur. Phys. J. B* **1998**, *1*, 295. [CrossRef]
61. Wanka, S.; Hagel, J.; Beckmann, D.; Wosnitza, J.; Schlueter, J.A.; Williams, J.M.; Nixon, P.G.; Winter, R.W.; Gard, G.L. Specific heat and critical fields of the organic superconductor β″−(BEDT−TTF)$_2$SF$_5$CH$_2$CF$_2$SO$_3$. *Phys. Rev. B* **1998**, *57*, 3084. [CrossRef]
62. Bangura, A.F.; Coldea, A.I.; Singleton, J.; Ardavan, A.; Akutsu-Sato, A.; Akutsu, H.; Turner, S.S.; Day, P.; Yamamoto, T.; Yakushi, K. Robust superconducting state in the low-quasiparticle-density organic metals β″−(BEDT−TTF)$_4$[(H$_3$O)M(C$_2$O$_4$)$_3$]·Y: Superconductivity due to proximity to a charge-ordered state. *Phys. Rev. B* **2005**, *72*, 014543. [CrossRef]

Article

Fermi Surface Structure and Isotropic Stability of Fulde-Ferrell-Larkin-Ovchinnikov Phase in Layered Organic Superconductor β''-(BEDT-TTF)$_2$SF$_5$CH$_2$CF$_2$SO$_3$

Shiori Sugiura [1,*,†], Hiroki Akutsu [2], Yasuhiro Nakazawa [2], Taichi Terashima [3], Syuma Yasuzuka [4], John A. Schlueter [5,6] and Shinya Uji [3,*,†]

1. Institute for Materials Research, Tohoku University, Sendai 980-8577, Japan
2. Department of Chemistry, Graduate School of Science, Osaka University, Toyonaka 560-0043, Japan; akutsu@chem.sci.osaka-u.ac.jp (H.A.); nakazawa@chem.sci.osaka-u.ac.jp (Y.N.)
3. Institute for Materials Science, Tsukuba 305-0003, Japan; TERASHIMA.Taichi@nims.go.jp
4. Research Center for Condensed Matter Physics, Hiroshima Institute of Technology, Hiroshima 731-5193, Japan; yasuzuka@cc.it-hiroshima.ac.jp
5. Materials Science Division, Argonne National Laboratory, Argonne, IL 60439, USA; JASchlueter@anl.gov
6. Division of Materials Research, National Science Foundation, Alexandria, VA 22314, USA
* Correspondence: shiori.sugiura.c5@tohoku.ac.jp (S.S.); UJI.shinya@nims.go.jp (S.U.); Tel.:+81-22-215-2028 (S.S.); +81-29-863-5512 (S.U.)
† These authors contributed equally to this work.

Abstract: The Fermi surface structure of a layered organic superconductor β''-(BEDT-TTF)$_2$SF$_5$CH$_2$CF$_2$SO$_3$ was determined by angular-dependent magnetoresistance oscillations measurements and band-structure calculations. This salt was found to have two small pockets with the same area: a deformed square hole pocket and an elliptic electron pocket. Characteristic corrugations in the field dependence of the interlayer resistance in the superconducting phase were observed at any in-plane field directions. The features were ascribed to the commensurability (CM) effect between the Josephson vortex lattice and the periodic nodal structure of the superconducting gap in the Fulde–Ferrell–Larkin–Ovchinnikov (FFLO) phase. The CM effect was observed in a similar field region for various in-plane field directions, in spite of the anisotropic nature of the Fermi surface. The results clearly showed that the FFLO phase stability is insensitive to the in-plane field directions.

Keywords: organic superconductor; resistance; FFLO phase; vortex dynamics

1. Introduction

The discovery of superconductivity has led to breakthroughs in a wide range of fields from fundamental research and applications [1]. In particular, since the discovery of high-temperature superconducting cuprates in 1980s, the search for new superconducting mechanisms has been one of the major trends in superconductivity basic research. Among the various superconductors, organic superconductors in the vicinity of metal-insulator transitions have brought about significant progress in basic research.

Organic conductors based on BEDT-TTF molecules are characterized by a stacked structure with anion molecule (insulating) layers and BEDT-TTF molecule (conducting) layers. These conductors have attracted significant interest because of the presence of various ground states, a dimer-Mott insulating phase, a charge-ordered phase, a density wave phase, and a superconducting phase, where the degree of dimerization of the BEDT-TTF molecules, the Fermi surface instability, and the strong electron correlation play important roles. In particular, the possibility of unconventional superconductivity, mediated by antiferromagnetic spin and/or charge fluctuations, is a central concern.

When the orbital effect is suppressed and the critical field (H_{c2}) is Pauli-limited, a unique superconducting phase, namely, the Fulde–Ferrell–Larkin–Ovchinnikov (FFLO)

superconducting phase is expected to emerge at high fields [2,3]. In conventional superconductors, the spin-singlet Cooper pairs formed by up and down spins are destroyed in a magnetic field by the Zeeman effect. This pair-breaking effect gives the Pauli limit, $H_{\text{Pauli}} = \Delta_0/\sqrt{2}\mu_B = 1.86 T_c$, where Δ_0 is the superconducting energy gap at 0 K, and μ_B is the Bohr magneton [4]. In the FFLO phase, the Cooper pairs are formed between up and down spins on the polarized Fermi surface. Therefore, the Cooper pairs have a finite center-of-mass momentum q and show a spatial modulation of the order parameter in real space; $\Delta(r) = \Delta\cos(qr)$, as shown in Figure 1a. As a result, the superconductivity can be stabilized even above H_{Pauli}. In recent years, experimental results suggesting its existence have been obtained in heavy fermion superconductors [5,6], oxide layered superconductors [7], ion-based superconductors [8], and organic superconductors [9–16]. In organic superconductors, the FFLO phase transition was first observed by a tuned-circuit differential susceptometer experiment for κ-(BEDT-TTF)$_2$Cu(NCS)$_2$ [10], and, since then, various measurements [17–24] have been performed to confirm the FFLO transition.

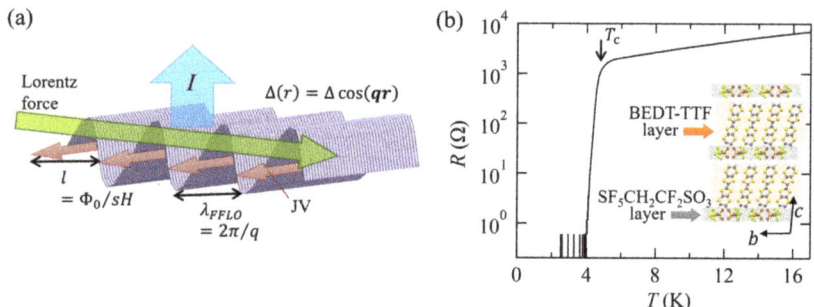

Figure 1. (a) Schematic illustration of the order parameter oscillation $\Delta(r)$ for a single q case in a FFLO phase and JV lattice in a layered superconductor. The JVs are easily driven by the Lorentz force in an interlayer current I, leading to nonzero interlayer resistance even in the superconducting phase. (b) Temperature dependence of the interlayer resistance of the β''-SF$_5$ salt. The onset of the superconducting transition can be defined as $T_c \approx 4.8$ K, consistent with the specific heat measurement [23]. Inset: crystal structure of β''-SF$_5$ salt.

Highly two-dimensional (2D) layered superconductors can be modeled as Josephson-coupled multi-layer systems. In such superconductors, magnetic flux lines penetrating the sample can be decomposed into two parts; the pancake vortices (PVs) penetrating the superconducting layers and the Josephson vortices (JVs) penetrating the insulating layers. The JVs are pinned more loosely than the PVs, since the order parameter vanishes in the insulating layers. Therefore, the JVs are easily driven by the Lorentz force in an interlayer current, and, consequently, nonzero interlayer resistance is observed even in the superconducting phase. In the FFLO phase, periodic nodal lines of the order parameter are formed by the finite center-of-mass momentum q of the Cooper pairs as depicted in Figure 1a. When the nodal lines are parallel to the JVs, they will work as pinning sites of the JVs. The wavelength of the order parameter oscillation is given by $\lambda_{\text{FFLO}} = 2\pi/q$, and the JV lattice spacing is $l = \Phi_0/sH$, where s is the interlayer spacing and Φ_0 is the flux quantum. It is expected that JVs are relatively strongly pinned by the nodal line structure for a commensurate condition $l/\lambda_{\text{FFLO}} = N$ (N : integer). Since λ_{FFLO} is also expected to decrease with an increasing field [25,26], the commensurate condition in the FFLO phase will be periodically satisfied, leading to fine structures in the interlayer resistance curves. This commensurability (CM) effect was first predicted by Bulaevskii et al. [27]. Thus far, the CM effect has been observed in the FFLO phases for various organic superconductors, λ-(BETS)$_2$FeCl$_4$ [15], β''-(BEDT-TTF)$_4$[(H$_3$O)Ga(C$_2$O$_4$)$_3$]C$_6$H$_5$NO$_2$ (β''-Ga salt) [28], and β''-(BEDT-TTF)$_2$SF$_5$CH$_2$CF$_2$SO$_3$ (β''-SF$_5$ salt) [29]. Among them, the highly 2D nature of the β''-Ga and β''-SF$_5$ salts with the large anion layers would provide an excellent platform for the FFLO studies, since JV dynamics play an essential role in the CM effects.

The β''-SF$_5$ salt with $T_c \approx 4.8$ K is composed of the conducting BEDT-TTF molecular layer and the large insulating SF$_5$CH$_2$CF$_2$SO$_3$ layer (inset of Figure 1b). A highly 2D electronic state has been realized, which is characterized by a large ratio of the intralayer to that of interlayer critical fields $H_{c2}^{\parallel}/H_{c2}^{\perp} \approx 11.5$. The specific-heat measurements show strongly coupled BCS-like behavior with a full gap given by $\Delta_0/k_B T = 2.18$ [30]. In a magnetic field parallel to the conducting layers, the critical field H_{c2} significantly exceeded the Pauli limit $H_{\text{Pauli}} \approx 10$ T, above which the FFLO superconductivity is realized [31–35]. Optical measurements of isostructural β''-(BEDT-TTF)$_2$SF$_5$RSO$_3$ (R = CH$_2$, CHFCF$_2$, CH$_2$CF$_2$, and CHF) compounds revealed that the superconducting phase is adjacent to a charge-ordered insulating phase [36]. This could indicate superconductivity mediated by charge fluctuations, which is another reason for the interest in the β'' salts.

In our previous studies, we clarified the FFLO phase boundary in terms of the magnetocaloric effect, torque [35], and resistance measurements [29]. We also observed the CM effect in fields almost parallel to the a-axis in the FFLO phase, above $\sim$9 T. The λ_{FFLO} values, ranging from $\sim$40 nm to $\sim$210 nm, were obtained under the assumption of a single q vector perpendicular to the field. The stability of the FFLO phase is closely related to the nesting instability of the Fermi surface, and the q vector leading to a large nesting part is favorable for the FFLO phase. Therefore, the optimum q vector depends on the anisotropic structure of the Fermi surface. Meanwhile, the orbital effect stabilizes the q vector parallel to the field. This situation can lead to complicated field-direction dependencies of the optimum q vector. Even multi-q-vector phases are theoretically predicted depending on the field strength and temperature [37].

In this study, we focused on the stability of the FFLO phase in the β''-SF$_5$ salt, with an anisotropic Fermi surface. Firstly, we clarify the Fermi surface structure from the measurements of angular-dependent magnetoresistance oscillations (AMROs), and then we report the CM effect in various in-plane field directions. The CM effect was surprisingly observed in a similar field region for various in-plane field directions, despite the anisotropic Fermi surface structure. Possible scenarios for explaining these results are presented.

2. Materials and Methods

Single crystals of the β''-SF$_5$ salt were synthesized using a standard electrochemical method [38]. Two gold wires of 10 μm diameter were attached to both sides of the single crystal using carbon paste. The interlayer resistance with an electric current perpendicular to the superconducting layers was measured using a conventional four-probe AC technique. The single crystals were mounted on a two-axis rotator in a ^{3}He cryostat with a 15 T superconducting magnet and cooled down to $\sim$0.5 K at a rate of $\sim$1 K/min. All measurements were performed at the Tsukuba Magnet Laboratories, NIMS.

3. Results

Figure 1b shows the temperature dependence of the interlayer resistance for the β''-SF$_5$ salt. The resistance decreased monotonically with decreasing temperature. At $\sim$4.8 K, a sudden drop in resistance was observed due to the superconducting transition. Below $\sim$4 K, the resistance was zero within the noise level. To investigate the 2D Fermi surface structure, we first measured the AMROs in various rotation planes. Typical AMRO data are presented in Figure 2a. The angles θ and φ are defined in the inset. The characteristic θ dependence of the interlayer resistance is shown in the upper part of Figure 2a. In the negative second derivative curves, we can observe AMROs, which are periodic with $\tan(\theta)$, as shown in the lower part of Figure 2a. The AMRO period δ directly yields the reciprocal lattice vector $k_{\parallel}$, $\delta(r) = \pi/s k_{\parallel}(\varphi)$ values. Figure 2b shows the polar plot of $k_{\parallel}$ obtained from the AMRO measurements at various φ. We can draw the cross-section of the 2D Fermi surface, inscribed in the $k_{\parallel}(\varphi)$ curves, by a solid curve, assuming an elliptical shape. The cross-section of the Fermi surface is very elongated, whose area was $\sim$6% of the first Brillouin zone. The AMRO results were almost consistent with those of previous reports [39].

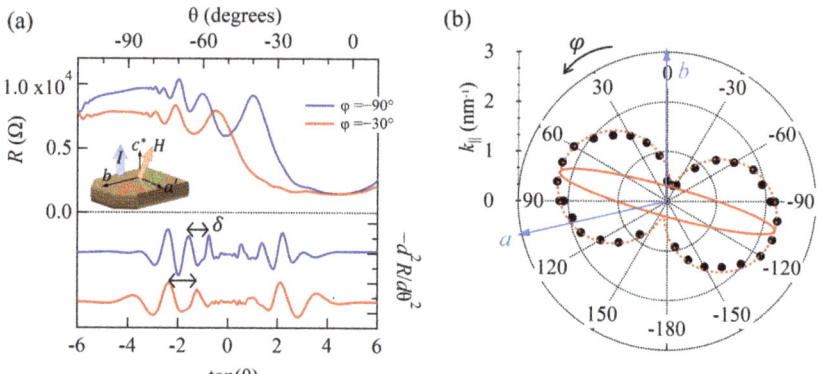

Figure 2. (a) Typical AMRO data and their negative second derivative curves at 1.5 K for 14 T. (b) Polar plot of $k_\parallel$ obtained from the AMRO measurements. The red solid curve shows the 2D Fermi surface obtained from the AMRO measurements.

In Figure 3a, we present the band calculations by an extended Huckel method [40], using lattice parameters obtained from X-ray crystallography [38]. The calculated 2D Fermi surface is depicted in Figure 3b. The results are different from the reported Fermi surface structure, with a pair of 1D Fermi surface and a closed Fermi surface [38,41], in which the Brillouin zone is apparently wrong. In our calculations, there were two pockets with different carriers: a deformed square electron pocket and an elliptic hole pocket. The areas were equal to each other, and a compensated metal was formed. This is consistent with a single frequency of the quantum oscillation [42–44]. For comparison, the Fermi surface obtained from the AMRO measurements is indicated by a red dotted curve, which is almost consistent with the electron pocket. In the AMRO measurements, the hole pocket was not observed. The reason for this is not clear at present.

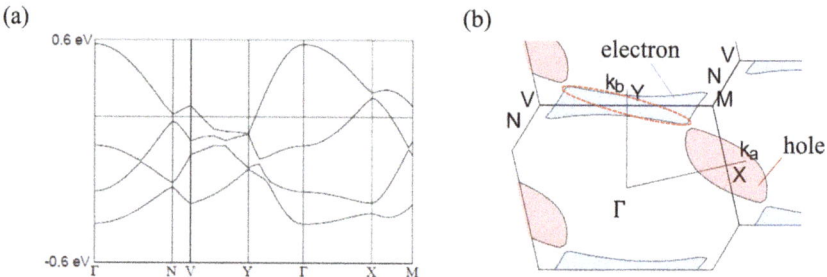

Figure 3. (a) Band structure by an extended Huckel method and (b) 2D Fermi surface structure for the β''-SF$_5$ salt. Deformed square hole and elliptic electron pockets were formed. The red dotted curve indicates the 2D Fermi surface determined by the AMRO measurements.

Figure 4a shows the magnetic field dependence of the interlayer resistance at various temperatures. The field was applied parallel to the b axis, in the superconducting a'–b plane, within the accuracy of 0.1°. At 0.5 K, the resistance increased with the field above 6.5 T, defined as H_{onset}. Characteristic corrugation was evident. The corrugation was reduced with increasing temperature. The critical field was determined as $H_{c2} \approx 13$ T at 0.7 K from the specific-heat measurements [23]. The finite resistance in the wide field region below H_{c2} can be ascribed to the motion of the JVs in the insulating layers, as has been observed in various organic superconductors [15,28,29]. To clearly see the corrugations in detail, negative second-derivative curves of the resistance are plotted in Figure 4b. At 0.5 K, we see a broad dip at ∼6.5 T, corresponding to the resistance increase from the noise level. Above ∼8 T, we see a quasi-periodic dip structure, which is most pronounced at ∼10 T. This structure can be ascribed to the CM effect, which is observable only in the FFLO phase, as discussed in the previous reports [29]. At higher fields, the CM effect is reduced and vanishes above ∼12 T, which corresponds to the melting transition of the JV lattice. As temperature increases, the dip structure is suppressed and shifts to a lower field region. Above 2.1 K, such astructure is not evident. As has been discussed [27,29], the dips mean relatively strong pinning of the JV lattice, ascribed to the CM effect between the JV lattice and the periodic nodal structure of the gap $\Delta(r)$.

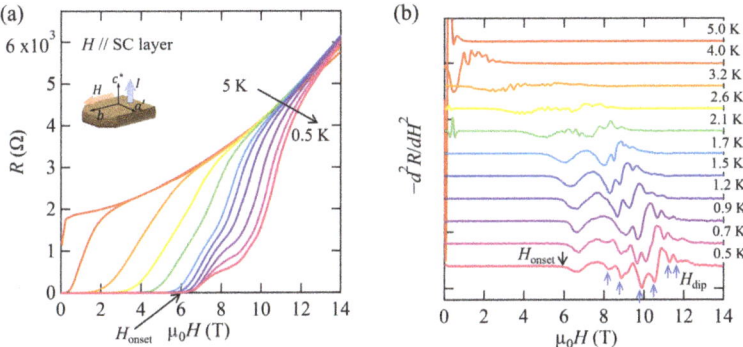

Figure 4. (a) Magnetic field dependence of the interlayer resistance at various temperatures. The field was applied parallel to the b axis in the superconducting a'–b plane within an accuracy of 0.1°. (b) Negative second-derivative curves of the resistance. Each curve was shifted for clarity.

From the above results, we obtained the temperature–field phase diagram shown in Figure 5. The blue squares indicate the dip fields H_{dip}, and the solid curve indicates H_{c2}, which was determined from specific-heat measurements [23]. The FFLO phase appears in a wide region above ∼8 T and below ∼2 K. The phase diagram is very similar to that for the $H \parallel a$-axis [29], although the Fermi surface was anisotropic.

Figure 6a shows the magnetic field dependence of the resistance at various field angles θ. For $\theta = 0°$ ($H \parallel b$-axis), the resistance increased with the field above $\mu_0 H_{onset} = 6$ T, which is indicated by an arrow. Figure 6b shows the negative second-derivative curves of the resistance. The low H_{onset} value for $\theta = 0°$, denoted by an arrow, indicates that only JVs were formed (no PVs), which were pinned very weakly in the insulating layers. When the field was tilted from the superconducting layer, H_{onset} increased. This behavior is explained by the stronger pinning of the flux lines in the superconducting layers, where PVs are formed. As the field was further tilted, H_{c2} was steeply reduced, leading to a decrease in H_{onset}. For $\theta = 0°$, small dips due to the CM effect can be seen above ∼9.5 T. As the field was tilted from the layer, the CM effect was suppressed, and no CM effect was observed for $|\theta| \gtrsim 0.6°$. The stability of the FFLO phase in such a small angle region is consistent with the results for the field almost parallel to the a axis [29].

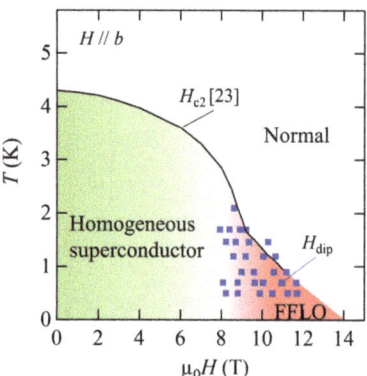

Figure 5. Temperature–field phase diagram for $H \parallel b$ axis. H_{c2} determined by the specific-heat measurements is indicated by a solid curve [23].

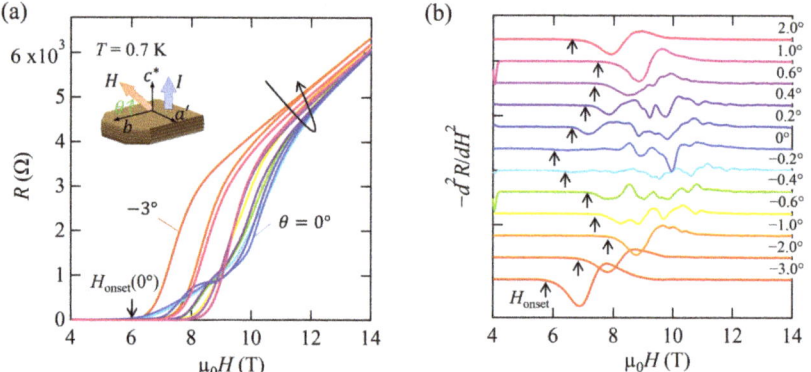

Figure 6. (**a**) Magnetic field dependence of the resistance at various field angles θ. Definition of θ is in the inset. (**b**) Negative second-derivative curves of the resistance. Each curve is shifted for clarity.

Figure 7a shows the field dependence of the negative second-derivative curves at various in-plane field directions φ, as shown in the inset of Figure 7b. For $\varphi = 0°$, we see the onset field $\mu_0 H_{\mathrm{onset}} = 6$ T (black arrow). The H_{onset} value had a nonmonotonic φ dependence. It should be noted that H_{onset} is the depinning field of the JV lattice [29], determined by the pinning strength at the sample edges and/or some other (impurity or defect) pinning sites, which is not related to the FFLO phase transition. The anisotropic behavior of H_{onset} is possibly due to the shape effect of the sample. In contrast, we observed many dips above ∼9 T, owing to the CM effect in a similar field region at any φ. This suggests that the FFLO phase stability was insensitive to the in-plane field direction. An important feature is that the largest dip was evident at $\mu_0 H_{\mathrm{dip}}^* \approx 9$ T (red arrow) in a wide-angle region, except for $\varphi = 0°$–$45°$. This φ dependence of the dip amplitude suggests some differences in the JV dynamics in the FFLO phase. The largest dip field H_{dip}^* corresponds to the strongest CM effect and is plotted as a function of φ in Figure 7b. We note that H_{dip}^* is almost isotropic, despite the anisotropic Fermi surface structure as presented in Figure 3b.

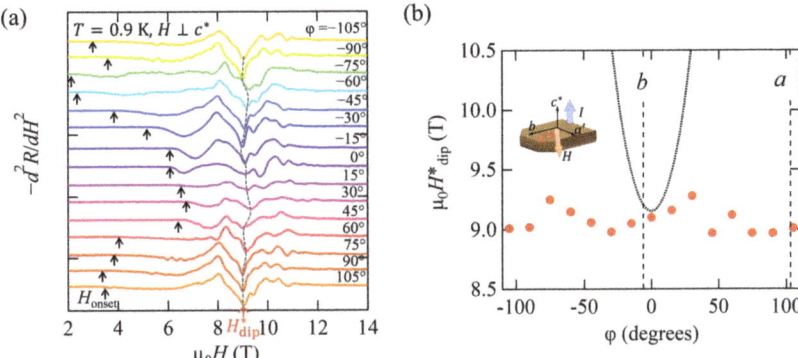

Figure 7. (a) Magnetic field dependence of the negative second-derivative curves at various in-plane field directions φ defined in the inset of (b). Each curve is shifted for clarity. The onset field H_{onset} and the largest dip field H^*_{dip} are indicated by black and red arrows, respectively. (b) H_{dip} as a function of φ. The dotted curve indicates the expected value for the CM condition $H_{dip} \propto 1/\cos(\varphi)$ with a single q vector.

4. Discussion

We observed the CM effect in various in-plane field directions, which is recognized as strong evidence of the FFLO phase characterized by the q vector. As pointed out, the q vector leading to a large nesting part is favorable for the FFLO phase, as schematically depicted in Figure 8a, where the largest number of Cooper pairs can be formed by the q vector, perpendicular to the flat part of the Fermi surface. On the other hand, in the presence of the orbital effect, the q vector parallel to the field is favorable for an isotropic Fermi surface, leading to no CM effect. The observation of the CM effect at any φ indicates that the q vector is not parallel to the field.

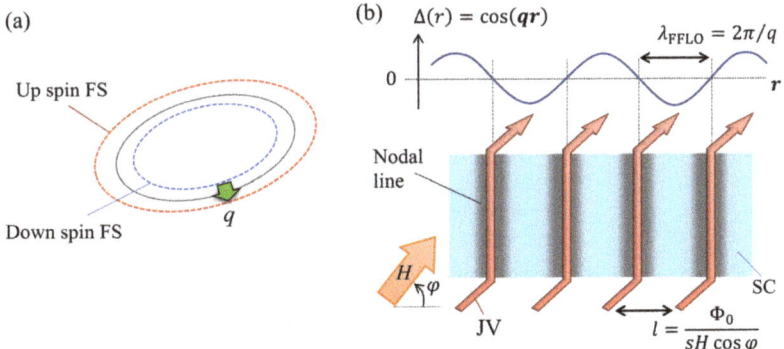

Figure 8. (a) Spin-polarized 2D Fermi surface in a magnetic field. Up and down Fermi surfaces are indicated by red and blue curves, respectively. (b) Schematic nodal line structure of a FFLO phase and JV lattice in a tilted field for the CM condition $l/\lambda_{FFLO} = 1$.

The largest dip at H^*_{dip} in Figure 7a suggest the strongest CM effect, $l/\lambda_{FFLO} = 1$, where all the flux lines can fit into the nodal lines completely. Assuming that the q vector is fixed to a certain direction (for instance, a'-axis), the JV lattice spacing is given by $l = \Phi_0/sH\cos(\varphi)$ as depicted in Figure 8b. This leads to large φ dependence of H^*_{dip} as indicated by the dotted curve in Figure 7b, which is inconsistent with the experimental result. This inconsistency requires another factor on the stability of the FFLO phase. At the nodal lines, the order parameter vanishes in the superconducting layers. Therefore, the most stable condition of the vortex structure will be that all the flux lines are almost parallel to the nodal lines; the q vector is almost perpendicular to the field. This suggests that the direction of the q vector changes with the field direction. In the anisotropic Fermi surface, the q vector length should also be anisotropic, depending on the energy dispersion. Although it is difficult to know the φ dependence of the q vector, this scenario could explain the lack of significant φ dependence of H_{dip}^* in Figure 7b.

Recent specific-heat measurements show that H_{c2} is the same at a few different in-plane field directions in a low temperature range [23]. The fact shows that the FFLO stability is independent of φ, and seems to be consistent with our results: no significant φ dependence of H_{dip}^*. Theoretically, the in-plane anisotropies of H_{c2} are led by Fermi surface structure and orbital effects in the FFLO phase [45]. Therefore, no in-plane anisotropy of H_{c2} in the specific-heat measurements suggests that the orbital effect is almost negligible, and the q vector is most likely pinned to an optimal direction independent of the in-plane field direction. The inconsistency with our results remains an open question.

Another possible scenario that could explain our results is that multi-q vectors [37] are formed in the β''-SF$_5$ salt, since the two different Fermi pockets are present as shown in Figure 3b. In this case, the q-dependent anisotropic stability of the FFLO phase could be smeared out; the FFLO phase may appear in a similar field range, independent of the in-plane field direction. This scenario may also explain the lack of a significant in-plane anisotropy of H_{dip}^* and H_{c2}.

Finally, we briefly mention the results for another FFLO superconductor λ-(BETS)$_2$FeCl$_4$, which had a pair of 1D and a 2D Fermi surfaces [15]. In this salt, the CM effect was first observed in the field-induced superconducting phase. The CM effect was clearly observed for $H \parallel c$ but not for $H \parallel a$. The results show that a single q vector was fixed to the a-axis in the whole FFLO phase. The different behavior of the CM effect between β''-SF$_5$ salt and λ-(BETS)$_2$FeCl$_4$ will be closely related to the Fermi surface structure. More detailed measurements of the CM effect in other FFLO superconductors will be required to clarify the correlation between the Fermi surface structure and the q vector.

5. Conclusions

The AMRO measurements and band-structure calculations in the β''-SF$_5$ salt show that the Fermi surface is composed of two small pockets, a deformed square electron pocket, and an elliptic hole pocket, which are different from the previous report. The CM effect in the interlayer resistance was observed in a similar field region at any in-plane field directions. This indicates that the stability of the FFLO phase is almost isotropic, which is consistent with the observations of precious specific-heat measurements. Two possible scenarios are proposed: (1) a single center-of-mass momentum q of the Cooper pairs, which changes with the in-plane field direction, and (2) multi-q vectors, originating from the two anisotropic Fermi surfaces.

Author Contributions: S.S., T.T. and S.U. designed the experiments. S.S. and S.Y. mainly performed the resistance measurements and analyzed the data. H.A. and Y.N. performed the band calculation. J.A.S. synthesized the single crystals. S.U. supervised the project. All authors have read and agreed to the published version of the manuscript.

Funding: This work was supported by KAKENHI 17H01144 and 20K14400.

Institutional Review Board Statement: Not applicable.

Informed Consent Statement: Not applicable.

Data Availability Statement: The data that support the findings of this study are available from the corresponding author on reasonable request.

Acknowledgments: Work at ANL was supported by U. Chicago Argonne, LLC, Operator of Argonne National Laboratory ("Argonne"). Argonne, a U.S. Department of Energy Office of Science labor s operated under Contract No. DE-AC02-06CH11357. J.A.S. acknowledges support from the Independent Research = Development program while serving at the National Science Foundation.

Conflicts of Interest: The authors declare no conflict of interest.

References

1. Martucciello, N.; Giubileo, F.; Grimaldi, G.; Corato, V. Introduction to the focus on superconductivity for energy. *Supercond. Sci. Technol.* **2015**, *28*, 070201. [CrossRef]
2. Fulde, P.; Ferrell, R.A. Superconductivity in a strong spin-exchange field. *Phys. Rev.* **1964**, *135*, A550. [CrossRef]
3. Larkin, A.I.; Ovchinnikov, Y.N. Nonuniform state of superconductors. *Zh. Eksp. Teor. Fiz* **1964**, *47*, 1136.
4. Clogston, A.M. Upper limit for the critical field in hard superconductors. *Phys. Rev. Lett.* **1962**, *9*, 266. [CrossRef]
5. Radovan, H.A.; Fortune, N.A.; Murphy, T.P.; Hannahs, S.T.; Palm, E.C.; Tozer, S.W.; Hall, D. Magnetic enhancement of superconductivity from electron spin domains. *Nature* **2003**, *425*, 51–55. [CrossRef] [PubMed]
6. Bianchi, A.R.; Movshovich, C.C.; Pagliuso, P.G.; Sarrao, J.L. Possible Fulde-Ferrell-Larkin-Ovchinnikov Superconducting State in CeCoIn$_5$. *Phys. Rev. Lett.* **2003**, *91*, 187004. [CrossRef] [PubMed]
7. Kikugawa, N.; Terashima, T.; Uji, S.; Sugii, K.; Maeno, Y.; Graf, D.; Baumbach, R.; Brooks, J. Superconducting subphase in the layered perovskite ruthenate Sr$_2$RuO$_4$ in a parallel magnetic field. *Phys. Rev. B* **2016**, *93*, 184513. [CrossRef]
8. Kasahara, S.; Sato, Y.; Licciardello, S.; Čulo, M.; Arsenijević, O.S.T.; Tominaga, T.; Böker, E.J.I.; Shibauchi, T.; Wosnitza, J.; Hussey, N.E.; et al. Evidence for an Fulde–Ferrell–Larkin–Ovchinnikov State with Segmented Vortices in the BCS–BEC–Crossover Superconductor FeSe. *Phys. Rev. Lett.* **2020**, *124*, 107001. [CrossRef]
9. Yonezawa, S.; Kusaba, S.; Maeno, Y.; Auban-Senzier, P.; Pasquier, C.; Bechgaard, K.; Jérome, D. Anomalous In-Plane Anisotropy of the Onset of Superconductivity in (TMTSF)$_2$ClO$_4$. *Phys. Rev. Lett.* **2008**, *100*, 117002. [CrossRef] [PubMed]
10. Singleton, J.; Symington, J.A.; Nam, M.S.; Ardavan, A.; Kurmoo, M.; Day, P. Observation of the Fulde-Ferrell-Larkin-Ovchinnikov state in the quasi-two-dimensional organic superconductor κ-(BEDT-TTF)$_2$Cu(NCS)$_2$ (BEDT-TTF= bis (ethylene-dithio) tetrathiafulvalene). *J. Phys. Condens. Matter* **2000**, *12*, L641. [CrossRef]
11. Tanatar, M.A.; Ishiguro, T.; Tanaka, H.; Kobayashi, H. Magnetic field–temperature phase diagram of the quasi-two-dimensional organic superconductor λ-(BETS)$_2$ GaCl$_4$ studied via thermal conductivity. *Phys. Rev. B* **2002**, *66*, 134503. [CrossRef]
12. Tanatar, M.A.; Ishiguro, T.; Tanaka, H.; Kobayashi, H. Superconducting phase diagram and FFLO signature in λ-(BETS)$_2$ GaCl$_4$ from rf penetration depth measurements. *Phys. Rev. B* **2011**, *83*, 224507.
13. Tanatar, M.A.; Ishiguro, T.; Tanaka, H.; Kobayashi, H. Vortex Dynamics and Diamagnetic Torque Signals in Two Dimensional Organic Superconductor λ-(BETS)$_2$ GaCl$_4$. *J. Phys. Soc. Jpn.* **2015**, *84*, 104709.
14. Balicas, L.; Brooks, J.S.; Storr, K.; Uji, S.; Tokumoto, M.; Tanaka, H.; Kobayashi, H.; Kobayashi, A.; Barzykin, V.; Gořkov, L.P. Superconductivity in an organic insulator at very high magnetic fields. *Phys. Rev. Lett.* **2001**, *87*, 067002. [CrossRef] [PubMed]
15. Uji, S.; Terashima, T.; Nishimura, M.; Takahide, Y.; Konoike, T.; Enomoto, K.; Cui, H.; Kobayashi, H.; Kobayashi, A.; Tanaka, H.; et al. Vortex dynamics and the Fulde-Ferrell-Larkin-Ovchinnikov state in a magnetic-field-induced organic superconductor. *Phys. Rev. Lett.* **2006**, *97*, 157001. [CrossRef] [PubMed]
16. Uji, S.; Kodama, K.; Sugii, K.; Terashima, T.; Takahide, Y.; Kurita, N.; Tsuchiya, S.; Kimata, M.; Kobayashi, A.; Zhou, B.; et al. Magnetic torque studies on FFLO phase in magnetic-field-induced organic superconductor λ-(BETS)$_2$ FeCl$_4$. *Phys. Rev. B* **2012**, *85*, 174530. [CrossRef]
17. Lortz, R.; Wang, Y.; Demuer, A.; Böttger, P.H.M.; Bergk, B.; Zwicknagl, G.; Nakazawa, Y.; Wosnitza, J. Calorimetric Evidence for a Fulde-Ferrell-Larkin-Ovchinnikov Superconducting State in the Layered Organic Superconductor κ-(BEDT-TTF)$_2$Cu(NCS)$_2$. *Phys. Rev. Lett.* **2007**, *99*, 187002. [CrossRef] [PubMed]
18. Bergk, B.; Demuer, A.; Sheikin, I.; Wang, Y.; Wosnitza, J.; Nakazawa, Y.; Lortz, R. Magnetic torque evidence for the Fulde-Ferrell-Larkin-Ovchinnikov state in the layered organic superconductor κ-(BEDT-TTF)$_2$Cu(NCS)$_2$. *Phys. Rev. B* **2011**, *83*, 064506. [CrossRef]
19. Wright, J.A.; Green, E.; Kuhns, P.; Reyes, A.; Brooks, J.; Schlueter, J.; Bkato, R.; Yamamoto, H.; Kobayashi, M.; Brown, S.E. Zeeman-Driven Phase Transition within the Superconducting State of κ-(BEDT-TTF)$_2$Cu(NCS)$_2$. *Phys. Rev. Lett.* **2011**, *107*, 087002. [CrossRef] [PubMed]
20. Agosta, C.C.; Jin, J.; Coniglio, W.A.; Smith, B.E.; Cho, K.; Stroe, I.; Martin, C.; Tozer, S.W.; Murphy, T.P.; Palm, E.C.; et al. Experimental and semiempirical method to determine the Pauli-limiting field in quasi-two-dimensional superconductors as applied to κ-(BEDT-TTF)$_2$Cu(NCS)$_2$: Strong evidence of a FFLO state. *Phys. Rev. B* **2012**, *85*, 214514. [CrossRef]
21. Mayaffre, H.; Krämer, S.; Horvatić, M.; Berthier, C.; Miyagawa, K.; Kanoda, K.; Mitrović, V.F. Evidence of Andreev bound states as a hallmark of the FFLO phase in κ-(BEDT-TTF)$_2$Cu(NCS)$_2$. *Nat. Phys.* **2014**, *10*, 928. [CrossRef]
22. Tsuchiya, S.; Yamada, J.I.; Sugii, K.; Graf, D.; Brooks, J.S.; Terashima, T.; Uji, S. Phase boundary in a superconducting state of κ-(BEDT-TTF)$_2$Cu(NCS)$_2$: Evidence of the Fulde–Ferrell–Larkin–Ovchinnikov phase. *J. Phys. Soc. Jpn.* **2015**, *84*, 034703. [CrossRef]
23. Wosnitza, J. FFLO states in layered organic superconductors. *Ann. Phys.* **2018**, *530*, 1700282. [CrossRef]

24. Agosta, C.C.; Fortune, N.A.; Hannahs, S.T.; Gu, S.; Liang, L.; Park, J.H.; Schleuter, J.A. Calorimetric Measurements of Magnetic-Field-Induced Inhomogeneous Superconductivity Above the Paramagnetic Limit. *Phys. Rev. Lett.* **2017**, *118*, 267001. [CrossRef]
25. Tachiki, M.; Takahashi, S.; Gegenwart, P.; Weiden, M.; Lang, M.; Geibel, C.; Steglich, F.; Modler, R.; Paulsen, C.; Ōnuki, Y. Generalized Fulde-Ferrell-Larkin-Ovchinnikov state in heavy-fermion and intermediate-valence systems. *Z. Phys. Condens. Matter* **1997**, *100*, 369–380. [CrossRef]
26. Shimahara, H. Fulde-Ferrell state in quasi-two-dimensional superconductors. *Z. Phys. Condens. Matter* **1994**, *50*, 12760. [CrossRef] [PubMed]
27. Bulaevskii, L.; Buzdin, A.; Maley, M. Intrinsic pinning of vortices as a direct probe of the nonuniform Larkin-Ovchinnikov-Fulde-Ferrell state in layered superconductors β''-(BEDT-TTF)$_2$SF$_5$CH$_2$CF$_2$SO$_3$. *Phys. Rev. Lett.* **2003**, *90*, 067003. [CrossRef]
28. Uji, S.; Iida, Y.; Sugiura, S.; Isono, T.; Sugii, K.; Kikugawa, N.; Terashima, T.; Yasuzuka, S.; Akutsu, H.; Nakazawa, Y.; et al. Fulde-Ferrell-Larkin-Ovchinnikov superconductivity in the layered organic superconductor β''-(BEDT-TTF)$_4$[(H$_3$O)Ga(C$_2$O$_4$)$_3$]C$_6$H$_5$NO$_2$. *Phys. Rev. B* **2018**, *97*, 144505. [CrossRef]
29. Sugiura, S.; Terashima, T.; Yasuzuka, S.; Schlueter, J.A.; Uji, S. Josephson vortex dynamics and Fulde-Ferrell-Larkin-Ovchinnikov superconductivity in the layered organic superconductor β''-(BEDT-TTF)$_2$SF$_5$CH$_2$CF$_2$SO$_3$. *Phys. Rev. B* **2019**, *100*, 014515. [CrossRef]
30. Wanka, S.; Hagel, J.; Beckmann, D.; Wosnitza, J.; Schlueter, J.A.; Williams, J.M.; Nixon, P.G.; Winter, R.W.; Gard, G.L. Specific heat and critical fields of the organic superconductor β''-(BEDT-TTF)$_2$SF$_5$CH$_2$CF$_2$SO$_3$. *Phys. Rev. B* **1998**, *57*, 3084. [CrossRef]
31. Cho, K.; Smith, B.E.; Coniglio, W.A.; Winter, L.E.; Agosta, C.C.; Schlueter, J.A. Upper critical field in the organic superconductor β''-(ET)$_2$SF$_5$CH$_2$CF$_2$SO$_3$: Possibility of Fulde-Ferrell-Larkin-Ovchinnikov state. *Phys. Rev. B* **2009**, *79*, 220507. [CrossRef]
32. Beyer, R.; Bergk, B.; Yasin, S.; Schlueter, J.A.; Wosnitza, J. Angle-dependent evolution of the Fulde-Ferrell-Larkin-Ovchinnikov state in an organic superconductor. *Phys. Rev. Lett.* **2012**, *109*, 027003. [CrossRef]
33. Beyer, R.; Wosnitza, J. Emerging evidence for FFLO states in layered organic superconductors. *Low Temp. Phys.* **2013**, *39*, 225–231. [CrossRef]
34. Koutroulakis, G.; Kühne, H.; Schlueter, J.A.; Wosnitza, J.; Brown, S.E. Microscopic study of the Fulde-Ferrell-Larkin-Ovchinnikov state in an all-organic superconductor. *Phys. Rev. Lett.* **2016**, *116*, 067003. [CrossRef] [PubMed]
35. Sugiura, S.; Isono, T.; Terashima, T.; Yasuzuka, S.; Schlueter, J.A.; Uji, S. Fulde–Ferrell–Larkin–Ovchinnikov and vortex phases in a layered organic superconductor. *J. Abbr.* **2008**, *10*, 142–149. [CrossRef]
36. Pustogow, A.; Saito, Y.; Rohwer, A.; Schlueter, J.A.; Dressel, M. Coexistence of charge order and superconductivity in organic superconductor β''-(BEDT-TTF)$_2$SF$_5$CH$_2$CF$_2$SO$_3$. *Phys. Rev. B* **2019**, *99*, 140509(R). [CrossRef]
37. Shimahara, H. Structure of the Fulde-Ferrell-Larkin-Ovchinnikov state in two-dimensional superconductors. *J. Phys. Soc. Jpn.* **1998**, *67*, 736–739. [CrossRef]
38. Geiser, U.; Schlueter, J.A.; Wang, H.H.; Kini, A.M.; Williams, J.M.; Sche, P.P.; Zakowicz, I.H.; Michael, V.; Dudek, D.J.; Nixon, P.G.; et al. Superconductivity at 5.2 K in an electron donor radical salt of bis (ethylenedithio) tetrathiafulvalene (BEDT-TTF) with the novel polyfluorinated organic anion SF$_5$CH$_2$CF$_2$SO$_3$. *J. Am. Chem. Soc.* **1996**, *118*, 9996–9997. [CrossRef]
39. Yasuzuka, S.; Uji, S.; Terashima, T.; Sugii, K.; Isono, T.; Iida, Y.; Schlueter, J.A. In-Plane Anisotropy of Upper Critical Field and Flux-Flow Resistivity in Layered Organic Superconductor β''-(BEDT-TTF)$_2$SF$_5$CH$_2$CF$_2$SO$_3$. *J. Phys. Soc. Jpn.* **2015**, *84*, 094709. [CrossRef]
40. Mori, T.; Kobayashi, A.; Sasaki, Y.; Kobayashi, H.; Saito, G.; Inokuchi, H. The Intermolecular Interaction of Tetrathiafulvalene and Bis(ethylenedithio)tetrathiafulvalene in Organic Metals. Calculation of Orbital Overlaps and Models of Energy-band Structures. *Bull. Chem. Soc. Jpn.* **1984**, *57*, 627. [CrossRef]
41. Beckmann, D.; Wanka, S.; Wosnitza, J.; Schlueter, J.A.; Williams, J.M.; Nixon, P.G.; Winter, R.W.; Gard, G.L.; Ren, J.; Whangbo, M.-H. Characterization of the Fermi surface of the organic superconductor-by measurements of Shubnikov-de Haas and angle-dependent magnetoresistance oscillations and by electronic band-structure calculations. *Eur. Phys. J. Condens. Matter Complex Syst.* **1998**, *3*, 295–300. [CrossRef]
42. Wosnitza, J.; Wanka, S.; Hagel, J.; Häussler, R.; Löhneysen, H.V.; Schlueter, J.A.; Geiser, U.; Nixon, P.G.; Winter, R.W.; Gard, G.L. Shubnikov–de Haas effect in the superconducting state of an organic superconductor. *Phys. Rev. B* **2000**, *62*, R11973. [CrossRef]
43. Wosnitza, J.; Hagel, J.; Meeson, P.J.; Bintley, D.; Schlueter, J.A.; Mohtasham, J.; Winter, R.W.; Gard, G.L. Enhanced magnetic quantum oscillations in the mixed state of a two-dimensional organic superconductor. *Phys. Rev. B* **2003**, *67*, 060504. [CrossRef]
44. Sugiura, S.; Terashima, T.; Uji, S.; Schlueter, J.A. Deformed Waveshape of Quantum Oscillation in Magnetocaloric Effect for Layered Organic Superconductor. *J. Phys. Soc. Jpn.* **2021**, *90*, 074601. [CrossRef]
45. Croitoru, M.D.; Houzet, M.; Buzdin, A.I. In-Plane Magnetic Field Anisotropy of the Fulde-Ferrell-Larkin-Ovchinnikov State in Layered Superconductors. *Phys. Rev. Lett.* **2012**, *108*, 207005. [CrossRef] [PubMed]

Article

Tuning Charge Order in (TMTTF)$_2$X by Partial Anion Substitution

Andrej Pustogow [1,2,*], Daniel Dizdarevic [1], Sebastian Erfort [1], Olga Iakutkina [1], Valentino Merkl [1], Gabriele Untereiner [1] and Martin Dressel [1,*]

[1] 1. Physikalisches Institut, Universität Stuttgart, Pfaffenwaldring 57, 70569 Stuttgart, Germany; danieldizdarevic@gmx.de (D.D.); erfort@theochem.uni-stuttgart.de (S.E.); olga.iakutkina@pi1.uni-stuttgart.de (O.I.); valentino.merkl@pi1.uni-stuttgart.de (V.M.); gabriele.untereiner@pi1.physik.uni-stuttgart.de (G.U.)

[2] Institute of Solid State Physics, TU Wien, 1040 Vienna, Austria

* Correspondence: pustogow@ifp.tuwien.ac.at (A.P.); dressel@pi1.physik.uni-stuttgart.de (M.D.)

Abstract: In the quasi-one-dimensional (TMTTF)$_2$X compounds with effectively quarter-filled bands, electronic charge order is stabilized from the delicate interplay of Coulomb repulsion and electronic bandwidth. The correlation strength is commonly tuned by physical pressure or chemical substitution with stoichiometric ratios of anions and cations. Here, we investigate the charge-ordered state through partial substitution of the anions in (TMTTF)$_2$[AsF$_6$]$_{1-x}$[SbF$_6$]$_x$ with $x \approx 0.3$, determined from the intensity of infrared vibrations, which is sufficient to suppress the spin-Peierls state. Our dc transport experiments reveal a transition temperature T_{CO} = 120 K and charge gap Δ_{CO} = 430 K between the values of the two parent compounds (TMTTF)$_2$AsF$_6$ and (TMTTF)$_2$SbF$_6$. Upon plotting the two parameters for different (TMTTF)$_2$X, we find a universal relationship between T_{CO} and Δ_{CO} yielding that the energy gap vanishes for transition temperatures $T_{CO} \leq 60$ K. While these quantities indicate that the macroscopic correlation strength is continuously tuned, our vibrational spectroscopy results probing the local charge disproportionation suggest that 2δ is modulated on a microscopic level.

Keywords: charge-transfer salts; (TMTTF)$_2$X; Fabre salts; charge order; strongly correlated electron systems; extended Hubbard model; bandwidth tuning; partial chemical substitution; negative chemical pressure; phase transitions; metal-insulator transitions; optical conductivity; vibrational spectroscopy; FTIR

PACS: 71.30.+h; 78.30.Jw; 75.25.Dk

1. Introduction

Organic charge-transfer salts are model systems realizing electronic correlations and Mott–Hubbard physics, yielding a plethora of metal–insulator transitions in many different compounds. Owing to their effectively quarter-filled bands with nominally one electron per two organic molecules, the quasi-one-dimensional Fabre salts (TMTTF)$_2$X are prone to charge-order instabilities [1]. Initially, charge order (CO) was suggested as a purely electronic effect due to intersite Coulomb repulsion. Within the extended Hubbard model, the ratio of nearest neighbor interaction V with respect to the bandwidth W is a measure of the correlation strength that can be varied by external pressure or by chemical means [2,3]. Eventually, CO is suppressed completely for sufficiently small V/W, and an insulator–metal transition takes place, stabilizing metallic and superconducting states. At lower temperatures, more complex phase transitions to anion-ordered and spin-Peierls phases result in modifications of the magnetic and structural degrees of freedom, such as the formation of a spin gap and tetramerization of the TMTTF molecules. Previous NMR and optical studies of electronically-driven charge order provided consistent results on (TMTTF)$_2$X with centrosymmetric anions $X = $ PF$_6^-$, AsF$_6^-$, SbF$_6^-$, and TaF$_6^-$ [1,4–13] and tetrahedral anions $X = $ BF$_4^-$, ReO$_4^-$ [14,15].

The wave function overlap $t \propto W$ and, hence, V/W, can be modified by changing the lattice parameter through introduction of bigger or smaller anions, or by the use of organic donor molecules with Se instead of S. Chemical substitution in a stoichiometric manner allows reaching distinct regions in the phase diagram, with a step size determined by the chemical properties of the respective molecules. Physical pressure, on the other hand, enables tuning in arbitrary steps through the phase diagram, at the cost of more difficult experiments that have to be carried out in a pressure cell. The advantages of both approaches—flexible tuning and ambient pressure experiments—can be achieved by partial substitution of the constituents, either the donor molecules [16–29] or the anions [30–33], as depicted in Figure 1. So far, partial substitution has remained poorly investigated compared to pressure tuning, but, in addition to bandwidth tuning, it also enables the study of disorder effects on metal–insulator transitions and superconductivity [29].

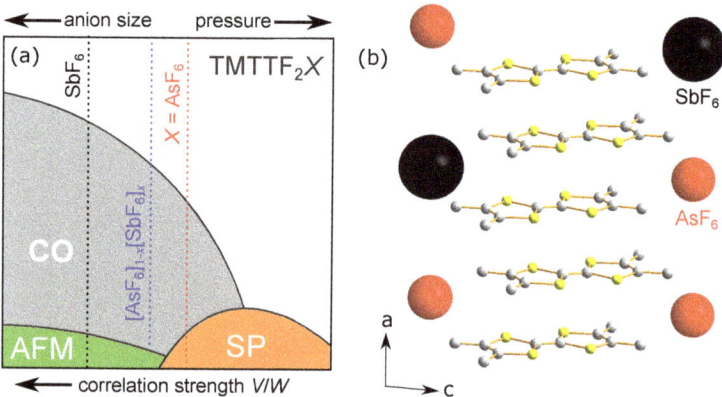

Figure 1. (Color online) (**a**) (TMTTF)$_2$X with X = SbF$_6$ exhibits larger electronic correlation strength V/W as the molecules are separated further apart than for X = AsF$_6$. Modifying the intersite spacing via physical pressure or chemical substitution allows tuning through charge-ordered (CO), antiferromagnetic (AFM), and spin-Peierls (SP) states in the phase diagram. (**b**) Partial substitution of AsF$_6$ anions (red) with larger SbF$_6$ (black) in (TMTTF)$_2$[AsF$_6$]$_{1-x}$[SbF$_6$]$_x$ yields a position between the two parent compounds, indicated by the dotted blue line in (**a**).

Here, we investigate charge order upon partial substitution of the anions in (TMTTF)$_2$-[AsF$_6$]$_{1-x}$[SbF$_6$]$_x$. The stoichiometry $x \approx 0.3$ is determined via the intensity of SbF$_6^-$ and AsF$_6^-$ vibration modes at 660 and 700 cm^{-1}, respectively. The lack of tetramerization and a spin gap deduced from optical and magnetic susceptibility measurements indicates the absence of a spin-Peierls state. Our dc transport results yield a transition temperature T_{CO} = 120 K and a charge gap Δ_{CO} = 430 K. These values line up with CO in the stoichiometric (TMTTF)$_2$X compounds, revealing that $\Delta_{CO} \to 0$ around $T_{CO} \approx 60$ K. Consistent with the resistivity data, our optical experiments on TMTTF vibrations in the infrared range yield a splitting of the charge-sensitive ν_{28} mode below T_{CO}. Despite the higher transition temperature, the charge disproportionation $2\delta = 0.21e$ (in the limit $T \to 0$) is similar to (TMTTF)$_2$AsF$_6$. Our findings indicate that the local amplitude of charge imbalance is linked closely to the nearest anions, while T_{CO} and Δ_{CO} are determined by the macroscopic mixture. This motivates more systematic studies of the role of the anions and the microscopic properties of CO via partial chemical substitution.

2. Materials and Experiments

CO in (TMTTF)$_2$X has been comprehensively studied in the stoichiometric compounds with octahedral and tetrahedral anions [1,4–15]. For the former, larger anion size $d(\text{TaF}_6^-) > d(\text{SbF}_6^-) > d(\text{AsF}_6^-) > d(\text{PF}_6^-)$ yields a bigger separation of the TMTTF molecules and, hence, an increase of electronic correlations V/W as the wave function overlap $t \propto W$ is reduced more strongly than the intersite Coulomb repulsion V (Figure 1a). This trend has been continued with the recently synthesized new member with NbF_6^- as counterion, which has similar T_{CO} as $(TMTTF)_2SbF_6$ [34]. Accordingly, the compounds with largest (smallest) anions have the highest (lowest) T_{CO}. Vice versa, physical pressure reduces the intermolecular distances, decreasing V/W and suppressing CO. However, the use of stoichiometric compounds confines the available phase space to a few distinct positions where suitable anions for single crystal growth are available. Reaching the regions in between two compounds, e.g., between $(TMTTF)_2SbF_6$ and $(TMTTF)_2AsF_6$, requires pressure tuning starting from the material with the larger anion, here SbF_6^- [1,6,35,36]; increasing V/W in small increments from the position of $X = AsF_6^-$ was not achieved so far. In principle, this region in the phase diagram could be reached by "negative" pressure which can be obtained by uniaxially applying tensile strain, but not through hydrostatic pressure.

Here, we perform continuous correlation tuning via partial chemical substitution in $(TMTTF)_2[AsF_6]_{1-x}[SbF_6]_x$. Single crystals were prepared following the standard electrochemical synthesis procedures [37]. While a mixture of approximately 1:1 SbF_6^- and AsF_6^- anions was used, the real stoichiometry upon crystallization can differ from this ratio. We determined the composition as $x \approx 0.3$ by comparing the infrared intensities of the respective anion vibration modes at 660 and 700 cm^{-1} in Figure 2. The position in the phase diagram in Figure 1a is further substantiated by measurements of the magnetic susceptibility. Our SQUID data in Figure 3 provide solid evidence that $(TMTTF)_2[AsF_6]_{1-x}[SbF_6]_x$ studied here does not exhibit a spin-Peierls ground state, because it lacks a pronounced drop of χ_s that would indicate a spin-singlet formation as in the case of $(TMTTF)_2AsF_6$ [7]. The data quality at the lowest temperatures prevents the assignment of a possible antiferromagnetic transition, which occurs at $T_N = 8$ K for $(TMTTF)_2SbF_6$ [6,35]. Further characterization of the magnetic ground state is a task for future investigations.

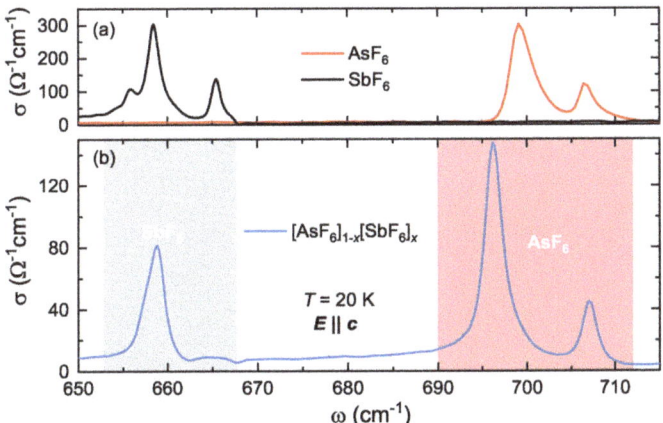

Figure 2. (a) The vibration modes of the anions SbF_6^- and AsF_6^- occur in the infrared spectra of $(TMTTF)_2SbF_6$ and $(TMTTF)_2AsF_6$ around 660 and 700 cm^{-1}, respectively. (b) The spectral features of both anion species are present in the spectrum of $(TMTTF)_2[AsF_6]_{1-x}[SbF_6]_x$. Through integrating the spectral weight SW in the grey and red shaded frequency ranges, we estimate a stoichiometry $x = 0.3$ from the ratios between the intensities of the SbF_6^- and AsF_6^- modes.

Standard optical spectroscopic experiments in the mid-infrared range (500–8000 cm^{-1}) were performed with the light polarized along the a, b, and c crystallographic axes in a temperature range from 300 K down to 5 K. We focused on evaluation of the charge-sensitive infrared active $\nu_{28}(b_{1u})$ mode probed for $E \parallel c$, the resonance frequency of which follows [1,38]

$$\nu_{28}(\rho) = 1632 \text{ cm}^{-1} - \rho \cdot 80 \text{ cm}^{-1}/e, \qquad (1)$$

where ρ is the molecular charge and e the charge of an electron. For a characterization of the conduction properties, plotted in Figure 4a, we measured the dc resistivity in situ during the optical experiments, revealing a considerable increase of $T_{CO} = 120$ K from the transition temperature of (TMTTF)$_2$AsF$_6$ ($T_{CO} = 102$ K) towards that of (TMTTF)$_2$SbF$_6$ ($T_{CO} = 156$ K).

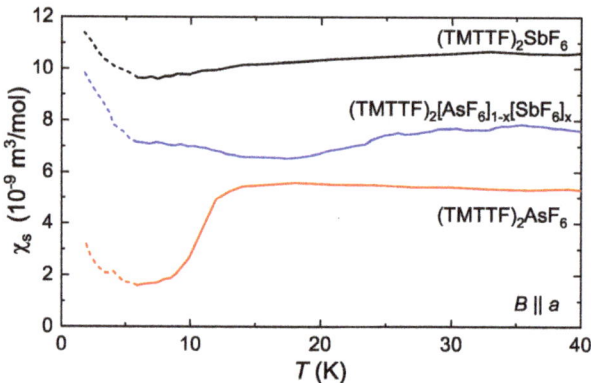

Figure 3. Temperature dependence of the magnetic susceptibility $\chi_s(T)$ for (TMTTF)$_2X$ with $X = $ SbF$_6$, [AsF$_6$]$_{1-x}$[SbF$_6$]$_x$ ($x = 0.3$) and AsF$_6$ measured by a SQUID magnetometer. In all cases, the pronounced Curie tail below 6 K has not been subtracted (dashed region). The magnetic field was $B = 0.5$ T for the alloy and 1 T for the two pure compounds; the single crystals are oriented with $B \parallel a$. The curves of $X = $ SbF$_6$ and [AsF$_6$]$_{1-x}$[SbF$_6$]$_x$ are vertically shifted by a positive offset for clarity reasons. The spin-Peierls transition in (TMTTF)$_2$AsF$_6$ can be clearly seen at $T_{SP} = 13$ K, whereas the antiferromagnetic transition at $T_N = 8$ K is barely visible for (TMTTF)$_2$SbF$_6$. For the alloy (TMTTF)$_2$[AsF$_6$]$_{1-x}$[SbF$_6$]$_x$, we cannot identify any transition within the accessible temperature range ($T > 1.8$ K).

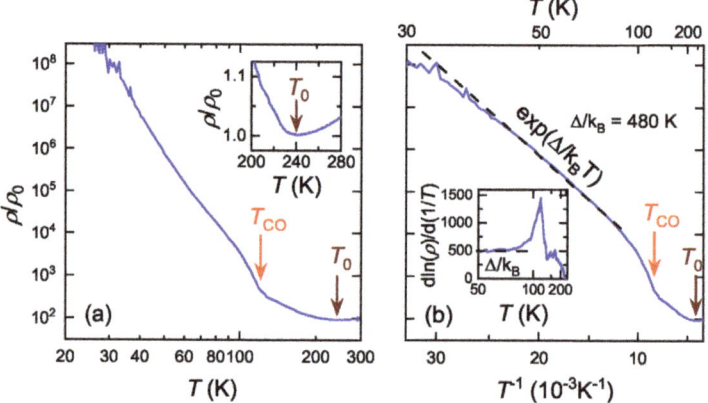

Figure 4. (**a**) Temperature-dependent electrical resistivity of (TMTTF)$_2$[AsF$_6$]$_{1-x}$[SbF$_6$]$_x$ with $x = 0.3$ (electrical current parallel to a-axis) measured in situ during optical reflection measurements in the ac-plane. The sharp increase at $T_{CO} = 120$ K indicates the CO transition. The localization temperature $T_0 = 240$ K is a little lower compared to the parent compounds (TMTTF)$_2$AsF$_6$ and (TMTTF)$_2$SbF$_6$ [37], but otherwise the transport properties are qualitatively similar. (**b**) The Arrhenius plot yields an approximately constant energy gap $\Delta/k_B = 480$ K in the temperature range 40–100 K, as illustrated in the transport derivative $d \ln \rho / d(1/T)$ in the inset.

3. Results and Analysis

The transition temperature $T_{CO} = 120$ K agrees with our expectations based on the stoichiometry, placing (TMTTF)$_2$[AsF$_6$]$_{1-x}$[SbF$_6$]$_x$ with $x = 0.3$ closer to (TMTTF)$_2$AsF$_6$ than to (TMTTF)$_2$SbF$_6$. Our comprehensive set of experimental results allows us to gain much deeper insight by evaluating distinctive quantities such as the energy gap associated with CO, as well as the charge disproportionation 2δ. In order to extract the transport gap, we plot the resistivity in an Arrhenius plot in Figure 4b. We find an approximately constant $\Delta/k_B = 480$ K between 40–100 K which is, again, in line with the gap size of the parent compounds [37]. The inset, showing the transport derivative $d \ln \rho / d(1/T)$, consistently yields a temperature-independent transport gap below the sharp peak that occurs in the vicinity of T_{CO}.

3.1. Universal Relation between Charge Gap and Transition Temperature

The bare value of the transport gap, however, does not reflect the CO state only, but involves also contributions from Mott localization that cause the upturn of resistivity below T_0. These individual contributions add in quadrature establishing the total value of $\Delta(T)$. To that end, we determine the CO contribution from the temperature-dependent energy gap $\Delta(T) = T \ln \rho$, following the procedure applied in [37]. $\Delta(T)$ of (TMTTF)$_2$[AsF$_6$]$_{0.7}$[SbF$_6$]$_{0.3}$ is shown in Figure 5a, together with the energy gaps of (TMTTF)$_2 X$ with $X = $ PF$_6^-$ (here we show the deuterated compound with $T_{CO} = 90$ K from [39]), AsF$_6^-$, and SbF$_6^-$ [37]. Note that the data have been shifted vertically and in all cases $\Delta(T_0) = 0$ by definition. Δ_{CO} is determined from the increase of $\Delta(T)$ below the respective T_{CO}, i.e., $\Delta_{CO} = \sqrt{\Delta_{max}^2 - \Delta^2(T_{CO})}$, as indicated by the double arrows and dashed lines in respective colors in Figure 5a. From our present transport results, we obtain $\Delta_{CO} = 430$ K for (TMTTF)$_2$[AsF$_6$]$_{0.7}$[SbF$_6$]$_{0.3}$.

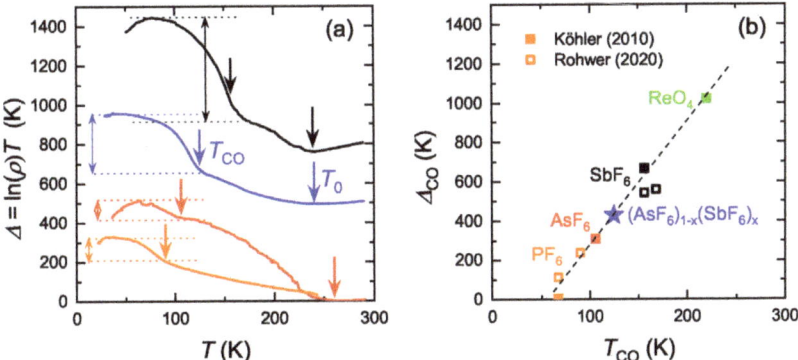

Figure 5. (a) Temperature-dependent energy gap of (TMTTF)$_2$[AsF$_6$]$_{1-x}$[SbF$_6$]$_x$ for $x = 0$ [37], 0.3, and 1 [37], and for deuterated (TMTTF)$_2$PF$_6$ with $T_{CO} = 90$ K [39]. The total gap size, measured by the resistivity, increases as Δ_{CO} adds up in quadrature $\Delta(T) = T \ln \rho = \sqrt{\Delta_{CO}^2 + \Delta^2(T_{CO})}$ in the CO state. The increase from $\Delta(T_{CO})$ to Δ_{max} is indicated by double arrows and dashed lines in respective colors. The curves have been vertically shifted; by definition $\Delta(T_0) = 0$. (b) Δ_{CO} of (TMTTF)$_2$[AsF$_6$]$_{1-x}$[SbF$_6$]$_x$ (blue star) is plotted versus T_{CO} together with the data (solid squares) of (TMTTF)$_2 X$ with $X = $ PF$_6^-$, AsF$_6^-$, SbF$_6^-$ and ReO$_4^-$ [37]. Open squares indicate measurements on pristine and deuterated PF$_6^-$ and SbF$_6^-$ compounds [39]. The data follow an approximately linear relationship until Δ_{CO} vanishes for $T_{CO} \leq 60$ K. Alternatively, this *universal* behavior can be interpreted as a negative Δ_{CO} for $T_{CO} \to 0$.

In Figure 5b, we plot Δ_{CO} as a function of T_{CO} for the data shown in (a), i.e., (TMTTF)$_2$-[AsF$_6$]$_{1-x}$[SbF$_6$]$_x$ with $x = 0.3$ (star) and PF$_6^-$, AsF$_6^-$, SbF$_6^-$, together with ReO$_4^-$ (solid squares) [37]. Included are also additional datasets for pristine and deuterated compounds from [39], indicated by the open squares. As we find, all data points fall on a *universal* line that yields $\Delta_{CO} = 0$ around $T_{CO} = 60$ K. It is tempting to relate this temperature scale with $D/k_B \approx 60$ K reported in Figure 8b of [14]. In that work, deuteration was found to contribute an energy D in quadrature to T_{CO}, suggesting that interactions between anions and the TMTTF donors via the hydrogen atoms interfere with the CO mechanism: T_{CO} is smaller in the protonated compounds compared to the heavier, less-mobile deuterium isotopes. Our results on (TMTTF)$_2$[AsF$_6$]$_{1-x}$[SbF$_6$]$_x$ provide an additional piece of evidence for the importance of anion-TMTTF interactions and motivate further research in this direction. More precise evaluation of the anion modes and vibrations of the methyl endgroups, possibly supplemented by calculations, may be a first step in this direction.

While assessing the *universal* behavior in Figure 5b, one could also consider that Δ_{CO} has a negative offset for $T_{CO} \to 0$, meaning that the repulsive interactions turn into attractive charge fluctuations as correlations diminish. This mechanism has been vividly discussed as a potential candidate for the pairing glue in the quasi-2D molecular superconductors nearby a CO instability [40–45]. This notion also calls for further scrutiny of the CO phenomenon by means of continuous correlation tuning, as presented here.

3.2. Charge Disproportionation Determined by Vibrational Spectroscopy

We investigate CO in (TMTTF)$_2$[AsF$_6$]$_{0.7}$[SbF$_6$]$_{0.3}$ in more detail by assessing the optical response of molecular vibrations in the midinfrared frequency range [1,38]. The charge-sensitive $\nu_{28}(b_{1u})$ mode probed for $E \parallel c$ exhibits a similar splitting in the CO phase ($T = 20$ K $< T_{CO}$) as in the two parent compounds (Figure 6). According to Equation (1), the separation of the two main peaks corresponds to the charge disproportionation 2δ between charge-rich and charge-poor molecular sites, which increases from $x = 0$ to $x = 1$. Overall, the ν_{28} mode for $x = 0.3$ very much resembles the spectrum of $x = 0$, in agreement with the chemical composition, placing the alloy closer to (TMTTF)$_2$AsF$_6$ in the phase diagram (Figure 1a). We also assessed the optical reflectivity parallel to the stacks ($E \parallel a$, see inset of Figure 6) and observe no significant activation of the ν_4 mode that is sensitive to tetramerization of the TMTTF molecules [1,36]. Therefore, we find no evidence for a spin-Peierls transition in (TMTTF)$_2$[AsF$_6$]$_{0.7}$[SbF$_6$]$_{0.3}$ based on our optical data, in line with the magnetic susceptibility measurements shown in Figure 3. The suppression of the spin-Peierls state is in agreement with pressure-dependent experiments on (TMTTF)$_2$SbF$_6$ and (TMTTF)$_2$AsF$_6$, where antiferromagnetism rapidly replaces the spin-gapped phase [6,35,36].

Figure 7a presents the optical conductivity of the $\nu_{28}(b_{1u})$ vibration in (TMTTF)$_2$[AsF$_6$]$_{0.7}$[SbF$_6$]$_{0.3}$ for various temperatures above and below T_{CO}. Consistent with our transport results in Figure 4, the line exhibits a splitting at $T \leq 120$ K. For a quantitative analysis, we determined the charge disproportionation 2δ from the resonance frequencies of the peak splitting according to Equation (1), as shown in panels (b) and (c). On first glance, 2δ exhibits a similar mean-field-like increase upon cooling below T_{CO}. While the transition temperature is higher than in the parent compound (TMTTF)$_2$AsF$_6$ [1,12], the charge disproportionation reaches a rather similar value of $2\delta = 0.21e$ in the limit $T \to 0$.

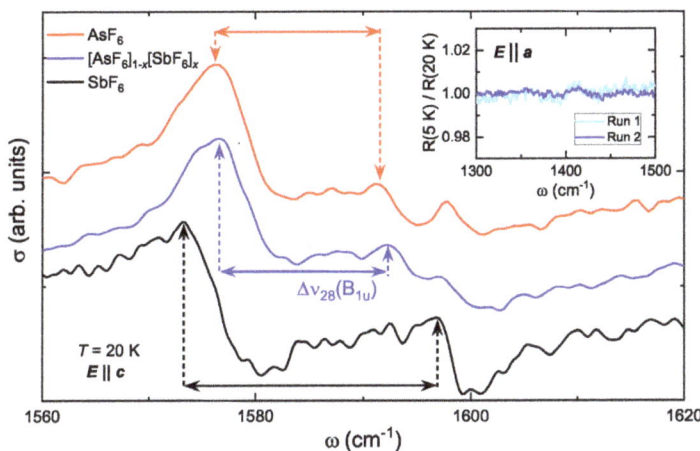

Figure 6. The ν_{28} mode in (TMTTF)$_2$[AsF$_6$]$_{1-x}$[SbF$_6$]$_x$ with $x = 0.3$ is plotted at $T = 20$ K and compared to the parent compounds (TMTTF)$_2$SbF$_6$ and (TMTTF)$_2$AsF$_6$ [1,12]. The overall low-temperature spectrum is very similar for $x = 0$ and 0.3, with a smaller line splitting $\Delta\nu_{28}$ than for $x = 1$. For better comparison of the peaks, the data were scaled and shifted by a vertical offset. Inset: Comparing the reflectivity for $E \parallel a$ at 5 K and 20 K yields no enhancement of the ν_4 mode [1,36]; two distinct measurement runs of the same (TMTTF)$_2$[AsF$_6$]$_{1-x}$[SbF$_6$]$_x$ sample are shown. The absence of tetramerization indicates that the spin-Peierls phase is suppressed for a substitution $x = 0.3$, which is taken into account in Figure 1a.

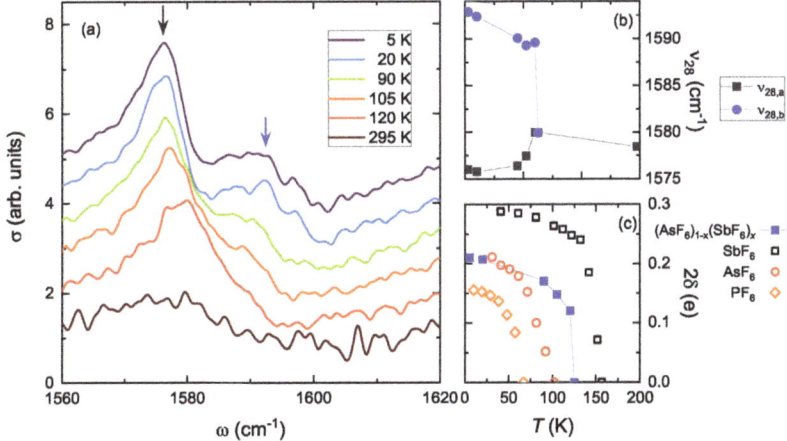

Figure 7. (a) Temperature dependence of the ν_{28} mode in (TMTTF)$_2$[AsF$_6$]$_{1-x}$[SbF$_6$]$_x$ for $x = 0.3$; the spectrum is very similar to (TMTTF)$_2$AsF$_6$ [1,12]. While the charge-rich, low-frequency peak can be identified clearly, the charge-poor, high-frequency peak is spread out more broadly. The data have been shifted vertically. (b) Peak frequencies from (a). (c) The charge disproportionation below $T_{CO} = 120$ K approaches a similar value towards $T \to 0$ as for (TMTTF)$_2$AsF$_6$. The temperature evolution of 2δ upon CO is compared among (TMTTF)$_2X$ with $X = $ PF$_6^-$, AsF$_6^-$ and SbF$_6^-$ (data taken from [1,12]).

4. Discussion and Outlook

The dichotomy of T_{CO} (and Δ_{CO}) and $2\delta(T = 0)$, where the former increases upon anion substitution x in (TMTTF)$_2$[AsF$_6$]$_{1-x}$[SbF$_6$]$_x$ while the latter remains unchanged (see Figure 6), is surprising in view of the monotonous increase of both quantities in the stoichiometric compounds. A priori, from the empirical trend (see Figure 4 in [14]), one would expect a 20% larger value of 2δ based on T_{CO} = 120 K. Possibly, the molecular charge arranges according to the closest anion: 2δ is larger nearby SbF_6^- anions and smaller around AsF_6^-, where the latter constitute the majority of anions. The local increase of charge disproportionation around SbF_6^- sites can be gradual as it depends on the structural extent of the lattice distortion, resulting in a distribution of 2δ. Indeed, the ν_{28} lines are broadened and more smeared in the spectrum of $x = 0.3$ compared to $x = 0$ (Figure 6). Note that, while our spectroscopic data are consistent with a microscopic modulation of the local CO amplitude, our transport data reveal only a single T_{CO} = 120 K, and the crystal as a whole does not exhibit two transitions. Structural phase separation is further ruled out as χ_s (Figure 3) does not exhibit a distinct drop at T_{SP} = 13 K expected for volume fractions of pure $(TMTTF)_2AsF_6$. This shows that the macroscopic CO transition is a 3D bulk phenomenon that requires substantial coupling among the one-dimensional TMTTF chains—through/with the anions. To that end, the abovementioned TMTTF–anion interactions via the hydrogen atoms are crucial ingredients to CO, opposite to the original notions about a *structureless* transition. Certainly, the microscopic charge disproportionation should be studied in more detail by assessing different values of the substitution x and by complementary methods, such as NMR or other local probes. Magnetic resonance measurements are also required to investigate the magnetic ground state: studying the borderland of antiferromagnetic and spin-Peierls phases is of particular interest to the field of frustrated magnetism and quantum spin liquids [46].

Using partial substitution of the anions, it will be interesting to inspect the length scales of long-range correlations and short-range CO modulations and associated disorder effects—in particular in the deuterated compounds where CO is affected from a change in TMTTF–anion coupling [14,39]. In addition, the interplay of CO and anion order can be studied via mixing of octahedral and tetrahedral anions, which can possibly stabilize novel forms of structural order at commensurate stoichiometries. We further suggest to partially introduce anions, that are not regularly used for single crystal growth, into "established" systems which provides another knob to tune through unexplored phase space.

Moreover, it is intriguing to compare the continuous correlation tuning methods of physical pressure and partial chemical substitution. While the former truly modifies the interactions locally, in the latter case, the microscopic mixture of distinct constituents yields a change of correlation strength on a macroscopic level, i.e., "on average". We expect fundamental differences between (super)conducting systems with itinerant charge carriers in the vicinity of the Mott transition [24,26–29,31] and the fully insulating Fabre salts inspected here. Clearly, partial chemical substitution is a powerful tool that enables us to put *new spin on metal-insulator transitions*.

5. Summary

We report transport and optical spectroscopy experiments on partially substituted $(TMTTF)_2[AsF_6]_{1-x}[SbF_6]_x$ with $x = 0.3$, which is equivalent to "negative" pressure applied to $(TMTTF)_2AsF_6$. The transition temperature T_{CO} = 120 K and charge gap Δ_{CO} = 430 K indicate that this alloy is between the parent compounds $(TMTTF)_2AsF_6$ and $(TMTTF)_2SbF_6$, a little closer to the former in agreement with the stoichiometry. This demonstrates the powerful capabilities of partial anion substitution for continuous bandwidth tuning. Upon plotting Δ_{CO} as a function of T_{CO} for various $(TMTTF)_2X$ salts exhibiting CO, all data points fall on a universal line. We find that Δ_{CO} vanishes around T_{CO} = 60 K—a value similar to the contribution to CO upon deuteration ($D \approx 60$ K reported in [14]). While the macroscopic CO transition and its underlying electronic correlation strength are tuned continuously, our measurements utilizing the local probe of vibrational spectroscopy indicate that the charge disproportionation adheres to the closest anion on a microscopic level, yielding a short-range modulation of the CO amplitude around the substituent sites.

Author Contributions: Conceptualization and supervision, A.P. and M.D.; crystal growth, G.U.; transport and optical investigation and analysis, A.P., D.D. and S.E.; magnetic characterization, O.I. and V.M.; writing and editing, A.P. All authors have read and agreed to the published version of the manuscript.

Funding: Deutsche Forschungsgemeinschaft.

Data Availability Statement: The authors declare that the data supporting the findings of this study are available within the paper. Further information can be provided by A.P. or M.D.

Acknowledgments: The project was supported by the Deutsche Forschungsgemeinschaft (DFG). The authors acknowledge TU Wien Bibliothek for financial support through its Open Access Funding Program.

Conflicts of Interest: The authors declare no conflicts of interest.

References

1. Dressel, M.; Dumm, M.; Knoblauch, T.; Masino, M. Comprehensive Optical Investigations of Charge Order in Organic Chain Compounds $(TMTTF)_2X$. *Crystals* **2012**, *2*, 528–578. [CrossRef]
2. Fukuyama, H. Physics of Molecular Conductors. *J. Phys. Soc. Jpn.* **2006**, *75*, 51001. [CrossRef]
3. Seo, H.; Merino, J.; Yoshioka, H.; Ogata, M. Theoretical Aspects of Charge Ordering in Molecular Conductors. *J. Phys. Soc. Jpn.* **2006**, *75*, 51009. [CrossRef]
4. Chow, D.S.; Zamborszky, F.; Alavi, B.; Tantillo, D.J.; Baur, A.; Merlic, C.A.; Brown, S.E. Charge Ordering in the TMTTF Family of Molecular Conductors. *Phys. Rev. Lett.* **2000**, *85*, 1698–1701. [CrossRef]
5. Zamborszky, F.; Yu, W.; Raas, W.; Brown, S.E.; Alavi, B.; Merlic, C.A.; Baur, A. Competition and coexistence of bond and charge orders in $(TMTTF)_2AsF_6$. *Phys. Rev. B* **2002**, *66*, 81103. [CrossRef]
6. Yu, W.; Zhang, F.; Zamborszky, F.; Alavi, B.; Baur, A.; Merlic, C.A.; Brown, S.E. Electron-lattice coupling and broken symmetries of the molecular salt $(TMTTF)_2SbF_6$. *Phys. Rev. B* **2004**, *70*, 121101. [CrossRef]
7. Dumm, M.; Salameh, B.; Abaker, M.; Montgomery, L.K.; Dressel, M. Magnetic and optical studies of spin and charge ordering in $(TMTTF)_2AsF_6$. *J. Phys. IV* **2004**, *114*, 57–60. [CrossRef]
8. Dumm, M.; Abaker, M.; Dressel, M. Mid-infrared response of charge-ordered quasi-1D organic conductors $(TMTTF)_2X$. *J. Phys. IV* **2005**, *131*, 55–58. [CrossRef]
9. Dumm, M.; Abaker, M.; Dressel, M.; Montgomery, L.K. Charge Order in $(TMTTF)_2PF_6$ Investigated by Infrared Spectroscopy. *J. Low Temp. Phys.* **2006**, *142*, 613–616. [CrossRef]
10. Pashkin, A.; Dressel, M.; Ebbinghaus, S.; Hanfland, M.; Kuntscher, C. Pressure-induced structural phase transition in the Bechgaard-Fabre salts. *Synth. Met.* **2009**, *159*, 2097–2100. [CrossRef]
11. Pashkin, A.; Dressel, M.; Hanfland, M.; Kuntscher, C.A. Deconfinement transition and dimensional crossover in the Bechgaard-Fabre salts: Pressure- and temperature-dependent optical investigations. *Phys. Rev. B* **2010**, *81*, 125109. [CrossRef]
12. Knoblauch, T.; Dressel, M. Charge disproportionation in $(TMTTF)_2X$ ($X = PF_6$, AsF_6 and SbF_6) investigated by infrared spectroscopy. *Phys. Status Solidi C* **2012**, *9*, 1158–1160. [CrossRef]
13. Oka, Y.; Matsunaga, N.; Nomura, K.; Kawamoto, A.; Yamamoto, K.; Yakushi, K. Charge Order in $(TMTTF)_2TaF_6$ by Infrared Spectroscopy. *J. Phys. Soc. Jpn.* **2015**, *84*, 114709. [CrossRef]
14. Pustogow, A.; Peterseim, T.; Kolatschek, S.; Engel, L.; Dressel, M. Electronic correlations versus lattice interactions: Interplay of charge and anion orders in $(TMTTF)_2X$. *Phys. Rev. B* **2016**, *94*, 195125. [CrossRef]
15. Rösslhuber, R.; Rose, E.; Ivek, T.; Pustogow, A.; Breier, T.; Geiger, M.; Schrem, K.; Untereiner, G.; Dressel, M. Structural and Electronic Properties of $(TMTTF)_2X$ Salts with Tetrahedral Anions. *Crystals* **2018**, *8*, 121. [CrossRef]
16. Parkin, S.S.P.; Coulon, C.; Jérome, D.; Fabre, J.M.; Giral, L. Substitution of TMTSeF with TMTTF in $(TMTSeF)_2ClO_4$: High pressure studies. *J. Phys. France* **1983**, *44*, 603–607. [CrossRef]
17. Ilakovac, V.; Ravy, S.; Pouget, J.P.; Lenoir, C.; Boubekeur, K.; Batail, P.; Babic, S.D.; Biskup, N.; Korin-Hamzic, B.; Tomic, S.; Bourbonnais, C. Enhanced charge localization in the organic alloys $[(TMTSF)_{1-x}(TMTTF)_x]_2ReO_4$. *Phys. Rev. B* **1994**, *50*, 7136–7139. [CrossRef]
18. Tomic, S.; Auban-Senzier, P.; Jérome, D. Charge localization in $[(TMTTF)_{0.5}(TMTSF)_{0.5}]_2ReO_4$: A pressure study. *Synth. Met.* **1999**, *103*, 2197–2198. [CrossRef]
19. Naito, T.; Miyamoto, A.; Kobayashi, H.; Kato, R.; Kobayashi1, A. Superconducting Transition Temperature of the Organic Alloy System. κ-$[(BEDT-TTF)_{1-x}(BEDT-STF)_x]_2Cu[N(CN)_2]Br$. *Chem. Lett.* **1992**, *21*, 119–122. [CrossRef]
20. Kawamoto, A.; Taniguchi, H.; Kanoda, K. Superconductor-Insulator Transition Controlled by Partial Deuteration in BEDT-TTF Salt. *J. Am. Chem. Soc.* **1998**, *120*, 10984–10985. [CrossRef]
21. Sasaki, J.; Kawamoto, A.; Kumagai, K.I. Magnetic Properties of the alloy of κ-$(BEDT-TTF)_{2(1-X)}(BEDSe-TTF)_{2X}Cu[N(CN)_2]Br$. *Synth. Met.* **2003**, *137*, 1249–1250. [CrossRef]
22. Sushko, Y.V.; Leontsev, S.O.; Korneta, O.B.; Kawamoto, A. SQUID-magnetometry study of the P-T phase diagram of κ-$[(BEDT-TTF)_{1-x}(BEDSe-TTF)_x]_2Cu[N(CN)_2]Br$. *J. Low Temp. Phys.* **2006**, *142*, 563–566. [CrossRef]

23. Kobayashi, T.; Ihara, Y.; Kawamoto, A. Modification of local electronic state by BEDT-STF doping to κ-(BEDT-TTF)$_2$Cu[N(CN)$_2$]Br salt studied by ^{13}C NMR spectroscopy. *Phys. Rev. B* **2016**, *93*, 94515. [CrossRef]
24. Saito, Y.; Minamidate, T.; Kawamoto, A.; Matsunaga, N.; Nomura, K. Site-specific ^{13}C NMR study on the locally distorted triangular lattice of the organic conductor κ-(BEDT-TTF)$_2$Cu$_2$(CN)$_3$. *Phys. Rev. B* **2018**, *98*, 205141. [CrossRef]
25. Yesil, E.; Imajo, S.; Nomoto, T.; Yamashita, S.; Akutsu, H.; Nakazawa, Y. Variation of Electronic Heat Capacity of κ-(BEDT-TTF)$_2$Cu[N(CN)$_2$]Br Induced by Partial Substitution of Donor Layers. *J. Phys. Soc. Jpn.* **2020**, *89*, 73701. [CrossRef]
26. Pustogow, A.; Saito, Y.; Löhle, A.; Sanz Alonso, M.; Kawamoto, A.; Dobrosavljević, V.; Dressel, M.; Fratini, S. Rise and fall of Landau's quasiparticles while approaching the Mott transition. *Nat. Commun.* **2021**, *12*, 1571. [CrossRef]
27. Pustogow, A.; Rösslhuber, R.; Tan, Y.; Uykur, E.; Böhme, A.; Wenzel, M.; Saito, Y.; Löhle, A.; Hübner, R.; Kawamoto, A.; et al. Low-temperature dielectric anomaly arising from electronic phase separation at the Mott insulator-metal transition. *NPJ Quantum Mater.* **2021**, *6*, 9. [CrossRef]
28. Saito, Y.; Rösslhuber, R.; Löhle, A.; Sanz Alonso, M.; Wenzel, M.; Kawamoto, A.; Pustogow, A.; Dressel, M. Chemical tuning of molecular quantum materials κ-[(BEDT-TTF)$_{1-x}$(BEDT-STF)$_x$]$_2$Cu$_2$(CN)$_3$: From the Mott-insulating quantum spin liquid to metallic Fermi liquid. *J. Mater. Chem. C* **2021**, *9*, 10841–10850. [CrossRef]
29. Saito, Y.; Löhle, A.; Kawamoto, A.; Pustogow, A.; Dressel, M. Pressure-Tuned Superconducting Dome in Chemically-Substituted κ-(BEDT-TTF)$_2$Cu$_2$(CN)$_3$. *Crystals* **2021**, *11*, 817. [CrossRef]
30. Joo, N.; Auban-Senzier, P.; Pasquier, C.R.; Jérome, D.; Bechgaard, K. Impurity-controlled superconductivity/spin density wave interplay in the organic superconductor: (TMTSF)$_2$ClO$_4$. *EPL* **2005**, *72*, 645–651. [CrossRef]
31. Faltermeier, D.; Barz, J.; Dumm, M.; Dressel, M.; Drichko, N.; Petrov, B.; Semkin, V.; Vlasova, R.; Mézière, C.; Batail, P. Bandwidth-controlled Mott transition in $\kappa-(BEDT-TTF)_2$Cu[N(CN)$_2$]Br$_x$Cl$_{1-x}$: Optical studies o. *Phys. Rev. B* **2007**, *76*, 165113. [CrossRef]
32. Dumm, M.; Faltermeier, D.; Drichko, N.; Dressel, M.; Mézière, C.; Batail, P. Bandwidth-controlled Mott transition in κ-(BEDT-TTF)$_2$Cu[N(CN)$_2$]Br$_x$Cl$_{1-x}$: Optical studies of correlated carriers. *Phys. Rev. B* **2009**, *79*, 195106. [CrossRef]
33. Yoshida, Y.; Maesato, M.; Tomeno, S.; Kimura, Y.; Saito, G.; Nakamura, Y.; Kishida, H.; Kitagawa, H. Partial Substitution of Ag(I) for Cu(I) in Quantum Spin Liquid κ-(ET)$_2$Cu$_2$(CN)$_3$, Where ET Is Bis(ethylenedithio)tetrathiafulvalene. *Inorg. Chem.* **2019**, *58*, 4820–4827. [CrossRef]
34. Kitou, S.; Zhang, L.; Nakamura, T.; Sawa, H. Complex changes in structural parameters hidden in the universal phase diagram of the quasi-one-dimensional organic conductors (TMTTF)$_2$X (X = NbF$_6$, AsF$_6$, PF$_6$, and Br). *Phys. Rev. B* **2021**, *103*, 184112. [CrossRef]
35. Yu, W.; Zamborszky, F.; Alavi, B.; Baur, A.; Merlic, C.A.; Brown, S.E. Influence of charge order on the ground states of TMTTF molecular salts. *J. Phys. IV France* **2004**, *114*, 35–40. [CrossRef]
36. Voloshenko, I.; Herter, M.; Beyer, R.; Pustogow, A.; Dressel, M. Pressure-dependent optical investigations of Fabre salts in the charge-ordered state. *J. Phys. Condens. Matter* **2017**, *29*, 115601. [CrossRef]
37. Köhler, B.; Rose, E.; Dumm, M.; Untereiner, G.; Dressel, M. Comprehensive transport study of anisotropy and ordering phenomena in quasi-one-dimensional (TMTTF)$_2$X salts (X = PF$_6$, AsF$_6$, SbF$_6$, BF$_4$, ClO$_4$, ReO$_4$). *Phys. Rev. B* **2011**, *84*, 035124. [CrossRef]
38. Meneghetti, M.; Bozio, R.; Zanon, I.; Pecile, C.; Ricotta, C.; Zanetti, M. Vibrational behavior of molecular constituents of organic superconductors: TMTSF, its radical cation and the sulphur analogs TMTTF and TMTTF$^+$. *J. Chem. Phys.* **1984**, *80*, 6210. [CrossRef]
39. Rohwer, A.; Dressel, M.; Nakamura, T. Deuteration Effects on the Transport Properties of (TMTTF)$_2$X Salts. *Crystals* **2020**, *10*, 1085. [CrossRef]
40. Merino, J.; McKenzie, R.H. Superconductivity Mediated by Charge Fluctuations in Layered Molecular Crystals. *Phys. Rev. Lett.* **2001**, *87*, 237002. [CrossRef]
41. Drichko, N.; Dressel, M.; Kuntscher, C.; Pashkin, A.; Greco, A.; Merino, J.; Schlueter, J. Electronic properties of correlated metals in the vicinity of a charge-order transition: Optical spectroscopy of α-(BEDT-TTF)$_2$MHg(SCN)$_4$ (M = NH$_4$, Rb, Tl). *Phys. Rev. B* **2006**, *74*, 235121. [CrossRef]
42. Kaiser, S.; Dressel, M.; Sun, Y.; Greco, A.; Schlueter, J.A.; Gard, G.L.; Drichko, N. Bandwidth Tuning Triggers Interplay of Charge Order and Superconductivity in Two-Dimensional Organic Materials. *Phys. Rev. Lett.* **2010**, *105*, 206402. [CrossRef]
43. Pustogow, A.; Saito, Y.; Rohwer, A.; Schlueter, J.A.; Dressel, M. Coexistence of charge order and superconductivity in $\beta''-(BEDT-TTF)_2$SF$_5$CH$_2$CF$_2$SO$_3$. *Phys. Rev. B* **2019**, *99*, 140509. [CrossRef]
44. Imajo, S.; Akutsu, H.; Akutsu-Sato, A.; Morritt, A.L.; Martin, L.; Nakazawa, Y. Effects of electron correlations and chemical pressures on superconductivity of β''-type organic compounds. *Phys. Rev. Res.* **2019**, *1*, 33184. [CrossRef]
45. Imajo, S.; Akutsu, H.; Kurihara, R.; Yajima, T.; Kohama, Y.; Tokunaga, M.; Kindo, K.; Nakazawa, Y. Anisotropic Fully Gapped Superconductivity Possibly Mediated by Charge Fluctuations in a Nondimeric Organic Complex. *Phys. Rev. Lett.* **2020**, *125*, 177002. [CrossRef]
46. Balents, L. Spin liquids in frustrated magnets. *Nature* **2010**, *464*, 199–208. [CrossRef]

Article

Electronic Heat Capacity and Lattice Softening of Partially Deuterated Compounds of κ-(BEDT-TTF)$_2$Cu[N(CN)$_2$]Br

Yuki Matsumura [1], Shusaku Imajo [2], Satoshi Yamashita [1], Hiroki Akutsu [1] and Yasuhiro Nakazawa [1,*]

[1] Department of Chemistry, Graduate School of Science, Osaka University, Machikaneyama 1-1, Toyonaka, Osaka 560-0043, Japan; matsumuray16@chem.sci.osaka-u.ac.jp (Y.M.); sayamash@chem.sci.osaka-u.ac.jp (S.Y.); akutsu@chem.sci.osaka-u.ac.jp (H.A.)
[2] Institute for Solid State Physics, University of Tokyo, Kashiwa, Chiba 277-8581, Japan; imajo@issp.u-tokyo.ac.jp
* Correspondence: nakazawa@chem.sci.osaka-u.ac.jp

Citation: Matsumura, Y.; Imajo, S.; Yamashita, S.; Akutsu, H.; Nakazawa, Y. Electronic Heat Capacity and Lattice Softening of Partially Deuterated Compounds of κ-(BEDT-TTF)$_2$Cu[N(CN)$_2$]Br. Crystals 2022, 12, 2. https://doi.org/10.3390/cryst12010002

Academic Editor: Andrei Vladimirovich Shevelkov

Received: 6 November 2021
Accepted: 19 December 2021
Published: 21 December 2021

Publisher's Note: MDPI stays neutral with regard to jurisdictional claims in published maps and institutional affiliations.

Copyright: © 2021 by the authors. Licensee MDPI, Basel, Switzerland. This article is an open access article distributed under the terms and conditions of the Creative Commons Attribution (CC BY) license (https://creativecommons.org/licenses/by/4.0/).

Abstract: Thermodynamic investigation by calorimetric measurements of the layered organic superconductors, κ-(BEDT-TTF)$_2$Cu[N(CN)$_2$]Br and its partially deuterated compounds of κ-(d[2,2]-BEDT-TTF)$_2$Cu[N(CN)$_2$]Br and κ-(d[3,3]-BEDT-TTF)$_2$Cu[N(CN)$_2$]Br, performed in a wide temperature range is reported. The latter two compounds were located near the metal–insulator boundary in the dimer-Mott phase diagram. From the comparison of the temperature dependences of their heat capacities, we indicated that lattice heat capacities of the partially deuterated compounds were larger than that of the pristine compound below about 40 K. This feature probably related to the lattice softening was discussed also by the sound velocity measurement, in which the dip-like structures of the $\Delta v/v$ were observed. We also discussed the variation of the electronic heat capacity under magnetic fields. From the heat capacity data at magnetic fields up to 6 T, we evaluated that the normal-state γ value of the partially deuterated compound, κ-(d[3,3]-BEDT-TTF)$_2$Cu[N(CN)$_2$]Br, was about 3.1 mJ K^{-2} mol^{-1}. Under the magnetic fields higher than 3.0 T, we observed that the magnetic-field insulating state was induced due to the instability of the mid-gap electronic state peculiar for the two-dimensional dimer-Mott system. Even though the volume fraction was much reduced, the heat capacity of κ-(d[3,3]-BEDT-TTF)$_2$Cu[N(CN)$_2$]Br showed a small hump structure probably related to the strong coupling feature of the superconductivity near the boundary.

Keywords: organic superconductor; strong electron correlations; heat capacity; Mott transition

1. Introduction

It is generally recognized that strong electron correlations in the half-filling state tend to induce a metal–insulator transition, which is called the Mott transition [1–3]. Unusual electronic phenomena such as high-T_c superconductivity [4,5], anomalous mass enhancement of electron carriers [6,7], non-Fermi liquid features [8], and charge disproportionation [9], are known to emerge around the transition. These phenomena are dominated by the Mott–Hubbard physics, which describes the competitive nature of the itineracy and localization of correlated electrons. The two-dimensional (2D) organic charge transfer compounds with the chemical formula of κ-(BEDT-TTF)$_2$X, where BEDT-TTF is bis(ethylenedithio)tetrathiafulvalene and X is the monovalent counter anion, are known as a typical electron correlation system of π-electrons originating from molecular HOMOs. These compounds give a phase relation determined by the ratio of the on-site or inter-site Coulomb repulsion U, V, and bandwidth W, which is tunable by pressure. Their physical properties are summarized as a so-called 2D dimer-Mott-type phase diagram, where the antiferromagnetic insulating phase and the metallic/superconductive phase are adjacent to each other [10,11]. The insulating phase and the superconductive phase are separated by the first-order Mott boundary. The existence of the critical endpoint at a finite temperature and the rounding of the Mott transition line [12] as well as the disorder-sensitive electronic states due to the glassy feature of ethylene groups make the curious merging of the electronic and phonon states near the boundary [13]. Furthermore, the peculiar criticality across the first-order boundary [14,15] and the unusual softening of lattice dynamics, known as critical elasticity, are now being discussed by the sound velocity and the thermal expansion measurements [16–18].

To further investigate the unusual features in physical properties just near the boundary in terms of multiple degrees of freedom, such as spin, charge, and lattice, information on thermodynamic properties is important. However, the experiments under pressure can detect only the relative change of thermal anomalies with the standard calorimetric technique under external pressures. Pursuing the systematic change of thermodynamic parameters from the accurate values of the heat capacity just near the boundary region is required for this purpose. The partial deuteration of donors [19–21] or solid solutions of halogen sites in the counter anions of $Cu[N(CN)_2]X$ (X = Br and Cl) [22,23] are performed for the purpose of chemical-pressure tuning to approach the critical boundary from the position of bulk superconductor, $\kappa\text{-(BEDT-TTF)}_2Cu[N(CN)_2]Br$ in the phase diagram. According to the results by Kawamoto et al., single crystals synthesized by the BEDT-TTF molecule, of which ethylene groups on both sides are partially deuterated, can tune chemical pressure systematically [19]. They labeled the deuterated ratio with the notation of d[n,n] to represent the number of the deuterium in each ethylene group. The increase of the substitution rate can change the position of the compound with the $Cu[N(CN)_2]Br^-$ anion to the Mott boundary gradually. $\kappa\text{-(d[0,0]-BEDT-TTF)}_2Cu[N(CN)_2]Br$ (d[0,0] compound, hereafter) shows bulk superconductivity below 11–12 K, while fully deuterated $\kappa\text{-(d[4,4]-BEDT-TTF)}_2Cu[N(CN)_2]Br$ (d[4,4] compound) located in the Mott-insulating region undergoes an antiferromagnetic transition around 15 K. According to the transport measurements, the compound with d[3,3]-BEDT-TTF (d[3,3] compound) is located very close to the boundary as is schematically illustrated in Figure 1 [19,24]. Using the partially deuterated d[2,2] and d[3,3] compounds, thermodynamic information related to the electronic and the lattice states just near the boundary are possible to be detected through the single-crystal calorimetry technique and sound velocity measurements.

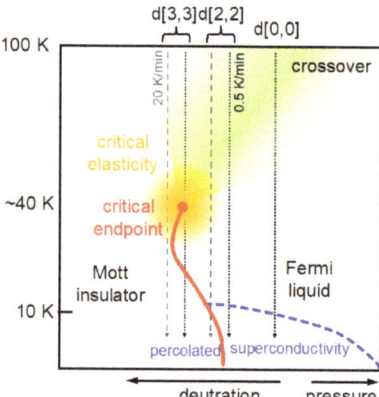

Figure 1. A schematic view of the two-dimensional (2D) dimer-Mott phase diagram for the $\kappa\text{-(BEDT-TTF)}_2X$ system. The positions of the d[0,0], d[2,2], and d[3,3] compounds in a slowly cooled case are shown by the dashed arrows. The possible positions for the rapidly cooled cases (20 K min^{-1}) for the latter two compounds are also shown by the gray dashed arrows. The position of the Mott boundary and that of the critical endpoint are shown in the red color.

2. Experiments

Low-temperature heat capacity measurements were performed by the thermal relaxation calorimeters designed for measuring single-crystal samples and pellet samples of molecule-based compounds. The plate-like single crystals of the d[2,2] and d[3,3] compounds weighing 2–3 mg were attached to the sample stage by a small amount of Apiezon N grease for low-temperature measurements with magnetic fields applied perpendicularly to the conducting plane. A ruthenium oxide sensor of which the resistance at room temperature is 10 kΩ was used for the sample-temperature sensor in the low-temperature experiments between 0.7 K and 3 K, and a Cernox1070 bare chip sensor (LakeShore, Westerville, OH, USA) was used for the experiments between 5 K and 100 K. The details of the relaxation calorimetry systems including measurements under magnetic fields are reported in the literature [25]. The higher-temperature heat capacity of the d[0,0] compound was measured by a adiabatic calorimeter using multiple pieces weighing about 25 mg. The thermometer for the adiabatic calorimetry cell was a Pt chip sensor which was calibrated between 20 K and 300 K. The accuracy of the thermal relaxation technique is within a few percentages of the absolute values, and the relative precision was about 0.5% for the present measurements. The sound velocity measurements were performed for block-type single-crystal samples of the d[2,2] and d[3,3] salts. The change of the sound velocities transmitted in the crystal was measured by the longitudinal ultrasound waves, of which the frequency was 30.5 MHz. They were generated and detected by $LiNbO_3$ piezoelectric transducers (thickness: 100 μm) attached to the side surfaces of the crystals of which details were similar to those in reference [26]. We applied ultrasonic sounds so as to propagate in the in-plane direction which was sensitive to the electronic states.

3. Results and Discussion

In Figure 2a, we show the temperature dependences of the heat capacity of the d[0,0] and d[3,3] compounds in a logarithmic plot. The red color represents the data of the d[0,0] compound from 0.8 K to 288 K. The low-temperature data below 20 K were obtained by the thermal relaxation technique as was already reported in references [27,28], and the higher temperature data were obtained by an adiabatic calorimeter. The black curve shown in the temperature range below 20 K is the normal state heat capacity obtained by fitting the data with 8 T, which was applied perpendicular to the conducting layer. The low-temperature data have already been reported by several groups [29–32].

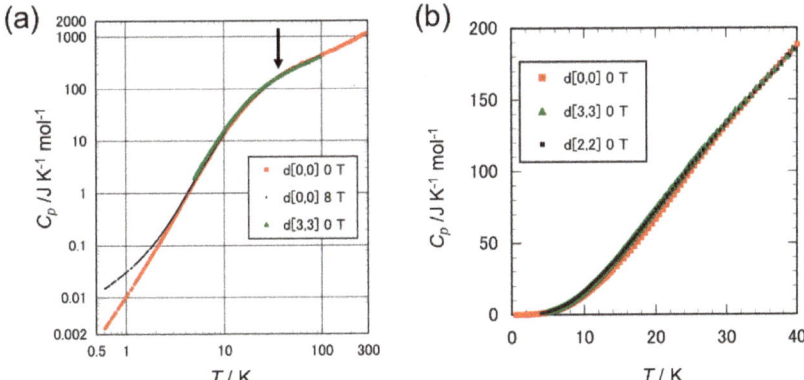

Figure 2. (a) Temperature dependences of the heat capacity of κ-(BEDT-TTF)$_2$Cu[N(CN)$_2$]Br (d[0,0] compound) shown by red squares and its partially deuterated compound (d[3,3] compound) denoted by green triangles shown in a logarithmic plot. The black dots show the data for the normal state heat capacity determined from the data in reference [27]. The arrow indicates the temperature of the critical endpoint of the Mott boundary. (b) C_p vs. T lot of the d[0,0], d[2,2], and d[3,3] compounds in a temperature range below 40 K, which shows the difference of the lattice heat capacity between the partially deuterated compound and the pristine compound.

From the overall temperature dependence of the d[0,0] compound, the phonon structure of the κ-type BEDT-TTF salts showed a Debye-like temperature dependence in the low-temperature region due to acoustic phonons. It showed a further increase above 100 K due to the multiple optical phonon modes and molecular vibrations. This feature is similar to the cases of typical molecular crystals [33]. We also showed the heat capacity data of the d[3,3] compound obtained for a single crystal between 4.9 K and 100 K in Figure 2a. The measurement of this sample was performed by slow cooling conditions from room temperature down to 4 K with a rate lower than 0.1 K min^{-1}. Although the data are shown in the logarithmic plot in the figure, the difference of the green and red curves observed below about 40 K means that the lattice heat capacity of the partially deuterated salt became larger than the pristine non-deuterated d[0,0] compound. From the figure, it is noted that the difference appeared below the region of the critical endpoint temperature that is shown by the arrow in the figure. To confirm the difference of the heat capacity between the two compounds, we showed the data below 40 K in C_p vs. T plot in Figure 2b. The difference was about 10–15% of the absolute values around 20 K, which exceeded the possible error bars related to the accuracy of the present heat capacity measurement. This difference was not originated from the spin entropy, since the κ−(BEDT-TTF)$_2$Cu[N(CN)$_2$]Cl having the bulk antiferromagnetic ordering showed a smaller heat capacity in this temperature region as was reported in references [10,11]. The increase of the heat capacity in the partially deuteration can be confirmed also in the data of the d[2,2] compound shown in Figure 2b, which were measured in slow cooling conditions. The difference from the d[0,0] data was slightly smaller, but a similar behavior to that of the d[3,3] compound seemed to exist. Although we do not have direct evidence at present, we speculate that the phonon structure shows a glassy feature due to the rounding of the phase boundary and merging of the Mott-insulating phase and the metallic phase. The lattice softening that gives a higher density of states of phonons occurs in the compounds just near the boundary can be understood as a kind of supercritical phenomenon that specifically appears near the boundary. Divergent compressibility has been reported by thermal expansion measurements [17,18,34].

The peculiar softening of the acoustic phonons of the partially deuterated d[2,2] and d[3,3] compounds below about 40 K, similar to the reported softening in κ-(BEDT-TTF)$_2$Cu[N(CN)$_2$]Cl [16] under pressures, was detected by the sound velocity measurement. The datasets shown in Figure 3 were the relative change in the ultrasonic sound velocity $\Delta v/v$ of the d[2,2] and d[3,3] compounds between 5 K and 100 K obtained for the cases with different cooling rates of 0.5 K min^{-1} and 20 K min^{-1}. $\Delta v/v$ showed a peculiar dip structure, indicating a tendency of significant lattice softening in a temperature range between 20 K and 50 K, and the temperatures showing the minima were 34 K for 20 K min^{-1} and 38 K for 0.5 K min^{-1} for the d[3,3] compound. Here, we emphasize that the temperature region where the decrease of the sound velocity took place was roughly corresponding to the region where the larger heat capacity appeared.

This fact demonstrated that these phenomena originated from the same origin. Although it is still a speculative discussion, the softening response of the ultrasonic compressional wave at this region implied that the coherence of acoustic phonons was disturbed by the microscopic merging of electronic phases and gave a kind of glassy state in a microscopic level. The observed experimental results are consistent with the experiment of κ-(BEDT-TTF)$_2$Cu[N(CN)$_2$]Cl with gas pressure control [16]. Indeed, although the accurate estimation of the absolute value of the ultrasonic attenuation is difficult owing to the significant lattice softening, the enhancement of the attenuation, which implies the growth of the phonon scattering, was observed in this region. By comparing the sound velocity data of the d[2,2] compound with those of the d[3,3] compound, we can find that the position of the dip structures shifted to a higher temperature and became much broader, since d[2,2] was located in the metallic region. The broadening of the hump agreed with the change in the position in the phase diagram. The shifts of the dip temperature by changing the cooling rate in both compounds should be related to the change of the positions of the d[2,2] and d[3,3] compounds in the phase diagram. As is known for this compound, the disorder in ethylene groups of BEDT-TTF molecules induced by rapid cooling made the volume large. The rapid cooling worked as a negative pressure effect. We roughly evaluated the possible positions of each cooling rate and displayed them in the phase diagram in Figure 1. The phonon softening near the boundary region is probably consistent with the unusual elasticity near the boundary reported by thermal expansion measurements [17,18]. It is important to measure the frequency dependences of $\Delta v/v$ in order to evaluate the glassy feature of the phonons. However, higher frequency measurements were not easy because of the strong reduction in the signals of ultrasonic echoes. The crossing of the 1st-order Mott boundary at low temperatures can also give an anomaly in sound velocity. Although we cannot see drastic discontinuity in $\Delta v/v$ at present, the small kink around 25 K observed in the slowly cooled d[3,3] salt (light blue) may be attributed to the 1st-order Mott transition, which was typically less significant than the critical elasticity around the critical endpoint as reported in reference [16].

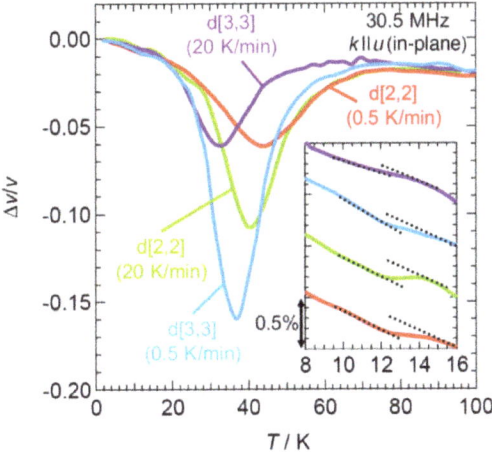

Figure 3. Temperature dependences of the relative change of the sound velocity ($\Delta v/v$) for the partially deuterated d[3,3] compound and d[2,2] compound obtained with two cooling rates of 0.5 K min^{-1} and 20 K min^{-1}. The inset shows the sound velocity anomaly around the superconducting transition temperature.

The inset of Figure 3 shows the sound velocity anomalies around the superconducting transition temperature of the d[3,3] and d[2,2] compounds. The sound velocity was sensitive to detect the superconductivity including the fluctuating superconductivity. Since the fluctuations of the superconductivity also made the lattice softened, the change in the sound velocity tended to occur at higher temperatures than T_c of the bulk superconductivity. The onset and the local minimum around 11–13 K of the dip typically correspond to the emergence of the fluctuating superconductivity and the superconducting transition, respectively. The present experiments were performed in the configuration in which ultrasonic sounds propagated in the in-plane direction. The absolute value of the change seemed to be larger than those in the previous measurements of interlayer ultrasonic responses for the d[0,0] compound [35] and κ-(BEDT-TTF)$_2$Cu(NCS)$_2$ [36], despite the percolative superconductivity originated from the macroscopic phase separation near the Mott transition in the present d[2,2] and d[3,3] compounds. The magnitude of the dip in the inset demonstrated that the volume fraction of the superconducting components changed due to the difference of deuterated numbers and cooling rates. This sensitive acoustic response to the superconductivity enabled us to discuss the details of the superconductivity including the fluctuating superconductivity and percolative superconductivity. In addition, it suggests a possibility to detect other superconducting state such as the Fulde–Ferrell–Larkin–Ovchinnikov (FFLO) superconductivity in the d[0,0] compound and other organic superconductors [37–40]. As a matter of fact, the FFLO state in the κ-(BEDT-TTF)$_2$Cu[N(CN)$_2$]Br was detected in the high-field ultrasound measurement by Imajo et al. [38].

The electronic properties related to the percolative feature of the superconductive and antiferromagnetic components near the boundary can be discussed through the analyses of the magnetic fields dependences of electronic heat capacity coefficient γ and lattice heat capacity coefficient β of the d[3,3] compound. It is known that the volume fraction of the superconductivity of this compound was significantly suppressed and the number of electrons related to the superconducting components became smaller than that of the d[0,0] compound.

In Figure 4a,b, we show the heat capacity data at extremely low temperatures between 0.7 K and 2.64 K (T^2 = 7 K^2) measured in magnetic fields up to 6 T. The data were plotted in $C_p T^{-1}$ vs. T^2. Here, the magnetic fields were applied perpendicularly to the superconducting layers. From the heat capacities at 0 T, 0.5 T, 1 T, and 1.5 T in Figure 4a, we can notice that the $C_p T^{-1}$ showed almost a linear dependence against T^2 in this low-energy region. The γ and the β for each magnetic field were determined using a linear fit $C_p T^{-1} = \gamma + \beta T^2$ to the data below 2 K. In Table 1, we show the two thermodynamic parameters γ and β obtained by the fitting analysis. The γ value at 0 T was about 1.3 ± 0.2 mJ K^{-2} mol^{-1}, reflecting on the gap formation in the superconductive ground state. A slight increase of γ in a low-filed region below 1.0 T indicated the gradual recovery of the electron density of states induced by the application of magnetic fields, although the values of γ were still very small. In fact, when applying magnetic fields with an out-of-plane configuration for various layered superconducting compounds, the pair-breaking mainly due to the orbital effect induced normal electrons. In the bulk superconductor of the d[0,0] compound, the recovery rate of the $C_p T^{-1}$ obeyed the square root of H, which was discussed as an evidence of the nodal superconductor explained by Volovik's theory. The d[0,0] compound showed a typical feature of the d-wave symmetry from the angle-resolved heat capacity measurements [27,28], and the normal state γ was evaluated as 22–25 mJ K^{-2} mol^{-1}. The pair-breaking feature was also reported in the d[1,1] and d[2,2] compounds in reference [41]. However, the change in γ by external magnetic fields was reported as negligibly small for the d[4,4] compound and κ-(BEDT-TTF)$_2$Cu[N(CN)$_2$]Cl located in the insulating region [41]. Although much smaller than d[2,2], the increase of the γ term observed below 1.0 T for d[3,3] demonstrated that the normal-state γ value detected was considered as ~3.1 mJ K^{-2} mol^{-1}.

Table 1. The thermodynamic parameters of the electronic heat capacity coefficient, γ, and the Debye term, β, for each magnetic field between 0 T and 6 T.

$\mu 0$ (H/T)	B (mJ K^{-4} mol^{-1})	Γ (mJ K^{-2} mol^{-1})
0	12.5	1.3
0.5	12.0	2.2
1	12.0	3.1
1.5	12.2	2.9
3.5	12.9	2.0
4	13.1	1.9
4.5	13.6	1.6
6	13.1	0.9

The heat capacity data obtained at higher magnetic fields of 3.5 T, 4.0 T, 4.5 T, and 6 T are shown in Figure 4 b. A curious tendency different from the pair breaking appeared in the temperature dependence of the electronic heat capacity. The $C_p T^{-1}$ vs. T^2 plots of 3.5 T, 4.0 T, and 4.5 T showed small hump-like structures around T = 1.4–1.7 K, and they were suppressed in the higher magnetic field of 6 T. The feature was different from the contribution of simple paramagnetic spins, which should give a Schottky-like heat capacity, since the Schottky anomaly gives the systematic change due to the increase of Zeeman splitting. The γ values in these fields were smaller than those of the data at H = 1.5 T. It is emphasized that the data at H = 6 T gave a much smaller γ value of 0.9 mJ K^{-2} mol^{-1}, which is a comparable value with that at H = 0 T.

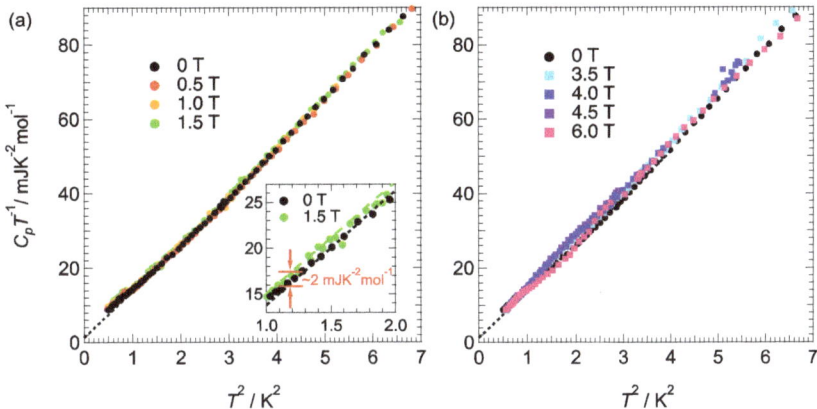

Figure 4. $C_p T^{-1}$ vs. T^2 plot of the partially deuterated d[3,3] compound obtained under magnetic fields applied perpendicular to the donor layer. The lower-field data of 0 T, 0.5 T, 1.0 T, and 1.5 T are shown in (**a**), and those at higher fields of 3.5 T, 4.0 T, 4.5 T, and 6.0 T are shown in (**b**) together with the data at the magnetic field of 0 T. The broken line represents the fitting result of 0 T using $C_p T^{-1} = \gamma + \beta T^2$. The inset in (**a**) is an enlarged plot of the datasets at magnetic fields of 0 T and 1.5 T around 1.0 K to make the field dependence clearer.

Figure 5 displays the change in these parameters in magnetic fields. It was confirmed that β was 12.5 ± 0.1 mJ K^{-4} mol^{-1} at 0 T, which is almost consistent with the previous report of κ-(BEDT-TTF)$_2$Cu[N(CN)$_2$]Br, and the change in β with the increase of the magnetic field was relatively small, as is shown in the figure. The change in the γ value with the magnetic field showed unusual features as the superconductive materials. From the variation of γ in the higher-field region, we should note that the shift of the Mott boundary, as was suggested by the transport measurements by Taniguchi et al., certainly occurs in this compound at this boundary region [24,42]. The normal state γ was evaluated by using the value at the peak in Figure 5 as 3.1 mJ K^{-2} mol^{-1}, although it must be an underestimate due to the field-dependent Mott boundary.

The change in the thermodynamic parameter γ inside the superconducting region in the dimer-Mott phase diagram was discussed from the theoretical and experimental viewpoints. According to the previous study of the low-temperature heat capacity of partial deuterated salts with Cu[N(CN)$_2$]Br, the normal state γ term is evaluated as 20 mJ K^{-2} mol^{-1} for the d[1,1] compound and 9–10 mJ K^{-2} mol^{-1} for the d[2,2] compound. The d[4,4] compound is located in the insulating region, and its γ value reaches almost zero [43]. The d[3,3] compound measured in this work revealed that the normal state γ value was 3.1 mJ K^{-2} mol^{-1}, although the insulating phase partially merged by applying magnetic fields and the gap-like structure seemed to be enhanced under magnetic fields. The systematic decrease of γ that occurs when approaching the boundary is understood by the enhancement of the Hubbard-like picture due to the increase of the U/W ratio [44,45]. This feature is interpreted by a theoretical suggestion for strongly correlated electron systems given by Kotliar et al. using the dynamical mean-field approach, as is shown in Figure 6 [46]. The superconducting component in the d[3,3] samples may be formed in the situation where the band-like and Mott–Hubbard features coexist, mainly produced by the electrons in the mid-gap states. The delicate balance of the magnetic insulating state and the superconductivity may induce the curious magnetic field dependence in the low-temperature thermal excitations for this material. It is considered that the bulk superconducting compounds, such as κ-(BEDT-TTF)$_2$I$_3$, κ-(BEDT-TTF)$_2$Ag(CN)$_2$H$_2$O, κ-

(BEDT-TTF)$_2$Cu(NCS)$_2$, are in the Fermi liquid region and the Brinkman–Rice enhancement emerges as was reported previously [47]. The crossover from the Brinkmann–Rice region to the Hubbard-gap region is considered as a specific feature of the 2D dime-Mott compound. Such a crossover can be explained by the picture schematically illustrated in Figure 6. The unconventional nature of the superconductivity was characterized by the four-fold oscillation of $C_p T^{-1}$ in the angle-resolved heat capacity measurements, which demonstrated that the antiferromagnetic spin fluctuations play an important role for relatively high transition temperatures of the dimer-Mott superconductors. The symmetry change of Cooper pairs from d_{xy} for κ-(BEDT-TTF)$_2$Ag(CN)$_2$H$_2$O to $d_{x2-y2} + s_\pm$ for the d[0,0] compound occurs inside the superconductive phase [28,48–53]. The relation with the symmetry change and the crossover inside the superconducting phase are interesting subjects to be solved by heat capacity measurements, although it is still a speculative discussion at present.

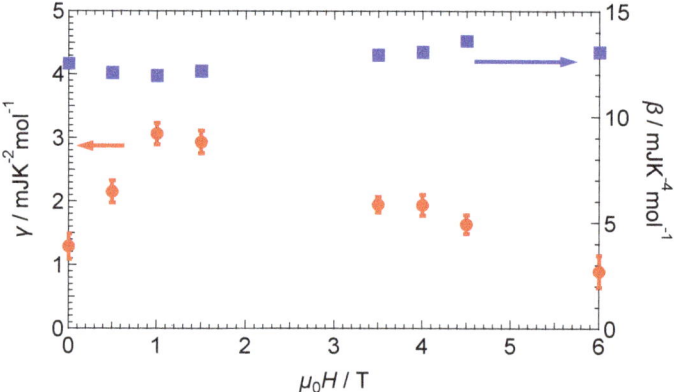

Figure 5. Magnetic fields dependence of the thermodynamic parameters γ and β of the d[3,3] compound. The peak structure at 1.0 T means that the normal state electronic heat capacity coefficient of this compound was about 3.0 mJ K^{-2} mol^{-1}. The decrease of γ above this field means that an insulating component was induced in the high-field region. The translucent curves are guides for the eye.

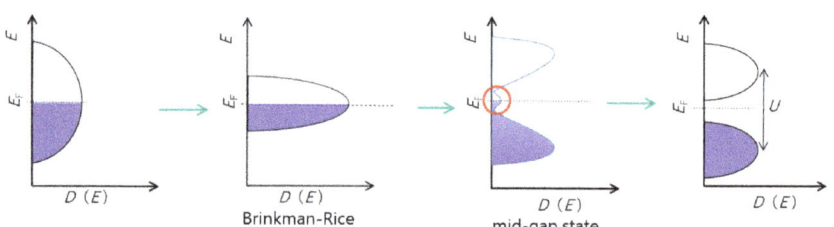

Figure 6. Schematic illustration of the electronic structure expected for the dimer-Mott compounds. Normal band metal on the left side gradually changed to the Mott–Hubbard state with the increase of U/W ratio. The variation from the Brinkman–Rice region to the mid-gap state region may be realized with the increase of partially deuteration.

Finally, we discussed the heat capacity jump, ΔC_p, around the superconductive transition of the d[3,3] compound. The non-deuterated d[0,0] compound showed a large heat capacity jump with about $\Delta C_p T^{-1}$ of 60 mJ K^{-2} mol^{-1}, and this anomaly was suppressed by applying magnetic fields above H_{c2}. The data of the d[0,0] compound showing the thermal anomaly are shown in Figure 7a. These results were already reported in references [27,28]. The temperature dependence of $C_p T^{-1}$ of the d[3,3] compound is shown in the green color in the same figure. The absolute value of C_pT^{-1} was 15% larger than that of d[0,0] at 10 K due to the difference in the lattice heat capacity as mentioned above. The heat capacity jump around T_c of the d[3,3] compound was quite small as compared with that of the d[0,0] compound but visible as a broad hump in the temperature dependence plot of C_pT^{-1} in Figure 7b. The suppression of this hump by the application of a magnetic field of 6 T indicated that the superconducting component certainly existed. The ΔC_pT^{-1} was roughly evaluated at most as 10 mJ K^{-2} mol^{-1} and the transition temperature was about 11 K, although the large lattice heat capacity gave ambiguity for the accurate determination of these electronic components. Since this compound was located just near the boundary, the much-reduced value of ΔC_pT^{-1} was of course consistent with the smaller normal state γ value. In spite of the smaller fraction in the d[3,3] compound, the large value of $\Delta C_p/\gamma T \sim 3.2$ indicated that the strong coupling feature of the superconductivity was retained. Here, we assumed the peak value of γ in Figure 5 was close to the normal state γ value. It may be underestimated, if we considered the relatively smaller magnetic field of 1T was used for the evaluation of the normal-state γ value. However, the tendency of the enhanced electron correlations near the boundary gives a strong attractive force for electron pairs [44], although the diminishing of electrons which contribute to the density of states in the mid-gap state is serious for stabilizing bulk superconductivity.

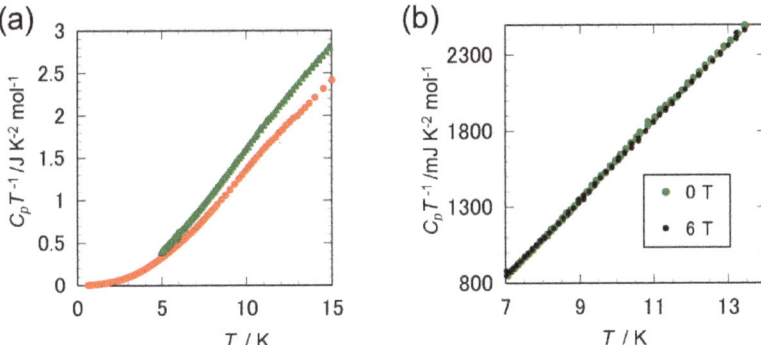

Figure 7. (a) C_pT^{-1} vs. T plot of the heat capacity of the d[0,0] compound (red color) and that of the d[3,3] compound (green color) obtained at 0 T; (b) the extended plot around the superconducting transition of the d[3,3] compound obtained at 0 T and 6 T. The data at magnetic fields of 0 T and 6 T are shown by the green circles and the black circles, respectively. A broad and small hump around the transition seemed to be suppressed by an external magnetic field.

4. Summary

The thermodynamic properties of κ-(BEDT-TTF)$_2$Cu[N(CN)$_2$]Br and partially deuterated d[3,3] compound have been studied by the heat capacity measurements with the thermal relaxation and the adiabatic calorimetry techniques. The d[3,3] compound located just near the boundary showed a larger heat capacity than the pristine compound at low temperatures below 40 K. This feature was confirmed by the sound velocity measurement as the significant lattice softening originating from the critical elasticity. The low-temperature heat capacity in magnetic fields demonstrated that the normal-state γ value was about 3.1 mJ K^{-2} mol^{-1}. The magnetic fields higher than 3.0 T were found to induce a gapped insulating state, and this field-induced feature was interpreted in terms of the instability of the mid-gap state that occurred due to the enhanced electronic correlations around the Mott boundary. Although the volume fraction was much reduced in the d[3,3] compound, the heat capacity data showed a small hump structure probably related to the strong-coupling feature of the superconductivity just near the boundary. This feature was also considered as the result of the larger U/W ratio near the boundary.

Author Contributions: Y.M., S.I. and S.Y. conducted the low-temperature heat capacity measurements. H.A. and Y.M. worked on sample characterization. The sound velocity measurements were performed by S.I., Y.M. and Y.N. designed the research plan, and all authors worked on the overall discussions throughout the work. Y.M. and Y.N. wrote the draft of the manuscript, and all authors commented on the manuscript. All authors have read and agreed to the published version of the manuscript.

Funding: This research was financially supported by JSPS KAKENHI (grant numbers: JP19K221690 and JP20H018620).

Institutional Review Board Statement: Not applicable.

Informed Consent Statement: Not applicable.

Data Availability Statement: The data presented in this study are available on request from the corresponding author.

Acknowledgments: The authors thank A. Kawamoto (Hokkaido University) and Hiromi Taniguchi (Saitama University) for valuable discussion in terms of crystal synthesis and sample characterization. Y.M. thanks JSPS for the fellowship for young researchers.

Conflicts of Interest: The authors declare no conflict of interest.

References

1. Mott, F.N. The transition to the metallic state. *Plilos. Mag.* **1961**, *6*, 287–309. [CrossRef]
2. Mcwhan, B.D.; Rice, M.T.; Tokura, Y. Metal-insulator transition in Cr-Doped V_2O_3. *Phys. Rev. Lett.* **1969**, *23*, 1384–1387. [CrossRef]
3. Imada, M.; Fujimori, A.; Tokura, Y. Metal-insulator transitions. *Rev. Mod. Phys.* **1998**, *70*, 1039–1263. [CrossRef]
4. Bednorz, J.G.; Müller, K.A. Possible high T_C superconductivity in the Ba-La-Cu-O System. *Z. Phys. B—Condens. Matter* **1986**, *64*, 189–193. [CrossRef]
5. Lee, A.P.; Nagaosa, N.; Wen, X.-G. Doping a Mott insulator: Physics of high temperature superconductivity. *Rev. Mod. Phys.* **2006**, *78*, 17–85. [CrossRef]
6. Stewart, R.G. Heavy-Fermion systems. *Rev. Mod. Phys.* **1984**, *56*, 755–787. [CrossRef]
7. Carter, A.S.; Rosenbaum, F.T.; Metcalf, P.; Honig, M.J.; Spalek, J. Mass enhancement and magnetic order at the Mott-Hubbard transition. *Phys. Rev. B* **1993**, *48*, 16841–16844. [CrossRef] [PubMed]
8. Von Löhneysen, H. Non-Fermi-liquid behaviour in the heavy-fermion system $CeCu_{6-x}Au_x$. *J. Phys. Condens. Matter* **1996**, *8*, 9689–9706. [CrossRef]
9. Hiraki, K.; Kanoda, K. Wigner Crystal Type of Charge Ordering in an Organic Conductor with a Quarter-Filled Band: (DI–DCNQI)$_2$Ag. *Phys. Rev. Lett.* **1998**, *80*, 4737–4740. [CrossRef]
10. Kanoda, K. Recent progress in NMR studies on organic conductors. *Hyperfine Interact.* **1997**, *104*, 235–249. [CrossRef]
11. Kanoda, K. Metal–Insulator Transition in κ-(ET)$_2$X and (DCNQI)$_2$M: Two Contrasting Manifestation of Electron Correlation. *J. Phys. Soc. Jpn.* **2006**, *75*, 051007. [CrossRef]
12. Kagawa, F.; Miyagawa, F.; Kanoda, K. Magnetic Mott criticality in a κ-type organic salt probed by NMR. *Nat. Phys.* **2005**, *5*, 880–884. [CrossRef]
13. Akutsu, H.; Saito, K.; Sorai, M. Phase Behavior of Organic Superconductors, κ-(BEDT-TTF)$_2$Cu[N(CN)$_2$]X (X = Br and Cl), Studied by ac Calorimetry. *Phys. Rev. B* **2000**, *61*, 4346–4352. [CrossRef]
14. Kagawa, F.; Miyagawa, K.; Kanoda, K. Unconventional critical behaviour in a quasi-two-dimensional organic conductor. *Nature* **2005**, *436*, 534–537. [CrossRef]
15. Sasaki, T.; Yoneyama, N.; Kobayashi, N. Mott transition and superconductivity in the strong correlated organic superconductor κ-(BEDT-TTF)$_2$Cu[N(CN)$_2$]Br. *Phys. Rev. B* **2008**, *77*, 054505. [CrossRef]
16. Fournier, D.; Poirier, M.; Castonguay, M.; Truong, D.K. Mott transition, Compressibility Divergence, and the P-T Phase Diagram of Layered Organic Superconductors: An Ultrasonic Investigation. *Phys. Rev. Lett.* **2003**, *90*, 127002. [CrossRef]
17. Gati, E.; Garst, M.; Manna, S.R.; Tutsch, U.; Bartosch, L.; Schubert, H.; Sasaki, T.; Sculueter, A.J.; Lang, M. Breakdown of Hookes law of elasticity at the Mott critical endpoint in an organic conductor. *Sci. Adv.* **2016**, *2*, e1601646. [CrossRef]
18. Gati, E.; Tutsch, U.; Naji, A.; Garst, M.; Köhler, S.; Schubert, H.; Sasaki, S.; Lang, M. Effects of Disorder on the Pressure-Induced Mott Transition in κ-(BEDT-TTF)$_2$Cu[N(CN)$_2$]Cl. *Crystals* **2018**, *8*, 38. [CrossRef]
19. Kawamoto, A.; Taniguchi, H.; Kanoda, K. Superconductor–Insulator Transition Controlled by Partial Deuteration in BEDT-TTF Salt. *J. Am. Chem. Soc.* **1998**, *120*, 10984–10985. [CrossRef]
20. Yoneyama, N.; Sasaki, T.; Kobayashi, N. Substitution Effect by Deuterated Donors on Superconductivity in κ-(BEDT-TTF)$_2$Cu[N(CN)$_2$]Br. *J. Phys. Soc. Jpn.* **2004**, *73*, 1434–1437. [CrossRef]
21. Müller, J.; Hartmann, B.; Sasaki, T. Fine-tuning the Mott metal–insulator transition and critical charge carrier dynamics in molecular conductors. *Philos. Mag.* **2017**, *97*, 3477–3494. [CrossRef]

22. Yasin, S.; Dumm, M.; Salameh, B.; Batail, P.; Mézière, C.; Dressel, M. Transport studies at the Mott transition of the two-dimensional organic metal κ-(BEDT-TTF)$_2$Cu[N(CN)$_2$]Br$_x$Cl$_{1-x}$. *Eur. Phys. J. B* **2011**, *79*, 383–390. [CrossRef]
23. Faltermeier, D.; Barz, J.; Dumm, M.; Dressel, M.; Drichko, N.; Petrov, B.; Semkin, V.; Vlasova, R.; Mézière, C.; Batail, P. Bandwidth-controlled Mott transition in κ−(BEDT−TTF)$_2$Cu[N(CN)$_2$]Br$_x$Cl$_{1-x}$: Optical studies of localized charge excitations. *Phys. Rev. B* **2007**, *76*, 165113. [CrossRef]
24. Taniguchi, H.; Kanoda, K.; Kawamoto, A. Field switching of superconductor-insulator bistability in artificially tuned organics. *Phys. Rev. B* **2003**, *67*, 014510. [CrossRef]
25. Sorai, M. *Comprehensive Handbook of Calorimetry and Thermal Analysis*; Wiley J. & Sons: West Susex, UK, 2004; pp. 68–74.
26. Imajo, S.; Akutsu, H.; Kurihara, R.; Yajima, T.; Kohama, Y.; Tokunaga, M.; Kindo, K.; Nakazawa, Y. Anisotropic Fully Gapped Superconductivity Possibly Mediated by Charge Fluctuations in a Nondimeric Organic Complex. *Phys. Rev. Lett.* **2020**, *125*, 177002. [CrossRef]
27. Imajo, S.; Yamashita, S.; Akutsu, H.; Nakazawa, Y. Quadratic temperature dependence of electronic heat capacities in the κ-type organic superconductors. *Int. J. Mod. Phys. B* **2016**, *30*, 1642014. [CrossRef]
28. Imajo, S.; Kindo, K.; Nakazawa, Y. Symmetry Change of d-wave Superconductivity in κ-type Organic Superconductors. *Phys. Rev. B* **2021**, *103*, L060508. [CrossRef]
29. Elsinger, H.; Wosnitza, J.; Wanka, S.; Hagel, J.; Schweitzer, D.; Strunz, W. κ-(BEDT−TTF)$_2$Cu[N(CN)$_2$]Br: A Fully Gapped Strong-Coupling Superconductor. *Phys. Rev. Lett.* **2000**, *84*, 6098–6101. [CrossRef]
30. Andraka, B.; Jee, S.C.; Kim, S.J.; Stewart, R.G.; Calson, D.K.; Wang, H.H.; Crouch SV, A.; Kini, M.A.; Williams, M.J. Specific heat of the high Tc organic superconductor κ-(ET)$_2$Cu[N(CN)$_2$]Br. *Solid State Commun.* **1991**, *79*, 57–59. [CrossRef]
31. Taylor, J.O.; Carrington, A.; Schlueter, A.J. Specific-Heat Measurements of the Gap Structure of the Organic Superconductors κ−(ET)$_2$Cu[N(CN)$_2$]Br and κ−(ET)$_2$Cu(NCS)$_2$. *Phys. Rev. Lett.* **2007**, *99*, 057001. [CrossRef]
32. Nakazawa, Y.; Kanoda, K. Low-temperature specific heat of κ-(BEDT-TTF)$_2$Cu[N(CN)$_2$]Br in the superconducting state. *Phys. Rev. B* **1997**, *55*, R8670–R8673. [CrossRef]
33. Sorai, M.; Nakazawa, Y.; Nakano, M.; Miyazaki, Y. Calorimetric Investigation of Phase Transitions Occurring in Molecule-Based Magnets. *Chem. Rev.* **2013**, *113*, 41–122. [CrossRef]
34. Souza, M.; Bartosch, L. Probing the Mott Physics in κ-(BEDT-TTF)$_2$X salts via thermal expansion. *J. Phys. Condens. Matt.* **2015**, *27*, 053203. [CrossRef]
35. Fournier, D.; Poirier, M.; Truong, K.D. Competition between magnetism and superconductivity in the organic metal κ-[BEDT-TTF]$_2$Cu[N(CN)$_2$]Br. *Phys. Rev. B* **2007**, *76*, 054509. [CrossRef]
36. Simizu, T.; Yoshimoto, N.; Nakamura, M.; Yoshizawa, M. High-frequency ultrasonic measurements of κ-(BEDT-TTF)$_2$Cu(NCS)$_2$ by devised transducer. *Phys. B* **2000**, *281&282*, 896–898. [CrossRef]
37. Imajo, S.; Nomura, T.; Kohama, Y.; Kindo, K. Nematic response of the Fulde-Ferrell-Larkin-Ovchinnikov state. *arXiv* **2021**, arXiv:2110.12774.
38. Imajo, S.; Kindo, K. The FFLO State in the Dimer Mott Organic Superconductor κ-(BEDT-TTF)$_2$Cu[N(CN)$_2$]Br. *Crystals* **2021**, *11*, 1358. [CrossRef]
39. Croitoru, M.D.; Buzdin, A.I. In search of unambiguous evidence of the Fulde-Ferrell-Larkin-Ovchinnikov state in quasi-low dimensional superconductors. *Condens. Matter* **2017**, *2*, 30. [CrossRef]
40. Croitoru, M.D.; Buzdin, A.I. The Fulde-Ferrell-Larkin-Ovchinnikov state in layered d-wave superconductors: In-plane anisotropy and resonance effects in the angular dependence of the upper critical field. *J. Phys. Condens. Matter* **2013**, *25*, 125702. [CrossRef]
41. Nakazawa, Y.; Taniguchi, H.; Kawamoto, A.; Kanoda, K. Electronic specific heat at the boundary region of the metal-insulator transition in the two-dimensional electronic system of κ−(BEDT−TTF)$_2$Cu[N(CN)$_2$]Br. *Phys. Rev. B* **2000**, *61*, R16295–R16298. [CrossRef]
42. Kagawa, F.; Itou, T.; Miyagawa, K.; Kanoda, K. Magnetic-Field-Induced Mott Transition in a Quasi-Two-Dimensional Organic Conductor. *Phys. Rev. Lett.* **2004**, *93*, 127001. [CrossRef]
43. Nakazawa, Y.; Kanoda, K. Thermodynamic investigation of the electronic states of deuterated κ−(BEDT−TTF)$_2$Cu[N(CN)$_2$]Br. *Phys. Rev. B* **1999**, *60*, 4263–4266. [CrossRef]
44. Nam, M.-S.; Mézière, C.; Batail, P.; Zorina, L.; Simonov, S.; Ardavan, A. Superconducting fluctuations in organic molecular metals enhanced by Mott criticality. *Sci. Rep.* **2013**, *3*, 3390. [CrossRef]
45. Powell, J.B.; McKenzie, H. Strong electronic correlations in superconducting organic charge transfer salts. *J. Phys. Condens. Matt.* **2006**, *18*, R827–R866. [CrossRef]
46. Kotliar, G.; Lange, E.; Rozenberg, M.J. Laudau Theory of the Finite Temperature Mott Transition. *Phys. Rev. Lett.* **2000**, *84*, 5180–5183. [CrossRef] [PubMed]
47. McWhan, B.D.; Remeika, P.J.; Rice, M.T.; Brinkman, F.M.; Maita, P.J.; Menth, A. Electronic Specific Heat of Metallic Ti-Doped V_2O_3. *Phys. Rev. Lett.* **1971**, *27*, 941. [CrossRef]
48. Malone, L.; Taylor, O.J.; Schlueter, A.J.; Carrington, A. Location of gap nodes in the organic superconductors κ-(ET)$_2$Cu(NCS)$_2$ and κ-(ET)$_2$Cu[N(CN)$_2$]Br determined by magnetocalorimetry. *Phys. Rev. B* **2010**, *82*, 014522. [CrossRef]
49. Kuroki, K.; Kimura, T.; Arita, R.; Tanaka, Y.; Matsuda, Y. $d_{x^2-y^2}$- versus d_{xy}- like pairings in organic superconductors κ−(BEDT−TTF)$_2$X. *Phys. Rev. B* **2002**, *65*, 100516. [CrossRef]
50. Kuroki, K. Pairing symmetry competition in organic superconductors. *J. Phys. Soc. Jpn.* **2006**, *75*, 051013. [CrossRef]

51. Guterding, D.; Altmeyer, M.; Jeschke, H.O.; Valentí, R. Near-degeneracy of extended s+$d_{x^2-y^2}$ and d_{xy} order parameters in quasi-two-dimensional organic superconductors. *Phys. Rev. B* **2016**, *94*, 024515. [CrossRef]
52. Watanabe, H.; Seo, H.; Yunoki, S. Mechanism of superconductivity and electron-hole doping asymmetry in κ-type molecular conductors. *Nat. Commun.* **2019**, *10*, 3167. [CrossRef] [PubMed]
53. Milbradt, S.; Bardin, A.A.; Truncik, C.J.S.; Huttema, W.A.; Jacko, A.C.; Burn, P.L.; Lo, S.-C.; Powell, B.J.; Broun, D.M. In-plane superfluid density and microwave conductivity of the organic superconductor κ-(BEDT-TTF)$_2$Cu[N(CN)$_2$]Br: Evidence for *d*-wave pairing and resilient quasiparticles. *Phys. Rev. B* **2013**, *88*, 064501. [CrossRef]

Article

Metallic Conduction and Carrier Localization in Two-Dimensional BEDO-TTF Charge-Transfer Solid Crystals

Hiroshi Ito [1,*], Motoki Matsuno [1], Seiu Katagiri [1], Shinji K. Yoshina [1], Taishi Takenobu [1], Manabu Ishikawa [2], Akihiro Otsuka [2,3], Hideki Yamochi [2,3], Yukihiro Yoshida [2,4], Gunzi Saito [4,5], Yongbing Shen [6] and Masahiro Yamashita [6,7]

1. Department of Applied Physics, Nagoya University, Nagoya 464-8603, Japan; moto315@me.com (M.M.); seiu.arsenal@gmail.com (S.K.); shinjiyoshina@gmail.com (S.K.Y.); takenobu@nagoya-u.jp (T.T.)
2. Division of Chemistry, Graduate School of Science, Kyoto University, Kyoto 606-8502, Japan; ishikawa.manabu.2s@kyoto-u.ac.jp (M.I.); otsuka@kuchem.kyoto-u.ac.jp (A.O.); yamochi@kuchem.kyoto-u.ac.jp (H.Y.); yoshiday@ssc.kuchem.kyoto-u.ac.jp (Y.Y.)
3. Research Center for Low Temperature and Materials Sciences, Kyoto University, Kyoto 606-8501, Japan
4. Faculty of Agriculture, Meijo University, Nagoya 468-8502, Japan; gunzi-s@mx2.canvas.ne.jp
5. Toyota Physical and Chemical Research Institute, Nagakute 480-1192, Japan
6. Department of Chemistry, Graduate School of Science, Tohoku University, Sendai 980-8578, Japan; shenyongbing17@gmail.com (Y.S.); yamasita.m@gmail.com (M.Y.)
7. School of Materials Science and Engineering, Nankai University, Tianjin 300350, China
* Correspondence: ito@nuap.nagoya-u.ac.jp

Abstract: Charge-transfer salts based on bis(ethylenedioxy)tetrathiafulvalene (BEDO-TTF or BO for short) provide a stable two-dimensional (2D) metallic state, while the electrical resistance often shows an upturn at low temperatures below ~10 K. Such 2D weak carrier localization was first recognized for BO salts in the Langmuir–Blodgett films fabricated with fatty acids; however, it has not been characterized in charge-transfer solid crystals. In this paper, we discuss the carrier localization of two crystalline BO charge-transfer salts with or without magnetic ions at low temperatures through the analysis of the weak negative magnetoresistance. The phase coherence lengths deduced with temperature dependence are largely dominated by the electron–electron scattering mechanism. These results indicate that the resistivity upturn at low temperatures is caused by the 2D weak localization. Disorders causing elastic scattering within the metallic domains, such as those of terminal ethylene groups, should be suppressed to prevent the localization.

Keywords: charge-transfer solid crystals; two-dimensional metal; carrier localization; negative magnetoresistance; phase coherence length

1. Introduction

Since the 1970s, molecular conductors with tetrathiafulvalene (TTF) derivatives have attracted attention for their rich variety of electronic phenomena such as metal–insulator transitions and superconductivity [1,2]. Bis(ethylenedithio)-substituted derivative BEDT-TTF (or ET for short) is especially well known for the production of a variety of superconductors (more than 40 kinds). Additionally, the bis(ethylenedioxy)-substituted derivative, BEDO-TTF (or BO for short, Figure 1a), in which the sulfur atoms in the outer six-membered ring of ET are replaced by oxygen atoms, has also attracted attention as a building molecule. This donor molecule shows the strong tendency to create complexes of two-dimensional (2D) layers due to the self-aggregation ability caused by the intermolecular π–π interactions, along with the C–H···O hydrogen bonding [2,3]. While most of the BO salts exhibit metallic conduction, only a few examples are reported to show superconductivity [4,5], in contrast to the ET complexes.

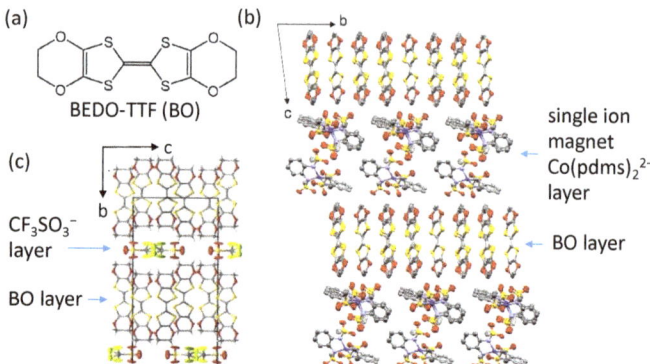

Figure 1. (**a**) The chemical structure of the donor molecule BEDO-TTF or BO, i.e., bis(ethylenedioxy) tetrathiafulvalene). Crystal structures of (**b**) β''-(BO)$_3$[Co(pdms)$_2$](CH$_3$CN)(H$_2$O)$_2$ and (**c**) κ-(BO)$_2$CF$_3$SO$_3$ viewed along a-axis. Each atom is colored as: gray, C; white, H; blue, N; green, F; navy, Co; red, O; yellow, S.

Applying this peculiarity, metallic Langmuir–Blodgett (LB) films were fabricated by mixing BO with fatty acids for potential applications in molecular electronics [4–6]. Self-organization provided a stable 2D BO layer which was sandwiched by the layers composed of fatty acids such as stearic [6,7] or arachidic [8] acid. Although the LB films exhibited room-temperature conductivity as high as 100 S cm^{-1}, the electrical resistance increased at low temperatures. Two-dimensional weak localization has been proposed as the origin of the behavior. The characteristic negative magnetoresistance (MR) under the magnetic field applied perpendicularly to the 2D layer was observed owing to the destruction of the constructive interference between waves along a closed loop in opposite senses with equal probabilities [9,10].

On the other hand, the carrier localization in crystalline charge-transfer salts at low temperatures [11] has not been mentioned much so far in BO complexes, although resistance upturn at low temperatures has often been found with [12] or without [3,13] magnetic ions. Clarification on the origin of the resistivity upturn at low temperatures is important to explore electronic phase transition, such as metal–insulator transition and/or superconductivity in charge-transfer solid crystals, whether they stem from magnetic interaction or not.

In this paper, we will discuss two crystalline BO salts with and without magnetic ions to explore the resistance upturn phenomena in single crystals, β''-(BO)$_3$[Co(pdms)$_2$](CH$_3$CN)(H$_2$O)$_2$ (abbreviated hereafter as β''-BO3) [14], where H$_2$pdms = 1,2-bis(methanesulfonamido) benzene, and κ-(BO)$_2$CF$_3$SO$_3$ (abbreviated hereafter as κ-BO2) [15]. The crystal structures of the two materials are shown in Figure 1b,c, respectively, consisting of 2D layers of BO molecules.

β''-BO3 has a 1/3 band filling with the β''-type molecular arrangement, according to the classification for ET salts, showing a metallic state down to 15 K [14]. This salt was synthesized in order to develop the possible interaction between the conduction electron and single ion magnet, exhibiting typical frequency dependence of ac magnetic susceptibility due to the bistable spin state of Co^{2+}. This is a sister compound of β''-(BO)$_4$[Co(pdms)$_2$]·3H$_2$O, (abbreviated hereafter as β''-BO4), which was grown using the same molecules but with a different solvent, which exhibited ferromagnetic interaction at low temperatures between BO and Co(pdms)$_2$ ions [16]. The MR of β''-BO4 cannot be understood by the weak localization model, showing contribution from the π–d interaction [16]. On the other hand, the π–d interaction with the BO layer in the β''-BO3 was considered to be weak in view of the band structure calculation [14].

κ-BO2 has a 1/4 band filling with the κ-type molecular arrangement according to the classification for ET salts. This is a BO analogue salt of the ET salt with the same structure, κ-(ET)$_2$CF$_3$SO$_3$ [17], which is a Mott insulator at ambient pressure but shows metallic behavior by applying a pressure above 1 GPa, and the superconducting transition occurs at 4.8 K under 1.1 GPa [18]. Although semiconducting behavior is observed from room temperature in the first report of κ-BO2 [15], here, a metallic conduction was observed down to around 10 K, below which electrical resistance increases. The change in the resistive behavior may be ascribed to the effect of grease (weak pressurization on crystals by cooling) and slow cooling in the present study.

In terms of common features, the low-temperature MR measurements showed that both β''-BO3 and κ-BO2 salts have a negative MR, similar to those observed in LB films. Here, we analyze the negative MR of the two different types of BO salt with the 2D weak localization model. The phase coherence lengths are deduced with temperature dependence largely dominated by the electron–electron scattering mechanism. These values are comparable to those of other type of materials, such as LB films and polymer poly(ethylenedioxythiophene):polystyrene sulfonate (PEDOT:PSS), implying a common origin of the inelastic scattering. These results indicate that the resistivity upturn at low temperatures is caused by the 2D weak localization, which confirms the 2D nature of the conduction layer made of BO molecules. Disorders causing elastic scattering within the metallic domains, such as those of terminal ethylene groups, should be suppressed to prevent the localization.

2. Materials and Methods

The single crystals were prepared by galvanostatic electrooxidation according to the literature [14,15]. The electrical resistivity was measured with the four-probe method by attaching annealed platinum wires on sample surfaces with carbon paste. We measured three samples for β''-BO3 (#1, #2, #3) and κ-BO2 (#1, #3 for intralayer direction and #2 for interlayer direction). For intralayer measurement, the current was applied along b-axis for β''-BO3 and along c-axis for κ-BO2. For κ-BO2, interlayer resistivity was also measured for the sample #2 along b-axis direction; however, β''-BO3 crystals were too thin to measure the interlayer resistivity reliably.

Electrical resistivity from room temperature to 2 K and MR under magnetic fields up to 9 T were measured using physical property measurement system, QUANTUM DESIGN, with a cooling rate of 0.2–0.3 K/min. A low measurement current of 10 μA was used to suppress self-heating. The sample was coated with Apiezon N grease to suppress the discontinuous jumping of resistance which has been often observed in charge-transfer complexes, possibly caused by microcracking [1]. A gentle pressure of 0.03 GPa was applied to the sample after being wrapped in Apiezon N grease and cooled down to low temperature [19]. The sample was rotated in the magnetic field to adjust the sample position using a horizontal rotator when the magnetic field was perpendicular and parallel to the 2D layer.

3. Results

Figure 2a shows the temperature dependence of the intralayer resistivity along the b-axis of β''-BO3 down to 2 K. The resistivity decreased at ambient pressure down to 10~15 K, below which it began to increase, as shown in Figure 2b [14]. Figure 2c shows the plot of the intralayer conductivity in the ln$T$ scale. In the temperature range of the resistivity increase, the conductivity followed the ln$T$ dependence asymptotically.

To investigate the origin of the low-temperature upturn of the resistivity, the angle dependence of MR was investigated in a magnetic field (B) oriented perpendicular ($\theta = 90°$) and parallel ($\theta = 0°$) with respect to the 2D layer. MR is defined by the increase of resistivity by the magnetic field: MR (%) = $[\rho(B, T) - \rho(0, T)]/\rho(0, T) \times 100\%$, where ρ is the resistivity. Figure 3a shows the MR of β''-BO3 at 2 K for $\theta = 0°$ and 90°. Positive MR of metals was observed for the high field region above 4 T for both magnetic field directions; however, clear negative MR was observed up to 4 T only when the magnetic field was applied perpendicular to the 2D layer ($\theta = 90°$). No negative MR was observed when the magnetic field was applied parallel to the 2D layer ($\theta = 0°$). This behavior is typical for the weak negative MR caused by weak localization within the 2D conducting layer [9,10], as was observed for the LB films. The MR can be understood without considering any contribution from the magnetic anion layers, in sharp contrast to the case of β''-BO4 salt [16]. As shown in Figure 3b, the negative MR at $\theta = 90°$ was observed up to ~15 K, where the ln$T$ dependence in the temperature dependence of resistivity disappeared and turned to show a metallic behavior.

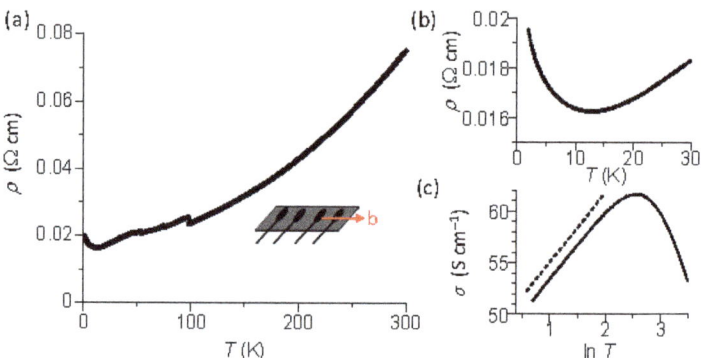

Figure 2. (a) The temperature (T) dependence of intralayer (b-axis) resistivity (ρ) of β''-BO3 (sample #1). (b) Intralayer resistivity in the low temperature region. (c) Intralayer conductivity as a function of ln$T$. The broken line represents the relationship of $\sigma \propto \ln T$.

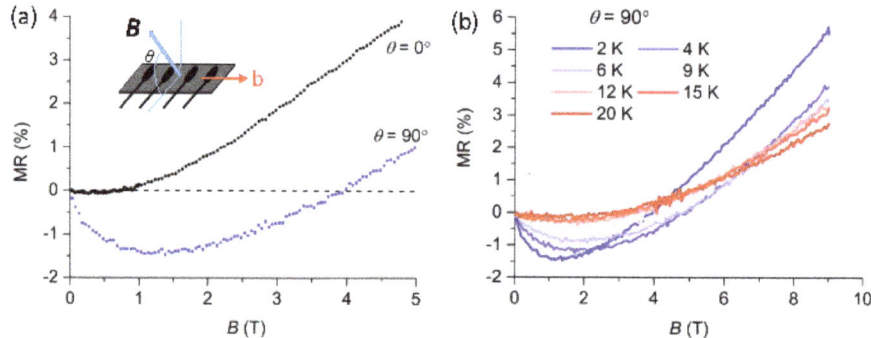

Figure 3. (a) The MR (current along b-axis) of β''-BO3 (sample #1) at 2 K under magnetic field B parallel ($\theta = 0°$) and perpendicular ($\theta = 90°$) to the 2D layer. Clear negative MR was observed when $\theta = 90°$. (b) The MR for the perpendicular field in the temperature range of 2–20 K.

Figure 4a shows the temperature dependence of the electrical conductivity of κ-BO2 in the direction parallel (sample #1, along c-axis) and perpendicular (sample #2, along b-axis) to the 2D layer. From room temperature down to ~10 K, the electrical resistivities decreased; however, the intralayer resistivity increased below 10 K, as shown in Figure 4b. Figure 4c shows the plot of the intralayer conductivity of the sample #1 in the ln$T$ scale; the conductivity follows the ln$T$ dependence asymptotically below 8 K, similar to β''-BO3. The interlayer resistivity (sample #2) also increased at low temperatures but very weakly below 4 K. The anisotropy of electrical conductivity reached 1×10^5, which is high compared to its ET counterpart, κ-(ET)$_2$CF$_3$SO$_3$ [18], implying the strong 2D nature of the conduction in κ-BO2.

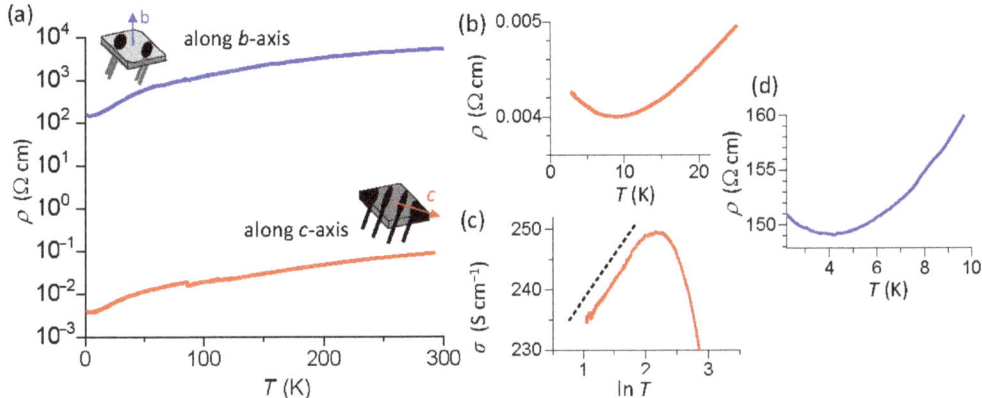

Figure 4. (a) The temperature (T) dependence of the electrical resistivity (ρ) of κ-BO2 in the direction parallel (sample #1, along c-axis) and perpendicular (sample #2, along b-axis) to the 2D layer. (b) Intralayer resistivity of the sample #1 in the low temperature region. (c) Intralayer conductivity of the sample #1 as a function of $\ln T$. The broken line represents the relationship of $\sigma \propto \ln T$. (d) Interlayer resistivity of sample #2 in the low temperature region.

Figure 5a shows the MR of κ-BO2 under magnetic fields parallel and perpendicular to the 2D layer. The negative MR for κ-BO2 was obvious at $B < 2$ T only under the perpendicular field, similar to β''-BO3. Figure 5b shows the MR of κ-BO2 at different temperatures (2–10 K) under the perpendicular field. The negative MR diminished at 10 K.

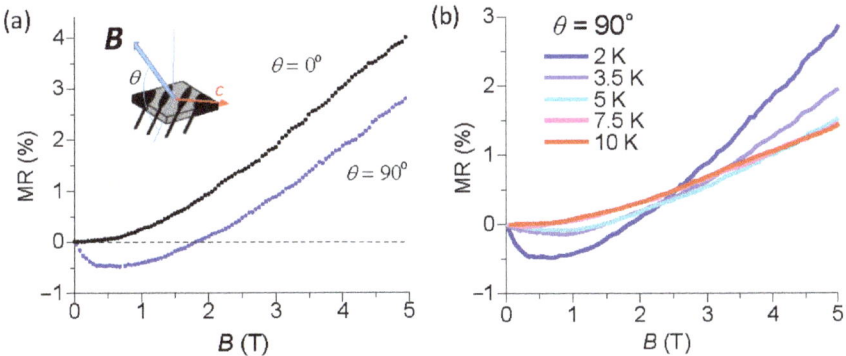

Figure 5. (a) The MR (current along c-axis) of κ-BO2 (sample #1) at 2 K under magnetic field B parallel ($\theta = 0°$) and perpendicular ($\theta = 90°$) to the 2D layer. Clear negative MR was observed when $\theta = 90°$. (b) The MR for the perpendicular field in the temperature range of 2–10 K.

4. Discussion

For both BO salt crystals, the intralayer resistivity increased asymptotically with $\ln T$ at low temperatures. Negative MR was observed under a magnetic field perpendicular to the 2D layer. Under the magnetic field parallel to the 2D layer, there manifested no negative MR, but only positive MR. These features are characteristic of 2D weak localization [9,10]. The elastic scatterings of electrons by disorders or imperfections occur several times within a metallic domain in which phase coherence of the wavefunction is preserved. Then, a closed loop can be formed within the metallic domain and the constructive interference between waves in opposite senses with equal probabilities causes the carrier localization [9,10].

To further investigate whether the negative MR showed weak 2D localization, the changes in MR with respect to temperature and magnetic field were examined. The time-reversal, symmetry-breaking perturbations, such as magnetic field, applied perpendicular to the closed loop destroyed the interference and, thus, suppressed the localization to cause the negative MR. The negative MR is formulated as the increase in the electrical conductivity σ due to the magnetic field (magnetoconductance, MC) based on the Hikami–Larkin–Nagaoka (HLN) expression [10]:

$$\frac{\sigma(B) - \sigma(B = 0\,\text{T})}{\sigma(B = 0\,\text{T})} = A_1 \left[\ln\left(\frac{B}{B_i}\right) + \psi\left(\frac{1}{2} + \frac{B_i}{B}\right) \right] - A_2 B^\alpha \qquad (1)$$

where A_1, A_2, B_i and α are fitting parameters. ψ is a digamma function. The first term represents the contribution from the 2D localization. The second term represents the negative MC (i.e., positive MR) of a metal which increases when the magnetic field increased. The exponent of α is in the range 1~2. The strength of the negative MR can be scaled with the ratio of A_1 and A_2. The characteristic magnetic field B_i is associated with the phase coherence length λ, in which the phase coherence of the carrier wavefunction is preserved between each inelastic scattering:

$$\lambda = \sqrt{\frac{\hbar}{4eB_i}} \qquad (2)$$

where $\hbar$ is the reduced Planck constant and e is the elementary charge.

We show typical examples of the fitting of MC under the perpendicular magnetic field with respect to Equation (1) in the supplementary materials, Figures S1–S3 for β''-BO3 and Figures S4 and S5 for κ-BO2. Corresponding fitting parameters are given in Tables S1–S5 in the supplementary materials, respectively. For β''-BO3, the curves are fitted to Equation (1) up to 6 T because the fitting to the metallic MC $A_2 B^\alpha$ with single exponent α up to 9 T is rather inappropriate and the exponent α seems to change to a smaller value above ~6 T, especially at the lowest temperature of 2 K.

Figure 6 shows the temperature dependence of λ deduced by the analysis in terms of Equations (1) and (2) on a logarithmic scale in the temperature range where the negative MR was observed for β''-BO3 and κ-BO2, together with those found for LB films [6–8]. The intralayer coherence length was far larger than the conducting 2D layer thickness of 1.5~2.0 nm, indicating the 2D nature of the conduction. The coherence length increased with lowering temperature. The temperature dependence of the phase coherence length λ followed the relation $\lambda \propto T^{-p/2}$, in which $p \sim 1$ and $p \sim 2$ were associated with the situations in which phase-breaking inelastic scatterings occurred due to electron–electron [20] and electron–phonon [21,22] interactions, respectively. In the present case of the BO solids and LB films [6–8], the exponent was rather close to $p \sim 1$. The mechanism of the inelastic scattering in these BO solids and LB films may be understood mainly as the electron–electron interactions.

The result indicates that the negative MR for the crystalline BO salts is well understood in terms of the weak 2D localization model, as in the case of the LB films reported previously [6–8]. The observation of the weak 2D localization is also shared by the recent report for polymer PEDOT:PSS, which shows high crystallinity and a 2D nature with a similar range of phase coherence length [23]. PEDOT also has terminal ethylene groups at the end of EDOT moiety, which may cause elastic scattering similar to the BO cases.

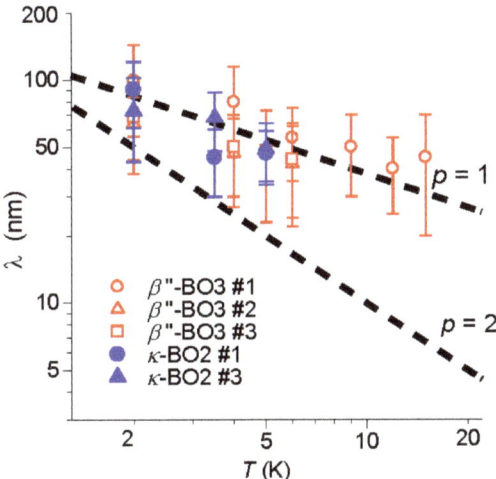

Figure 6. Temperature dependence of the phase coherence length λ for β''-BO3 (sample #1, #2, #3), κ-BO2 (sample #1, #3). The broken lines represent the temperature dependence following $\lambda \propto T^{-p/2}$ with $p = 1$ or $p = 2$.

5. Conclusions

We have studied the low-temperature carrier localization of the charge-transfer salts of BO. The resistivity upturn in the two BO salt crystals can be ascribed to the 2D weak localization with the observation of the negative MR. The present study implies that the 2D weak localization was universally observed in BO charge-transfer salts, regardless of the BO packing and band filling. The electron–electron interaction is considered to be the dominant mechanism for the inelastic scattering. The phenomena demonstrate the formation of the robust 2D metallic state in the BO salts. However, the carrier localization may mask the possible electronic transition such as superconductivity of BO salts. Disorders causing elastic scattering within the metallic domains, such as those of terminal ethylene groups, should be suppressed to prevent the localization.

Supplementary Materials: The following are available online at https://www.mdpi.com/article/10.3390/cryst12010023/s1, Figure S1: A typical fitting result for the magnetoconductance of β''-BO3 #1 along the weak 2D localization model under the perpendicular magnetic field. Figure S2: A typical fitting result for the magnetoconductance of β''-BO3 #2 along the weak 2D localization model under the perpendicular magnetic field. Figure S3: A typical fitting result for the magnetoconductance of β''-BO3 #3 along the weak 2D localization model under the perpendicular magnetic field. Figure S4: A typical fitting result for the magnetoconductance of κ-BO2 #1 along the weak 2D localization model under the perpendicular magnetic field. Figure S5: A typical fitting result for the magnetoconductance of κ-BO2 #3 along the weak 2D localization model under the perpendicular magnetic field. Table S1: Parameters obtained by the fitting shown in Figure S1. Table S2: Parameters obtained by the fitting shown in Figure S2. Table S3: Parameters obtained by the fitting shown in Figure S3. Table S4: Parameters obtained by the fitting shown in Figure S4. Table S5: Parameters obtained by the fitting shown in Figure S5.

Author Contributions: BO molecule synthesis, M.I., A.O. and H.Y.; β''-BO3 crystals preparation, Y.S. and M.Y.; κ-BO2 crystals preparation, Y.Y. and G.S.; measurements and analysis, H.I., M.M., S.K., S.K.Y. and T.T.; manuscript preparation, H.I. All authors have read and agreed to the published version of the manuscript.

Funding: This research was funded by JSPS KAKENHI (grant nos. 19K22127, 20H05867, 20H05664, 20H05862, 20K05448 and 21K04865) and JST CREST (grant no. JPMJCR17I5). M.Y. acknowledges the support by the 111 project (B18030) from China.

Institutional Review Board Statement: Not applicable.

Informed Consent Statement: Not applicable.

Data Availability Statement: Data are available from H.I. upon request.

Conflicts of Interest: The authors declare no conflict of interest.

References

1. Ishiguro, T.; Yamaji, K.; Saito, G. *Organic Superconductors*, 2nd ed.; Springer: Berlin, Germany, 1998.
2. Yamada, J.; Sugimoto, T. *TTF Chemistry*; Kodansha & Springer: Tokyo, Japan, 2004.
3. Horiuchi, S.; Yamochi, H.; Saito, G.; Sakaguchi, K.; Kusunoki, M. Nature and Origin of Stable Metallic State in Organic Charge-Transfer Complexes of Bis(ethylenedioxy)tetrathiafulvalene. *J. Am. Chem. Soc.* **1996**, *118*, 8604–8622. [CrossRef]
4. Beno, M.A.; Wang, H.H.; Kini, A.M.; Carlson, K.D.; Geiser, U.; Kwok, W.K.; Thompson, J.E.; Williams, J.M.; Ren, J.; Whangbo, M.H. The first ambient pressure organic superconductor containing oxygen in the donor molecule, βm-(BEDO-TTF)$_3$Cu$_2$(NCS)$_3$, Tc = 1.06 K. *Inorg. Chem.* **1990**, *29*, 1599–1601. [CrossRef]
5. Kahlich, S.; Schweitzer, D.; Heinen, I.; Lan, S.E.; Nuber, B.; Keller, H.J.; Winzer, K.; Helberg, H.W. (BEDO-TTF)$_2$ReO$_4$(H$_2$O): A new organic superconductor. *Solid State Comm.* **1991**, *80*, 191–195. [CrossRef]
6. Ishizaki, Y.; Izumi, M.; Ohnuki, H.; Lipinska, K.K.; Imakubo, T.; Kobayashi, K. Formation of two-dimensional weak localization in conducting Langmuir-Blodgett films. *Phys. Rev. B* **2001**, *63*, 134201. [CrossRef]
7. Ishizaki, Y.; Izumi, M.; Ohnuki, H.; Imakubo, T.; Kalita-Lipinska, K. Observation of two-dimensional weak localization as a sign of coherent carrier transport in the conducting Langmuir–Blodgett films of BEDO-TTF and stearic acid. *Colloids Surf. A* **2002**, *198–200*, 723–728. [CrossRef]
8. Ito, H.; Tamura, H.; Kuroda, S.; Yamochi, H. Structural and Transport Studies of BEDO-TTF-Arachidic-Acid Conducting Langmuir-Blodgett Films. *Trans. MRS-J* **2005**, *30*, 131–134.
9. Lee, P.A.; Ramakrishnan, T.V. Disordered electronic systems. *Rev. Mod. Phys.* **1985**, *57*, 287–337. [CrossRef]
10. Hikami, S.; Larkin, A.I.; Nagaoka, Y. Spin-Orbit Interaction and Magnetoresistance in the Two Dimensional Random System. *Prog. Theor. Phys.* **1980**, *63*, 707–710. [CrossRef]
11. Ulmet, J.P.; Bachere, L.; Askenazy, S.; Ousset, J.C. Negative magnetoresistance in some dimethyltri-methylene-tetraselenafulvalenium salts: A signature of weak-localization effects. *Phys. Rev. B* **1988**, *38*, 7782–7788. [CrossRef]
12. Prokhorova, T.G.; Simonov, S.V.; Khasanov, S.S.; Zorina, L.V.; Buravov, L.I.; Shibaeva, R.P.; Yagubskii, E.B.; Morgunov, R.B.; Foltynowiczc, D.; S'wietlikc, R. Bifunctional molecular metals based on BEDO-TTF radical cation salts with paramagnetic [M^{III}(CN)$_6$]$^{3-}$ anions, M= Fe, Cr, (Fe$_{0.5}$Co$_{0.5}$). *Synth. Met.* **2008**, *158*, 749–757. [CrossRef]
13. Dubrovakii, A.D.; Spitsina, N.G.; Buravov, L.I.; Shilov, G.V.; Dyachenko, O.A.; Yagubskii, E.B.; Laukhin, V.N.; Canadell, E. New molecular metals based on BEDO radical cation salts with the square planar Ni(CN)$_4{}^{2-}$ anion. *J. Mater. Chem.* **2005**, *15*, 1248–1254.
14. Shen, Y.; Ito, H.; Zhang, H.; Yamochi, H.; Cosquer, G.; Herrmann, C.; Ina, T.; Yoshina, S.K.; Breedlove, B.K.; Otsuka, A.; et al. Emergence of Metallic Conduction and Cobalt(II)-Based Single-Molecule Magnetism in the Same Temperature Range. *J. Am. Chem. Soc.* **2021**, *143*, 4891–4895. [CrossRef] [PubMed]
15. Fettouhi, M.; Ouahab, L.; Serhani, D.; Fabre, J.-M.; Ducasse, L.; Arniell, J.; Canet, R.; Delhaesd, P. Structural and physical properties of BEDO-TTF charge-transfer salts: κ-phase with CF$_3$SO$_3{}^-$. *J. Mater. Chem.* **1993**, *3*, 1101–1107. [CrossRef]
16. Shen, Y.; Ito, H.; Zhang, H.; Yamochi, H.; Katagiri, S.; Yoshina, S.K.; Otsuka, A.; Ishikawa, M.; Cosquer, G.; Uchida, K.; et al. Simultaneous manifestation of metallic conductivity and single-molecule magnetism in a layered molecule-based compound. *Chem. Sci.* **2020**, *11*, 11154–11161. [CrossRef] [PubMed]
17. Fettouhi, M.; Ouahab, L.; Gomez-Garcia, C.; Ducasse, L.; Delhaes, P. Structural and physical properties of κ-(BEDT-TTF)$_2$(CF$_3$SO$_3$). *Synth. Met.* **1995**, *70*, 1131–1132. [CrossRef]
18. Ito, H.; Asai, T.; Shimizu, Y.; Hayama, H.; Yoshida, Y.; Saito, G. Pressure-induced superconductivity in the antiferromagnet κ-(ET)$_2$CF$_3$SO$_3$ with quasi-one-dimensional triangular spin lattice. *Phys. Rev. B* **2016**, *94*, 020503(R). [CrossRef]
19. Ito, H.; Suzuki, D.; Yokochi, Y.; Kuroda, S.; Umemiya, M.; Miyasaka, H.; Sugiura, K.I.; Yamashita, M.; Tajima, H. Charge Carriers in Divalent BEDT-TTF Conductor (BEDT-TTF)Cu$_2$Br$_4$. *Phys. Rev. B* **2005**, *71*, 085202. [CrossRef]
20. Altshuler, B.L.; Aronov, A.G.; Khmelnitsky, D.E. Effects of electron-electron collisions with small energy transfers on quantum localization. *J. Phys. C Solid State Phys.* **1982**, *15*, 7367–7386. [CrossRef]
21. Belitz, D.; Das Sarma, S. Inelastic phase-coherence time in thin metal films. *Phys. Rev. B* **1987**, *36*, 7701–7704. [CrossRef]
22. Sergeev, A.; Mitin, V. Electron-phonon interaction in disordered conductors: Static and vibrating scattering potentials. *Phys. Rev. B* **2000**, *61*, 6041–6047. [CrossRef]
23. Homma, Y.; Itoh, K.; Masunaga, H.; Fujiwara, A.; Nishizaki, T.; Iguchi, S.; Sasaki, T. Mesoscopic 2D Charge Transport in Commonplace PEDOT:PSS Films. *Adv. Electron. Mater.* **2018**, *4*, 1700490.

Review

Simultaneous Control of Bandfilling and Bandwidth in Electric Double-Layer Transistor Based on Organic Mott Insulator κ-(BEDT-TTF)$_2$Cu[N(CN)$_2$]Cl

Yoshitaka Kawasugi [1,*] and Hiroshi M. Yamamoto [2,*]

[1] Department of Physics, Toho University, Funabashi 274-8510, Chiba, Japan
[2] Institute for Molecular Science, National Institutes of Natural Sciences, Okazaki 444-8585, Aichi, Japan
* Correspondence: yoshitaka.kawasugi@sci.toho-u.ac.jp (Y.K.); yhiroshi@ims.ac.jp (H.M.Y.)

Citation: Kawasugi, Y.; Yamamoto, H.M. Simultaneous Control of Bandfilling and Bandwidth in Electric Double-Layer Transistor Based on Organic Mott Insulator κ-(BEDT-TTF)$_2$Cu[N(CN)$_2$]Cl. *Crystals* **2022**, *12*, 42. https://doi.org/10.3390/cryst12010042

Academic Editor: Andrej Pustogow

Received: 4 December 2021
Accepted: 26 December 2021
Published: 28 December 2021

Publisher's Note: MDPI stays neutral with regard to jurisdictional claims in published maps and institutional affiliations.

Copyright: © 2021 by the authors. Licensee MDPI, Basel, Switzerland. This article is an open access article distributed under the terms and conditions of the Creative Commons Attribution (CC BY) license (https://creativecommons.org/licenses/by/4.0/).

Abstract: The physics of quantum many-body systems have been studied using bulk correlated materials, and recently, moiré superlattices formed by atomic bilayers have appeared as a novel platform in which the carrier concentration and the band structures are highly tunable. In this brief review, we introduce an intermediate platform between those systems, namely, a band-filling- and bandwidth-tunable electric double-layer transistor based on a real organic Mott insulator κ-(BEDT-TTF)$_2$Cu[N(CN)$_2$]Cl. In the proximity of the bandwidth-control Mott transition at half filling, both electron and hole doping induced superconductivity (with almost identical transition temperatures) in the same sample. The normal state under electric double-layer doping exhibited non-Fermi liquid behaviors as in many correlated materials. The doping levels for the superconductivity and the non-Fermi liquid behaviors were highly doping-asymmetric. Model calculations based on the anisotropic triangular lattice explained many phenomena and the doping asymmetry, implying the importance of the noninteracting band structure (particularly the flat part of the band).

Keywords: organic conductor; Mott insulator; electric double-layer transistor; uniaxial strain

1. Introduction

The Mott transition, one of the core subjects in condensed matter physics, allows for the observation of intriguing phenomena, such as high-temperature superconductivity, exotic magnetism, pseudogap, and bad-metal behavior [1]. Although the Hubbard model is thought to include the essential physics of these phenomena, a detailed comparison of the model and real materials is lacking because the pristine Mott state is commonly obscured by a complicated band structure. In addition, the control parameter in a real Mott insulator is usually limited to either bandfilling or bandwidth. Moiré superlattices formed by atomic bilayers have recently emerged as a novel platform for correlated electron systems. Twisted bilayer graphene exhibits superconductivity and correlated insulating states [2], and transition metal dichalcogenide heterobilayers provide a correlation-tunable, Mott-insulating state on the triangular lattice [3]. These artificial systems are a powerful tool to understand the fundamental physics of quantum many-body systems. However, the electronic energy scale of these systems is quite different from that of bulk correlated materials, and an intermediate platform between the artificial and highly tunable moiré superlattices and the bulk correlated materials, such as high-T_C cuprates, is invaluable.

In this brief review, we introduce bandfilling- and bandwidth-control measurements in a transistor device based on an organic antiferromagnetic Mott insulator (Figure 1) [4–6]. We fabricated an electric double layer (EDL) transistor [7], which is a type of field-effect transistor, using an organic Mott insulator. Gate voltages induced extra charges on the Mott insulator surface, which resulted in bandfilling shifts. On the other hand, by bending the EDL transistor, the Mott insulator was subjected to strain, which resulted in bandwidth changes. The (gate voltage)-(strain) phase diagram corresponded to the conceptual phase diagram of the Mott insulator in bandfilling-bandwidth 2D space. We experimentally mapped the insulating, metallic, and superconducting phases in the phase diagram. The experimental phase diagram showed that the superconducting phase surrounded the insulating phase with a particularly doping-asymmetric shape. The asymmetry was partly reproducible by calculations based on the Hubbard model on an anisotropic triangular lattice, implying the importance of the noninteracting band structure. We also showed that the normal states in the doped Mott-insulating state exhibited non-Fermi liquid behaviors, probably due to the partial disappearance of the Fermi surface (FS), similarly to the high-T_C cuprates.

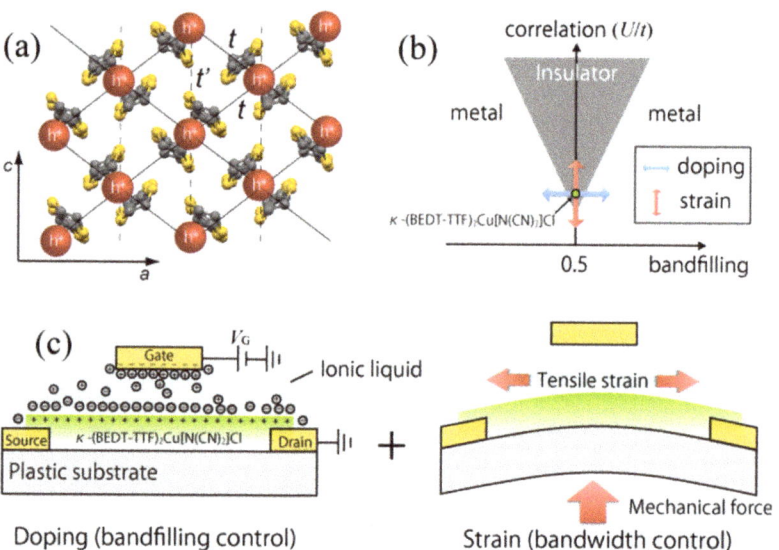

Figure 1. (a) Conducting BEDT-TTF layer in κ-(BEDT-TTF)$_2$Cu[N(CN)$_2$]Cl. (b) Conceptual phase diagram based on the Hubbard model [1]. The vertical axis denotes the strength of the electron correlation. κ-(BEDT-TTF)$_2$Cu[N(CN)$_2$]Cl is located near the tip of the insulating region. (c) Schematic side view of the device structure. Doping concentration and bending strain are controlled by EDL gating and substrate bending with a piezo nanopositioner, respectively.

2. Materials and Methods

2.1. Subject Material: Organic Mott Insulator κ-(BEDT-TTF)$_2$Cu[N(CN)$_2$]Cl

κ-(BEDT-TTF)$_2$Cu[N(CN)$_2$]Cl (BEDT-TTF: bisethylenedithio-tetrathiafulvalene, abbreviated κ-Cl hereinafter) is a quasi-two-dimensional molecular conductor in which the conducting (BEDT-TTF)$_2^+$ layer and the insulating Cu[N(CN)$_2$]Cl$^-$ layer are stacked alternately [8]. The unit cell contains four BEDT-TTF molecules forming four energy bands based on the molecular orbital approximation. Two electrons are transferred to the anion layer, resulting in a 3/4 filled system [9]. However, the BEDT-TTF molecules are strongly dimerized, and the upper two energy bands are sufficiently apart from the remaining two bands, resulting in an effective half-filled system. If we regard the two BEDT-TTF dimers in the unit cell (with different orientations) as equivalent, κ-Cl can be modeled as a half-filled, single-band Hubbard model on an anisotropic triangular lattice: $t'/t = -0.44$ [10], where t is the nearest-neighbor hopping, and t' is the next-nearest-neighbor hopping. Similar to the high-T_C cuprates, the sign of t'/t is negative so that the van Hove singularity lies below the Fermi energy (hole-doped side). However, t' exists only for one diagonal of the dimer sites and accordingly, the FS is elliptical (Figure 2).

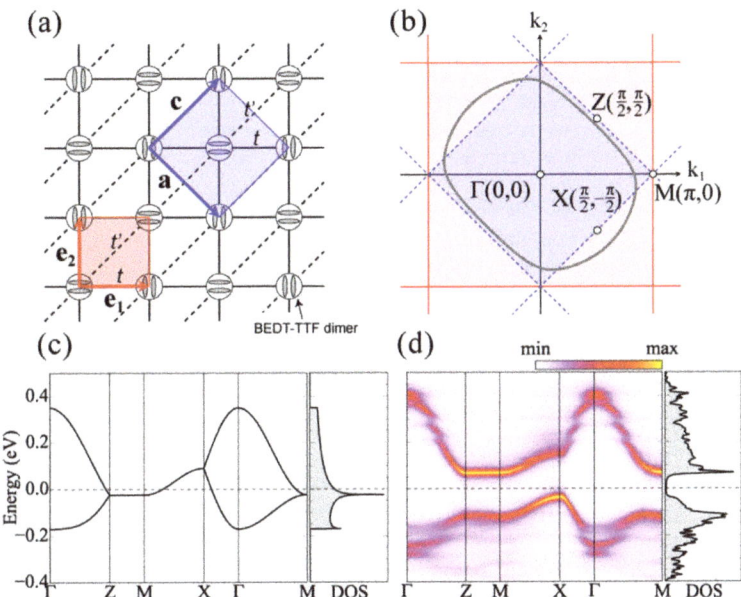

Figure 2. Unit cells, Brillouin zones, band structure, and single-particle spectral functions of κ-Cl. Note that the calculations are based on the one-band model. However, the band structure and the spectral functions are shown in the two-site Brillouin zone [blue shaded area in (**b**)] because the adjacent BEDT-TTF dimers are not completely equivalent in the material. Accordingly, the X, Z, and M points in (**c**) and (**d**) correspond to points $(\pi/2, -\pi/2)$, $(\pi/2, \pi/2)$, and $(\pi, 0)$ in the Brillouin zone of the one-site unit cell. (**a**) Schematic of the anisotropic triangular lattice of κ-Cl. Translational vectors e_1 and e_2 (*a* and *c*) are represented by red (blue) arrows. The red (blue) shaded region represents the unit cell containing one site (two sites). The ellipses on the sites denote the conducting BEDT-TTF dimers. (**b**) The momentum space for the anisotropic triangular lattice. The Brillouin zones of the one- (two-) site unit cell are represented by the red (blue) shaded region bounded by the red solid (blue-dashed) lines. The solid gray line indicates the FS. (**c**) Noninteracting, tight-binding band structure along highly symmetric momenta and density of states (DOS) of κ-Cl ($t'/t = -0.44$ with $t = 65$ meV). The Fermi level for half filling is set to zero and denoted by the dashed lines. (**d**) Single-particle spectral functions and DOS of κ-Cl at half filling in the antiferromagnetic state at zero temperature, calculated by variational cluster approximation [5]. The Fermi level is denoted by the dashed lines at zero energy. Reproduced with permission from [6].

Because of the narrow bandwidth and the half filling, κ-Cl ($U/t = 5.5$ [10]) is an antiferromagnetic Mott insulator at low temperatures due to the on-site Coulomb repulsion [11]. The material is in close proximity to the bandwidth-control Mott transition. When low hydrostatic pressure is applied (~20 MPa), κ-Cl exhibits the first-order Mott transition to a Fermi liquid/superconductor ($T_C \sim 13$ K). κ-(BEDT-TTF)$_2$Cu[N(CN)$_2$]Br ($U/t = 5.1$ [10]), which is a derivative with a slightly larger t, is also a Fermi liquid/superconductor. The transition has been thoroughly investigated using precise pressure control, such as continuously controllable He gas pressure [12] and chemical pressure by deuterated BEDT-TTF in κ-(BEDT-TTF)$_2$Cu[N(CN)$_2$]Br [13]. As T_C is relatively high for the low Fermi temperatures, the bandwidth-control Mott/superconductor transition in κ-Cl is sometimes regarded as a counterpart to the bandfilling-control Mott-insulator/superconductor transitions in the high-T_C cuprates [14].

κ-Cl has a few hole-doped derivatives. κ-(BEDT-TTF)$_4$Hg$_{2.89}$Br$_8$ has a large U/t (nearly twice that of κ-Cl) but shows metallic conduction and superconductivity because of ~11% hole doping [15]. The transport properties [16,17] are reminiscent of high-T_C cuprates;

they show linear-in-temperature resistivity above the Mott–Ioffe–Regel limit (bad-metal behavior) and Hall coefficients inconsistent with the noninteracting FS. Applying pressure reduces U/t, and the temperature dependence of the resistivity approaches that of a Fermi liquid. If the doping concentration is precisely controllable, we would be able to obtain the desired bandwidth–bandfilling phase diagram. However, the doping concentration is fixed, and the doped derivatives are limited (only 11% [15] and 27% [18]).

2.2. Experimental Method for Bandfilling Control: EDL Doping

To control the bandfilling of κ-Cl, we employed a doping method based on the EDL transistor [Figure 3a]. The EDL transistor is a type of field-effect transistor in which the gate-insulating film is replaced by a liquid electrolyte such as an ionic liquid. EDL doping enables a higher doping concentration than the typical field-effect doping using a solid gate insulator due to the strong electric fields by the liquid electrolyte. First, we prepared polyethylene terephthalate (PET) substrates and patterned Au electrodes (source, drain, voltage-measuring electrodes, and side-gate electrodes) using photolithography. Next, we synthesized thin single crystals of κ-Cl by electrolysis of a 1,1,2-trichloroethane [10% (v/v) ethanol] solution in which BEDT-TTF (20 mg), TPP[N(CN)$_2$] [tetraphenylphosphonium (TPP), 200 mg], CuCl (60 mg), and TPP-Cl (100 mg) were dissolved. We applied 8 μA current overnight and obtained tiny thin crystals of κ-Cl. However, we could not easily remove the thin crystals from the solution because the surface tension of the solution easily broke the crystals. We therefore moved the crystals together with a small amount of the solution by pipetting them into 2-propanol (an inert liquid). Then, using the tip of a hair shaft, we manipulated one crystal and placed it on the substrate in 2-propanol. After the substrate with the κ-Cl single crystal was taken out from 2-propanol and dried, the crystal tightly adhered to the substrate (probably via electrostatic force). The κ-Cl single crystal was shaped into a Hall bar using a pulsed laser beam at the wavelength of 532 nm [Figure 3b]. Lastly, we added a droplet of 1-ethyl-3-methylimidazolium 2-(2-methoxyethoxy) ethyl sulfate ionic liquid on the sample and the Au side-gate electrode and placed a 1.2-μm-thick polyethylene naphthalate (PEN) film on it to make the liquid phase thin. The thinning of the gate electrolyte using the PEN film reduced the thermal stress at low temperatures. We immediately cooled the sample to 220 K (~3 K/min), where the ionic liquid was less reactive. At lower temperatures, the ionic liquid solidified.

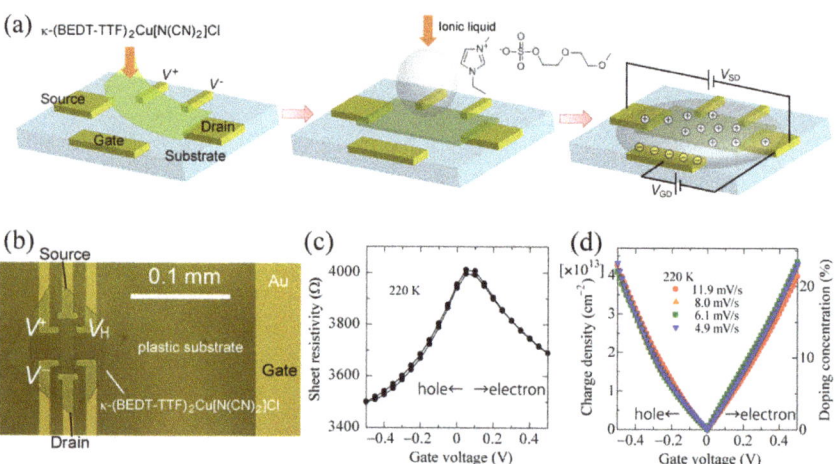

Figure 3. (**a**) Schematic view of device fabrication procedure. (**b**) Optical top view of an EDL transistor device. The κ-Cl crystal is laser-shaped into a Hall bar. (**c**) Gate voltage dependence of sheet resistivity at 220 K. (**d**) Gate-voltage dependence of accumulated charge density and doping concentration.

We controlled the doping concentration of the κ-Cl crystal surface by varying the gate voltage, V_G, at 220 K. Both the positive and negative gate voltages, corresponding to the electron and hole doping, reduced the sample resistance, implying the deviation of the bandfilling from 1/2 [Figure 3c]. According to the charge displacement current measurements [19], the doping concentration reached approximately ±20% at V_G of ±0.5 V [Figure 3d]. Increasing the gate voltage ($|V_G| > 0.7$ V) led to the irreversible increase of resistance, indicating sample degradation due to chemical reactions. The choice of ionic liquid was important; the crystal immediately disappeared when we employed ionic liquids that were too reactive or too good for the solubilization of κ-Cl. Diethylmethyl(2-methoxyethyl)ammonium bis(trifluoromethylsulfonyl)imide [DEME-TFSI], a typical ionic liquid for EDL doping, was suitable for electron doping but not for hole doping at low temperatures. At the moment, 1-ethyl-3-methylimidazolium 2-(2-methoxyethoxy) ethyl sulfate is the best choice. We focused on the doping effect on this ionic liquid.

2.3. Experimental Method for Bandwidth Control: Uniaxial Bending Strain via Substrate

We usually control the bandwidth of a molecular conductor by applying hydrostatic pressure using a pressure medium oil and a pressure cell. However, because the heterogeneous device structure was unsuitable for hydrostatic pressure application, we adopted the strain effect caused by substrate bending, as shown in Figure 4. This method required no liquid pressure medium (the ionic liquid is already on the crystal) and enabled precise strain control by fine tuning the piezo nanopositioner that bent the substrate (Figure 4). Assuming that the bent substrate is an arc of a circle (angle: 2θ, curvature radius: r), strain S is estimated as

$$S = \frac{2\theta(r + d/2) - 2\theta r}{2\theta r} = \frac{\theta d}{l}$$

where d and l are the thickness and the length of the substrate, respectively. The relationship between the sides of the shaded triangle in Figure 4 gives

$$r \sin\theta \sim r\theta = \sqrt{r^2 - (r-x)^2}, \therefore \theta = \frac{4lx}{l^2 + 4x^2}$$

using the small angle approximation, where x is the displacement of the piezo nanopositioner. As a result,

$$S = 4dx/\left(l^2 + 4x^2\right).$$

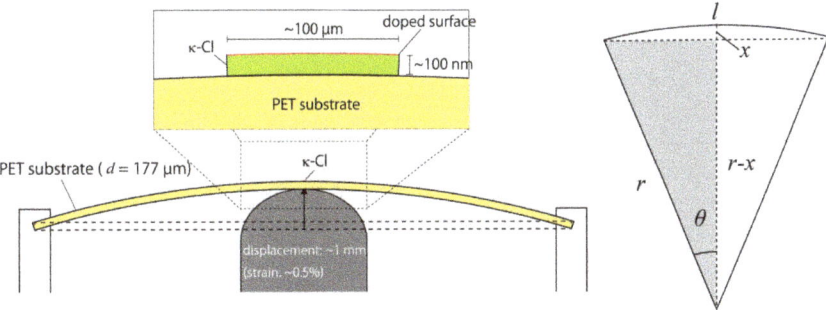

Figure 4. Schematic illustration for the application of uniaxial tensile strain to the κ-Cl sample.

We employed PET substrates with $d = 177$ μm and $l = 12$ mm, and x was up to 2.5 mm so that the typical value of S was ~1%. Note that the strain in this experimental setup was tensile and uniaxial. As the strain generated by bending was tensile, we employed a PET substrate with a large thermal expansion coefficient to start the bandwidth scanning from the superconducting region. Biaxial compression of the κ-Cl crystal by the substrate at low temperatures resulted in the superconducting state without bending strain. Therefore, we applied the bending tensile strain to enhance U/t and induce the bandwidth-control Mott transition from the metallic/superconducting side to the insulating side. The strain effect should be dependent on the strain direction. However, we leave the detailed direction dependence to future work because the strain-induced Mott transition could be observed regardless of the strain direction at the moment.

3. Results

First, we introduced the superconducting phase transitions around the tip of the Mott-insulating state in the bandwidth–bandfilling phase diagram in Section 3.1 [4]. Then, we showed the transport properties under EDL doping at a large U/t in Section 3.2 [5,6] (we did not apply the bending strain shown in Section 2.3 here).

3.1. Superconducting Phase around the Mott-Insulating Phase

3.1.1. Strain Effect without Gate Voltage

First, we showed the strain effect without the gate voltage (Figure 5). As mentioned in the Methods section, κ-Cl became a superconductor without bending strain owing to the thermal contraction of the substrate. T_C (~12 K) was similar to that of the bulk κ-Cl crystal under low hydrostatic pressure. The superconducting state disappeared upon applying the uniaxial tensile strain, S, and the insulating state appeared at low temperatures. Despite the uniaxiality, the temperature dependence of the resistivity was qualitatively similar to that in the bulk κ-Cl crystal under hydrostatic pressure [Figure 5c]. The slope of the metallic/insulating phase in the phase diagram implied that the insulating state at the lowest temperatures had low entropy and was the antiferromagnetic Mott-insulating state. Thus, we could control the bandwidth (and consequently U/t) of the sample across the bandwidth-control Mott transition at half filling.

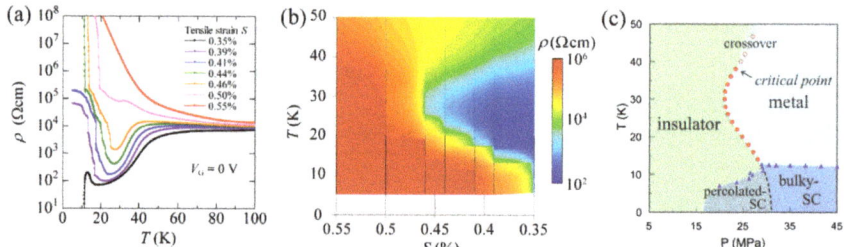

Figure 5. (a) Resistivity vs. temperature plots under different tensile strains at gate voltage $V_G = 0$ V, and (b) contour plots of the resistivity data. (c) Pressure-temperature phase diagram of bulk κ-Cl. Reproduced with permission from [12].

3.1.2. Doping Effect at Fixed Strain

Next, we fixed the uniaxial tensile strain at the very tip of the insulating state in the bandwidth–bandfilling phase diagram ($S = 0.41\%$) and applied gate voltages (with warming of the sample to 220 K, changing V_G, and cooling again). Both electron and hole doping reduced the resistivity and induced the superconducting state, as shown in Figure 6. T_C was similar (~12 K) among the electron-doped and hole-doped states (and the undoped metallic state). However, the doping effect was highly asymmetric against the polarity of V_G. By hole doping, the resistivity monotonically decreased, and superconductivity emerged for $V_G \leq -0.3$ V [approximately 10% hole doping according to Figure 3d]. On the other hand, the resistivity abruptly dropped, and a superconducting state emerged with low electron doping ($+0.14$ V $\leq V_G \leq +0.22$ V, approximately $4 \sim 7\%$ electron doping). At the phase boundary, the resistivity discretely fluctuated [5]. After further electron doping, the resistivity increased again, and superconductivity disappeared. Interestingly, the normal-state resistivity at $T > T_C$ also decreased first and increased thereafter with electron doping.

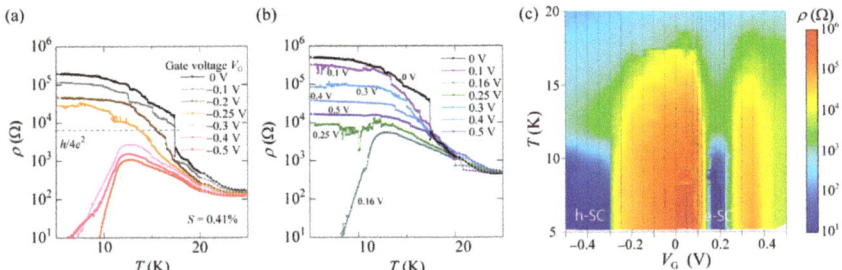

Figure 6. Sheet resistivity vs. temperature plots under (**a**) hole doping and (**b**) electron doping at tensile strain $S = 0.41\%$. The dashed line indicates pair quantum resistance $h/4e^2$. (**c**) Contour plots of the data in (**a**,**b**). h-SC and e-SC denote hole-doped superconductor and electron-doped superconductor, respectively.

3.1.3. Gate Voltage vs. Strain Phase Diagram

After obtaining the resistivity vs. gate voltage data at the fixed strain, we slightly increased the strain and repeated the same cycles, as shown in Figure 7a. These measurements resulted in the gate voltage vs. strain phase diagram at low temperatures, which corresponded (although not proportionally) to the bandfilling–bandwidth phase diagram, as shown in Figure 7b. The insulating phase was triangular on the hole-doped side (left), similar to the conceptual phase diagram of a Mott insulator, and the superconducting phase surrounded the insulating phase. On the other hand, the superconducting phase appeared to "penetrate" into the insulating phase on the electron-doped side (right).

As shown in Figure 8, the doping asymmetry was qualitatively reproduced by variational cluster approximation (VCA) calculations of the antiferromagnetic and superconducting order parameters in a Hubbard model on an anisotropic triangular lattice. At low U/t, the superconducting (antiferromagnetic) order parameter more drastically increased (decreased) by electron doping than hole doping [Figure 8a]. The doping dependence of the chemical potential was nonmonotonic only on the electron-doped side [Figure 8b], implying the possibility of a phase separation between the Mott-insulating and superconducting phases [Figure 8c]. The nonmonotonic behavior of the chemical potential seemed to have originated from the flat part at the bottom of the upper Hubbard band (along the Z–M axis), which was originally located below the Fermi energy in the noninteracting energy band. The flat part of the energy band was caused by the absence of t' along the crystallographic a-axis, namely, by the nature of the triangular lattice.

Notice that the experimental results are not understood uniquely within the half-filled band scenario. Calculations on a more detailed quarter-filled band model for the κ-BEDT-TTF salts also predict the doping-induced superconductivity, where the doping polarity alters the pairing symmetry (electron doping: extended $s + d_{x^2-y^2}$, hole doping: d_{xy}) [20]. In addition, many quantum Monte Carlo calculations on the Hubbard model indicate the absence of superconductivity at near half filling [21,22], while superconductivity is predicted near quarter filling [23].

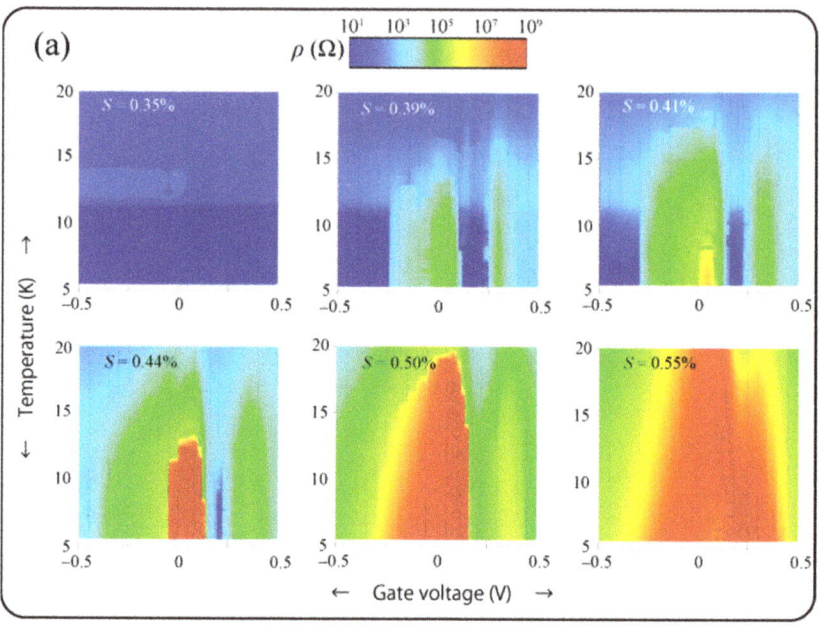
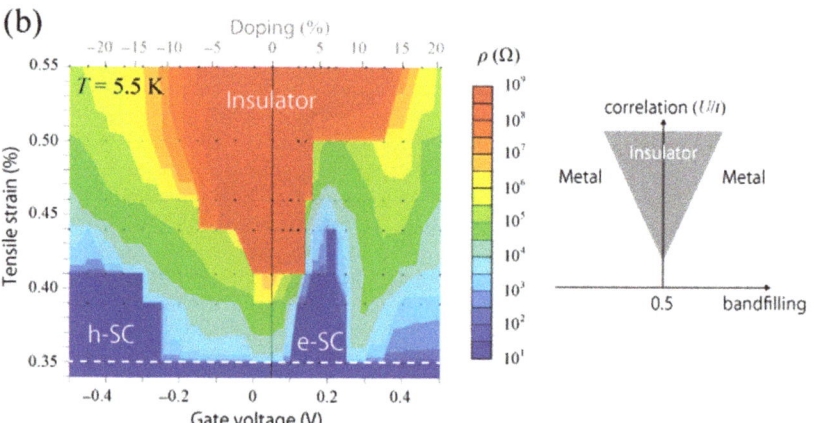

Figure 7. (**a**) Contour plots of sheet resistivity, ρ, under tensile strains, S, of 0.35%, 0.39%, 0.41%, 0.44%, 0.50%, and 0.55% as a function of temperature and gate voltage. (**b**) Contour plots of sheet resistivity, ρ, at 5.5 K as a function of gate voltage and tensile strain (left) and the corresponding conceptual phase diagram (right). Black dots in all figures indicate the data points where the sheet resistivity was measured. The doping concentration estimated from the average density of charge accumulated in the charge displacement current measurement [Figure 3d] is shown for reference on the upper horizontal axis in (**b**).

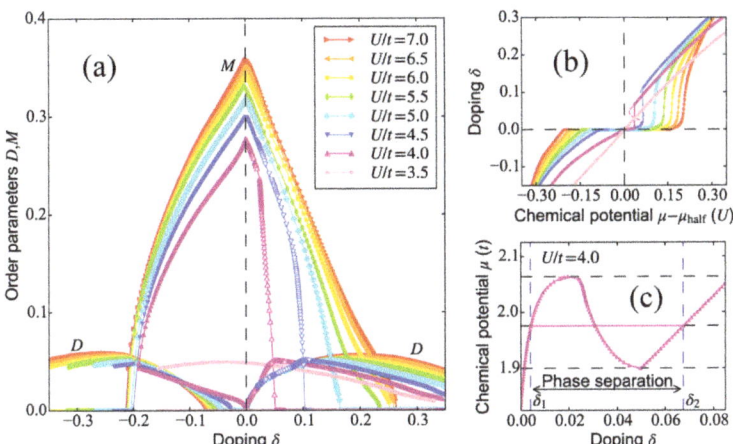

Figure 8. VCA calculations. (**a**) Antiferromagnetic and $dx^2 - dy^2$ superconducting order parameters, M and D, respectively, vs. doping concentration, δ, for several values of U/t. M and D for metastable and unstable solutions (empty symbols) at $U/t = 4$ and 4.5 under electron doping (corresponding to positive d) are also shown. (**b**) Doping concentration, δ, vs. chemical potential, μ, relative to that at half filling (μ_{half}) for several values of U/t. The results for metastable and unstable solutions at $U/t = 4$ and 4.5 are indicated by dashed lines, and the results obtained by the Maxwell construction are denoted by solid vertical lines. The results imply the presence of phase separation and a first-order phase transition. It is noteworthy that there is a steep (nearly vertical) increase in δ with increasing μ for larger values of U/t under electron doping, suggesting a strong tendency toward phase separation. (**c**) Chemical potential, μ, vs. doping concentration, δ, for $U/t = 4$. δ_1 and δ_2 are the doping concentrations of the two extreme states in the phase separation. All results in (**a**) to (**c**) were calculated using VCA for the single-band Hubbard model on an anisotropic triangular lattice ($t'/t = -0.44$) with a 4×3 cluster.

3.2. Non-Fermi Liquid Behaviors in the Normal State under Doping

Non-Fermi liquid behaviors, such as the metallic-like resistivity above the Mott–Ioffe–Regel limit and the Hall coefficient inconsistent with the volume of the FS, are ubiquitous features of the normal state of many strongly correlated materials. Here we show that our Mott EDL transistor also exhibited such behaviors.

Due to the principle of the EDL transistor, only the sample surface was doped. However, the nondoped region of the sample was also conductive at high temperatures in κ-Cl. Therefore, to discuss the non-Fermi liquid behaviors at high temperatures, we extracted surface resistivity, ρ_s, and surface Hall coefficient, R_{Hs}. Assuming a simple summation of the conductivity tensors of two parallel layers (surface monolayer and remaining bulk layers), we derived ρ_s and R_{Hs} from

$$\rho_s = \frac{L}{W}\left(\frac{1}{\rho_{measured}} - \frac{1}{\rho_{0V}} \times \frac{N-1}{N}\right)^{-1}$$

$$R_{Hs} = \rho_s^2 \left(\frac{R_{H\,measured}^2}{\rho_{measured}^2} - \frac{R_{H\,0V}^2}{\rho_{0V}^2} \times \frac{N-1}{N}\right)$$

where L, W, and N denote the length, width, and number of conducting layers, respectively. Suffixes "measured" and "0 V" stand for the actual measured (combined) values and the values at 0 V (nondoped values), respectively. A sample with a larger U/t (using the PEN substrate that had less thermal contraction) than the previous superconducting sample was measured, and uniaxial tensile strain was not applied.

3.2.1. Temperature Dependence of the Resistivity

Figure 9a shows the temperature dependence of the surface resistivity, ρ_s, under electron doping. Without the gate voltage, the system was insulating at all measured temperatures (2–200 K). Upon low electron doping ($V_G \sim 0.1$ V, >3% electron doping), metallic-like conduction ($d\rho_s/dT > 0$) above the Mott–Ioffe–Regel limit, ρ_{MIR}, ($\sim h/e^2$, assuming a two-dimensional isotropic FS) appeared at high temperatures even though the system remained insulating at the lowest temperatures. Although the resistivity was not linear-in-temperature (between linear and quadratic) in this temperature range, this was a bad-metal behavior in the sense that the mean free path of carriers was shorter than the site distance. At $V_G = 0.34$ V, the resistivity below 50 K also exhibited an insulator-metal crossover across ρ_{MIR}. For $V_G > 0.5$ V, the temperature dependence of ρ_s approached a Fermi liquid (quadratic in temperature) below 20 K.

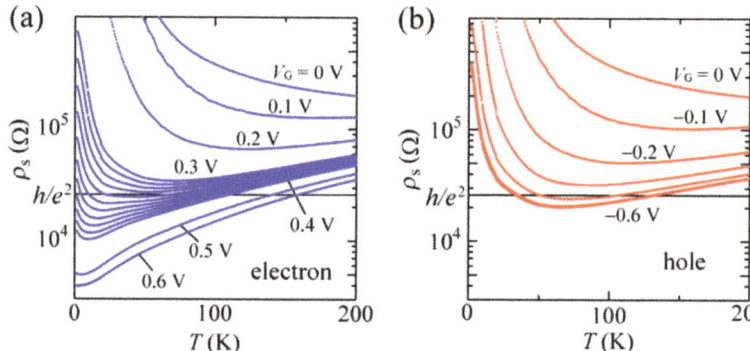

Figure 9. Temperature T dependence of surface resistivity, ρ_s, under (**a**) electron doping and (**b**) hole doping.

Hole doping also induced the bad-metal behavior at high temperatures, as shown in Figure 9b. Although an accurate estimation of the power-law exponent was difficult, the temperature dependence of ρ_s appeared more linear in temperature than in the case of electron doping. The temperature dependence was consistent with the linear-in-temperature resistivity in the hole-doped compound, κ-(BEDT-TTF)$_4$Hg$_{2.89}$Br$_8$, at high temperatures [16,17]. However, we could not observe metallic conduction or the Fermi liquid behavior at low temperatures down to V_G of -0.6 V. Thus, at high temperatures, the bad-metal behavior emerged in a wide doping range except at $V_G = 0$ V, whereas the Fermi liquid state at low temperatures appeared only under high electron doping.

3.2.2. Hall Coefficient

In the case of a Fermi liquid with a single type of carrier, $1/e|R_H|$ (R_H: Hall coefficient) would denote the carrier density corresponding to the volume enclosed by FS [24] and should be independent of temperature. On the contrary, temperature-dependent R_H, which was inconsistent with the volume of the noninteracting FS, often appeared in the normal state of strongly correlated materials. In the noninteracting single-band picture, κ-Cl had a large hole-like FS so that $1/eR_H$ in the metallic state should be $+p(1-\delta)$, where p and δ are the half-filled hole density per layer and the electron doping concentration, respectively. Figure 10b shows the V_G dependence of $1/eR_H$ at 40 K. Near the charge-neutrality point (Mott-insulating state), we could not observe the distinct Hall effect due to the high resistivity. Upon electron doping, the Hall coefficients became measurable and were positive despite the doped electrons. $1/eR_H$ appeared to obey $+p(1-\delta)$ under sufficient electron doping, indicating that the Mott-insulating state collapsed, and the system approached the metallic state based on the noninteracting band structure. However, upon hole doping, $1/eR_H$ became much less than $+p(1-\delta)$. The values also differed from the externally doped hole density, $-p\delta$, implying that the Mott-insulating state collapsed by hole doping, but the system approached a different electronic state with a smaller FS than the noninteracting case.

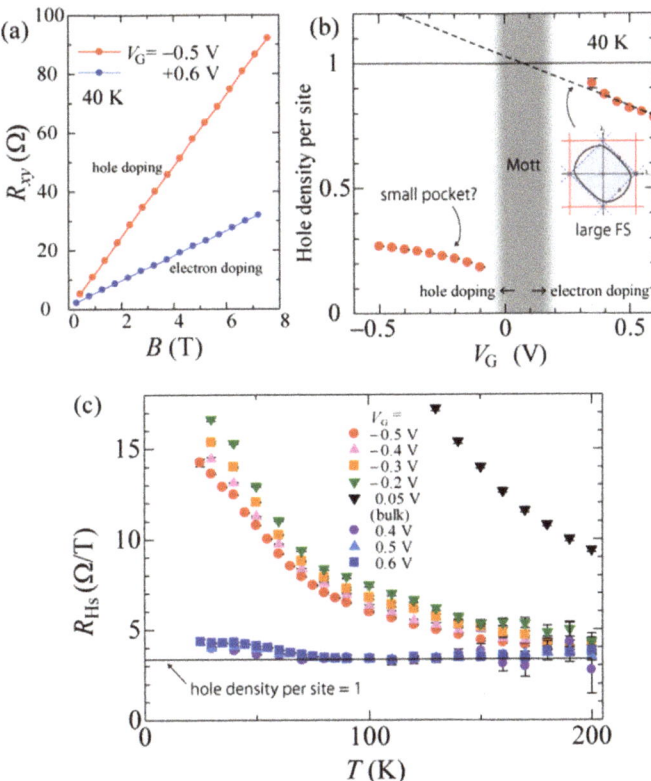

Figure 10. (**a**) Hall resistance vs. magnetic field at 40 K. (**b**) Gate-voltage dependence of the hole density per site (estimated from $1/eR_H$) at 40 K. The dashed line denotes the hole density per site estimated from the volume bounded by the noninteracting FS assuming that doping concentration is proportional to V_G (20% doping at 0.5 V). The center of the shaded insulating region corresponds to the charge neutrality point (resistivity peak). (**c**) Temperature dependence of R_{Hs}. The solid line indicates the value where the hole density per site becomes one (half filling).

The temperature dependence of R_{Hs} also revealed the peculiarity of the hole-doped state, as shown in Figure 10c. R_{Hs} was almost temperature-independent under electron doping, as expected for a conventional metal. By contrast, R_{Hs} under hole doping monotonically decreased with an increasing temperature, approaching values similar to those under electron doping.

3.2.3. Resistivity Anisotropy

In-plane conductivity anisotropy also reflected the anomalous state under hole doping. Figure 11 shows the in-plane anisotropy of the surface resistivity, ρ_c/ρ_a, up to 200 K. Here, ρ_c (ρ_a) denotes the surface resistivity along the c axis [a axis; the short axis of the elliptical FS is parallel to the c axis, as shown in Figure 2b]. Under electron doping, the resistivity was almost isotropic ($\rho_c/\rho_a \sim 1$) and independent of temperature. Under hole doping, by contrast, ρ_c/ρ_a was distinctly larger than one and increased with cooling. The conduction along the c-axis diminished in the hole-doped state, and its origin was weakened at high temperatures.

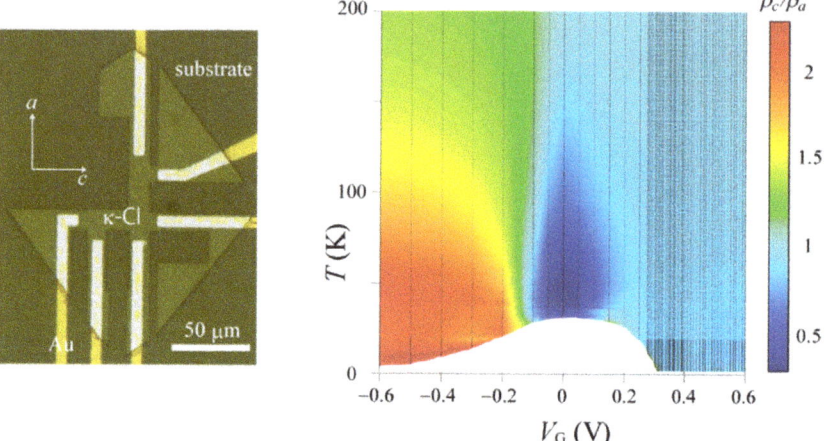

Figure 11. Temperature dependence of in-plane anisotropy of surface resistivity (note that both a and c axes are parallel to the conducting plane in this material). Data are missing at low temperatures and low doping (white region in the right panel) due to the high resistance.

3.2.4. Single-Particle Spectral Functions

We compared the above experimental results with model calculations. Figure 12 shows the single-particle spectral functions (corresponding to the DOS) of the Hubbard model on an anisotropic triangular lattice at 30 K using the cluster perturbation theory (CPT). The Mott-insulating state [Figure 12b,e] was reproduced at half filling, where the energy gap opened at all the k-points. When 17% of the electrons were doped [Figure 12c,f], the noninteracting-like FS emerged. On the other hand, the topology of FS under 17% hole doping [Figure 12a,d] appeared different from the noninteracting case. The spectral weight near the Z point was strongly suppressed (pseudogap), and a lens-like small hole pocket remained. The partial disappearance of FS was notable in the lightly hole-doped cuprates.

The calculated FS provides insights into the origin of the doping asymmetry of the Hall effect and the resistivity anisotropy. Sufficient electron doping reconstructed the noninteracting-like large hole FS, resulting in $1/eR_H \sim p(1-\delta)$. On the other hand, hole doping induced the partially suppressed lens-like small FS. In this state, Luttinger's theorem seemed violated, and R_H could no longer be simply estimated. However, it was possible that R_H was predominantly governed by quasiparticles with a relatively long lifetime (bright points of the spectral function in the reciprocal space in Figure 12), resulting in similar values of $1/eR_H$ corresponding to the area of the lens-like hole pocket. In addition, the large resistivity anisotropy under hole doping could be simply explained by the suppression of the quasi-one-dimensional FS along the Z–M line, which contributed to the conduction along the c-axis. As shown in Figure 12g, the spectral density on the Z–M line was recovered at high temperatures, consistent with the tendency in the Hall and anisotropy measurements.

The suppression of the spectral weight near the Z–M line under hole doping seemed to be related to the van Hove singularity (the van Hove critical points lie on the Z–M axis). With the doping of holes, FS approached the van Hove singularity, and the effect of the interaction was expected to be enhanced. It was also revealed that the spin fluctuation was stronger in the hole-doped state because of the van Hove singularity. By contrast, FS departed from the van Hove singularity with the doping of electrons, resulting in a weaker interaction effect and a more noninteracting-like FS.

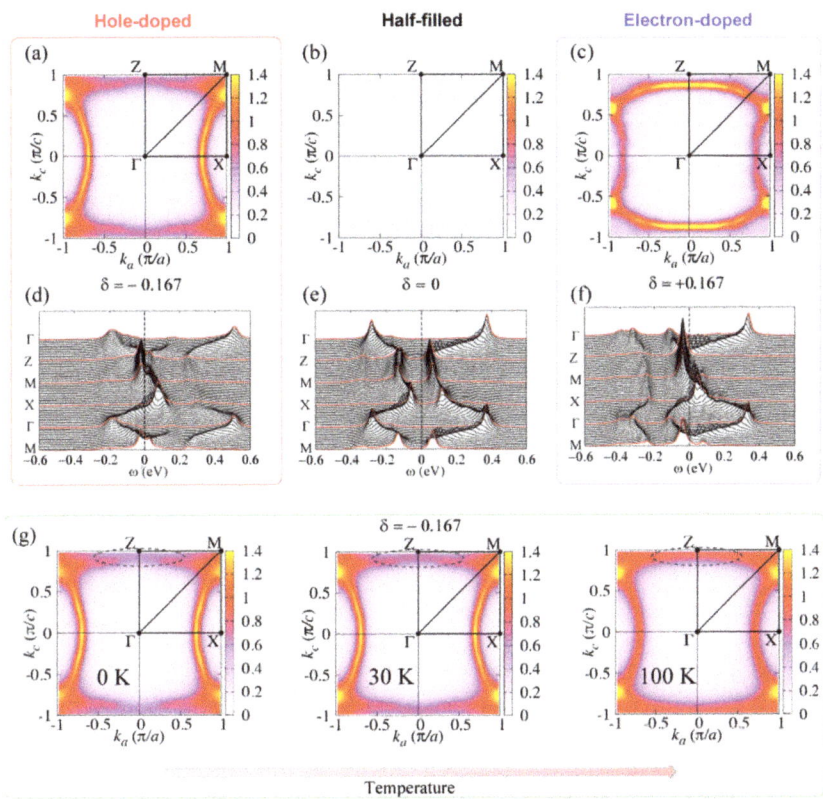

Figure 12. Fermi surfaces and single-particle spectral functions of the Hubbard model on an anisotropic triangular lattice at 30 K, calculated using the cluster perturbation theory (CPT). (**a–c**) Fermi surfaces for (**a**) 17% hole doping, (**b**) half filling, and (**c**) 17% electron doping, determined by the largest spectral intensity at the Fermi energy. (**d–f**) Single-particle spectral functions for (**d**) 17% hole doping, (**e**) half filling, and (**f**) 17% electron doping. The Fermi energy is located at $\omega = 0$ and the parameter set of this model is $t'/t = -0.44$, $U/t = 5.5$, and $t = 65$ meV. (**g**) Temperature evolution of the spectral density at 17% hole doping. The suppression near the Z–M line diminishes at high temperatures.

4. Summary

We fabricated bandfilling- and bandwidth-tunable EDL transistor devices using a flexible organic Mott insulator with a simple band structure. As shown in Section 3.1, drastic resistivity changes ranging from superconducting to highly insulating ($\rho > 10^9$ Ω) states occurred in the proximity of the tip of the Mott-insulating phase in the phase diagram. The superconducting phase surrounded the Mott-insulating phase in the bandfilling−bandwidth phase diagram. The superconducting transition temperature, T_C, was almost identical among the electron-doped, hole-doped, and nondoped states, in contrast to those of the high-T_C cuprates. Model calculations based on the anisotropic triangular lattice qualitatively reproduced the doping asymmetry on the doping levels for superconductivity and the tendency towards phase separation under electron doping, implying the significance of the flat part of the upper Hubbard band (originating from the noninteracting band structure). However, the calculations did not reproduce the reentrance into the slightly insulating state beyond the electron-doped superconducting state. One possibility was that a magnetic or charge-ordered state emerged at specific doping levels (for example, ~12.5%), as in the case of the stripe order in the cuprates [25–28].

At high U/t where no superconductivity was observed, the transport properties exhibited non-Fermi liquid behaviors, as shown in Section 3.2. At high temperatures (above $T \sim 100$ K), the metallic-like conduction above the Mott–Ioffe–Regel limit (the bad-metal behavior) widely emerged regardless of the doping polarity, supporting the universality of the bad-metal behavior near the Mott transitions. At lower temperatures, the anomalously large, temperature-dependent Hall coefficient and in-plane resistivity anisotropy appeared under sufficient hole doping. Model calculations of the spectral density explained the anomaly under hole doping in terms of the partial disappearance of FS (pseudogap) due to the approach of the Fermi energy to the flat part of the energy band (same part as the flat part of the upper Hubbard band before doping). The pseudogap state also appeared in the lightly doped high-T_C cuprates. However, the location of the pseudogap in k-space differed, owing to the difference of the noninteracting band structure, resulting in the different doping asymmetry.

The experimental methods shown here are applicable to other molecular conductors, including κ-(BEDT-TTF)$_2$Cu$_2$(CN)$_3$, which is a genuine Mott insulator without antiferromagnetic ordering [29]. The same experiments on this material may reveal more universal behaviors of a doped Mott insulator. Similar experiments on the molecular Dirac fermion system [30] are also possible and of great interest.

Author Contributions: Conceptualization, Y.K. and H.M.Y.; methodology, Y.K. and H.M.Y.; data curation, Y.K.; writing—original draft preparation, Y.K.; writing—review and editing, H.M.Y.; visualization, Y.K.; supervision, H.M.Y.; funding acquisition, Y.K. and H.M.Y. All authors have read and agreed to the published version of the manuscript.

Funding: This research was funded by MEXT and JSPS KAKENHI, Grant Numbers JP16H06346, JP19K03730, JP19H00891.

Data Availability Statement: All data needed to draw the conclusions in the paper are presented in the paper. Additional data related to this paper may be requested from the authors.

Acknowledgments: We would like to acknowledge R. Kato, K. Seki, S. Yunoki, J. Pu, T. Takenobu, S. Tajima, N. Tajima, and Y. Nishio for the collaborations and valuable discussions. The VCA and CPT calculations shown here were performed by K. Seki and S. Yunoki. This research was supported by MEXT and JSPS KAKENHI, Grant Numbers JP16H06346, JP19K03730, JP19H00891.

Conflicts of Interest: The authors declare no conflict of interest.

References

1. Imada, M.; Fujimori, A.; Tokura, Y. Metal-insulator transitions. *Rev. Mod. Phys.* **1998**, *70*, 1039–1263. [CrossRef]
2. Cao, Y.; Fatemi, V.; Fang, S.; Watanabe, K.; Taniguchi, K.; Kaxiras, E.; Jarillo-Herrero, P. Unconventional superconductivity in magic-angle graphene superlattices. *Nature* **2018**, *556*, 43–50. [CrossRef]
3. Li, T.; Jiang, S.; Li, L.; Zhang, Y.; Kang, K.; Zhu, J.; Watanabe, K.; Taniguchi, T.; Chowdhury, D.; Fu, L.; et al. Continuous Mott transition in semiconductor moiré superlattices. *Nature* **2021**, *597*, 350–354. [CrossRef] [PubMed]
4. Kawasugi, Y.; Seki, K.; Pu, J.; Takenobu, T.; Yunoki, S.; Yamamoto, H.M.; Kato, R. Electron–hole doping asymmetry of Fermi surface reconstructed in a simple Mott insulator. *Nat. Commun.* **2016**, *7*, 12356. [CrossRef]
5. Kawasugi, Y.; Seki, K.; Tajima, S.; Pu, J.; Takenobu, T.; Yunoki, S.; Yamamoto, H.M.; Kato, R. Two-dimensional ground-state mapping of a Mott-Hubbard system in a flexible field-effect device. *Sci. Adv.* **2019**, *5*, eaav7278. [CrossRef] [PubMed]
6. Kawasugi, Y.; Seki, K.; Pu, J.; Takenobu, T.; Yunoki, S.; Yamamoto, H.M.; Kato, R. Non-Fermi-liquid behavior and doping asymmetry in an organic Mott insulator interface. *Phys. Rev. B* **2019**, *100*, 115141. [CrossRef]
7. Ueno, K.; Shimotani, H.; Yuan, H.; Ye, J.; Kawasaki, M.; Iwasa, Y. Field-induced superconductivity in electric double layer transistors. *J. Phys. Soc. Jpn.* **2014**, *83*, 032001. [CrossRef]
8. Williams, J.M.; Kini, A.M.; Wang, H.H.; Carlson, K.D.; Geiser, U.; Montgomery, L.K.; Pyrka, G.J.; Watkins, D.M.; Kommers, J.M.; Boryschuk, S.J.; et al. From semiconductorsemiconductor transition (42 K) to the highest-T_c organic superconductor, κ-(ET)$_2$Cu[N(CN)$_2$]Cl (T_c = 12.5 K). *Inorg. Chem.* **1990**, *29*, 3272–3274. [CrossRef]
9. Mori, T.; Mori, H.; Tanaka, S. Structural Genealogy of BEDT-TTF-Based Organic Conductors II. Inclined Molecules: θ, α, and κ Phases. *Bull. Chem. Soc. Jpn.* **1999**, *72*, 179–197. [CrossRef]
10. Kandpal, H.C.; Opahle, I.; Zhang, Y.-Z.; Jeschke, H.O.; Valentí, R. Revision of Model Parameters for κ-Type Charge Transfer Salts: An Ab Initio Study. *Phys. Rev. Lett.* **2009**, *103*, 067004. [CrossRef]

11. Kanoda, K. Metal–Insulator Transition in κ-(ET)$_2$X and (DCNQI)$_2$M: Two Contrasting Manifestation of Electron Correlation. *J. Phys. Soc. Jpn.* **2006**, *75*, 051007. [CrossRef]
12. Kagawa, F.; Itou, T.; Miyagawa, K.; Kanoda, K. Transport criticality of the first-order Mott transition in the quasi-two-dimensional organic conductor κ-(BEDT-TTF)$_2$Cu[N(CN)$_2$]Cl. *Phys. Rev. B* **2004**, *69*, 064511. [CrossRef]
13. Kawamoto, A.; Miyagawa, K.; Kanoda, K. Deuterated κ-(BEDT-TTF)$_2$Cu[N(CN)$_2$]Br: A system on the border of the superconductor–magnetic-insulator transition. *Phys. Rev. B* **1997**, *55*, 14140–14143. [CrossRef]
14. Mckenzie, R.H. Similarities Between Organic and Cuprate Superconductors. *Science* **1997**, *278*, 820–821. [CrossRef]
15. Lyubovskaya, R.N.; Zhilyaeva, E.I.; Pesotskii, S.I.; Lyubovskii, R.B.; Atovmyan, L.O.; D'yachenko, O.A.; Takhirov, T.G. Superconductivity of (ET)$_4$Hg$_{2.89}$Br$_8$ at atmospheric pressure and T_c = 4.3 K and the critical-field anisotropy. *JETP Lett.* **1987**, *46*, 149–152.
16. Taniguchi, H.; Okuhata, T.; Nagai, T.; Satoh, K.; Môri, N.; Shimizu, Y.; Hedo, M.; Uwatoko, Y. Anomalous Pressure Dependence of Superconductivity in Layered Organic Conductor, κ-(BEDT-TTF)$_4$Hg$_{2.89}$Br$_8$. *J. Phys. Soc. Jpn.* **2007**, *76*, 113709. [CrossRef]
17. Oike, H.; Miyagawa, K.; Taniguchi, H.; Kanoda, K. Pressure-Induced Mott Transition in an Organic Superconductor with a Finite Doping Level. *Phys. Rev. Lett.* **2015**, *114*, 067002. [CrossRef] [PubMed]
18. Lyubovskaya, R.N.; Lyubovskii, R.B.; Shibaeva, R.P.; Aldoshina, M.Z.; Gol'denberg, L.M.; Rozenberg, L.P.; Khidekel', M.L.; Shul'pyakov, Y.F. Superconductivity in a BEDT-TTF organic conductor with a chloromercurate anion. *JETP Lett.* **1985**, *42*, 468–472.
19. Xie, W.; Frisbie, C.D. Organic electrical double layer transistors based on rubrene single crystals: Examining transport at high surface charge densities above 1013 cm^{-2}. *J. Phys. Chem. Rev.* **2011**, *115*, 14360–14368. [CrossRef]
20. Watanabe, H.; Seo, H.; Yunoki, S. Mechanism of superconductivity and electron-hole doping asymmetry in κ-type molecular conductors. *Nat. Commun.* **2019**, *10*, 3167. [CrossRef]
21. Huang, Z.B.; Lin, H.Q. Quantum Monte Carlo study of Spin, Charge, and Pairing correlations in the t-t'-U Hubbard model. *Phys. Rev. B* **2001**, *64*, 205101. [CrossRef]
22. Qin, M.; Chung, C.-M.; Shi, H.; Vitali, E.; Hubig, C.; Schollwöck, U.; White, S.R.; Zhang, S. Absence of Superconductivity in the Pure Two-Dimensional Hubbard Model. *Phys. Rev. X* **2020**, *10*, 031016. [CrossRef]
23. Gomes, N.; Silva, W.; Dutta, T.; Clay, R.; Mazumdar, S. Coulomb-enhanced superconducting pair correlations and paired-electron liquid in the frustrated quarter-filled band. *Phys. Rev. B* **2016**, *93*, 165110. [CrossRef]
24. Luttinger, J.M. Fermi Surface and Some Simple Equilibrium Properties of a System of Interacting Fermions. *Phys. Rev.* **1960**, *119*, 1153. [CrossRef]
25. Tranquada, J.M.; Sternlieb, B.J.; Axe, J.D.; Nakamura, Y.; Uchida, S. Evidence for stripe correlations of spins and holes in copper oxide superconductors. *Nature* **1995**, *375*, 561–563. [CrossRef]
26. Ghiringhelli, G.; Le Tacon, M.; Minola, M.; Blanco-Canosa, S.; Mazzoli, C.; Brookes, N.B.; De Luca, G.M.; Frano, A.; Hawthorn, D.G.; He, F.; et al. Long-range incommensurate charge fluctuations in (Y,Nd)Ba$_2$Cu$_3$O$_{6+x}$. *Science* **2012**, *337*, 821–825. [CrossRef] [PubMed]
27. Comin, R.; Frano, A.; Yee, M.M.; Yoshida, Y.; Eisaki, H.; Schierle, E.; Weschke, E.; Sutarto, R.; He, F.; Soumyanarayanan, A.; et al. Charge order driven by Fermi-arc instability in Bi$_2$Sr$_{2-x}$La$_x$CuO$_{6+\delta}$. *Science* **2014**, *343*, 390–392. [CrossRef] [PubMed]
28. Campi, G.; Poccia, N.; Bianconi, G.; Barba, L.; Arrighetti, G.; Innocenti, D.; Karpinski, J.; Zhigadlo, N.D.; Kazakov, S.M.; Burghammer, M.; et al. Inhomogeneity of charge-density-wave order and quenched disorder in a high-Tc superconductor. *Nature* **2015**, *525*, 359–362. [CrossRef] [PubMed]
29. Shimizu, Y.; Miyagawa, K.; Kanoda, K.; Maesato, M.; Saito, G. Spin liquid state in an organic Mott insulator with a triangular lattice. *Phys. Rev. Lett.* **2003**, *91*, 107001. [CrossRef]
30. Tajima, N.; Sugawara, S.; Tamura, M.; Nishio, Y.; Kajita, K. Electronic phases in an organic conductor α-(BEDT-TTF) 2I3: Ultra narrow gap semiconductor, superconductor, metal, and charge-ordered insulator. *J. Phys. Soc. Jpn.* **2006**, *75*, 051010. [CrossRef]

Article

A Discrepancy in Thermal Conductivity Measurement Data of Quantum Spin Liquid β′-EtMe₃Sb[Pd(dmit)₂]₂ (dmit = 1,3-Dithiol-2-thione-4,5-dithiolate)

Reizo Kato *, Masashi Uebe, Shigeki Fujiyama and Hengbo Cui

Condensed Molecular Materials Laboratory, RIKEN, Wako, Saitama 351-0198, Japan; masaliam020902@gmail.com (M.U.); fujiyama@riken.jp (S.F.); hengbocui@gmail.com (H.C.)
* Correspondence: reizo@riken.jp

Abstract: A molecular Mott insulator β′-EtMe₃Sb[Pd(dmit)₂]₂ is a quantum spin liquid candidate. In 2010, it was reported that thermal conductivity of β′-EtMe₃Sb[Pd(dmit)₂]₂ is characterized by its large value and gapless behavior (a finite temperature-linear term). In 2019, however, two other research groups reported opposite data (much smaller value and a vanishingly small temperature-linear term) and the discrepancy in the thermal conductivity measurement data emerges as a serious problem concerning the ground state of the quantum spin liquid. Recently, the cooling rate was proposed to be an origin of the discrepancy. We examined effects of the cooling rate on electrical resistivity, low-temperature crystal structure, and ^{13}C-NMR measurements and could not find any significant cooling rate dependence.

Keywords: molecular conductors; quantum spin liquid; thermal conductivity; cooling rate; electrical resistivity; low-temperature crystal structure; ^{13}C-NMR

1. Introduction

Quantum spin liquid (QSL) in a strongly frustrated spin system on a triangular lattice is characterized by the absence of long-range magnetic order or valence bond solid order among entangled quantum spins even at zero temperature [1,2]. Although theoretical works indicated that the ideal nearest-neighbor Heisenberg antiferromagnet on the triangular lattice has the long-range Néel ordered ground state (120-degree-structured state), a possibility of this third fundamental state for magnetism in the general $S = 1/2$ antiferromagnetic triangular lattice systems has attracted much attention. Indeed, the number of QSL candidates of real materials is increasing since the beginning of the 2000 [2].

An isostructural series of anion radical salts of a metal complex Pd(dmit)₂ (dmit = 1,3-dithiol-2-thione-4,5-dithiolate), β′-Et$_x$Me$_{4-x}$Z[Pd(dmit)₂]₂ (Et: C₂H₅, Me: CH₃, Z = P, As, and Sb; x = 0–2), are Mott insulators at ambient pressure [3]. In crystals of β′-Pd(dmit)₂ salts with the space group C2/c, Pd(dmit)₂ anion radicals are strongly dimerized to form a dimer with spin 1/2, [Pd(dmit)₂]₂⁻ (Figure 1). The dimers are arranged in an approximately isosceles-triangular lattice parallel to the *ab* plane, which leads to a frustrated $S = 1/2$ Heisenberg spin system. The anion radical layers and the non-magnetic cation layers are arranged alternately along the *c* axis. The ground state of the Pd(dmit)₂ salts is found to change among antiferromagnetic long-range order (AFLO), QSL, and charge order (CO), depending on the anisotropy of the triangular lattice that can be tuned by the choice of the counter cation [3]. The cation effect on the degree of frustration is associated with the arch-shaped distortion of the Pd(dmit)₂ molecule [4]. The QSL phase found in β′-EtMe₃Sb[Pd(dmit)₂]₂ is situated between AFLO and CO phases [5,6].

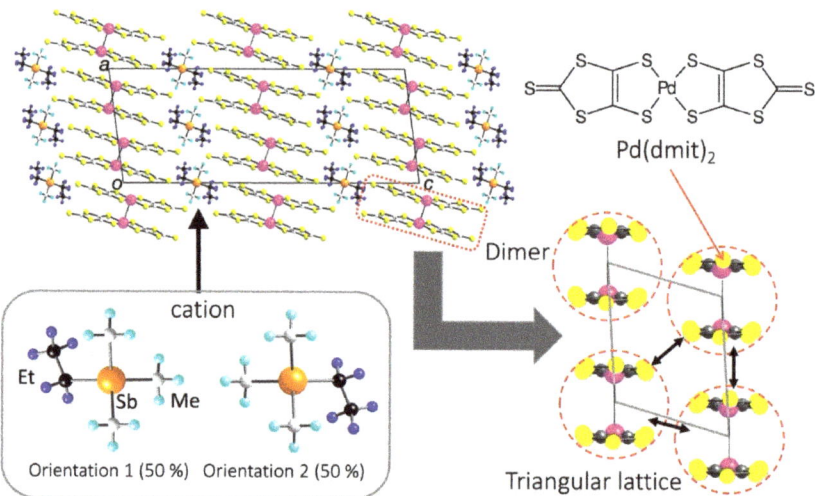

Figure 1. Crystal structure of β′-EtMe$_3$Sb[Pd(dmit)$_2$]$_2$. The cation on the two-fold axis shows an orientational disorder.

In β′-EtMe$_3$Sb[Pd(dmit)$_2$]$_2$, no magnetic order is detected down to a very low temperature (~19 mK) that corresponds to $J/12,000$, where J (~250 K) is the nearest-neighbor spin interaction energy [7]. Although ^{13}C-NMR spectra show an inhomogeneous broadening at low temperatures, the observed local static fields are too small to be explained by AFLO or spin glass state. Low-energy excitations in the QSL state of the β′-EtMe$_3$Sb[Pd(dmit)$_2$]$_2$ are open to debate even now. Heat capacity and magnetization indicate gapless fermion-like excitations, while ^{13}C-NMR indicates an existence of a nodal gap [7–9]. The thermal conductivity κ is free from the contribution from the rotation of the methyl (Me-) group that disturbs the heat capacity analysis below 1 K [8]. In addition, thermal conductivity measurements can detect the spin-mediated heat transport. In 2010, Yamashita and Matsuda reported that κ/T is finite as temperature T goes to zero, which indicates the presence of gapless excitations [10]. The finite T-linear term as well as largely enhanced κ values led to a proposal of contributions across the spinon Fermi surface. In 2019, however, two research groups reported that κ values are much smaller and κ/T is vanishingly small at 0 K, which caused a serious problem concerning the ground state of QSL [11,12].

In order to explain this sharp discrepancy in the thermal conductivity measurement data, Yamashita claimed that there were two kinds of crystals (large-κ and small-κ groups) in [13] published earlier than [11,12]. Yamashita pointed out the domain formation associated with the cation disorder or the micro cracks as an origin. It should be noted that in the context of "two kinds of crystals", the words "domain" and "micro cracks" are read as intrinsic properties that emerge in a crystal growth process or in a low-temperature phase, that is, they should be distinguished from extrinsic ones induced by improper sample handling. Although Yamashita did not disclose experimental evidence to justify the claim in [13], the claim had enough impact [14,15]. In response to the Yamashita's claim, the existence of two kinds of crystals was verified using X-ray diffraction (XRD), scanning electron microscope, and electrical resistivity measurements. The conclusion is that there is only one kind of crystal [11,12]. For example, no difference was found between the small-κ sample (sample G1) in [11] and the large-κ sample (sample C) in [13], both of which come from the very same growth batch (No. 752).

Meanwhile, in 2020, Yamashita et al. reported that one kind of crystal gives different results in the κ measurements depending on an experimental condition, "cooling rate" [16,17]. In their measurements, very slow cooling (−0.4 K/h) led to a finite linear residual thermal conductivity. In contrast, when the sample was cooled down rapidly

(-13 K/h), κ/T vanished at the zero-temperature limit, and the phonon thermal conductivity was strongly suppressed. These results suggest the existence of random scatterers introduced during the cooling process as another origin of the discrepancy. This proposal has raised a problem about effects of the newly proposed experimental parameter on other kinds of measurements. Herein, we investigated effects of the cooling rate on electrical resistivity, low-temperature crystal structure, and ^{13}C-NMR measurements with relevance to the discrepancy in the thermal conductivity data.

2. Materials and Methods

For electrical resistivity and XRD measurements, we used single crystals from the same growth batch (No. 898) as that used in [13] (small-κ samples E and F) and [16] (crystals 1–3). The procedure of the crystal growth was as follows: (EtMe$_3$Sb)$_2$[Pd(dmit)$_2$] (60 mg) was dissolved in acetone (100 mL). After addition of acetic acid (9.5 mL), the resultant solution was allowed to stand at -11 °C for 3 months. The β'-type crystals (black hexagonal plates) were obtained as a single phase. The ^{13}C-enriched dmit ligand for ^{13}C-NMR measurements was synthesized from tetrachloroethylene-^{13}C$_1$ (99 atom %, Sigma-Aldrich) that was converted to tetrathiooxalate for the reaction with CS$_2$ [18]. The ^{13}C-enriched single crystals of β'-EtMe$_3$Sb[Pd(dmit)$_2$]$_2$ (No. 899 for slow cooling and No. 923 for rapid cooling) were also obtained by the above-mentioned procedure.

The temperature-dependent electrical resistivities along the a and b axes ($\rho_{//a}$ and $\rho_{//b}$) were measured by the standard four-probe method from room temperature to 1.8 K using a physical property measurement system (Quantum Design Inc., San Diego, CA, USA). The electric leads were ϕ10 μm gold wires connected by carbon paste. Probe sizes are 325 × (1100 × 100) μm^3 for $\rho_{//a}$ and 150 × (1350 × 80) μm^3 for $\rho_{//b}$, respectively. For each current direction, the pristine crystal was cooled down to 1.8 K with different cooling rates of -0.6 K/h, -1.2 K/h, -6 K/h, and -150 K/h, in this order. In each thermal cycle, the warming rate was $+6$ K/h, except for the final cycle ($+150$ K/h).

Single crystal X-ray diffraction data were collected by a Weisenberg-type imaging plate system (R-AXIS RAPID/CS, Rigaku Corp., Tokyo, Japan) with monochromated Mo Kα radiation (UltraX6-E, Rigaku Corp., Tokyo, Japan). Low-temperature experiments were carried out in the cryostat cooled by a closed-cycle helium refrigerator (HE05/UV404, ULVAC CRYOGENICS Inc., Chigasaki, Japan). The temperature was controlled by Model 22C Cryogenic Temperature Controller (Cryogenic Control Systems Inc., Rancho Santa Fe, CA, USA). The pristine crystal was cooled down with a rate of -0.6 K/h. After the data collection at 5 K, the crystal was warmed up to room temperature with a rate of $+60$ K/h. The next cooling process was performed with a cooling rate of -120 K/h. All the diffraction data were processed using the CrystalStructure 3.8 crystallographic software package [19]. The structures were solved by the direct method (SIR92) [20] and refined by the full-matrix least-squares method (SHELXL-2018/3) [21]. The H atom coordinates were placed on calculated positions and refined with the riding model. Due to the orientational disorder of the EtMe$_3$Sb cation on the two-fold axis (Figure 1), we assumed that the ethyl group and the corresponding methyl group share two equivalent positions with 50% occupation factor for the refinements.

^{13}C-NMR spectra and nuclear relaxation rates were obtained by standard pulse Fourier transform technique using a single crystal. The magnetic field was applied along the direction 11 degrees tilted from the c^* axis to avoid an accidental cancellation of hyperfine fields originating from $2p$ and $2s$ electrons of ^{13}C. The temperature was controlled by Model 32 Cryogenic Temperature Controller. The crystals were cooled at -40 K/h (rapid cooling) and -0.6 K/h (slow cooling), respectively.

3. Results

3.1. Electrical Resistivity

The electrical resistivity is sensitive to the crack formation. In addition, when the emergence of an electronic phase depends on the cooling rate as is the case of θ-(BEDT-

TTF)$_2$RbZn(SCN)$_4$ (BEDT-TTF = Bis(ethylenedithio)tetrathiafulvalene) that possesses the charge-glass-forming ability, the electrical resistivity can detect a change of the electronic state [22]. Figure 2 shows temperature-dependent resistivities along the a and b axes measured with four different cooling rates, −0.6, −1.2, −6, and −150 K/h in this order. β'-EtMe$_3$Sb[Pd(dmit)$_2$]$_2$ was a semiconductor and the resistivity became too high to measure below 28 K. Anisotropy within the ab plane was small, including the activation energy (~41 meV). As shown in Figure 2, temperature-resistivity curves for each cooling rate overlap almost completely, which means that there was no cooling rate dependence. In addition, no crack formation (indicated by an abrupt jump of the resistivity) and no thermal cycle dependence was observed.

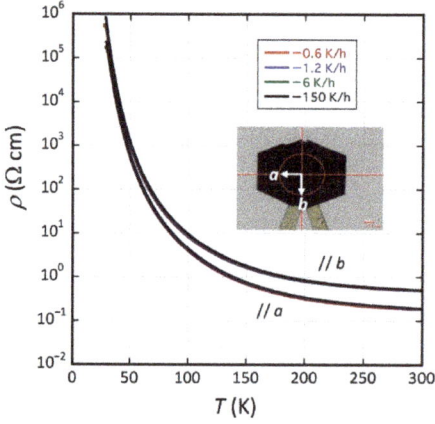

Figure 2. Temperature-dependent resistivities measured along the a axis ($\rho_{//a}$) and the b axis ($\rho_{//b}$) with four different cooling rates for β'-EtMe$_3$Sb[Pd(dmit)$_2$]$_2$. The inset is a photo of a single crystal. For both $\rho_{//a}$ and $\rho_{//b}$, four ρ–T curves (cooling process) overlap almost completely.

3.2. Low-Temperature Crystal Structure

Using the same single crystal, the crystal structure of β'-EtMe$_3$Sb[Pd(dmit)$_2$]$_2$ at 5 K was determined with two different cooling rates, −0.6 (slow) and −120 (rapid) K/h in this order. In both cases, the space group remained $C2/c$ and no additional diffraction peak was observed. Determined crystal structures were identical with the previous result [23], and did not show any significant effects of the cooling rate on temperature factors and differential Fourier synthesis (Table 1).

Table 1. Crystal data for β'-EtMe$_3$Sb[Pd(dmit)$_2$]$_2$ at 5 K obtained using two different cooling rates.

Cooling Rate	−120 K/h (Rapid)	−0.6 K/h (Slow)
a (Å)	14.346 (8)	14.342 (8)
b (Å)	6.317 (3)	6.315 (3)
c (Å)	37.07 (2)	37.07 (2)
β (Å)	97.624 (14)	97.640 (13)
V (Å^3)	3330 (3)	3328 (3)
R factor	0.0370	0.0357
G.O.F.	1.079	1.049
$\Delta\rho_{max}$ (e Å^{-3})	1.363	1.314
$\Delta\rho_{min}$ (e Å^{-3})	−1.089	−0.960

In the crystal, the EtMe$_3$Sb$^+$ cation is located on the 2-fold axis (//b). Since the cation does not have the 2-fold symmetry, the cation has two possible orientations (1 and 2 in Figure 1), which could work as an origin of the domain formation. However, our analysis, where the ethyl group is assumed to be overlaid with the methyl group with 50% occupation factor, did not find significant difference in the average cation structure for both cooling rates (Figure 3).

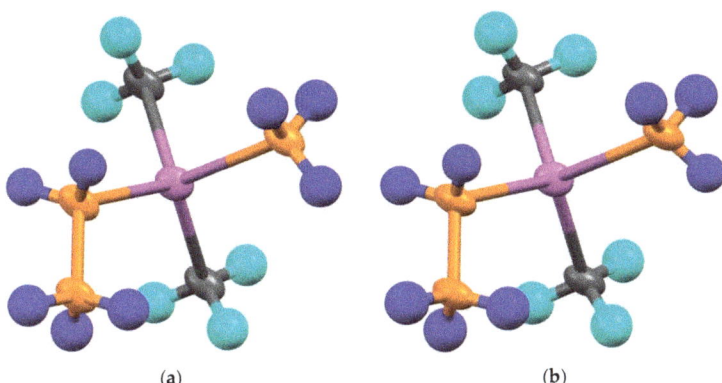

Figure 3. Average structure of the cation at 5 K viewed from the b axis (the two-fold axis) for two different cooling rates: (**a**) Rapid cooling (-120 K/h); and (**b**) slow cooling (-0.6 K/h).

3.3. ^{13}C-NMR

^{13}C-NMR enables us to investigate the microscopic electronic states of a crystal. Compared with the electrical resistivity, both NMR spectra and nuclear relaxations are insensitive to microcracks in a crystal. However, they would be able to detect possible domain formations in a cooling procedure.

Figure 4a shows the spectra at 5K. The ^{13}C atom is introduced into one of the carbon sites in the C=C bond in the dmit ligand. The four independent ^{13}C sites at which hyperfine coupling constants distribute up to 10% cause asymmetric spectra [24]. Two spectra with contrasting cooling rates nearly overlap with each other, by which we conclude negligible cooling rate variation in the static electronic states.

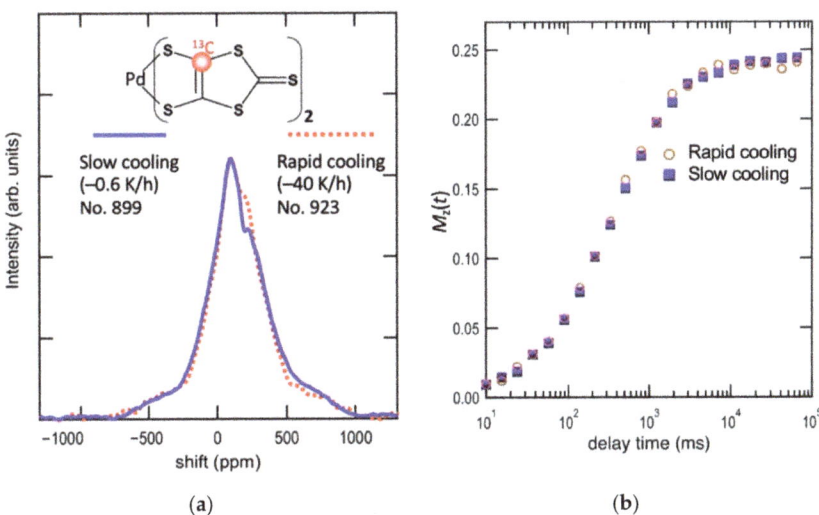

Figure 4. ^{13}C-NMR of β'-EtMe$_3$Sb[Pd(dmit)$_2$]$_2$ at 5 K with rapid and slow cooling rates. (**a**) Spectra; and (**b**) nuclear magnetization curves.

Nuclear magnetization (M_z) curves of the $1/T_1$ (nuclear spin-lattice relaxation rate) measurements at 5 K by the rapid and slow cooling procedures are shown in Figure 4b. Two curves agree with each other, and we conclude that the spin dynamics of the quantum spin liquid state is insensitive to the cooling rate.

It should be noted that the previous ^{13}C-NMR data obtained using randomly orientated crystals suggest no cooling rate dependence either [5,7]. In [5], the cooling process at a rate of ca. −10 K/h and measurements at a constant temperature (during 1–2 days) were repeated alternately, and the sample was cooled from room temperature down to 1.4 K spending about one month. In [7], on the other hand, the sample was cooled from room temperature down to 1.8 K within 10 h before the measurements in very low-temperature region (1.8 K–20 mK). The ^{13}C-NMR data with these two different experimental conditions coincide with each other in the same temperature region [25].

4. Discussion

In this work, we could not observe the effect of the cooling rate on resistivity, low-temperature crystal structure, and ^{13}C-NMR. Of course, we must be careful in discussing the relation between these physical/structural properties and the thermal conductivity. The problem we are facing is the thermal conductivity below 1 K. In the low temperature region, the electrical resistivity of the present material is very high, and the mean free path of a charge carrier becomes shorter than lattice lengths. In such a case, conduction electrons would be unaffected by an event in the whole crystal. On the other hand, the low-temperature crystal structures we determined are average ones and do not provide direct information about the possible domain or defect formation.

Nevertheless, the cooling rate engages in a process in the whole temperature region. If random scatterers are generated during the cooling process, they would be detected by physical/structural properties other than the thermal conductivity even in the high temperature region. In addition, local changes in a crystal could affect an average crystal structure and ^{13}C-NMR. As we mentioned before, the cooling rate dependence was not detected in the high temperature region (>~5 K) in this work. In the lower temperature region, the smaller heat capacity provides more homogeneous temperature distribution, and thus it is less plausible that the cooling rate plays an important role.

In this sense, the results of this work suggest that further analysis is necessary before concluding that the cooling rate is an essential experimental condition. Let us reconsider two different sets of thermal conductivity data from [10,16] (Figure 5). Measurements with different cooling rates indicated that the slower cooling rate gave the larger κ and finite κ/T [16]. However, even with the slowest cooling rate (−0.4 K/h), the κ values are much smaller than the first reported ones [10]. That is, the large-κ data have never been reproduced. In addition, the measurements in [10] performed with the rapid cooling rate of −10 K/h show larger κ values than those for Crystal 1 measured with the slowest cooling rate of −0.4 K/h in [16]. This is quite puzzling and suggests that the cooling rate is not essential.

In conclusion, the present situation is that one kind of crystal provides two different thermal conductivity data and a role of the newly proposed experimental parameter, cooling rate, remains to be seen. In addition, the large-κ data have never been reproduced. It is an urgent matter to clarify the intrinsic thermal conductivity. From this point of view, the description *"the crystals often do not recover to the initial state after a thermal cycle"* in [16] suggests that the stress from experimental environments including a setup may enhance or suppress the thermal conductivity. Indeed, crystals used in [10,16] frequently fell apart when the leads were removed by rinsing out the paste with diethyl succinate. This suggests the existence of the stress from leads on a crystal. In contrast, we did not observe any thermal cycle dependence in this work. In order to clarify this point, monitoring of electrical resistivity during thermal conductivity measurements will be valuable, because the resistivity is sensitive to the crack formation and pressure in the wide range of temperature (>~28K).

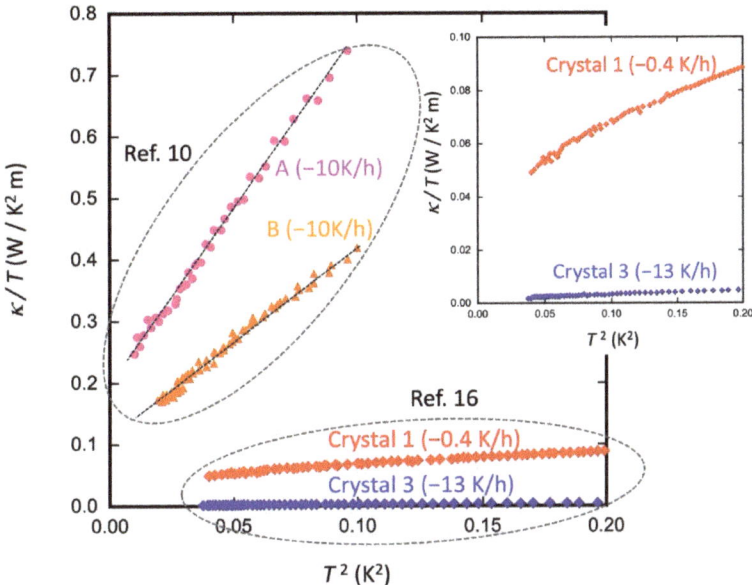

Figure 5. Two different sets of the thermal conductivity (κ) data for β'-EtMe$_3$Sb[Pd(dmit)$_2$]$_2$ from [10,16]. The cooling rate in each measurement is indicated in a parenthesis. The inset is an enlarged view of the data from [16]. The behavior of κ reported in [11,12] is similar to that of crystal 3 (-13 K/h) in [16].

Author Contributions: Conceptualization, funding acquisition, project administration, sample preparation, and writing—original draft preparation, R.K.; investigation and data curation, M.U., S.F. and H.C. All authors have read and agreed to the published version of the manuscript.

Funding: This research was partially supported by JSPS KAKENHI [grant number JP16H06346].

Institutional Review Board Statement: Not applicable.

Informed Consent Statement: Not applicable.

Data Availability Statement: Crystallographic information files are available from the CCDC, reference numbers 2125994-2125995. These data can be obtained free of charge via https://www.ccdc.cam.ac.uk/structures/.

Acknowledgments: We deeply thank Daisuke Hashizume (RIKEN Center for Emergent Matter Science) for technical help with the crystal structure analysis of β'-EtMe$_3$Sb[Pd(dmit)$_2$]$_2$ at 5 K.

Conflicts of Interest: The authors declare no conflict of interest.

References

1. Anderson, P.W. Resonating valence bonds: A new kind of insulator? *Mater. Res. Bull.* **1973**, *8*, 153–160. [CrossRef]
2. Balents, L. Spin liquids in frustrated magnets. *Nature* **2010**, *464*, 199–208. [CrossRef]
3. Kato, R. Development of π-electron systems based on [M(dmit)$_2$] (M= Ni and Pd; dmit: 1,3-dithiole-2-thione-4,5-dithiolate) Anion Radicals. *Bull. Chem. Soc. Jpn.* **2014**, *87*, 355–374. [CrossRef]
4. Kato, R.; Hengbo, C. Cation dependence of crystal structure and band parameters in a series of molecular conductors, β'-(Cation)[Pd(dmit)$_2$]$_2$ (dmit = 1,3-dithiol-2-thione-4,5-dithiolate). *Crystals* **2012**, *2*, 861–874. [CrossRef]
5. Itou, T.; Oyamada, A.; Maegawa, S.; Tamura, M.; Kato, R. Quantum spin liquid in the spin-1/2 triangular antiferromagnet EtMe$_3$Sb[Pd(dmit)$_2$]$_2$. *Phys. Rev. B* **2008**, *77*, 104413. [CrossRef]
6. Kanoda, K.; Kato, R. Mott physics in organic conductors with triangular lattices. *Annu. Rev. Condens. Matter Phys.* **2011**, *2*, 167–188. [CrossRef]

7. Itou, T.; Oyamada, A.; Maegawa, S.; Kato, R. Instability of a quantum spin liquid in an organic triangular-lattice antiferromagnet. *Nat. Phys.* **2010**, *6*, 673–676. [CrossRef]
8. Yamashita, S.; Yamamoto, T.; Nakazawa, Y.; Tamura, M.; Kato, R. Gapless spin liquid of an organic triangular compound evidenced by thermodynamic measurements. *Nat. Commun.* **2011**, *2*, 1–6. [CrossRef]
9. Watanabe, D.; Yamashita, M.; Tonegawa, S.; Oshima, Y.; Yamamoto, H.M.; Kato, R.; Sheikin, I.; Behnia, K.; Terashima, T.; Uji, S.; et al. Novel Pauli-paramagnetic quantum phase in a Mott insulator. *Nat. Commun.* **2012**, *3*, 1–6. [CrossRef]
10. Yamashita, M.; Nakata, N.; Senshu, Y.; Nagata, M.; Yamamoto, H.M.; Kato, R.; Shibauchi, T.; Matsuda, Y. Highly mobile gapless excitations in a two-dimensional candidate quantum spin liquid. *Science* **2010**, *328*, 1246–1248. [CrossRef]
11. Bourgeois-Hope, P.; Laliberté, F.; Lefrançois, E.; Grissonnanche, G.; René de Cotret, S.; Gordon, R.; Kitou, S.; Sawa, H.; Cui, H.; Kato, R.; et al. Thermal conductivity of the quantum spin liquid candidate EtMe$_3$Sb[Pd(dmit)$_2$]$_2$: No evidence of mobile gapless excitations. *Phys. Rev. X* **2019**, *9*, 041051. [CrossRef]
12. Ni, J.M.; Pan, B.L.; Song, B.Q.; Huang, Y.Y.; Zeng, J.Y.; Yu, Y.J.; Cheng, E.J.; Wang, L.S.; Dai, D.Z.; Kato, R.; et al. Absence of magnetic thermal conductivity in the quantum spin liquid candidate EtMe$_3$Sb[Pd(dmit)$_2$]$_2$. *Phys. Rev. Lett.* **2019**, *123*, 247204. [CrossRef] [PubMed]
13. Yamashita, M. Boundary-limited and glassy-like phonon thermal conduction in EtMe$_3$Sb[Pd(dmit)$_2$]$_2$. *J. Phys. Soc. Jpn.* **2019**, *88*, 083702. [CrossRef]
14. Fukuyama, H. Comment on "Boundary-limited and glassy-like phonon thermal conduction in EtMe$_3$Sb[Pd(dmit)$_2$]$_2$. *J. Phys. Soc. Jpn.* **2020**, *89*, 086001. [CrossRef]
15. Yamashita, M. Reply to comment on "Boundary-limited and glassy-like phonon thermal conduction in EtMe$_3$Sb[Pd(dmit)$_2$]$_2$". *J. Phys. Soc. Jpn.* **2020**, *89*, 086002. [CrossRef]
16. Yamashita, M.; Sato, Y.; Tominaga, T.; Kasahara, Y.; Kasahara, S.; Cui, H.; Kato, R.; Shibauchi, T.; Matsuda, Y. Presence and absence of itinerant gapless excitations in the quantum spin liquid candidate EtMe$_3$Sb[Pd(dmit)$_2$]$_2$. *Phys. Rev. B* **2020**, *101*, 140407. [CrossRef]
17. Yamashita, M. Erratum: Boundary-limited and glassy-like phonon thermal conduction in EtMe$_3$Sb[Pd(dmit)$_2$]$_2$. *J. Phys. Soc. Jpn.* **2020**, *89*, 068001. [CrossRef]
18. Bretizer, J.G.; Chou, J.-H.; Rauchfuss, T.B. A new synthesis of tetrathiooxalate and its conversion to $C_3S_5^{2-}$ and $C_4S_6^{2-}$. *Inorg. Chem.* **1998**, *37*, 2080–2082. [CrossRef]
19. *Crystal Structure*, Version 3.8; Crystal Structure Analysis Package; Rigaku and Rigaku Americas: The Woodlands, TX, USA, 2006.
20. Altomare, A.; Cascarano, G.; Giacovazzo, C.; Guagliardi, A.; Burla, M.C.; Polidori, G.; Camalli, M. SIR92: A program for automatic solution of crystal structures by direct methods. *J. Appl. Cryst.* **1994**, *27*, 435. [CrossRef]
21. Sheldrick, G.M. Crystal structure refinement with SHELXL. *Acta Crystallogr. Sect. C* **2015**, *C71*, 3–8.
22. Kagawa, F.; Sato, T.; Miyagawa, K.; Kanoda, K.; Tokura, Y.; Kobayashi, K.; Kumai, R.; Murakami, Y. Charge-cluster glass in an organic conductor. *Nat. Phys.* **2013**, *9*, 419–422. [CrossRef]
23. Ueda, K.; Tsumuraya, T.; Kato, R. Temperature dependence of crystal structures and band parameters in quantum spin liquid β'-EtMe$_3$Sb[Pd(dmit)$_2$]$_2$ and related materials. *Crystals* **2018**, *8*, 138. [CrossRef]
24. Fujiyama, S.; Kato, R. Fragmented electronic spins with quantum fluctuations in organic Mott insulators near a quantum spin liquid. *Phys. Rev. Lett.* **2019**, *122*, 147204. [CrossRef] [PubMed]
25. Itou, T. (Tokyo University of Science, Tokyo, Japan). Personal communication, 2021.

Perspective

Are Heavy Fermion Strange Metals Planckian?

Mathieu Taupin and Silke Paschen *

Institute of Solid State Physics, Vienna University of Technology, Wiedner Hauptstr. 8-10, 1040 Vienna, Austria; taupin@ifp.tuwien.ac.at
* Correspondence: paschen@ifp.tuwien.ac.at; Tel.: +43-1-58801-13716

Abstract: Strange metal behavior refers to a linear temperature dependence of the electrical resistivity that is not due to electron–phonon scattering. It is seen in numerous strongly correlated electron systems, from the heavy fermion compounds, via transition metal oxides and iron pnictides, to magic angle twisted bi-layer graphene, frequently in connection with unconventional or "high temperature" superconductivity. To achieve a unified understanding of these phenomena across the different materials classes is a central open problem in condensed matter physics. Tests whether the linear-in-temperature law might be dictated by Planckian dissipation—scattering with the rate $\sim k_B T/\hbar$—are receiving considerable attention. Here we assess the situation for strange metal heavy fermion compounds. They allow to probe the regime of extreme correlation strength, with effective mass or Fermi velocity renormalizations in excess of three orders of magnitude. Adopting the same procedure as done in previous studies, i.e., assuming a simple Drude conductivity with the above scattering rate, we find that for these strongly renormalized quasiparticles, scattering is much weaker than Planckian, implying that the linear temperature dependence should be due to other effects. We discuss implications of this finding and point to directions for further work.

Keywords: heavy fermion compounds; strange metals; Planckian dissipation; quantum criticality; Kondo destruction

Citation: Taupin, M.; Paschen, S. Are Heavy Fermion Strange Metals Planckian? *Crystals* **2022**, *12*, 251. https://doi.org/10.3390/cryst12020251

Academic Editor: Andrej Pustogov

Received: 24 December 2021
Accepted: 10 February 2022
Published: 12 February 2022

Publisher's Note: MDPI stays neutral with regard to jurisdictional claims in published maps and institutional affiliations.

Copyright: © 2022 by the authors. Licensee MDPI, Basel, Switzerland. This article is an open access article distributed under the terms and conditions of the Creative Commons Attribution (CC BY) license (https://creativecommons.org/licenses/by/4.0/).

1. Introduction

A first step in understanding matter is to delineate the different phases in which it manifests. To do so, a characteristic that uniquely identifies a phase must be found, and using its order has worked a long way. How this classification should be extended to also incorporate topological phases [1] is a matter of current research. Here, we focus on topologically trivial matter and thus take order-parameter descriptions [2] as a starting point and consider the case of second-order phase transitions. As an order parameter develops below a transition (or critical) temperature, the system's symmetry is lowered (or broken). Cornerstones are the power law behavior of physical properties near the critical temperature, with universal critical exponents, and the associated scaling relationships. Combined with renormalization-group ideas [3], this framework is now referred to as the Landau–Ginzburg–Wilson (LGW) paradigm. It has also been extended to zero temperature. Here, phase transitions—now called *quantum* phase transitions [4]—can occur as the balance between competing interactions is tipped. To account for the inherently dynamical nature of the $T = 0$ case, a dynamical critical exponent needs to be added. This increases the effective dimensionality of the system, which may then surpass the upper critical dimension for the transition, so that the system behaves as noninteracting, or "Gaussian". Interestingly, however, cases have been identified where this expectation is violated [5–8], evidenced for instance by the observation of dynamical scaling relationships [9] that should be absent according to the above rationale. We will refer to this phenomenon as "beyond order parameter" quantum criticality. It appears to be governed by new degrees of freedom specific to the quantum critical point (QCP). This is a topic of broad interest both in condensed matter physics and beyond, but a general framework is lacking. We will here discuss it from the perspective of heavy fermion compounds, where it can manifest as Kondo destruction quantum criticality [5,6]. We will in particular discuss materials that display linear-in-temperature "strange metal" electrical resistivity, as well as the proposed relation [10,11] to Planckian dissipation. We will allude to similar phenomena in other material platforms and point to directions for further research to advance the field.

2. Simple Models for Strongly Correlated Electron Systems

Strongly correlated electron systems host electrons at the brink of localization. The simplest model that can capture this physics is the Hubbard model

$$H = -t \sum_{\langle ij \rangle, \sigma} (d_{i\sigma}^\dagger d_{j\sigma} + d_{j\sigma}^\dagger d_{i\sigma}) + U \sum_i d_{i\uparrow}^\dagger d_{i\uparrow} d_{i\downarrow}^\dagger d_{i\downarrow} . \qquad (1)$$

The hopping integral t transfers electrons from site to site and thus promotes itinerancy, whereas the onsite Coulomb repulsion U penalizes double occupancy of any site, thereby promoting localization. Thus, with increasing U/t, a (Mott) metal–insulator transition is expected. This simple model is suitable for materials where transport is dominated by one type of orbital with moderate nearest neighbor overlap, leading to one relatively narrow band. Well-known examples are found in transition metal oxides, for instance the cuprates. Here, the relevant orbitals are copper d orbitals, kept at distance by oxygen atoms. The creation and annihilation operators are called d and $d^\dagger$ here.

If two different types of orbitals interplay—one much more localized than the other—a better starting point for a theoretical description is the (periodic) Anderson model that, for the one-dimensional case, reads [12,13]

$$H = \sum_{k,\sigma} \epsilon_k c_{k\sigma}^\dagger c_{k\sigma} + \sum_{j,\sigma} \epsilon_f f_{j\sigma}^\dagger f_{j\sigma} + U \sum_j f_{j\uparrow}^\dagger f_{j\uparrow} f_{j\downarrow}^\dagger f_{j\downarrow} + \sum_{j,k,\sigma} V_{jk}(e^{ikx_j} f_{j\sigma}^\dagger c_{k\sigma} + e^{-ikx_j} c_{k\sigma}^\dagger f_{j\sigma}) . \qquad (2)$$

Orbitals with large overlap, with the associated creation and annihilation operators c and $c^\dagger$, form a conduction band with dispersion ϵ_k. Orbitals with vanishing overlap situated at the positions x_j are associated with the operators f and $f^\dagger$. They are assumed to be separated by a distance greater than the f orbital diameter and thus no hopping between them is considered. However, the hybridization term V allows the f electrons to interact. This model is particularly well suited for the heavy fermion compounds, which contain lanthanide (with partially filled $4f$ shells) or actinide elements (with partially filled $5f$ shells) in addition to s, p, and d electrons. For the so-called Kondo regime, where f orbitals effectively act as local moments, the Anderson model can be transformed into the Kondo (lattice) model

$$H = \sum_{k,\sigma} \epsilon_k c_{k\sigma}^\dagger c_{k\sigma} - J \sum_i \vec{S}_i \cdot c_{i,\sigma}^\dagger \vec{\sigma}_{\sigma,\sigma'} c_{i\sigma'} , \qquad (3)$$

where the interaction between the localized and itinerant electrons is expressed in terms of an antiferromagnetic exchange coupling J. $\vec{S}$ is the local magnetic moment of the f orbital and $\vec{\sigma}_{\sigma,\sigma'}$ are the Pauli spin matrices. One of the possible ground states of this model is a paramagnetic heavy Fermi liquid with a large Fermi surface, which contains both the local moment and the conduction electrons. The resonant elastic scattering at each site generates a renormalized band at the Fermi energy. Its width is of the order of the Kondo temperature T_K, which can be orders of magnitude smaller than the noninteracting band width. In the (typically considered) simplest case (with a uniform and k independent hybridization), this band extends across essentially the entire Brillouin zone.

In popular terms, this heavy fermion band could be seen as the realization of a nearly perfect "flat band" (an early description of an interaction-driven truly flat band, with zero energy, is given in [14] and its relevance for strange metal physics is discussed in [15,16]). Flat bands have also been predicted [17] and later identified in magic angle twisted bi-layer graphene (MATBG) [18] as a result of moiré band formation, and are expected in lattices with specific geometries [19,20] such as the kagome lattice [21,22] through destructive phase interference of certain hopping paths. Whereas the theoretical description of these latter flat band systems may be simpler than solving even the simplest Hamiltonians for strongly correlated electron systems, such as (1)–(3), the inverse might be true for the challenge on the experimental side. Heavy fermion compounds with a large variety of chemical compositions and structures [23–25] can be quite readily synthesized as high-quality (bulk) single crystals; the heavy fermion "flat bands" are robust (not fine tuned), naturally extend essentially across the entire Brillouin zone, and are pinned to the Fermi energy. Albeit, they form in the Kondo coherent ground state of the system, which is typically only fully developed at low temperatures. To realize such physics via a complementary route that might bring these properties to room temperature is an exciting perspective. Bringing together these different approaches bears enormous potential for progress. Indeed, for both twisted trilayer graphene [26] and MATBG [27] the connection to heavy fermion physics has very recently been pointed out. Another topic discussed across the various platforms is "strange metal" physics, which we address next.

3. Strange Metal Phase Diagrams

Metals usually obey Fermi liquid theory, even in the limit of strong interactions. This is impressively demonstrated by the large body of heavy fermion compounds that, at sufficiently low temperatures, display the canonical Fermi liquid forms of the electronic specific heat

$$C_p = \gamma T, \qquad (4)$$

the Pauli susceptibility

$$\chi = \chi_0, \qquad (5)$$

and the electrical resistivity

$$\rho = \rho_0 + AT^2, \qquad (6)$$

where ρ_0 is the residual (elastic) resistivity. Theoretically, the prefactors γ, χ_0, and A all depend on the renormalized electronic density of states $N^* = N/N_0$, or the related renormalized (density-of-states) effective mass $m^* = m/m_0 \sim N^*$, to first approximation as $\gamma \sim m^*$, $\chi_0 \sim m^*$, and $A \sim (m^*)^2$. N_0 and m_0 are the free electron quantities. Indeed, in double-logarithmic plots of γ vs. χ_0 (Sommerfeld-Wilson) and A vs. γ (Kadowaki-Woods), experimental data of a large number of heavy fermion compounds fall on universal lines, thereby confirming the theoretically expected universal ratios [28]. The scaling works close to perfectly if corrections due to different ground state degeneracies [29] and effects of dimensionality, electron density, and anisotropy [30] are taken into account.

More surprising, then, was the discovery that this very robust Fermi liquid behavior can nevertheless cease to exist. This can have multiple reasons, but the predominant and best investigated one is quantum criticality [4,25,31,32]. In the standard scenario for quantum criticality of itinerant fermion systems [33–35], a continuously vanishing Landau order parameter (typically of a density wave) governs the physical properties. Its effect on the electrical resistivity is expected to be modest because (i) the long-wavelength critical modes of the bosonic order parameter can only cause small-angle scattering, which does not degrade current efficiently, and (ii) critical density wave modes only scatter those areas on the Fermi surface effectively that are connected by the ordering wavevector. Fermions from the rest of the Fermi surface will short circuit these hot spots [36]. For itinerant ferromagnets, $\rho \sim T^{5/3}$ is theoretically predicted [4] and experimentally observed [37]. For itinerant antiferromagnets, this type of order-parameter quantum criticality should result in $\rho \sim T^\epsilon$ with $1 \leq \epsilon \leq 1.5$, depending on the amount of disorder [36]. Whereas this may be consistent with experiments on a few heavy fermion compounds, a strong dependence of ϵ with the degree of disorder has, to the best of our knowledge, not been reported. More importantly, for relatively weak disorder, the current is dominated by the contributions from the cold regions of the Fermi surface which stay as quasiparticles and the resistivity would have the T^2 dependence of a Fermi liquid [38].

Instead, a number of heavy fermion compounds exhibit a linear-in-temperature electrical resistivity

$$\rho = \rho'_0 + A'T,\qquad(7)$$

a dependence dubbed "strange metal" behavior from the early days of high-temperature superconductivity on [39]. In Figure 1a–d we show four examples, in the form of temperature–magnetic field (a,b,d) or temperature–pressure (c) phase diagrams with color codings that reflect the exponent ϵ of the temperature-dependent inelastic electrical resistivity, $\Delta\rho \propto T^\epsilon$, determined locally as $\epsilon = \partial(\ln \Delta\rho)/\partial(\ln T)$. In all cases, fans of non-Fermi liquid behavior ($\epsilon \neq 2$) appear to emerge from QCPs, with ϵ close to 1 in the center of the fan and extending to the lowest accessed temperatures (at least in a,c,d).

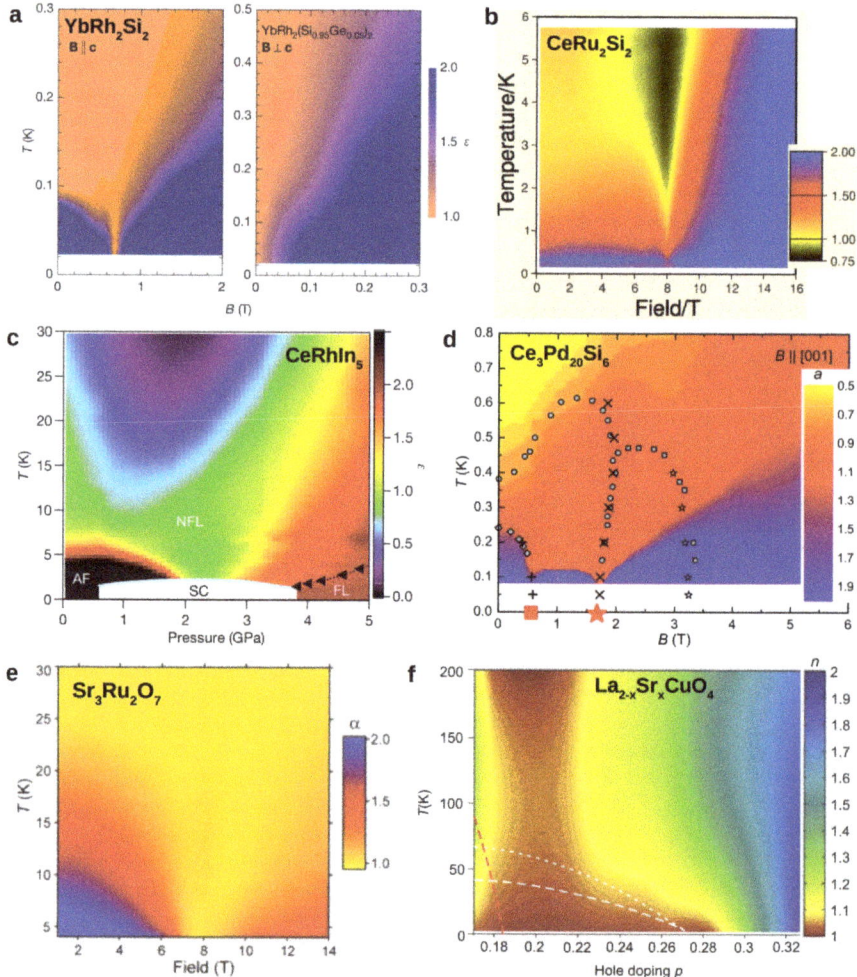

Figure 1. Cont.

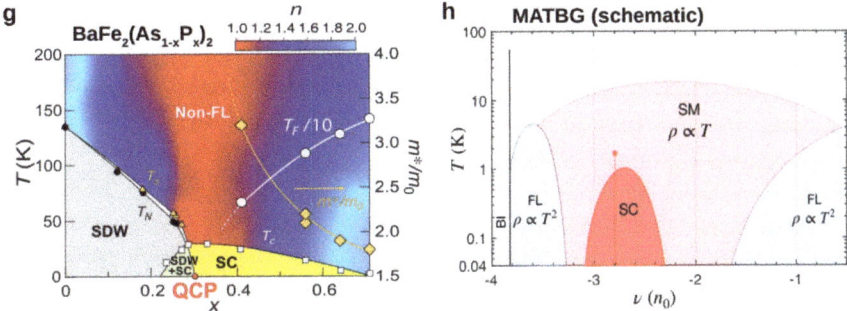

Figure 1. Color-coded phase diagrams featuring strange metal behavior in various materials platforms. (**a**) YbRh$_2$Si$_2$ (**left**) and YbRh$_2$(Si$_{0.95}$Ge$_{0.05}$)$_2$ (**right**), from [40]. (**b**) CeRu$_2$Si$_2$, from [41]. (**c**) CeRhIn$_5$, from [42]. (**d**) Ce$_3$Pd$_{20}$Si$_6$, from [43]. (**e**) SrRu$_3$O$_7$. Note that the temperature scale is cut at 4.5 K. At lower temperatures, deviations from linear behavior towards larger powers are observed; from [44]. (**f**) La$_{2-x}$Sr$_x$CuO$_4$, from [45]. (**g**) BaFe$_2$(As$_{1-x}$P$_x$)$_2$, from [46]. (**h**) Magic-angle twisted bi-layer graphene, adapted from [47].

The most pronounced such behavior is found in YbRh$_2$Si$_2$ (Figure 1a, left). Below 65 mK, the system orders antiferromagnetically [48]. As magnetic field (applied along the crystallographic c axis) continuously suppresses the order to zero at 0.66 T [40], linear-in-temperature resistivity, with $A' = 1.8\,\mu\Omega$cm/K and $\rho'_0 = 2.43\,\mu\Omega$cm, extends from about 15 K [48] down to the lowest reached temperature (below 25 mK) [40]. Recently, this range was further extended down to 5 mK, showing $A' = 1.17\,\mu\Omega$cm/K for a higher-quality single crystal ($\rho'_0 = 1.23\,\mu\Omega$cm) [49], thus spanning in total 3.5 orders of magnitude in temperature. This happens in a background of Fermi liquid behavior away from the QCP. A linear-in-temperature resistivity is also seen in the substituted material YbRh$_2$(Si$_{0.95}$Ge$_{0.05}$)$_2$. Its residual resistivity is about five times larger than that of the stoichiometric compound. That this sizeably enhanced disorder does not change the power ϵ indicates that the order-parameter-fluctuation description of an itinerant antiferromagnetic quantum critical point [36] is not appropriate here. This point will be further discussed in Section 7.

For CeRu$_2$Si$_2$ (Figure 1b), the situation is somewhat more ambiguous. Linear-in-temperature resistivity does not cover the entire core region of the fan; both above 2 K and below 0.5 K, crossovers to other power laws can be seen [41]. In CeRhIn$_5$ (Figure 1c), at the critical pressure of 2.35 GPa, linear-in-temperature resistivity extends from about 15 K down to 2.3 K, the maximum critical temperature of a dome of unconventional superconductivity [42]. That the formation of emergent phases such as unconventional superconductivity tends to be promoted by quantum critical fluctuations is, of course, of great interest in its own right even if, pragmatically, it can be seen as hindering the investigation of the strange metal state. Finally, Ce$_3$Pd$_{20}$Si$_6$ exhibits two consecutive magnetic field-induced QCPs, with linear-in-temperature resistivity emerging from both [43]. Other heavy fermion systems show similar behavior, though color-coded phase diagrams may not have been produced. A prominent example is CeCoIn$_5$. Its electrical resistivity was first broadly characterized as being linear-in-temperature below 20 K down to the superconducting transition temperature of 2.3 K [50]. Both magnetic field [51,52] and pressure [53] suppress the linear-in-temperature dependence and stabilize Fermi liquid behavior, in agreement with temperature over magnetic field scaling of the magnetic Grüneisen ratio indicating that a quantum critical point is situated at zero field [54]. Indeed, small Cd doping stabilizes an antiferromagnetic state [55].

In Figure 1e–h, we show resistivity-exponent color-coded phase diagrams of other classes of strongly correlated materials, a ruthenate, a cuprate, an iron pnictide, and a schematic phase diagram of MATBG. Extended regions of linear-in-temperature resistivity are also observed.

Before we discuss this strange metal behavior in more detail in Section 5, we take a closer look at the Fermi liquid regions of the heavy fermion phase diagrams.

4. Fermi Liquid Behavior near Quantum Critical Points

The low energy scales and associated low magnetic ordering temperatures typically found in heavy fermion compounds call for investigations of these materials at very low temperatures. Indeed, since early on, measurements down to dilution refrigerator temperatures have been the standard. Because scattering from phonons is strongly suppressed at such low temperatures, this is ideal to study non-Fermi liquid and Fermi liquid behavior alike. The phase diagrams in Figure 1a–d all feature Fermi liquid regions, at least on the paramagnetic side of the QCPs. The fan-like shape of the quantum critical regions dictates that the upper bound of the Fermi liquid regions shrinks upon approaching the QCP. Nevertheless, high-resolution electrical resistivity measurements still allow to extract the evolution of the Fermi liquid A coefficient upon approaching the QCP. In Figure 2 we show such dependencies for four different heavy fermion compounds. In all cases, the A coefficient is very strongly enhanced towards the QCP. In fact, within experimental uncertainty, the data are even consistent with a divergence of A at the QCP, as indicated by the power law fits, $A \sim 1/(B - B_c)^a$, with a close to 1, in Figure 2a,c,d, suggesting that the effective mass diverges at the QCP.

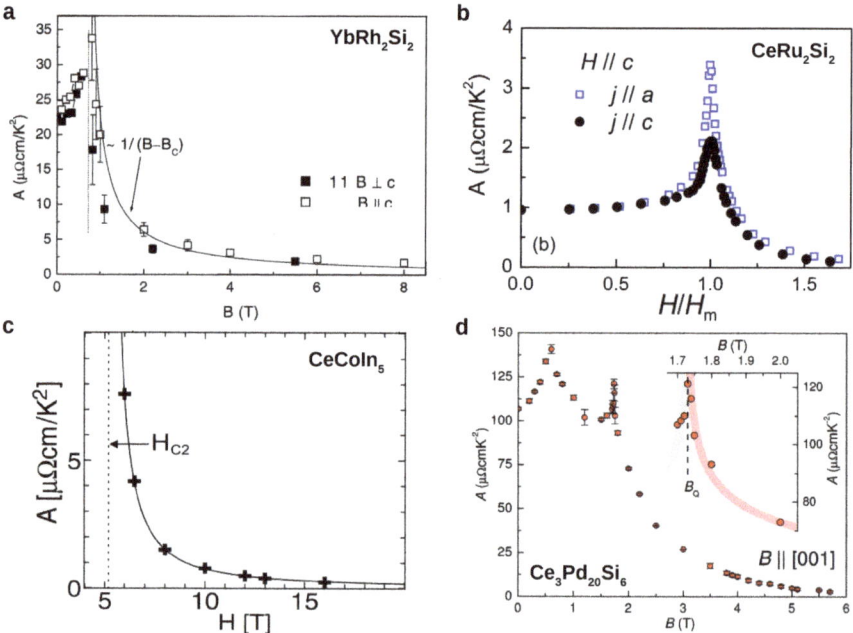

Figure 2. Variation of the A coefficient of the Fermi liquid form of the electrical resistivity, $\rho = \rho_0 + AT^2$, across QCPs in various heavy fermion compounds. (**a**) YbRh$_2$Si$_2$, from [40]. (**b**) CeRu$_2$Si$_2$, from [56]. (**c**) CeCoIn$_5$, from [51]. (**d**) Ce$_3$Pd$_{20}$Si$_6$, from [43].

This finding challenges the classification of heavy fermion compounds into lighter and heavier versions, that has been so popular in the early days of heavy fermion studies and that had culminated in the celebrated Kadowaki–Woods and Sommerfeld–Wilson plots, with each heavy fermion compound represented by a single point. Which A (γ, χ_0) value should now be used in these graphs? In [32] the use of lines instead of points was suggested, using the largest and smallest actually measured values (and not extrapolations beyond them) as end points. The question that remains is whether there is a "background" value, away from a quantum critical point, that is characteristic of a given compound. We will get back to this question in the next section.

5. Strange Metal Behavior and Planckian Dissipation

The occurrence of fans or, in some cases, differently shaped regions of linear-in-temperature resistivity in the phase diagrams of a broad range of correlated electron systems, as highlighted in Figure 1, raises the question whether a universal principle may be behind it. A frequently made argument is that linear-in-temperature resistivity is a natural consequence of the systems' energy scales vanishing at a quantum critical point and thus temperature becoming the only relevant scale. However, both the experimental observation of power laws $\Delta \rho \sim T^{\epsilon}$ with $\epsilon \neq 1$ in quantum critical heavy fermion compounds [57–60] and predictions from order-parameter-fluctuation theories of such laws [36] tell us that this argument cannot hold in general. We thus have to be more specific and ask whether for quantum critical systems that *do* exhibit linear-in-temperature resistivities and, apparently, require description beyond this order-parameter framework, a universal understanding can be achieved.

A direction that is attracting considerable attention [10,11,61] is to test whether the transport scattering rate $1/\tau$ of such systems may be dictated by temperature via

$$\frac{1}{\tau} = \alpha \frac{k_B T}{\hbar} \tag{8}$$

with $\alpha \approx 1$. Should this be the case and τ be the only temperature-dependent quantity in the electrical resistivity, then a linear-in-temperature resistivity would follow naturally. Conceptually, this roots in the insight, gained from the study of models without quasiparticles [4,62–65], that a local equilibration time (after the action of a local perturbation) of any many-body quantum system cannot be faster than the Planckian time

$$\tau_P = \frac{\hbar}{k_B T} \tag{9}$$

associated with the energy $k_B T$ via the Heisenberg uncertainty principle [65]. The question then is how to experimentally test this scenario. The simplest starting point is the Drude form for the electrical resistivity which, in the dc limit, reads

$$\rho = \frac{m}{ne^2} \frac{1}{\tau}, \tag{10}$$

with a temperature-independent effective mass m and charge carrier concentration n, and (8) for the scattering rate $1/\tau$, leading to

$$\rho = \alpha \frac{m}{ne^2} \frac{k_B T}{\hbar}. \tag{11}$$

Interpreting this as the inelastic part of the linear-in-temperature electrical resistivity (7), with $d\rho/dT = A'$, one obtains

$$\alpha = \frac{n}{m} \frac{e^2 \hbar}{k_B} A' \tag{12}$$

or, in convenient units format,

$$\alpha = 2.15 \cdot \frac{n(\text{nm}^{-3})}{m/m_0} \cdot A'(\mu\Omega\text{cm/K}), \tag{13}$$

where m_0 is the free electron mass. When this results in $\alpha \approx 1$, the dissipation is said to be "Planckian". Before looking at experiments, let's contemplate this for a moment. Relation (12) is based on the simple Drude model, and combines properties of well defined quasiparticles (n and m) with a property that characterizes a non-Fermi liquid (A')—possibly one without quasiparticles—that is unlikely to follow the Drude model. Furthermore, as shown in Section 4, the Fermi liquid A coefficient, which is a measure of m, varies strongly with the distance to the QCP. Another defining property of at least some of these strange metals are Fermi surface jumps at the QCP (see Section 7). This adds a nontrivial temperature and tuning parameter dependence to n. One should thus bear in mind that choosing a simple Drude model as starting point holds numerous pitfalls. If still doing so, it is unclear which m and n value to use.

In [10], published quantum oscillation data, in part combined with results from density functional theory (DFT), were used to estimate m and n for a range of different materials, including also "bad metals" (see Section 6) and simple metals in the regime where their resistivity is linear-in-temperature due to scattering from phonons. As an example, for $Sr_3Ru_2O_7$, de Haas–van Alphen (dHvA) data [66] measured at dilution refrigerator temperatures on the low-field side of the strange metal fan (Figure 1e) were used. Contributions from the different bands, assumed as strictly 2D, were summed up as

$$\sigma = \tau \frac{e^2}{\hbar} \sum_i \frac{n_i}{m_i}, \qquad (14)$$

i.e., a constant relaxation time was assumed for all bands. Then, the heavy bands with small carrier concentration play only a minor role. In this way, $\alpha = 1.6$ was obtained. The dHvA effective masses of all bands were found to be modest (at most $10 m_0$) and essentially field-independent [66], even though the A coefficient increases by more than a factor of 8 on approaching the strange metal regime from the low field side [66]. The dHvA experiments may thus not have detected all mass enhancement [10,66]. As shown below, using a larger effective mass would reduce α.

Similar analyses were performed for the other materials [10] and we replot the results as black points in Figure 3. The x axis of this plot is the Fermi velocity v_F which, for a 3D system, can be brought into the form

$$v_F(m/s) = 3.58 \cdot 10^5 \cdot \frac{[n(\mathrm{nm}^{-3})]^{1/3}}{m/m_0}. \qquad (15)$$

The y axis is the inverse of v_F multiplied by α (13) which, again for a 3D system, can be written as

$$\frac{\alpha}{v_F}(s/m) = 6.01 \cdot 10^{-6} \cdot A'(\mu\Omega\mathrm{cm/K}) \cdot [n(\mathrm{nm}^{-3})]^{2/3}. \qquad (16)$$

To further assess how the results for α depend on the choice of the quasiparticle parameters m and n, we here take a different approach. Instead of quantum oscillation data, we use global (effective) properties, namely, the A coefficient and the Hall coefficient R_H, and estimate α for a number of strange metal heavy fermion compounds. Because of the extreme mass renormalizations observed in this class of materials (see Section 4), it is particularly well suited for this test. Combining

$$\frac{m}{m_0} \cdot n^{1/3} = \frac{\gamma_{\mathrm{mole-f.u.}}}{V_{\mathrm{f.u.}}} \frac{3\hbar^2}{N_A m_0 k_B^2 (3\pi^2)^{1/3}} \qquad (17)$$

with the Kadowaki–Woods ratio $A/\gamma^2 = 10^{-5}\,\mu\Omega\mathrm{cm}(\mathrm{mole\,K/mJ})^2$, which is known to be very well obeyed in heavy fermion compounds [28], we obtain

$$\frac{m}{m_0} \cdot [n(\mathrm{nm}^{-3})]^{1/3} = 3.26 \cdot 10^4 \frac{\sqrt{A(\mu\Omega\mathrm{cm/K}^2)}}{V_{\mathrm{f.u.}}(\text{Å}^3)}. \qquad (18)$$

The rationale for using A instead of γ is that precise resistivity measurements are most abundant in the literature (also under challenging conditions such as high pressure and magnetic field) and that the resistivity is much less sensitive to extra contributions from phase transitions than the specific heat. In addition, and unlike γ, the A coefficient picks up effective mass anisotropies, which further improves our analysis. In all cases where reliable γ values were available [43,67–69], the agreement with our A coefficient γ was satisfactory.

A note is due on the determination of the charge carrier concentration n. It is commonly extracted from the Hall coefficient R_H, using the simple one-band relation $R_H = 1/ne$. Heavy fermion compounds are typically multiband systems, and thus compensation effects from electron and hole contributions can occur [70]. To limit the effect of anomalous Hall contributions, low-temperature data should be used [71]. Quantum oscillation experiments can determine the carrier concentration of single bands. However, heavy bands are hard to detect and it is unclear how to sum up contributions from different bands. An alternative is to determine n via the superfluid density [72], as was done previously [49,73], using the relation (in cgs units)

$$n = \left(\frac{\xi_0 \cdot T_c \cdot \gamma}{7.95 \cdot 10^{-24}} \right)^{3/2}, \tag{19}$$

where ξ_0 is the superconducting coherence length, T_c is the superconducting transition temperature, and γ is the normal-state Sommerfeld coefficient, which can be rewritten as

$$n(\mathrm{nm}^{-3}) = 3020 \cdot \left(\frac{\xi_0(\mathrm{nm}) \cdot T_c(\mathrm{K}) \cdot \gamma(\mathrm{Jmol}^{-1}\mathrm{K}^{-2})}{V_{\mathrm{f.u.}}(\mathrm{\AA}^3)} \right)^{3/2}. \tag{20}$$

This may be used as a lower bound of the carrier concentration in the normal state.

Table 1 lists the materials we inspected, with their A coefficients (or, when unavailable, γ), the best estimate of the charge carrier concentration n following the above discussion (see Table 2 for details), and the strange metal A' coefficient. m/m_0 as calculated via (18), or (17), is also listed.

Table 1. Parameters used for Figures 3 and 4. The red (or blue) square represents the largest A coefficient (measured closest to the QCP), the shaded red (or blue) lines the range of A coefficient measured upon moving away from the QCP. The Sommerfeld coefficient γ is estimated from A via the Kadowaki–Wood ratio, unless A data are unavailable. The charge carrier concentrations n and their error bars (where applicable) are taken from Table 2. For CeCoIn$_5$, several values are listed because the A coefficient is different for in-plane (H_a) and out-of-plane (H_c) field, and the A' coefficient is different for in-plane (j_a) and out-of-plane (j_c) currents. For YbAgGe, the A' coefficient changes with field; the two extreme A' values are denoted by the two red squares. For CeCoIn$_5$ ($j \perp c$), Figure 3 shows the range $A' = (0.8 \pm 0.2)$ $\mu\Omega\mathrm{cm/K}$ from [74]. Data for Ce$_3$Pd$_{20}$Si$_6$ refer to the second QCP (near 2 T, see Figure 1d) because for the lower field QCP no full data set on single crystals is published [43,75].

Compound	A ($\mu\Omega\mathrm{cm/K^2}$)	γ (J/molK^2)	m/m_0	n (nm^{-3})	A' ($\mu\Omega\mathrm{cm/K}$)
Ce$_2$IrIn$_8$	–	0.65 [76]	183	2.5	8.8 [76]
Ce$_3$Pd$_{20}$Si$_6$	5–120 [43]	0.707–3.46	136–665	1.7	18.3 [43]
CeCoIn$_5$ (j_a, H_a)	12.4–28.3 [67]	1.11–1.68	310–470	12.4	0.8 [77]
CeCoIn$_5$ (j_a, H_c)	1.72–11.5 [67]	0.414–1.07	116–300	12.4	0.8 [77]
CeCoIn$_5$ (j_c, H_c)	1.72–11.5 [67]	0.414–1.07	116–300	12.4	2.475 [77]
CeRu$_2$Si$_2$	0.1–3.4 [56]	0.1–0.583	53–310	11.6	0.91 [41]
UPt$_3$	–	0.425–0.625 [78]	223–329	21.4	1.1 [10]
YbAgGe ($H//a$)	–	0.87–1.4 [79]	1300–2100	1.6	27–59 [80]
YbRh$_2$Si$_2$	1.7–33.8 [68]	0.41–1.85	250–1100	10	1.83 [68]

All these data are then included in Figure 3 in the following way. The v_F (15) and α/v_F (16) value resulting from the largest measured A coefficient (or γ value) for each compound is shown as red square. The shaded red lines represent the published ranges of A coefficient (or γ value). The error bars represent uncertainties in the determination of the charge carrier concentration (see Table 1). Lines for $\alpha = 1, 0.1$, and 0.01 are also shown. It is clear that none of the shaded red lines overlaps with the $\alpha = 1$ line. The discrepancy with the points extracted from quantum oscillation experiments [10] is quite striking.

Table 2. Charge carrier concentrations (in nm^{-3}) determined as follows: (i) n_{sc} from the superconducting coherence length ξ_0, the superconducting transition temperature T_c, and the normal-state Sommerfeld coefficient γ, all in zero field, via (20); (ii) n_H from the Hall coefficient at the lowest temperatures, where anomalous contributions are minimal, via $R_H = 1/ne$; (iii) n_{qo} from quantum oscillation experiments reviewed in [10], by summing up the carrier concentrations from all detected bands. For CeCoIn$_5$, the γ coefficient is taken at 2.5 K, without taking into account the logarithmic divergence. The error bar in n used for CeCoIn$_5$ ($j \perp c$) in Figure 3 reflects the range of the parameters given in [74]. YbRh$_2$Si$_2$ is close to being a compensated metal, resulting in a strong sensitivity of n to small differences in the residual resistivity. The largest reported R_H value, which corresponds to $n_H = 26.0$ [71], has the lowest compensation and is thus most accurate. Nevertheless, the R_H value of LuRh$_2$Si$_2$ is even larger, corresponding to $n_H = 11.6 \, nm^{-3}$ [70], suggesting that there is still some degree of compensation in the sample of [71]. We list the average of both values, $18.8 \, nm^{-3}$, as best n_H estimate. For the plots, we use the approximate average of n_{sc} and n_H, i.e., $10 \, nm^{-3}$, with an asymmetric error bar $\delta n_+ = 10 \, nm^{-3}$ and $\delta n_- = -5 \, nm^{-3}$ (see Table 1). Similar compensation effects are also encountered in UPt$_3$ [81]. Bold fonts indicate the values used for the α estimates (see Table 1).

Compound	ξ_0 (nm)	T_c (K)	γ (J/molK^2)	n_{sc}	n_H	n_{qo}
Ce$_2$IrIn$_8$	-	-	-	-	2.5 [82]	-
Ce$_3$Pd$_{20}$Si$_6$	-	-	-	-	1.7 [43]	-
CeCoIn$_5$	5.6 [83]	2.3 [50]	290 [50]	10.8	10.1 [84]–12.5 [74]	12.4 [10]
CeRu$_2$Si$_2$	-	-	-	-	3.1 [41]–7.8 [85]	11.6 [10]
UPt$_3$	12 [86]	0.52 [86]	0.43 [78]	22.4	9 [85]	21.4 [10]
YbAgGe	-	-	-	-	1.6 [87]	-
YbRh$_2$Si$_2$	97 [49]	0.0079 [49]	1.42 [49]	4.86	18.8 [70,71]	-

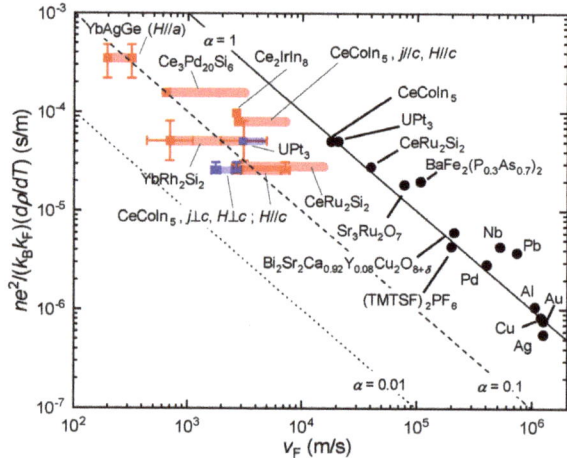

Figure 3. Planckian dissipation plot of [10] revisited. Double-logarithmic plot of Fermi velocity v_F vs. $ne^2/(k_B k_F)(d\rho/dT) = \alpha/v_F$ with data from [10] (black points) and data of the heavy fermion compounds listed in Table 1 and analyzed here. The red squares result from the largest measured A coefficient (or γ value) for each compound near the strange metal regime, the shaded red lines from the published ranges of A coefficient (or γ value), and the error bars from uncertainties in the determination of the charge carrier concentration n and sometimes other parameters (see Table 1). The full, dashed, and dotted line represent $\alpha = 1, 0.1$, and 0.01, respectively.

In Figure 4 we present these results in a different form, as α vs. $(m/m_0)/n$. The red squares and red shaded lines have the same meaning as in Figure 3. The dashed lines are extrapolations of the shaded lines to $\alpha = 1$. We can thus directly read off the values of $(m/m_0)/n$ for which a given compound would, in this simple framework, be a Planckian scatterer. In all cases, this is for effective masses significantly smaller than even the smallest measured ones in the Fermi liquid regime.

What are the implications of this finding? We first comment on the discrepancy with the results from [10]. Apparently, averaging the contributions from different bands detected in quantum oscillation experiments via (14) leads to sizeably larger Fermi velocities (sizeably smaller effective masses) than our A coefficient approach. In heavy fermion compounds, a coherent heavy fermion state forms at low temperatures, and the Fermi liquid A coefficient is known to be a pertinent measure thereof. It is thus either the use of (14) that should be reconsidered or the reliance in quantum oscillation experiments to detect the heaviest quasiparticles. Clearly, if dissipation in strange metal heavy fermion compounds is to be Planckian, this would hold only for the very weakly renormalized quasiparticles, as argued for in [88]. To us, this is a rather puzzling result as heavy fermion bands get successively renormalized with decreasing temperature and thus one would have expected that the "background" to effects of quantum critical fluctuations already contains a sizeable non-critical Kondo renormalization.

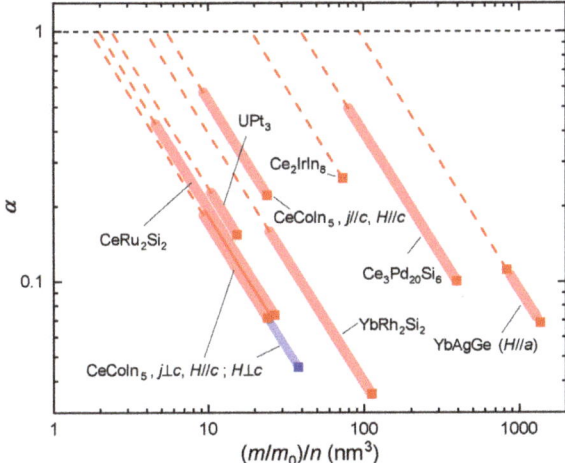

Figure 4. No Planckian dissipation from heavy quasiparticles in heavy fermion compounds. Double-logarithmic plot of α vs. $(m/m_0)/n$ for various strange metal heavy fermion compounds, as given in Table 1. Red squares and shaded lines have the same meaning as in Figure 3. The dashed lines are to help reading off the values of $(m/m_0)/n$ for which the linear-in-temperature electrical resistivity in these compounds could be governed by Planckian dissipation. Note that in all cases the "Planckian dissipation" effective masses obtained in this way are sizeably smaller than even the smallest values experimentally accessed by tuning the systems away from the strange metal regime (top end of full shaded lines).

6. Strange Metal Behavior and the Mott–Ioffe–Regel Limit

In a number of strongly correlated electron systems, including quasi-2D conductors such as the high-T_c cuprates but also 3D transition metal oxides and alkali-doped fullerides, linear-in-temperature resistivity is observed beyond the Mott–Ioffe–Regel (MIR) limit [89,90]. At this limit, the electron mean free path approaches certain microscopic length scales such as the interatomic spacing or the wavelength $2\pi/k_F$ [65,91–94]. Semiclassical transport of long-lived quasiparticles might then, at least naively, be expected not to exist and the resistivity should saturate, in 3D systems of interest to us here to

$$\rho_{\text{MIR}} = \frac{h}{e^2} \cdot L, \tag{21}$$

where L is the relevant microscopic length scale. Using the Drude resistivity (10) with the Fermi velocity $v_\text{F} = \hbar k_\text{F}/m$, the Fermi wave vector $k_\text{F} = (3\pi^2 n)^{1/3}$, and the mean free path $\ell = \tau v_\text{F}$ one obtains

$$\rho = \frac{h}{e^2} \cdot \frac{3\pi}{2} \frac{1}{k_\text{F}^2 \ell} = \frac{h}{e^2} \cdot L \cdot C, \tag{22}$$

where the value of the constant C depends on details of the electronic and crystal structure. Assuming $C = 1$, one gets

$$\rho_{\text{MIR}}(\mu\Omega\text{cm}) = 258 \cdot L(\text{Å}), \tag{23}$$

In heavy fermion compounds, linear-in-temperature resistivities are limited to low temperatures (Figure 1a–d) and the A' coefficients (Table 1) typically result in inelastic resistivities of the order of 10 µΩcm at the upper bound of the linear regime. This is well below the MIR limit. For instance, for YbRh$_2$Si$_2$, using the lattice parameters $a = 4.007$ Å and $c = 9.858$ Å [48] for L in (23) gives $\rho_{\text{MIR}} \approx 1000$ µΩcm and ≈ 2500 µΩcm, respectively, much larger than even the total resistivity at 15 K (which is about 30 µΩcm for YbRh$_2$Si$_2$ [48]), the upper bound of linear-in-temperature resistivity for that compound. In this case, a confusion with a linear-in-temperature resistivity due to electron-phonon scattering [65,95] can be safely ruled out.

7. Strange Metal Behavior and Fermi Surface Jumps

In Section 5, a simple Drude form was used for the electrical resistivity and all temperature dependence was attributed to the scattering rate. Then, the question was asked which quasiparticles (with which m/n) to take if the scattering were to be Planckian. The answer was that this would have to be very weakly interacting quasiparticles, certainly not the ones close to the QCP from which the strange metal behavior emerges. Here we address another phenomenon that may challenge a Planckian scattering rate picture: Fermi surface jumps across these QCPs.

This phenomenon was first detected by Hall effect measurements on YbRh$_2$Si$_2$ [71,96] (Figure 5a). Let us first recapitulate the experimental evidence for a Fermi surface jump across a QCP, as put forward in these works. Hall coefficient R_H (or Hall resistivity ρ_H) isotherms are measured as function of a tuning parameter δ (in case of YbRh$_2$Si$_2$ the magnetic field) across the QCP. A phenomenological crossover function, $R_\text{H}^\infty - (R_\text{H}^\infty - R_\text{H}^0)/[1 + (\delta/\delta_0)^p]$ [71], is fitted to $R_\text{H}(\delta)$ [or to $d\rho_\text{H}/dB(\delta)$] and its full width at half maximum (FWHM) is determined as a reliable measure of the crossover width. Only if this width extrapolates to zero in the zero-temperature limit a Hall coefficient jump is established. Of course, the jump size must remain finite in the zero temperature limit. To identify a Fermi surface jump, other origins of Hall effect changes must be ruled out, for instance anomalous Hall contributions from abrupt magnetization changes at a metamagnetic/first order transition [97]. All this was done for YbRh$_2$Si$_2$ [71,96]. For Ce$_3$Pd$_{20}$Si$_6$, using a very similar procedure, two Fermi surface jumps were found at the two consecutive QCPs (Figure 1d) [43,75]. The crossover at the first QCP [75] is shown in Figure 5b. It is also important to remind oneself that no Fermi surface discontinuity is expected at a conventional antiferromagnetic QCP as described by the spin-density-wave/order-parameter scenario [6]. Band folding of the (even at $T = 0$) continuously onsetting order parameter can in that case only lead to a continuously varying Hall coefficient, as seen for instance in the itinerant antiferromagnet Cr upon the suppression of the order by doping or pressure (see [98] for more details and the original references).

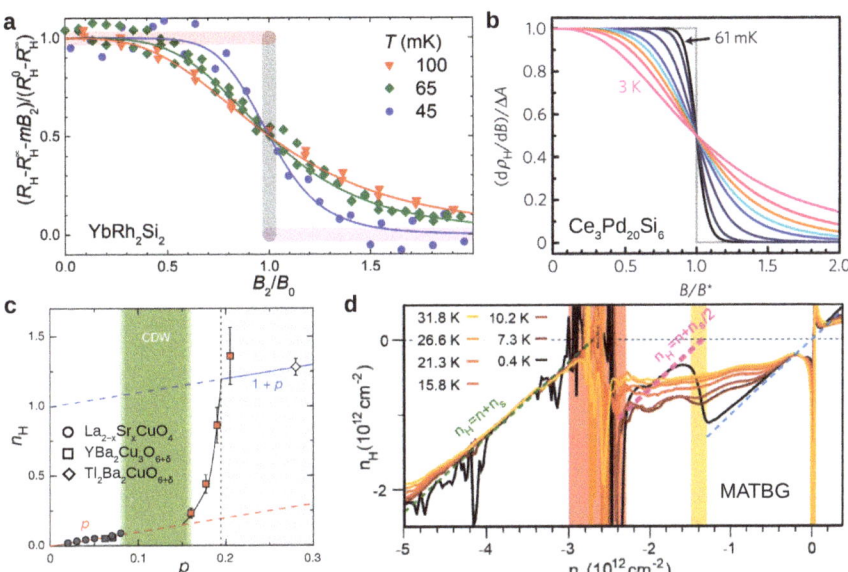

Figure 5. Fermi surface jumps as evidenced by Hall effect measurements in several strange metals. (**a**) YbRh$_2$Si$_2$, from [32,96]. (**b**) Ce$_3$Pd$_{20}$Si$_6$, from [75]. (**c**) Substitution series of three high-T_c cuprates, from [99]. (**d**) MATBG, from [100].

These jumps are understood as defining signatures of a Kondo destruction QCP, first proposed theoretically [5,6,101] in conjunction with inelastic neutron scattering experiments on CeCu$_{5.9}$Au$_{0.1}$ [9]. At such a QCP, the heavy quasiparticles, composites with f and conduction electron components, disintegrate. The Fermi surface jumps because the local moment, which is part of the Fermi surface in the paramagnetic Kondo coherent ground state [102], drops out as the f electrons localize. As such, Kondo destruction QCPs are sometimes referred to as f-orbital selective Mott transitions. More recently, THz time-domain transmission experiments on YbRh$_2$Si$_2$ thin films grown by molecular beam epitaxy revealed dynamical scaling of the optical conductivity [103]. This shows that the charge carriers are an integral part of the quantum criticality, and should not be seen as a conserved quantity that merely undergo strong scattering (as in order-parameter-fluctuation descriptions with intact quasiparticles). We also note that a Drude description of the optical conductivity fails rather drastically in the quantum critical regime [103]. It is thus unclear how this physics could be captured by the simple Planckian scattering approach described above.

Interestingly, Hall effect experiments in other strange metal platforms also hint at Fermi surface reconstructions. Two examples are included in Figure 5: a series of substituted high-T_c cuprates [99] (panel c) and MATBG as function of the total charge density induced by the gate [100] (panel d). Evidence for related physics has also been found in the pnictides [104]. The physics here appears to be related to the presence of d orbitals with a different degree of localization, with one of them undergoing a Mott transition, such as described by multi-orbital Hubbard models [105,106]. It may well be that Fermi surface jumps are an integral part of strange metal physics, and should be included as a starting point in its description.

8. Summary and Outlook

We have revisited the question whether the strange metal behavior encountered in numerous strongly correlated electron materials may be the result of Planckian dissipation. For this purpose, we have examined strange metal heavy fermion compounds. Their temperature–tuning parameter phase diagrams are particularly simple: Fans of strange metal behavior emerge from quantum critical points, in a Fermi liquid background. This, together with the extreme mass renormalizations found in these materials, makes them a particularly well-suited testbed.

As done previously, we use the Drude form of the electrical conductivity as a starting point, but complementary to a previous approach based on quantum oscillation data, we here rely on the Fermi liquid A coefficient as precise measure of the quasiparticle renormalization. We find that for any of the measured A coefficients, the slope of the linear-in-temperature strange metal resistivity A' is much smaller than the value expected from Planckian dissipation. We also propose a new plot that allows to read off the ratio of effective mass to carrier concentration that one would have to attribute to the quasiparticles for their scattering to be Planckian. It corresponds to very modest effective masses. While this could be something like a smooth background to quantum critical phenomena, the fact that the strange metal regime occurs entirely below the temperature for the initial onset of the dynamical Kondo correlations suggests that this background should already incorporate the non-critical Kondo correlations and thus correspond to a relatively heavy mass.

We have also pointed out that several heavy fermion compounds exhibit Fermi surface jumps across strange metal quantum critical points and that this challenges the Drude picture underlying the Planckian analysis. Indications for such jumps are also seen in other platforms and may thus be a common feature of strange metals. Further careful studies that evidence a sharp Fermi surface change in the zero temperature limit, such as providing for some of the heavy fermion compounds, are called for. On the theoretical side, approaches that discuss the electrical resistivity as an entity and do not single out a scattering rate as the only origin of strangeness, are needed.

Author Contributions: Conceptualization and original draft preparation, S.P.; data analysis, M.T. and S.P. All authors have read and agreed to the published version of the manuscript.

Funding: This research has received funding from the European Union's Horizon 2020 Research and Innovation Programme under Grant Agreement no 824109, and from the Austrian Science Fund (FWF Grant 29296-N27).

Institutional Review Board Statement: Not applicable.

Informed Consent Statement: Not applicable.

Data Availability Statement: No new data were created in this study. Data sharing is thus not applicable to this article.

Acknowledgments: Open Access Funding by the Austrian Science Fund (FWF). We acknowledge fruitful discussions with Joe Checkelsky, Piers Coleman, Pablo Jarillo-Herrero, Stefan Kirchner, Jose Lado, Patrick Lee, Xinwei Li, Doug Natelson, Aline Ramires, T. Senthil, Vasily R. Shaginyan, Qimiao Si, Chandra Varma, and Grigory Volovik.

Conflicts of Interest: The authors declare no conflict of interest.

References

1. Bhattacharjee, S.M.; Mj, M.; Bandyopadhyay, A. *Topology and Condensed Matter Physics*; Springer: Singapore, 2017. [CrossRef]
2. Landau, L.D. On the theory of phase transitions. I. *Phys. Z. Soviet.* **1937**, *11*, 26. Available online: http://cds.cern.ch/record/480039 (accessed on 8 February 2022).
3. Wilson, K.G. The renormalization group: Critical phenomena and the Kondo problem. *Rev. Mod. Phys.* **1975**, *47*, 773–840. [CrossRef]
4. Sachdev, S. *Quantum Phase Transitions*; Cambridge University Press: Cambridge, UK, 1999. [CrossRef]
5. Si, Q.; Rabello, S.; Ingersent, K.; Smith, J. Locally critical quantum phase transitions in strongly correlated metals. *Nature* **2001**, *413*, 804. [CrossRef] [PubMed]
6. Coleman, P.; Pépin, C.; Si, Q.; Ramazashvili, R. How do Fermi liquids get heavy and die? *J. Phys. Condens. Matter* **2001**, *13*, R723–R738. [CrossRef]
7. Senthil, T.; Vishwanath, A.; Balents, L.; Sachdev, S.; Fisher, M. Deconfined quantum critical points. *Science* **2004**, *303*, 1490–1494. [CrossRef]
8. Senthil, T.; Balents, L.; Sachdev, S.; Vishwanath, A.; Fisher, M.P.A. Quantum criticality beyond the Landau-Ginzburg-Wilson paradigm. *Phys. Rev. B* **2004**, *70*, 144407. [CrossRef]

9. Schröder, A.; Aeppli, G.; Coldea, R.; Adams, M.; Stockert, O.; v. Löhneysen, H.; Bucher, E.; Ramazashvili, R.; Coleman, P. Onset of antiferromagnetism in heavy-fermion metals. *Nature* **2000**, *407*, 351–355. [CrossRef]
10. Bruin, J.A.N.; Sakai, H.; Perry, R.S.; Mackenzie, A.P. Similarity of scattering rates in metals showing *T*-linear resistivity. *Science* **2013**, *339*, 804. [CrossRef]
11. Legros, A.; Benhabib, S.; Tabis, W.; Laliberté, F.; Dion, M.; Lizaire, M.; Vignolle, B.; Vignolles, D.; Raffy, H.; Li, Z.Z.; et al. Universal *T*-linear resistivity and Planckian dissipation in overdoped cuprates. *Nat. Phys.* **2019**, *15*, 142. [CrossRef]
12. Hewson, A.C. *The Kondo Problem to Heavy Fermions*; Cambridge University Press: Cambridge, UK, 1997. [CrossRef]
13. Coleman, P. *Introduction to Many-Body Physics*; Cambridge University Press: Cambridge, UK, 2015. [CrossRef]
14. Khodel, V.A.; Shaginyan, V.R. Superfluidity in system with fermion condensate. *JETP Lett.* **1990**, *51*, 553.
15. Volovik, G.E. Flat band and Planckian metal. *JETP Lett.* **2019**, *110*, 352–353. [CrossRef]
16. Shaginyan, V.R.; Amusia, M.Y.; Msezane, A.Z.; Stephanovich, V.A.; Japaridze, G.S.; Artamonov, S.A. Fermion condensation, *T*-linear resistivity, and Planckian limit. *JETP Lett.* **2019**, *110*, 290–295. [CrossRef]
17. Bistritzer, R.; MacDonald, A.H. Moiré bands in twisted double-layer graphene. *Proc. Natl. Acad. Sci. USA* **2011**, *108*, 12233. [CrossRef] [PubMed]
18. Cao, Y.; Fatemi, V.; Demir, A.; Fang, S.; Tomarken, S.L.; Luo, J.Y.; Sanchez-Yamagishi, J.D.; Watanabe, K.; Taniguchi, T.; Kaxiras, E.; et al. Correlated insulator behaviour at half-filling in magic-angle graphene superlattices. *Nature* **2018**, *556*, 80–84. [CrossRef]
19. Derzhko, O.; Richter, J.; Maksymenko, M. Strongly correlated flat-band systems: The route from Heisenberg spins to Hubbard electrons. *Int. J. Mod. Phys. B* **2015**, *29*, 1530007. [CrossRef]
20. Leykam, D.; Andreanov, A.; Flach, S. Artificial flat band systems: From lattice models to experiments. *Adv. Phys. X* **2018**, *3*, 1473052. [CrossRef]
21. Kang, M.; Ye, L.; Fang, S.; You, J.S.; Levitan, A.; Han, M.; Facio, J.I.; Jozwiak, C.; Bostwick, A.; Rotenberg, E.; et al. Dirac fermions and flat bands in the ideal kagome metal FeSn. *Nat. Mater.* **2020**, *19*, 163. [CrossRef]
22. Ye, L.; Fang, S.; Kang, M.G.; Kaufmann, J.; Lee, Y.; Denlinger, J.; Jozwiak, C.; Bostwick, A.; Rotenberg, E.; Kaxiras, E.; et al. A flat band-induced correlated kagome metal. *arXiv* **2021**, arXiv:2106.10824.
23. Stewart, G.R. Heavy-fermion systems. *Rev. Mod. Phys.* **1984**, *56*, 755. [CrossRef]
24. Stewart, G.R. Non-Fermi-liquid behavior in *d*- and *f*-electron metals. *Rev. Mod. Phys.* **2001**, *73*, 797. [CrossRef]
25. v. Löhneysen, H.; Rosch, A.; Vojta, M.; Wölfle, P. Fermi-liquid instabilities at magnetic quantum critical points. *Rev. Mod. Phys.* **2007**, *79*, 1015. [CrossRef]
26. Ramires, A.; Lado, J.L. Emulating heavy fermions in twisted trilayer graphene. *Phys. Rev. Lett.* **2021**, *127*, 026401. [CrossRef] [PubMed]
27. Song, Z.D.; Bernevig, B.A. MATBG as topological heavy fermion: I. Exact mapping and correlated insulators, *arXiv* **2021**, arXiv:2111.05865.
28. Kadowaki, K.; Woods, S.B. Universal relationship of the resistivity and specific heat in heavy-fermion compounds. *Solid State Commun.* **1986**, *58*, 507–509. [CrossRef]
29. Tsujii, N.; Kontani, H.; Yoshimura, K. Universality in heavy fermion systems with general degeneracy. *Phys. Rev. Lett.* **2005**, *94*, 057201. [CrossRef]
30. Jacko, A.C.; Fjaerestad, J.O.; Powell, B.J. A unified explanation of the Kadowaki–Woods ratio in strongly correlated metals. *Nat. Phys.* **2009**, *5*, 422. [CrossRef]
31. Kirchner, S.; Paschen, S.; Chen, Q.; Wirth, S.; Feng, D.; Thompson, J.D.; Si, Q. Colloquium: Heavy-electron quantum criticality and single-particle spectroscopy. *Rev. Mod. Phys.* **2020**, *92*, 011002. [CrossRef]
32. Paschen, S.; Si, Q. Quantum phases driven by strong correlations. *Nat. Rev. Phys.* **2021**, *3*, 9–26. [CrossRef]
33. Hertz, J.A. Quantum critical phenomena. *Phys. Rev. B* **1976**, *14*, 1165. [CrossRef]
34. Millis, A.J. Effect of a nonzero temperature on quantum critical points in itinerant fermion systems. *Phys. Rev. B* **1993**, *48*, 7183. [CrossRef]
35. Moriya, T. *Spin Fluctuations in Itinerant Electron Magnetism*; Springer: Berlin, Germany, 1985; Volume 56, pp. 44–81. [CrossRef]
36. Rosch, A. Interplay of disorder and spin fluctuations in the resistivity near a quantum critical point. *Phys. Rev. Lett.* **1999**, *82*, 4280. [CrossRef]
37. Smith, R.P.; Sutherland, M.; Lonzarich, G.G.; Saxena, S.S.; Kimura, N.; Takashima, S.; Nohara, M.; Takagi, H. Marginal breakdown of the Fermi-liquid state on the border of metallic ferromagnetism. *Nature* **2008**, *455*, 1220–1223. [CrossRef]
38. Hlubina, R.; Rice, T.M. Resistivity as a function of temperature for models with hot spots on the Fermi surface. *Phys. Rev. B* **1995**, *51*, 9253–9260. [CrossRef] [PubMed]
39. Nagaosa, N. RVB vs Fermi liquid picture of high-T_c superconductors. *J. Phys. Chem. Solids* **1992**, *53*, 1493–1498. [CrossRef]
40. Custers, J.; Gegenwart, P.; Wilhelm, H.; Neumaier, K.; Tokiwa, Y.; Trovarelli, O.; Geibel, C.; Steglich, F.; Pépin, C.; Coleman, P. The break-up of heavy electrons at a quantum critical point. *Nature* **2003**, *424*, 524. [CrossRef] [PubMed]
41. Daou, R.; Bergemann, C.; Julian, S.R. Continuous evolution of the Fermi surface of $CeRu_2Si_2$ across the metamagnetic transition. *Phys. Rev. Lett.* **2006**, *96*, 026401. [CrossRef]
42. Park, T.; Sidorov, V.A.; Ronning, F.; Zhu, J.X.; Tokiwa, Y.; Lee, H.; Bauer, E.D.; Movshovich, R.; Sarrao, J.L.; Thompson, J.D. Isotropic quantum scattering and unconventional superconductivity. *Nature* **2008**, *456*, 366–368. [CrossRef]

43. Martelli, V.; Cai, A.; Nica, E.M.; Taupin, M.; Prokofiev, A.; Liu, C.C.; Lai, H.H.; Yu, R.; Ingersent, K.; Küchler, R.; et al. Sequential localization of a complex electron fluid. *Proc. Natl. Acad. Sci. USA* **2019**, *116*, 17701. [CrossRef]
44. Grigera, S.A.; Perry, R.S.; Schofield, A.J.; Chiao, M.; Julian, S.R.; Lonzarich, G.G.; Ikeda, S.I.; Maeno, Y.; Millis, A.J.; Mackenzie, A.P. Magnetic field-tuned quantum criticality in the metallic ruthenate $Sr_3Ru_2O_7$. *Science* **2001**, *294*, 329–332. [CrossRef]
45. Cooper, R.A.; Wang, Y.; Vignolle, B.; Lipscombe, O.J.; Hayden, S.M.; Tanabe, Y.; Adachi, T.; Koike, Y.; Nohara, M.; Takagi, H.; et al. Anomalous criticality in the electrical resistivity of $La_{2-x}Sr_xCuO_4$. *Science* **2009**, *323*, 603–607. [CrossRef]
46. Hashimoto, K.; Cho, K.; Shibauchi, T.; Kasahara, S.; Mizukami, Y.; Katsumata, R.; Tsuruhara, Y.; Terashima, T.; Ikeda, H.; Tanatar, M.A.; et al. A sharp peak of the zero-temperature penetration depth at optimal composition in $BaFe_2(As_{1-x}P_x)_2$. *Science* **2012**, *336*, 1554. [CrossRef] [PubMed]
47. Jaoui, A.; Das, I.; Battista, G.D.; Díez-Mérida, J.; Lu, X.; Watanabe, K.; Taniguchi, T.; Ishizuka, H.; Levitov, L.; Efetov, D.K. Quantum-critical continuum in magic-angle twisted bilayer graphene, *arXiv* **2021**, arXiv:2108.07753.
48. Trovarelli, O.; Geibel, C.; Mederle, S.; Langhammer, C.; Grosche, F.M.; Gegenwart, P.; Lang, M.; Sparn, G.; Steglich, F. $YbRh_2Si_2$: Pronounced non-Fermi-liquid effects above a low-lying magnetic phase transition. *Phys. Rev. Lett.* **2000**, *85*, 626–629. [CrossRef] [PubMed]
49. Nguyen, D.H.; Sidorenko, A.; Taupin, M.; Knebel, G.; Lapertot, G.; Schuberth, E.; Paschen, S. Superconductivity in an extreme strange metal. *Nat. Commun.* **2021**, *12*, 4341. [CrossRef]
50. Petrovic, C.; Pagliuso, P.G.; Hundley, M.F.; Movshovich, R.; Sarrao, J.L.; Thompson, J.D.; Fisk, Z.; Monthoux, P. Heavy-fermion superconductivity in $CeCoIn_5$ at 2.3 K. *J. Phys. Condens. Matter* **2001**, *13*, L337–L342. [CrossRef]
51. Paglione, J.; Tanatar, M.A.; Hawthorn, D.G.; Boaknin, E.; Hill, R.W.; Ronning, F.; Sutherland, M.; Taillefer, L.; Petrovic, C.; Canfield, P.C. Field-induced quantum critical point in $CeCoIn_5$. *Phys. Rev. Lett.* **2003**, *91*, 246405. [CrossRef]
52. Bianchi, A.; Movshovich, R.; Vekhter, I.; Pagliuso, P.G.; Sarrao, J.L. Avoided antiferromagnetic order and quantum critical point in $CeCoIn_5$. *Phys. Rev. Lett.* **2003**, *91*, 257001. [CrossRef]
53. Sidorov, V.A.; Nicklas, M.; Pagliuso, P.G.; Sarrao, J.L.; Bang, Y.; Balatsky, A.V.; Thompson, J.D. Superconductivity and quantum criticality in $CeCoIn_5$. *Phys. Rev. Lett.* **2002**, *89*, 157004. [CrossRef]
54. Tokiwa, Y.; Bauer, E.D.; Gegenwart, P. Zero-field quantum critical point in $CeCoIn_5$. *Phys. Rev. Lett.* **2013**, *111*, 107003. [CrossRef]
55. Pham, L.D.; Park, T.; Maquilon, S.; Thompson, J.D.; Fisk, Z. Reversible tuning of the heavy-fermion ground state in $CeCoIn_5$. *Phys. Rev. Lett.* **2006**, *97*, 056404. [CrossRef]
56. Boukahil, M.; Pourret, A.; Knebel, G.; Aoki, D.; Ōnuki, Y.; Flouquet, J. Lifshitz transition and metamagnetism: Thermoelectric studies of $CeRu_2Si_2$. *Phys. Rev. B* **2014**, *90*, 075127. [CrossRef]
57. Mathur, N.; Grosche, F.; Julian, S.; Walker, I.; Freye, D.; Haselwimmer, R.; Lonzarich, G. Magnetically mediated superconductivity in heavy fermion compounds. *Nature* **1998**, *394*, 39–43. [CrossRef]
58. Gegenwart, P.; Kromer, F.; Lang, M.; Sparn, G.; Geibel, C.; Steglich, F. Non-Fermi-liquid effects at ambient pressure in a stoichiometric heavy-fermion compound with very low disorder: $CeNi_2Ge_2$. *Phys. Rev. Lett.* **1999**, *82*, 1293–1296. [CrossRef]
59. Knebel, G.; Braithwaite, D.; Canfield, P.C.; Lapertot, G.; Flouquet, J. Electronic properties of $CeIn_3$ under high pressure near the quantum critical point. *Phys. Rev. B* **2001**, *65*, 024425. [CrossRef]
60. Nakatsuji, S.; Kuga, K.; Machida, Y.; Tayama, T.; Sakakibara, T.; Karaki, Y.; Ishimoto, H.; Yonezawa, S.; Maeno, Y.; Pearson, E.; et al. Superconductivity and quantum criticality in the heavy-fermion system β-$YbAlB_4$. *Nat. Phys.* **2008**, *4*, 603–607. [CrossRef]
61. Grissonnanche, G.; Fang, Y.; Legros, A.; Verret, S.; Laliberté, F.; Collignon, C.; Zhou, J.; Graf, D.; Goddard, P.A.; Taillefer, L.; et al. Linear-in temperature resistivity from an isotropic Planckian scattering rate. *Nature* **2021**, *595*, 667–672. [CrossRef]
62. Zaanen, J. Why the temperature is high. *Nature* **2004**, *430*, 512–513. [CrossRef]
63. Hartnoll, S.A.; Lucas, A.; Sachdev, S. Holographic quantum matter. *arXiv* **2016**, arXiv:1612.07324.
64. Hartnoll, S.A.; Mackenzie, A.P. Planckian dissipation in metals. *arXiv* **2021**, arXiv:2107.07802.
65. Chowdhury, D.; Georges, A.; Parcollet, O.; Sachdev, S. Sachdev-Ye-Kitaev models and beyond: A window into non-Fermi liquids. *arXiv* **2021**, arXiv:2109.05037.
66. Mercure, J.F.; Rost, A.W.; O'Farrell, E.C.T.; Goh, S.K.; Perry, R.S.; Sutherland, M.L.; Grigera, S.A.; Borzi, R.A.; Gegenwart, P.; Gibbs, A.S.; et al. Quantum oscillations near the metamagnetic transition in $Sr_3Ru_2O_7$. *Phys. Rev. B* **2010**, *81*, 235103. [CrossRef]
67. Ronning, F.; Capan, C.; Bianchi, A.; Movshovich, R.; Lacerda, A.; Hundley, M.F.; Thompson, J.D.; Pagliuso, P.G.; Sarrao, J.L. Field-tuned quantum critical point in $CeCoIn_5$ near the superconducting upper critical field. *Phys. Rev. B* **2005**, *71*, 104528. [CrossRef]
68. Gegenwart, P.; Custers, J.; Geibel, C.; Neumaier, K.; Tayama, T.; Tenya, K.; Trovarelli, O.; Steglich, F. Magnetic-field induced quantum critical point in $YbRh_2Si_2$. *Phys. Rev. Lett.* **2002**, *89*, 056402. [CrossRef] [PubMed]
69. Aoki, Y.; Matsuda, T.; Sugawara, H.; Sato, H.; Ohkuni, H.; Settai, R.; Onuki, Y.; Yamamoto, E.; Haga, Y.; Andreev, A.; et al. Thermal properties of metamagnetic transition in heavy-fermion systems. *J. Magn. Magn. Mater.* **1998**, *177–181*, 271–276. [CrossRef]
70. Friedemann, S.; Wirth, S.; Oeschler, N.; Krellner, C.; Geibel, C.; Steglich, F.; MaQuilon, S.; Fisk, Z.; Paschen, S.; Zwicknagl, G. Hall effect measurements and electronic structure calculations on $YbRh_2Si_2$ and its reference compounds $LuRh_2Si_2$ and $YbIr_2Si_2$. *Phys. Rev. B* **2010**, *82*, 035103. [CrossRef]
71. Paschen, S.; Lühmann, T.; Wirth, S.; Gegenwart, P.; Trovarelli, O.; Geibel, C.; Steglich, F.; Coleman, P.; Si, Q. Hall-effect evolution across a heavy-fermion quantum critical point. *Nature* **2004**, *432*, 881. [CrossRef]

72. Orlando, T.P.; McNiff, E.J., Jr.; Foner, S.; Beasleya, M.R. Critical fields, Pauli paramagnetic limiting, and material parameters of Nb$_3$Sn and V$_3$Si. *Phys. Rev. B* **1979**, *19*, 4545. [CrossRef]
73. Rauchschwalbe, U.; Lieke, W.; Bredl, C.D.; Steglich, F.; Aarts, J.; Martini, K.M.; Mota, A.C. Critical fields of the "heavy-fermion" superconductor CeCu$_2$Si$_2$. *Phys. Rev. Lett.* **1982**, *49*, 1448–1451. [CrossRef]
74. Maksimovic, N.; Eilbott, D.H.; Cookmeyer, T.; Wan, F.; Rusz, J.; Nagarajan, V.; Haley, S.C.; Maniv, E.; Gong, A.; Faubel, S.; et al. Evidence for a delocalization quantum phase transition without symmetry breaking in CeCoIn$_5$. *Science* **2021**, *375*, 76–81. [CrossRef]
75. Custers, J.; Lorenzer, K.; Müller, M.; Prokofiev, A.; Sidorenko, A.; Winkler, H.; Strydom, A.M.; Shimura, Y.; Sakakibara, T.; Yu, R.; et al. Destruction of the Kondo effect in the cubic heavy-fermion compound Ce$_3$Pd$_{20}$Si$_6$. *Nat. Mater.* **2012**, *11*, 189. [CrossRef]
76. Kim, J.S.; Moreno, N.O.; Sarrao, J.L.; Thompson, J.D.; Stewart, G.R. Field-induced non-Fermi-liquid behavior in Ce$_2$IrIn$_8$. *Phys. Rev. B* **2004**, *69*, 024402. [CrossRef]
77. Tanatar, M.A.; Paglione, J.; Petrovic, C.; Taillefer, L. Anisotropic violation of the Wiedemann-Franz law at a quantum critical point. *Science* **2007**, *316*, 1320–1322. [CrossRef] [PubMed]
78. van der Meulen, H.P.; Tarnawski, Z.; de Visser, A.; Franse, J.J.M.; Perenboom, J.A.A.J.; Althof, D.; van Kempen, H. Specific heat of UPt$_3$ in magnetic fields up to 24.5 T. *Phys. Rev. B* **1990**, *41*, 9352–9357. [CrossRef] [PubMed]
79. Tokiwa, Y.; Pikul, A.; Gegenwart, P.; Steglich, F.; Bud'ko, S.L.; Canfield, P.C. Low-temperature thermodynamic properties of the heavy-fermion compound YbAgGe close to the field-induced quantum critical point. *Phys. Rev. B* **2006**, *73*, 094435. [CrossRef]
80. Niklowitz, P.G.; Knebel, G.; Flouquet, J.; Bud'ko, S.L.; Canfield, P.C. Field-induced non-Fermi-liquid resistivity of stoichiometric YbAgGe single crystals. *Phys. Rev. B* **2006**, *73*, 125101. [CrossRef]
81. Kambe, S.; Huxley, A.; Flouquet, J.; Jansen, A.G.M.; Wyder, P. Hall resistivity in the heavy Fermion normal state of UPt$_3$ up to 26 T. *J. Phys. Condens. Matter* **1999**, *11*, 221–227. [CrossRef]
82. Sakamoto, I.; Shomi, Y.; Ohara, S. Anomalous Hall effect in heavy fermion compounds Ce$_2$MIn$_8$ (M=Rh or Ir). *Physica B* **2003**, *329–333*, 607–609. [CrossRef]
83. Zhou, B.B.; Misra, S.; da Silva Neto, E.H.; Aynajian, P.; Baumbach, R.E.; Thompson, J.D.; Bauer, E.D.; Yazdani, A. Visualizing nodal heavy fermion superconductivity in CeCoIn$_5$. *Nat. Phys.* **2013**, *9*, 474–479. [CrossRef]
84. Singh, S.; Capan, C.; Nicklas, M.; Rams, M.; Gladun, A.; Lee, H.; DiTusa, J.F.; Fisk, Z.; Steglich, F.; Wirth, S. Probing the quantum critical behavior of CeCoIn$_5$ via Hall effect measurements. *Phys. Rev. Lett.* **2007**, *98*, 057001. [CrossRef]
85. Hadžić-Leroux, M.; Hamzić, A.; Fert, A.; Haen, P.; Lapierre, F.; Laborde, O. Hall effect in heavy-fermion systems: UPt$_3$, UAl$_2$, CeAl$_3$, CeRu$_2$Si$_2$. *Europhys. Lett.* **1986**, *1*, 579–584. [CrossRef]
86. Chen, J.W.; Lambert, S.E.; Maple, M.B.; Fisk, Z.; Smith, J.L.; Stewart, G.R.; Willis, J.O. Upper critical magnetic field of the heavy-fermion superconductor UPt$_3$. *Phys. Rev. B* **1984**, *30*, 1583–1585. [CrossRef]
87. Bud'ko, S.; Morosan, E.; Canfield, P. Anisotropic Hall effect in single-crystal heavy-fermion YbAgGe. *Phys. Rev. B* **2005**, *71*, 054408. [CrossRef]
88. Varma, C.M. Colloquium: Linear in temperature resistivity and associated mysteries including high temperature superconductivity. *Rev. Mod. Phys.* **2020**, *92*, 031001. [CrossRef]
89. Hussey, N.E.; Takenaka, K.; Takagi, H. Universality of the Mott–Ioffe–Regel limit in metals. *Philos. Mag.* **2004**, *84*, 2847–2864. [CrossRef]
90. Gunnarsson, O.; Calandra, M.; Han, J.E. Colloquium: Saturation of electrical resistivity. *Rev. Mod. Phys.* **2003**, *75*, 1085–1099. [CrossRef]
91. Ioffe, A.F.; Regel, A.R. Non-crystalline, amorphous and liquid electronic semiconductors. *Prog. Semicond.* **1960**, *4*, 237–291.
92. Mott, N.F. Conduction in non-crystalline systems IX. the minimum metallic conductivity. *Philos. Mag.* **1972**, *26*, 1015–1026. [CrossRef]
93. Abrahams, E.; Si, Q. Quantum criticality in the iron pnictides and chalcogenides. *J. Phys. Condens. Matter* **2011**, *23*, 223201. [CrossRef]
94. Werman, Y.; Kivelson, S.A.; Berg, E. Non-quasiparticle transport and resistivity saturation: A view from the large-N limit. *Npj Quantum Mater.* **2017**, *2*, 7. [CrossRef]
95. Emery, V.J.; Kivelson, S.A. Superconductivity in bad metals. *Phys. Rev. Lett.* **1995**, *74*, 3253–3256. [CrossRef]
96. Friedemann, S.; Oeschler, N.; Wirth, S.; Krellner, C.; Geibel, C.; Steglich, F.; Paschen, S.; Kirchner, S.; Si, Q. Fermi-surface collapse and dynamical scaling near a quantum-critical point. *Proc. Natl. Acad. Sci. USA* **2010**, *107*, 14547. [CrossRef] [PubMed]
97. Chu, J.H.; Analytis, J.G.; Kucharczyk, C.; Fisher, I.R. Determination of the phase diagram of the electron-doped superconductor Ba(Fe$_{1-x}$Co$_x$)$_2$As$_2$. *Phys. Rev. B* **2009**, *79*, 014506. [CrossRef]
98. Si, Q.; Paschen, S. Quantum phase transitions in heavy fermion metals and Kondo insulators. *Phys. Status Solidi B* **2013**, *250*, 425. [CrossRef]
99. Badoux, S.; Tabis, W.; Laliberté, F.; Grissonnanche, G.; Vignolle, B.; Vignolles, D.; Béard, J.; Bonn, D.A.; Hardy, W.N.; Liang, R.; et al. Change of carrier density at the pseudogap critical point of a cuprate superconductor. *Nature* **2016**, *531*, 210–214. [CrossRef]
100. Cao, Y.; Fatemi, V.; Fang, S.; Watanabe, K.; Taniguchi, T.; Kaxiras, E.; Jarillo-Herrero, P. Unconventional superconductivity in magic-angle graphene superlattices. *Nature* **2018**, *556*, 43–50. [CrossRef]
101. Senthil, T.; Vojta, M.; Sachdev, S. Weak magnetism and non-Fermi liquids near heavy-fermion critical points. *Phys. Rev. B* **2004**, *69*, 035111. [CrossRef]

102. Oshikawa, M. Topological approach to Luttinger's theorem and Fermi surface of a Kondo lattice. *Phys. Rev. Lett.* **2000**, *84*, 3370–3373. [CrossRef]
103. Prochaska, L.; Li, X.; MacFarland, D.C.; Andrews, A.M.; Bonta, M.; Bianco, E.F.; Yazdi, S.; Schrenk, W.; Detz, H.; Limbeck, A.; et al. Singular charge fluctuations at a magnetic quantum critical point. *Science* **2020**, *367*, 285. [CrossRef]
104. Yi, M.; Liu, Z.K.; Zhang, Y.; Yu, R.; Zhu, J.X.; Lee, J.; Moore, R.; Schmitt, F.; Li, W.; Riggs, S.; et al. Observation of universal strong orbital-dependent correlation effects in iron chalcogenides. *Nat. Commun.* **2015**, *6*, 7777. [CrossRef]
105. Yu, R.; Si, Q. Orbital-selective Mott phase in multiorbital models for iron pnictides and chalcogenides. *Phys. Rev. B* **2017**, *96*, 125110. [CrossRef]
106. Komijani, Y.; Kotliar, G. Analytical slave-spin mean-field approach to orbital selective Mott insulators. *Phys. Rev. B* **2017**, *96*, 125111. [CrossRef]

Article
Fingerprints of Topotactic Hydrogen in Nickelate Superconductors

Liang Si [1,2,*], Paul Worm [2] and Karsten Held [2,*]

[1] School of Physics, Northwest University, Xi'an 710127, China
[2] Institute for Solid State Physics, Vienna University of Technology, 1040 Vienna, Austria; p.worm@a1.net
* Correspondence: liang.si@ifp.tuwien.ac.at (L.S.); held@ifp.tuwien.ac.at (K.H.)

Abstract: Superconductivity has entered the nickel age marked by enormous experimental and theoretical efforts. Notwithstanding, synthesizing nickelate superconductors remains extremely challenging, not least due to incomplete oxygen reduction and topotactic hydrogen. Here, we present density-functional theory calculations for nickelate superconductors with additional topotactic hydrogen or oxygen, namely $La_{1-x}Sr_xNiO_2H_\delta$ and $LaNiO_{2+\delta}$. We identify a phonon mode as a possible indication for topotactic hydrogen and discuss the charge redistribution patterns around oxygen and hydrogen impurities.

Keywords: superconductivity; nickelates; strongly correlated electron systems

1. Introduction

Computational materials calculations have predicted superconductivity in nickelates [1] and the heterostructures thereof [2–4] since many decades, mainly based on apparent similarly to cuprate superconductors. Three years ago, superconductivity in nickelates was finally discovered in an experiment by Li, Hwang, and coworkers [5], breaking the grounds for a new age of superconductivity, the nickel age. It is marked by an enormous theoretical and experimental activity, including but not restricted to [5–31]. Superconductivity has been found by now, among others, in $Nd_{1-x}Sr_xNiO_2$ [5,6], $Pr_{1-x}Sr_xNiO_2$ [7], $La_{1-x}Ca_xNiO_2$ [8], $La_{1-x}Sr_xNiO_2$ [9], and most recently in the pentalayer nickelate $Nd_6Ni_5O_{12}$ [10]. Figure 1 shows some of the hallmark experimental critical temperatures (T_c's) for the nickelates in comparison with the preceding copper [32] and iron age [33] of unconventional superconductivity. Also shown are some other noteworthy superconductors, including the first superconductor, solid Hg, technologically relevant NbTi, and hydride superconductors [34]. The last are superconducting at room temperature [35], albeit only at a pressure of 267GPa exerted in a diamond anvil cell. All of these compounds are marked in gray in Figure 1 as they are conventional superconductors. That is, the pairing of electrons originates from the electron-phonon coupling, as described in the theory of Bardeen, Cooper, and Schrieffer (BCS) [36].

In contrast, cuprates, nickelates, and, to a lesser extent, iron pnictides are strongly correlated electron systems with a large Coulomb interaction between electrons because of their narrow transition metal orbitals. Their T_c is too high for BCS theory [37,38], and the origin of superconductivity in these strongly correlated systems is still hotly debated. One prospective mechanism is antiferromagnetic spin fluctuations [39–43] stemming from strong electronic correlations. Another mechanism is based on charge density wave fluctuations and received renewed interest with the discovery of charge density wave ordering in cuprates [44,45]. Dynamical vertex approximation [46–49] calculations for nickelates [27], which are unbiased with respect to charge and spin fluctuations, found that spin fluctuations dominated and successfully predicted the superconducting dome prior to experiment in $Nd_{1-x}Sr_xNiO_2$ [6,50,51].

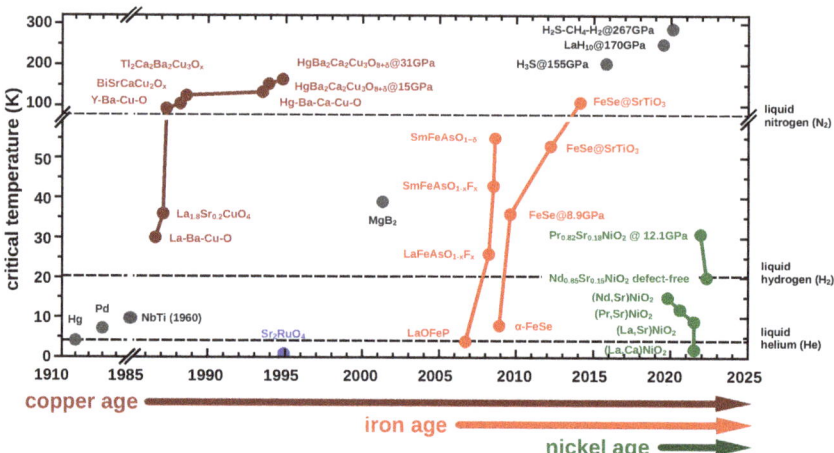

Figure 1. Superconducting T_c vs. year of discovery for selected superconductors. The discovery of cuprates, iron pnictides and nickelates led to enormous experimental and theoretical activities. Hence, one also speaks of the copper, iron and nickel age of superconductivity.

Why did it take 20 years to synthesize superconducting nickelates that have been so seemingly predicted on a computer? To mimic the cuprate Cu $3d^9$ configuration, as in NdNiO$_2$, nickel has to be in the uncommon oxidation state Ni^{1+}, which is rare and prone to oxidize further. Only through a complex two-step procedure, Lee, Hwang, and coworkers [52] were able to synthesize superconducting nickelates. In a first step, modern pulsed laser deposition (PLD) was used to grow a Sr$_x$Nd$_{1-x}$NiO$_3$ film on a SrTiO$_2$ substrate. This nickelate is still in the 3D perovskite phase—see Figure 2 (left)—with one oxygen atom too many and will thus not show superconductivity. Hence, this additional oxygen between the layers needs to be removed in a second step. The reducing agent CaH$_2$ is used to this end within a quite narrow temperature window [52]. If all goes well, one arrives at the superconducting Sr$_x$Nd$_{1-x}$NiO$_2$ film (top center). However, this process is prone to incomplete oxidation or to intercalate hydrogen topotactically, i.e., at the position of the removed oxygen; see Figure 2 (bottom center). Both of those unwanted outcomes are detrimental for superconductivity.

In [21,53,54], it was shown by density functional theory (DFT) calculations that NdNiO$_2$H is indeed energetically favorable to NdNiO$_2$ + 1/2 H. For the doped system, on the other hand, Nd$_{0.8}$Sr$_{0.2}$NiO$_2$ without the hydrogen intercalated is energetically favorable. The additional H or likewise an incomplete oxidation to SrNdNiO$_{2.5}$ alters the physics completely. Additional H or O$_{0.5}$ will remove an electron from the Ni atoms, resulting in Ni^{2+} instead of Ni^{1+}. The formal electronic configuration is hence $3d^8$ instead of $3d^9$, or two holes instead of one hole in the Ni d-shell. Dynamical mean-field theory (DMFT) calculations [21] evidence that the basic atomic configuration is the one of Figure 2 (lower right). That is, because of Hund's exchange, the two holes in NdNiO$_2$H occupy two different orbitals, $3d_{x^2y^2}$ and $3d_{3z^2-r^2}$, and form a spin-1. A consequence of this is that DMFT calculations predict NdNiO$_2$H to be a Mott insulator, whereas NdNiO$_2$ is a strongly correlated metal with a large mass enhancement of about five [21].

To the best of our knowledge, such a two-orbital, more 3D electronic structure is unfavorable for high-T_c superconductivity. The two-dimensionality of cuprate and nickelate superconductors helps to suppress long-range antiferromagnetic order, while at the same time retaining strong antiferromagnetic fluctuations that can act as a pairing glue for superconductivity. In experiment, we cannot expect ideal NdNiO$_2$, NdNiO$_2$H or NdNiO$_{2.5}$ films, but most likely some H or additional O will remain in the NdNiO$_2$ film, after the CaH$_2$ reduction. Additional oxygen can be directly evidenced in standard X-ray diffraction analysis after the synthesis step. However, hydrogen, being very light, evades such an X-ray analysis. It has been evidenced in nickelates only by nuclear magnetic resonance (NMR) experiments [55] which, contrary to X-ray techniques, are very sensitive to hydrogen. Ref. [56] suggested hydrogen in LaNiO$_2$ to be confined at grain boundaries or secondary-phase precipitates. Given these difficulties, it is maybe not astonishing that it took almost one year before a second research group [6] was able to reproduce superconductivity in nickelates. Despite enormous experimental efforts, only a few groups have succeeded hitherto.

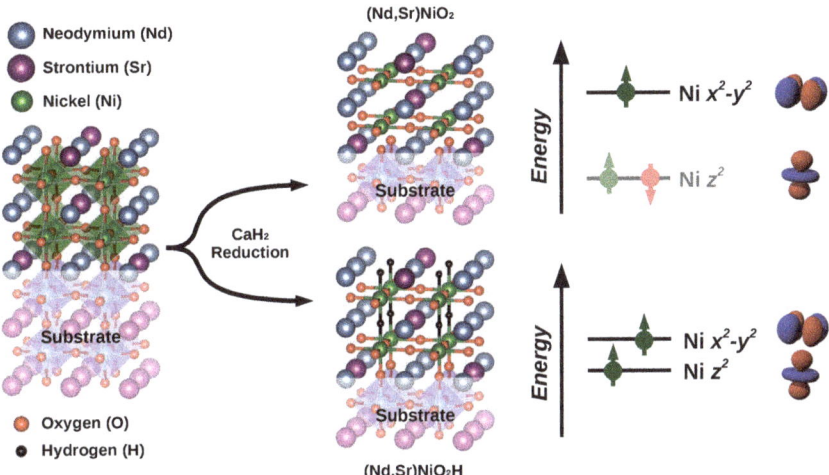

Figure 2. For synthesizing superconducting nickelates (1, **left**), a perovskite film of Nd(La)$_{1-x}$Sr$_x$NiO$_3$ is grown on a SrTiO$_3$ substrate, and (2, **center**) the O atoms between the planes are removed by reduction with CaH$_2$. Besides the pursued nickelate Nd(La)$_{1-x}$Sr$_x$NiO$_2$ (top center), also excess oxygen or topotactic H may remain in the film, yielding Nd(La)$_{1-x}$Sr$_x$NiO$_2$H (**bottom center**). The excess hydrogen results in two holes instead of one hole within the topmost two Ni 3d orbitals (**right**). Adapted from Ref. [57].

In this paper, we present additional DFT results for topotactic hydrogen and incomplete oxygen reduction in nickelate superconductors: In Section 3, we provide technical information on the DFT calculations. In Section 3, we analyze the energy gain to topotactically intercalate hydrogen in LaNiO$_2$ and NdNiO$_2$. In Section 4, we analyze the phonon spectrum and identify a high-energy mode originating from the Ni-H-Ni bond as a characteristic feature of intercalated hydrogen. In Section 5, we show the changes of the charge distribution caused by topotactic hydrogen or oxygen. Finally, Section 6 provides a summary and outlook.

2. Method

Computational details on E_b. In both our previous theoretical study [21] and this article, the binding energy E_b of hydrogen atoms is computed as:

$$E_b = E[ABO_2] + \mu[H] - E[ABO_2H]. \tag{1}$$

Here, $E[ABO_2]$ and $E[ABO_2H]$ are the total energy of infinite-layer ABO_2 and hydride-oxides ABO_2H, while $\mu[H] = E[H_2]/2$ is the chemical potential of H. Note that H$_2$ is a typical byproduct for the reduction with CaH$_2$ and also emerges when CaH$_2$ is in contact with H$_2$O. Hence, it can be expected to be present in the reaction. A positive (negative) E_b indicates the topotactic H process is energetically favorable (unfavorable) to obtain ABO_2H instead of ABO_2 and H$_2$/2.

In the present paper, we go beyond [21] that reported E_b of various ABO_2 compounds by investigating E_b of $La_{1-x}Ca_xNiO_2$ systems for many different doping levels. Here, the increasing Ca-doping is achieved by using the virtual crystal approximation (VCA) [58,59] from $LaNiO_2$ ($x = 0$) to $CaNiO_2$ ($x = 1$). For each Ca concentration, structure relaxation and static total energy calculation is carried out for $La_{1-x}Ca_xNiO_2$ and $La_{1-x}Ca_xNiO_2H$ within the tetragonal space group $P4/mmm$. To this end, we use density-functional theory (DFT) [60,61] with the VASP code [62,63] and the generalized gradient approximations (GGA) of Perdew, Burke, and Ernzerhof (PBE) [64], and PBE revised for solids (PBEsol) [65]. For undoped $LaNiO_2$, the GGA-PBEsol relaxations predict its in-plane lattice constant as 3.890 Å, which is close to that of the STO substrate: 3.905 Å. The computations for $La_{1-x}Ca_xNiO_2$ and $LaCoO_2$, $LaCuO_2$, $SrCoO_2$, and $SrNiO_2$ are performed without spin-polarization and a DFT+U treatment [66], as the inclusion of Coulomb U and spin-polarization only slightly decrease the E_b by ~5% for $LaNiO_2$ [57]. For $NdNiO_2$, an inevitably computational issue is the localized Nd-$4f$ orbitals. These f-orbitals are localized around the atomic core, leading to strong correlations. In non-spin-polarized DFT calculations, this generates flat bands near the Fermi level E_F and leads to unsuccessful convergence. To avoid this, we employed DFT+U [U_f(Nd) = 7 eV and U_d(Ni) = 4.4 eV] and initialized a G-type anti-ferromagnetic ordering for both Nd- and Ni-sublattice in a $\sqrt{2} \times \sqrt{2} \times 2$ supercell of $NdNiO_2$. For the $Nd_{0.75}Sr_{0.25}NiO_2$ case, 25% Sr-doping is achieved by replacing one out of the four Nd atoms by Sr in a $\sqrt{2} \times \sqrt{2} \times 2$ $NdNiO_2$ supercell.

Computational details on phonons. The phonon computations for $LaNiO_2$, $LaNiO_2H$, $LaNiO_2H_{0.125}$, and $LaNiO_{2.125}$ are performed with the frozen phonon method using the PHONONY [67] code interfaced with VASP. Computations with density functional perturbation theory (DFPT) method [68] are also carried out for double checking. For $LaNiO_2$ and $LaNiO_2H$, the unit cells shown in Figure 3a,b are enlarged to a $2 \times 2 \times 2$ supercell, while for $LaNiO_2H_{0.125}$ and $LaNiO_{2.125}$, the phonons are directly computed with the supercell of Figure 3c,d.

Computational details on electron density. The electron density distributions of $LaNiO_2$, $LaNiO_2H$, $LaNiO_2H_{0.125}$, and $LaNiO_{2.125}$ are computed using the WIEN2K code [69] while taking the VASP-relaxed crystal structure as input. The isosurfaces are plotted from 0.1 (yellow lines) to 2.0 (center of atoms) with spacing of 0.1 in units of e/Å^2.

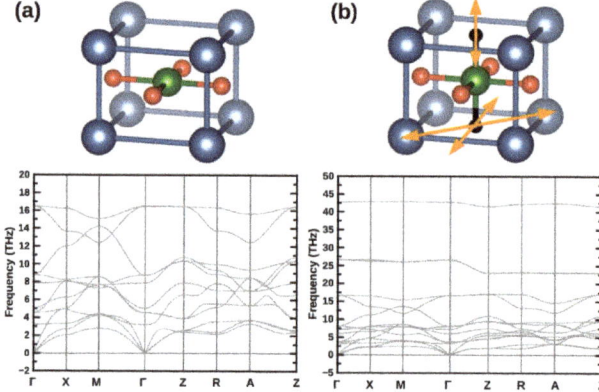

Figure 3. *Cont.*

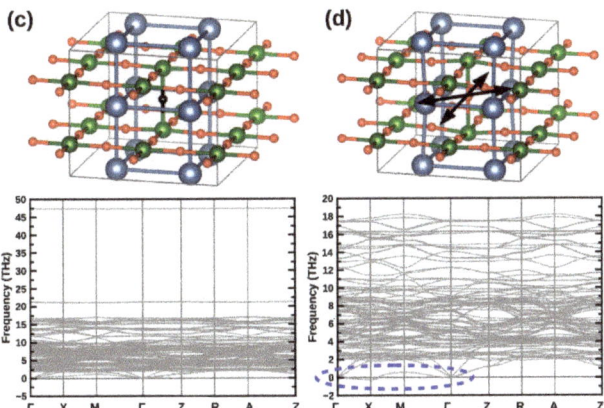

Figure 3. Phonon spectra of (**a**) LaNiO$_2$ and (**b**) LaNiO$_2$H and in a 2 × 2 × 2 LaNiO$_2$ supercell doped with a single (**c**) H and (**d**) O atom (i.e., LaNiO$_2$H$_{0.125}$ in (**c**) and LaNiO$_{2.125}$ in (**d**)). The orange and black arrows in (**b**,**d**) represent vibrations of H and O atoms. The blue dashed oval in (**d**) labels the unstable phonon modes induced by intercalating additional O atoms in LaNiO$_2$.

3. Energetic Stability

Figure 4 shows the results of the hydrogen binding energy E_b for the infinite layer nickelate superconductors Nd$_{1-x}$Sr$_x$NiO$_2$ [5,6,50] and La$_{1-x}$Ca$_x$NiO$_2$ [8]. To reveal the evolution of E_b when the B-site band filling deviates from their original configurations ($3d^9$ in LaNiO$_2$ when $x = 0$ and $3d^8$ in CaNiO$_2$ when $x = 1$), we also show the binding energy of LaCoO$_2$ ($3d^8$), LaCuO$_2$ ($3d^{10}$), SrCoO$_2$ ($3d^7$), and SrNiO$_2$ ($3d^8$).

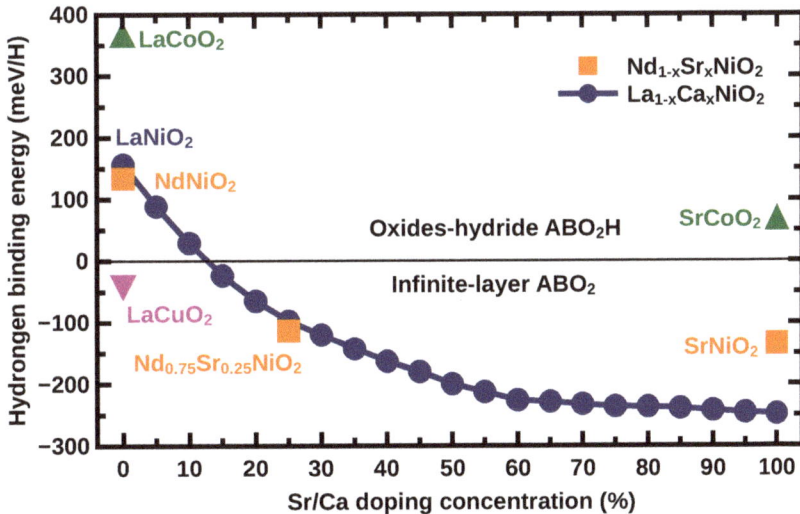

Figure 4. Hydrogen binding energy (E_b) per hydrogen in two nickelate superconductors, La$_{1-x}$Ca$_x$NiO$_2$ and Nd$_{1-x}$Sr$_x$NiO$_2$ vs. Sr/Ca doping concentration x; LaCuO$_2$, LaCoO$_2$, and SrCoO$_2$ are shown for comparison. Slightly above 10% (Sr,Ca)-doping infinite layer nickelates are energetically more stable. Note that the doping changes the filling of the B-3d orbital. To study the relationship between E_b and the types of B-site elements, E_b of several other ABO_2 compounds is computed: LaCoO$_2$, LaCuO$_2$, SrCoO$_2$ and SrNiO$_2$. Note that this changes the filling of B-3d orbital within a large range: e.g., $3d^8$ for LaCoO$_2$ and $3d^9$ for LaNiO$_2$.

Let us start with the case of $La_{1-x}Ca_xNiO_2$ [8]. Here, the unoccupied La-$4f$ orbitals make the computation possible even without spin-polarization and Coulomb U for La-$4f$, whereas for $NdNiO_2$, this is not practicable due to Nd-$4f$ flat bands near E_F. Positive (negative) E_b above (below) the horizontal line in Figure 4 indicates topotactic H is energetically favorable (unfavorable). When $x = 0$, i.e., for bulk $LaNiO_2$, the system tends to confine H atoms, resulting in oxide-hydride ABO_2H with $E_b = 157$ meV/H. As the concentration of Ca increases, E_b monotonously decreases, reaching -248 meV for the end member of the doping series, $CaNiO_2$. The turning point between favorable and unfavorable topotactic H inclusion is around 10% to 15% Ca-doping. Let us note that $E_b = 0$ roughly agrees with the onset of superconductivity, which for Ca-doped $LaNiO_2$ emerges for $x > 15$% Ca-doping [8].

To obtain E_b in $NdNiO_2$ a much higher computational effort is required: firstly, the Nd-$4f$ orbitals must be computed with either treating them as core-states or including spin-splitting. Secondly, for the spin-polarized DFT(+U) calculations, an appropriate (anti-)ferromagnetic ordering has to be arranged for both Ni- and Nd-sublattices. In oxide-hydride ABO_2H compounds, the δ-type bond between Ni and H stabilizes a G-type anti-ferromagnetic order by driving the system from a quasi two-dimensional (2D) system to a three dimensional (3D) one [21]. Given the large computational costs of E_b for $Nd_{1-x}Sr_xNiO_2$ by using anti-ferromagnetic DFT+U calculations for both Nd-$4f$ ($U \sim 7$ eV) and Ni-$3d$ ($U = 4.4$ eV) orbitals, we merely show here the results of $NdNiO_2$ ($x = 0$), $Nd_{0.75}Sr_{0.25}NiO_2$ ($x = 0.25$), and $SrNiO_2$ ($x = 1$), which are adopted from [21]. With 25% Sr-doping, the E_b of $NdNiO_2$ is reduced from 134 meV to -113 meV. Please note that E_b of (Nd,Sr)NiO_2 is slightly smaller than in (La,Ca)NiO_2, at least in the low doping range. This can be explained by shorter lattice constants in $NdNiO_2$, in agreement with the finding [21] that compressive strain plays an important role in reducing E_b.

One can speculate that this suppression of topotactic hydrogen may also play a role when comparing the recently synthesized (Nd,Sr)NiO_2 films on a $(LaAlO_3)_{0.3}(Sr_2TaAlO_6)_{0.7}$ (LSAT) substrate [51] with the previously employed $SrTiO_3$ (STO) substrate [50]. Lee et al. [51] reported cleaner films without defects and also a higher superconducting transition temperature $T_c \sim 20$ K for the LSAT film as compared to $T_c = 15$ K and plenty of stacking fault defects for the STO substrate [50]. As for (La,Ca)NiO_2, $E_b = 0$ falls in the region of the onset of the superconductivity for (Sr,Nd)NiO_2, which is $x \sim 10$% Sr-doping in LSAT-strained defect-free films [51] and $x \sim 12.5$% at $SrTiO_3$-substrate states [50]. Topotactic hydrogen might play a role in suppressing superconductivity in this doping region.

In Figure 4, we further show additional infinite layer compounds $LaCoO_2$, $LaCuO_2$, $SrCoO_2$, and $SrNiO_2$ for comparison. Their E_b is predicted to be 367, -42, 69, and -134 meV, respectively. Combining the results of $LaNiO_2$ and $CaNiO_2$, we summarize several tendencies on how to predict E_b of ABO_2: (1) the strongest effect on E_b is changing the B-site element. However, this seems unpractical for nickelate superconductors as the band filling is strictly restricted to be $3d^{9-x}$ ($x \sim 0.2$). For both trivalent (La, Nd) and bivalent (Sr, Ca) cations, E_b decreases when the B-site cation goes from early to late transition metal elements, e.g., from $LaCoO_2$ ($3d^8$) to $LaNiO_2$ ($3d^9$) to $LaCuO_2$ ($3d^{10}$). (2) Compressive strains induced by either substrate or external pressure can effectively reduce E_b, and we believe that this might be used for growing defect-free films. (3) According to our theoretical calculations, E_b mainly depends on lattice parameters and band filling of the B-site $3d$-orbitals, but much less on magnetic ordering and Coulomb interaction U.

4. Phonon Dispersion

As revealed by previous DFT phonon spectra calculations [16], $NdNiO_2$ is dynamically stable. One of the very fundamental questions would be whether topotactic H from over-reacted reduction and/or O from unaccomplished reductive reactions affect the lattice stability. To investigate this point, we perform DFT phonon calculations and analyze the lattice vibration induced by H/O intercalation, as shown in Figure 3.

The phonon spectrum of LaNiO$_2$ (Figure 3a) is essentially the same as in Ref. [16]; all the phonon frequencies are positive, indicating it is dynamically stable. Its upmost optical phonon at around 14 to 16 THz can be identified with the recent experimental resonant inelastic x-ray scattering (RIXS) data [70] showing a weakly dispersing optical phonon at ~60 meV ≈ 15 THz. In Figure 3b, the oxides-hydride LaNiO$_2$H is also predicted to be dynamically stable. Please note that the phonon dispersions between 0 and 20 THz are basically the same as those in LaNiO$_2$ (Figure 3a; note the different scale of the y-axis). However, one can see new, additional vibration modes from the light H-atoms at frequencies of ~27 THz and ~43 THz. Among these vibrations, the double degenerate mode at lower frequency is generated by an in-plane (xy-plane) vibration of the topotactic H atom. There are two such in-plane vibrations of H atoms, either along the (100) or (110) direction (and symmetrically related directions), as indicated by the orange arrows in Figure 3b. The mode located at the higher frequency ~43 THz is, on the other hand, formed by an out-of-plane (z-direction) vibration and is singly degenerate.

We explain these phonon modes in detail by computing the bonding strength between H-1s–Ni-d_{z^2} and H-1s–La-d_{xy} orbitals. Our tight-binding calculations yield an electron hopping term of -1.604 eV between H-1s and Ni-d_{z^2}, while it is -1.052 eV from La-d_{xy} to H-1s. That is, the larger H-1s–Ni-d_{z^2} overlap leads to a stronger δ-type bonding and, together with the shorter c-lattice constant, to a higher phonon energy. Additionally, the shorter c-lattice in LaNiO$_2$ should also play a role at forming a stronger H-1s–Ni-d_{z^2} bond.

In our previous analysis of the band character for LaNiO$_2$H [21], the H-1s bands were mainly located at two energy regions: a very flat band that is mostly from the H-1s itself at ~-7 to -6 eV, and a hybridized band between H-1s and Ni-d_{z^2} at ~-2 eV. Together with the higher phonon energy, this indicates that the topotactic H atoms are mainly confined by a Ni sub-lattice via bonding and anti-bonding states formed by H-1s and Ni-d_{z^2} orbitals, instead of the La(Nd) sub-lattice.

The complete (full) topotactic inclusion of H, where all vacancies induced by removing oxygen are filled by H, is an ideal limiting case. Under varying experimental conditions, such as chemical reagent, substrate, temperature, and strain, the H-topotactic inclusion may be incomplete, and thus ABO_2H_δ ($\delta < 1$) may be energetically favored. Hence, we also compute the phonon spectrum at a rather low H-topotactic density: LaNiO$_2$H$_{0.125}$, achieved by including a single H into $2 \times 2 \times 2$ LaNiO$_2$ supercells as shown in Figure 3c. Moreover, such a local H defect, as revealed by the positive frequency at all q-vectors in the lower panel of Figure 3c, does not destroy the dynamical stability of the LaNiO$_2$ crystal. In fact, the only remarkable qualitative difference between the complete and 12.5% topotactic H case is the number of phonon bands at 0 THz to 20 THz. This is just a consequence of the larger $2 \times 2 \times 2$ LaNiO$_2$ supercell, with eight times more phonons. Some quantitative differences can be observed with respect to the energy of the phonon mode: The out-of-plane vibration energy is enhanced from ~43 THz in LaNiO$_2$H (Figure 3b) to ~47 THz in LaNiO$_2$H$_{0.125}$ (Figure 3b), and the in-plane vibration mode frequency is reduced from ~27 THz in LaNiO$_2$H (Figure 3b) to ~21 THz in LaNiO$_2$H$_{0.125}$ (Figure 3c). This is because the H-intercalation shrinks the local c-lattice, i.e., the distance between two Ni atoms separated by topotactic H, from 3.383 Å in (LaNiO$_2$H: Figure 3b) to 3.327 Å (LaNiO$_2$H$_{0.125}$: Figure 3c). The bond length between H and La is, on the other hand, slightly increased from 2.767 Å in (LaNiO$_2$H: Figure 3b) to 2.277 Å (LaNiO$_2$H$_{0.125}$: Figure 3c). This lattice compression (enlargement) explains the enhancement (reduction) for the out-of-plane (in-plane) phonon frequencies (energies).

These results pave a new way to detect the formation of topotactic H in infinite nickelate superconductors: by measuring the phonon modes. The existence of localized phonon modes with little dispersion at ~25 THz and ~45 THz indicates the presence of topotactic hydrogen, which otherwise would be extremely hard to detect. These frequencies correspond to energies of 103 meV and 186 meV, respectively, beyond the range <80 meV measured for La$_{1-x}$Sr$_x$NiO$_2$ in [71].

Lastly, we further study the case representing an incompleted reduction process: $LaNiO_{2.125}$, achieved by intercalating a single O into a $2 \times 2 \times 2$ $LaNiO_2$ supercell ($LaNiO_{2.125}$: Figure 3d). As the same consequence of employing a supercell in phonon computation, the number of phonon bands is multiplied by a factor of 8 in the frequency region between 0 THz to 20 THz. One obvious difference between undoped $LaNiO_2$ (Figure 3a) and $LaNiO_{2.125}$ (Figure 3d) is that the additional O leads to an unstable phonon mode near $q = X(\pi,0,0)$ (blue region in Figure 3d). This phonon mode is formed by an effective vibration of the additional O along the xy plane in the (001) or (110) direction (and symmetrically related directions depending on the exact q-vector) of locally cubic coordinate. Such a mode is related to the structural transition from cubic Pm-$3m$ to a R-$3c$ rhombohedral phase as in bulk $LaNiO_3$, with the Ni-O-Ni bond along the z-direction deviating from $180°$. Our simulations for other concentrations of additional O atoms (not shown) also indicate that incomplete oxygen reduction reactions generally result in local instabilities of $LaNiO_{2+\delta}$ with $\delta > 0$.

5. Charge Distribution

In this section, we perform electron density calculations for $LaNiO_2$, $LaNiO_2H$, $LaNiO_2H_{0.125}$, and $LaNiO_{2.125}$ compounds to investigate the bond types resulting from intercalated H and O atoms. Figure 5a,b show the electron density of $LaNiO_2$ at the NiO-plane and La-plane (light green planes of the top panels). In Figure 5a, a strong Ni-O bond is observed, while the low electron density between each Ni-O layer reveals a very weak inter-layer coupling, indicating the strong quasi-2D nature of the infinite layer nickelates. In Figure 5b, no bonds are formed between the La (Nd) atoms. The A-site rare-earth elements merely play the role of electron donors.

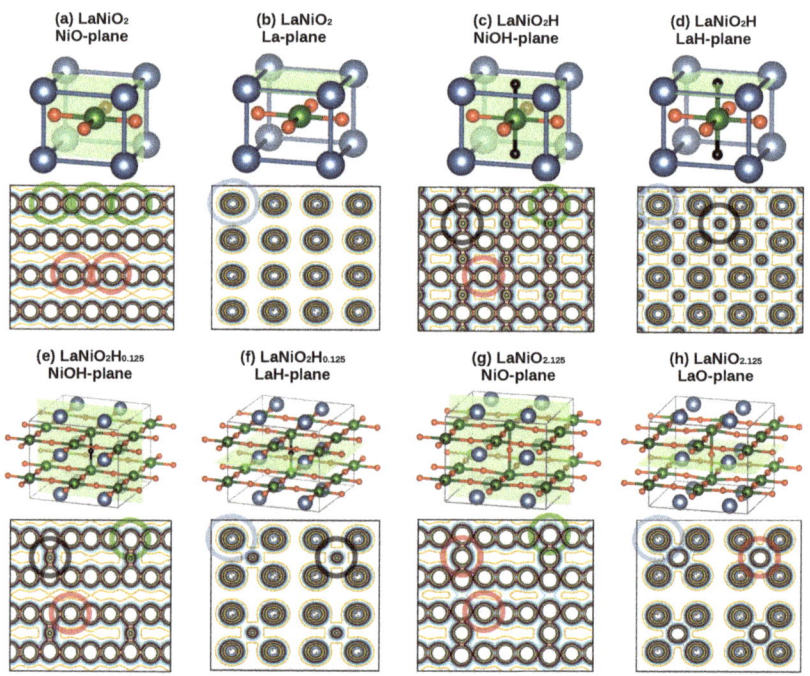

Figure 5. DFT calculated valence charge density of (**a**,**b**) $LaNiO_2$, (**c**,**d**) $LaNiO_2H$, and a $LaNiO_2$ supercell doped with a single (**e**,**f**) H and (**g**,**h**) O atom. For each compound, the charge density of (020) and (001) planes are shown in panels (**a**,**c**,**e**,**g**) and (**b**,**d**,**f**,**h**), respectively. The La, Ni, O, and H atoms are labeled by blue, green, red, and black circles, respectively.

Figure 5c,d present the electron density of LaNiO$_2$H along the same planes. In the NiOH-plane of Figure 5c, the comparison to Figure 5c shows that intercalated H boosts a 3D picture with an additional δ-type bond formed by Ni-d_{z^2} and H-1s orbitals (black circle). Along the LaH-plane (Figure 5d), δ-type bonds are formed by the orbital overlap between La-d_{xy} and H-1s orbitals. For LaNiO$_2$ with partial topotactic H (LaNiO$_2$H$_{0.125}$ in Figure 5e,f), the additional H atoms play similar roles at the Ni-H and the La-H bonds as in LaNiO$_2$H. The Ni and La atoms without H in between are similar to those in Figure 5a,b, and those with H are akin to Figure 5c,d. This indicates that the effects induced by topotactic H are indeed very local, i.e., they only affect the the nearest Ni and La atoms.

In Figure 5g,h, for LaNiO$_{2.125}$, the additional O increases the local c-lattice (Ni-Ni bond length via the additional O) from the LaNiO$_2$ value of 3.338 Å to 4.018 Å which is even larger than the DFT-relaxed value of LaNiO$_3$: 3.80 Å. This lattice expansion can be clearly seen in Figure 5g. The large electron density between Ni and O along the z-direction indicates the strength of this Ni-O bond in the z-direction is comparable with the ones along x/y directions. From Figure 5h, we conclude that similar La-O bonds are formed after intercalating additional O atoms, and the La-La distance is shrunken by the additional O atom from 3.889 Å (LaNiO$_2$) to 3.746 Å between the La atoms pointing to the additional O. However, from the electron density plot, the La-O bond strength seems not stronger than the La-H bonding in Figure 5c,e. This can be explained by the fact that both O-p_x and -p_y orbitals do not point to orbital lobes of La-d_{xy}, leading to a comparable bond strength as the La-H bond in LaNiO$_2$H$_\delta$.

6. Conclusions and Outlook

Our theoretical study demonstrates that the parent compounds of infinite-layer nickelate superconductors, LaNiO$_2$ and NdNiO$_2$, are energetically unstable with respect to topotactic H in the reductive process from perovskite La(Nd)NiO$_3$ to La(Nd)NiO$_2$. The presence of H, which reshapes the systems from ABO_2 to the hydride-oxide ABO_2H, triggers a transition from a quasi-2D strongly correlated single-band ($d_{x^2-y^2}$) metal, to a two-band ($d_{x^2-y^2}+d_{z^2}$) anti-ferromagnetic 3D Mott insulator. Our predictions [21] have been reproduced by other groups using DFT+U calculations for other similar ABO_2 systems [53,54]. The recent experimental observation [70] of Ni^{2+} ($3d^8$) in nickelates indicates the existence of topotactic H, as do NMR experiments [55]. The presence of H and its consequence of a 3D Mott-insulator is unfavorable for the emergence of superconductivity in nickelates. However, it is difficult to detect topotactic H in experiment. Three factors contribute to this difficulty: (1) the small radius of H makes it hard to be detected by commonly employed experimental techniques such as X-ray diffraction and scanning transmission electron microscopy (STEM). (2) As revealed by our phonon calculations, the dynamical stability of La(Nd)NiO$_2$ does not rely on the concentration of intercalated H atoms. Hence, the same infinite-layer structures should be detected by STEM even in the presence of H. (3) As revealed by electron density distributions, the topotactic H does not break the local crystal structure either (e.g., bond length and angle); the H atoms merely affect the most nearby Ni atoms via a Ni-d_{z^2}-H-1s δ-bond. This is different if we have additional O atoms instead of H: O atoms do not only induce a dynamical instability but also obviously change the local crystal by enlarging the Ni-Ni bond length and angle visibly. Oxygen impurities also lead to unstable phonon modes in LaNiO$_{2+\delta}$ and thus a major lattice reconstruction.

The ways to avoid topotactic H revealed by our calculations are: in-plane compressive strains and bivalent cation doping with Sr or Ca. This draws our attention to the recently synthesized (Nd,Sr)NiO$_2$ films [51], which were grown on a (LaAlO$_3$)$_{0.3}$(Sr$_2$TaAlO$_6$)$_{0.7}$ (LSAT) instead of a SrTiO$_3$ (STO) substrate, inducing an additional 0.9% compressive strain. These new films were shown to be defect-free and with a considerably larger superconducting dome from 10% to 30% Sr-doping and a higher maximal T_c $\sim$20 K [51], compared to 12.5%–25% Sr-doping and T_c $\sim$15 K for nickelate films grown on STO which show many stacking faults [5,6,50]. The compressive strain induced by replacing the STO substrate (a = 3.905 Å) by LSAT (a = 3.868 Å) may turn the positive E_b to negative, thus contributing to suppressing defects and recovering a single $d_{x^2-y^2}$-band picture.

Besides avoiding topotactic H, compressive strain is also predicted as an effective way to enhance T_c. Previous dynamical vertex approximation calculations [27,57] reveal that the key to enhance T_c in nickelates is to enhance the bandwidth W and reduce the ratio of Coulomb interaction U to W. Based on this prediction, we have proposed [27,57] three experimental ways to enhance T_c in nickelates: (1) In-plane compressive strain, which can indeed be achieved by using other substrates having a smaller lattice than STO, such as LSAT (3.868 Å), LaAlO$_3$ (3.80 Å), or SrLaAlO$_4$ (3.75 Å). The smaller in-plane lattice shrinks the distance between Ni atoms and thus increases their orbital overlap, leading to a larger W and a smaller U/W. Recent experimental reports have confirmed the validity of this approach by growing (Nd,Sr)NiO$_2$ on LSAT [51] and Pr$_{0.8}$Sr$_{0.2}$NiO$_2$ on LSAT [72]. (2) Applying external pressure on the films plays the same role as in-plane strain for the, essentially 2D, nickelates. This has been experimentally realized in [73]: under 12.1 GPa pressure, T_c can be enhanced monotonously to 31 K without yet showing a saturation. (3) Replacing 3d Ni by 4d Pd. In infinite-layer palladates such as NdPdO$_2$ or LaPdO$_2$ and similar compounds with 2D PdO$_2$ layers and separating layers between them, the more extended 4d orbitals of Pd are expected to reduce U/W from U/W ~7 for nickelates to U/W ~6 for palladates. Further experimental and theoretical research on the electronic and magnetic structure and the superconductive properties of palladates are thus worth performing.

Author Contributions: L.S. and K.H. conceptionalized the study; L.S. performed the DFT calculations; L.S., K.H. and P.W. contributed to the writing. All authors have read and agreed to the published version of the manuscript.

Funding: This research has been supported by the Austrian Science Funds (FWF) through project P 32044.

Institutional Review Board Statement: Not applicable.

Informed Consent Statement: Not applicable.

Data Availability Statement: Data will be made available upon reasonable request.

Acknowledgments: We thank M. Kitatani, J. Tomczak, O. Janson and Z. Zhong for valuable discussions and the Austrian Science Funds (FWF) for funding through project P 32044. L.S. also thanks the starting funds from Northwest University. Calculations have been done on the Vienna Scientific Clusters (VSC).

Conflicts of Interest: The authors declare no competing interests.

References

1. Anisimov, V.I.; Bukhvalov, D.; Rice, T.M. Electronic structure of possible nickelate analogs to the cuprates. *Phys. Rev. B* **1999**, *59*, 7901–7906. [CrossRef]
2. Chaloupka, J.; Khaliullin, G. Orbital Order and Possible Superconductivity in LaNiO$_3$/LaMO$_3$ Superlattices. *Phys. Rev. Lett.* **2008**, *100*, 016404. [CrossRef] [PubMed]
3. Hansmann, P.; Yang, X.; Toschi, A.; Khaliullin, G.; Andersen, O.K.; Held, K. Turning a Nickelate Fermi Surface into a Cupratelike One through Heterostructuring. *Phys. Rev. Lett.* **2009**, *103*, 016401. [CrossRef] [PubMed]
4. Hansmann, P.; Toschi, A.; Yang, X.; Andersen, O.; Held, K. Electronic structure of nickelates: From two-dimensional heterostructures to three-dimensional bulk materials. *Phys. Rev. B* **2010**, *82*, 235123. [CrossRef]
5. Li, D.; Lee, K.; Wang, B.Y.; Osada, M.; Crossley, S.; Lee, H.R.; Cui, Y.; Hikita, Y.; Hwang, H.Y. Superconductivity in an infinite-layer nickelate. *Nature* **2019**, *572*, 624–627. [CrossRef]
6. Zeng, S.; Tang, C.S.; Yin, X.; Li, C.; Li, M.; Huang, Z.; Hu, J.; Liu, W.; Omar, G.J.; Jani, H.; et al. Phase Diagram and Superconducting Dome of Infinite-Layer Nd$_{1-x}$Sr$_x$NiO$_2$ Thin Films. *Phys. Rev. Lett.* **2020**, *125*, 147003. [CrossRef]
7. Osada, M.; Wang, B.Y.; Lee, K.; Li, D.; Hwang, H.Y. Phase diagram of infinite layer praseodymium nickelate Pr$_{1-x}$Sr$_x$NiO$_2$ thin films. *Phys. Rev. Mater.* **2020**, *4*, 121801. [CrossRef]
8. Zeng, S.; Li, C.; Chow, L.E.; Cao, Y.; Zhang, Z.; Tang, C.S.; Yin, X.; Lim, Z.S.; Hu, J.; Yang, P.; et al. Superconductivity in infinite-layer nickelate La1- xCaxNiO2 thin films. *Sci. Adv.* **2022**, *8*, eabl9927.

9. Osada, M.; Wang, B.Y.; Goodge, B.H.; Harvey, S.P.; Lee, K.; Li, D.; Kourkoutis, L.F.; Hwang, H.Y. Nickelate Superconductivity without Rare-Earth Magnetism: (La, Sr)NiO$_2$. *Adv. Mater.* **2021**, *33*, 2104083. [CrossRef]
10. Pan, G.A.; Segedin, D.F.; LaBollita, H.; Song, Q.; Nica, E.M.; Goodge, B.H.; Pierce, A.T.; Doyle, S.; Novakov, S.; Carrizales, D.C.; et al. Superconductivity in a quintuple-layer square-planar nickelate. *Nat. Mater.* **2022**, *21*, 160–164. [CrossRef]
11. Botana, A.S.; Norman, M.R. Similarities and Differences between LaNiO$_2$ and CaCuO$_2$ and Implications for Superconductivity. *Phys. Rev. X* **2020**, *10*, 011024. [CrossRef]
12. Sakakibara, H.; Usui, H.; Suzuki, K.; Kotani, T.; Aoki, H.; Kuroki, K. Model Construction and a Possibility of Cupratelike Pairing in a New d^9 Nickelate Superconductor (Nd, Sr)NiO$_2$. *Phys. Rev. Lett.* **2020**, *125*, 077003. [CrossRef]
13. Hirayama, M.; Tadano, T.; Nomura, Y.; Arita, R. Materials design of dynamically stable d^9 layered nickelates. *Phys. Rev. B* **2020**, *101*, 075107. [CrossRef]
14. Hu, L.H.; Wu, C. Two-band model for magnetism and superconductivity in nickelates. *Phys. Rev. Res.* **2019**, *1*, 032046. [CrossRef]
15. Wu, X.; Di Sante, D.; Schwemmer, T.; Hanke, W.; Hwang, H.Y.; Raghu, S.; Thomale, R. Robust $d_{x^2-y^2}$-wave superconductivity of infinite-layer nickelates. *Phys. Rev. B* **2020**, *101*, 060504. [CrossRef]
16. Nomura, Y.; Hirayama, M.; Tadano, T.; Yoshimoto, Y.; Nakamura, K.; Arita, R. Formation of a two-dimensional single-component correlated electron system and band engineering in the nickelate superconductor NdNiO$_2$. *Phys. Rev. B* **2019**, *100*, 205138. [CrossRef]
17. Zhang, G.M.; Yang, Y.F.; Zhang, F.C. Self-doped Mott insulator for parent compounds of nickelate superconductors. *Phys. Rev. B* **2020**, *101*, 020501. [CrossRef]
18. Jiang, M.; Berciu, M.; Sawatzky, G.A. Critical Nature of the Ni Spin State in Doped NdNiO$_2$. *Phys. Rev. Lett.* **2020**, *124*, 207004. [CrossRef]
19. Werner, P.; Hoshino, S. Nickelate superconductors: Multiorbital nature and spin freezing. *Phys. Rev. B* **2020**, *101*, 041104. [CrossRef]
20. Lechermann, F. Late transition metal oxides with infinite-layer structure: Nickelates versus cuprates. *Phys. Rev. B* **2020**, *101*, 081110. [CrossRef]
21. Si, L.; Xiao, W.; Kaufmann, J.; Tomczak, J.M.; Lu, Y.; Zhong, Z.; Held, K. Topotactic Hydrogen in Nickelate Superconductors and Akin Infinite-Layer Oxides ABO_2. *Phys. Rev. Lett.* **2020**, *124*, 166402. [CrossRef]
22. Lechermann, F. Multiorbital Processes Rule the Nd$_{1-x}$Sr$_x$NiO$_2$ Normal State. *Phys. Rev. X* **2020**, *10*, 041002. [CrossRef]
23. Petocchi, F.; Christiansson, V.; Nilsson, F.; Aryasetiawan, F.; Werner, P. Normal State of Nd$_{1-x}$Sr$_x$NiO$_2$ from Self-Consistent GW + EDMFT. *Phys. Rev. X* **2020**, *10*, 041047. [CrossRef]
24. Adhikary, P.; Bandyopadhyay, S.; Das, T.; Dasgupta, I.; Saha-Dasgupta, T. Orbital-selective superconductivity in a two-band model of infinite-layer nickelates. *Phys. Rev. B* **2020**, *102*, 100501. [CrossRef]
25. Bandyopadhyay, S.; Adhikary, P.; Das, T.; Dasgupta, I.; Saha-Dasgupta, T. Superconductivity in infinite-layer nickelates: Role of f orbitals. *Phys. Rev. B* **2020**, *102*, 220502. [CrossRef]
26. Karp, J.; Botana, A.S.; Norman, M.R.; Park, H.; Zingl, M.; Millis, A. Many-Body Electronic Structure of NdNiO$_2$ and CaCuO$_2$. *Phys. Rev. X* **2020**, *10*, 021061. [CrossRef]
27. Kitatani, M.; Si, L.; Janson, O.; Arita, R.; Zhong, Z.; Held, K. Nickelate superconductors—A renaissance of the one-band Hubbard model. *NPJ Quantum Mater.* **2020**, *5*, 59. [CrossRef]
28. Worm, P.; Si, L.; Kitatani, M.; Arita, R.; Tomczak, J.M.; Held, K. Correlations turn electronic structure of finite-layer nickelates upside down. *arXiv* **2021**, arXiv:2111.12697.
29. Geisler, B.; Pentcheva, R. Correlated interface electron gas in infinite-layer nickelate versus cuprate films on SrTiO$_3$(001). *Phys. Rev. Res.* **2021**, *3*, 013261. [CrossRef]
30. Klett, M.; Hansmann, P.; Schäfer, T. Magnetic Properties and Pseudogap Formation in Infinite-Layer Nickelates: Insights From the Single-Band Hubbard Model. *Front. Phys.* **2022**, *10*, 834682. [CrossRef]
31. LaBollita, H.; Botana, A.S. Correlated electronic structure of a quintuple-layer nickelate. *Phys. Rev. B* **2022**, *105*, 085118. [CrossRef]
32. Bednorz, J.G.; Müller, K.A. Possible high T$_C$ superconductivity in the Ba-La-Cu-O system. *Z. Phys. B Condens. Matter* **1986**, *64*, 189–193. [CrossRef]
33. Kamihara, Y.; Watanabe, T.; Hirano, M.; Hosono, H. Iron-Based Layered Superconductor La[O$_{1-x}$F$_x$]FeAs ($x = 0.05 - 0.12$) with $T_c = 26K$. *J. Am. Chem. Soc.* **2008**, *130*, 3296. [CrossRef] [PubMed]
34. Drozdov, A.P.; Eremets, M.I.; Troyan, I.A.; Ksenofontov, V.; Shylin, S.I. Conventional superconductivity at 203 kelvin at high pressures in the sulfur hydride system. *Nature* **2015**, *512*, 73. [CrossRef]
35. Snider, E.; Dasenbrock-Gammon, N.; McBride, R.; Debessai, M.; Vindana, H.; Vencatasamy, K.; Salamat, K.V.L.A.; Dias, R.P. Room-temperature superconductivity in a carbonaceous sulfur hydride. *Nature* **2020**, *586*, 373. [CrossRef]
36. Bardeen, J.; Cooper, L.N.; Schrieffer, J.R. Microscopic Theory of Superconductivity. *Phys. Rev.* **1957**, *106*, 162–164. [CrossRef]
37. Savrasov, S.Y.; Andersen, O.K. Linear-Response Calculation of the Electron-Phonon Coupling in Doped CaCuO$_2$. *Phys. Rev. Lett.* **1996**, *77*, 4430–4433. [CrossRef]
38. Boeri, L.; Dolgov, O.V.; Golubov, A.A. Is LaFeAsO$_{1-x}$F$_x$ an Electron-Phonon Superconductor? *Phys. Rev. Lett.* **2008**, *101*, 026403. [CrossRef]
39. Scalapino, D.J. A common thread: The pairing interaction for unconventional superconductors. *Rev. Mod. Phys.* **2012**, *84*, 1383–1417. [CrossRef]

40. Vilardi, D.; Bonetti, P.M.; Metzner, W. Dynamical functional renormalization group computation of order parameters and critical temperatures in the two-dimensional Hubbard model. *Phys. Rev. B* **2020**, *102*, 245128. [CrossRef]
41. Sordi, G.; Sémon, P.; Haule, K.; Tremblay, A.M.S. Strong Coupling Superconductivity, Pseudogap, and Mott Transition. *Phys. Rev. Lett.* **2012**, *108*, 216401. [CrossRef]
42. Gull, E.; Millis, A. Numerical models come of age. *Nat. Phys.* **2015**, *11*, 808. [CrossRef]
43. Kitatani, M.; Schäfer, T.; Aoki, H.; Held, K. Why the critical temperature of high-T_c cuprate superconductors is so low: The importance of the dynamical vertex structure. *Phys. Rev. B* **2019**, *99*, 041115. [CrossRef]
44. Comin, R.; Damascelli, A. Resonant X-Ray Scattering Studies of Charge Order in Cuprates. *Annu. Rev. Condens. Matter Phys.* **2016**, *7*, 369–405. [CrossRef]
45. Wu, T.; Mayaffre, H.; Krämer, S.; Horvatić, M.; Berthier, C.; Hardy, W.N.; Liang, R.; Bonn, D.A.; Julien, M.H. Magnetic-field-induced charge-stripe order in the high-temperature superconductor YBa2Cu3Oy. *Nature* **2011**, *477*, 191. [CrossRef]
46. Toschi, A.; Katanin, A.A.; Held, K. Dynamical vertex approximation; A step beyond dynamical mean-field theory. *Phys Rev. B* **2007**, *75*, 045118. [CrossRef]
47. Held, K.; Katanin, A.; Toschi, A. Dynamical Vertex Approximation—An Introduction. *Prog. Theor. Phys. (Suppl.)* **2008**, *176*, 117. [CrossRef]
48. Katanin, A.A.; Toschi, A.; Held, K. Comparing pertinent effects of antiferromagnetic fluctuations in the two- and three-dimensional Hubbard model. *Phys. Rev. B* **2009**, *80*, 075104. [CrossRef]
49. Rohringer, G.; Hafermann, H.; Toschi, A.; Katanin, A.A.; Antipov, A.E.; Katsnelson, M.I.; Lichtenstein, A.I.; Rubtsov, A.N.; Held, K. Diagrammatic routes to nonlocal correlations beyond dynamical mean field theory. *Rev. Mod. Phys.* **2018**, *90*, 025003. [CrossRef]
50. Li, D.; Wang, B.Y.; Lee, K.; Harvey, S.P.; Osada, M.; Goodge, B.H.; Kourkoutis, L.F.; Hwang, H.Y. Superconducting Dome in $Nd_{1-x}Sr_xNiO_2$ Infinite Layer Films. *Phys. Rev. Lett.* **2020**, *125*, 027001. [CrossRef]
51. Lee, K.; Wang, B.Y.; Osada, M.; Goodge, B.H.; Wang, T.C.; Lee, Y.; Harvey, S.; Kim, W.J.; Yu, Y.; Murthy, C.; et al. Character of the "normal state" of the nickelate superconductors. *arXiv* **2022**, arXiv:2203.02580.
52. Lee, K.; Goodge, B.H.; Li, D.; Osada, M.; Wang, B.Y.; Cui, Y.; Kourkoutis, L.F.; Hwang, H.Y. Aspects of the synthesis of thin film superconducting infinite-layer nickelates. *APL Mater.* **2020**, *8*, 041107. [CrossRef]
53. Malyi, O.I.; Varignon, J.; Zunger, A. Bulk $NdNiO_2$ is thermodynamically unstable with respect to decomposition while hydrogenation reduces the instability and transforms it from metal to insulator. *Phys. Rev. B* **2022**, *105*, 014106. [CrossRef]
54. Bernardini, F.; Bosin, A.; Cano, A. Geometric effects in the infinite-layer nickelates. *arXiv* **2021**, arXiv:2110.13580.
55. Cui, Y.; Li, C.; Li, Q.; Zhu, X.; Hu, Z.; feng Yang, Y.; Zhang, J.; Yu, R.; Wen, H.H.; Yu, W. NMR Evidence of Antiferromagnetic Spin Fluctuations in $Nd_{0.85}Sr_{0.15}NiO_2$. *Chin. Phys. Lett.* **2021**, *38*, 067401. [CrossRef]
56. Puphal, P.; Pomjakushin, V.; Ortiz, R.A.; Hammoud, S.; Isobe, M.; Keimer, B.; Hepting, M. Investigation of Hydrogen Incorporations in Bulk Infinite-Layer Nickelates. *Front. Phys.* **2022**, *10*, 842578. [CrossRef]
57. Held, K.; Si, L.; Worm, P.; Janson, O.; Arita, R.; Zhong, Z.; Tomczak, J.M.; Kitatani, M. Phase Diagram of Nickelate Superconductors Calculated by Dynamical Vertex Approximation. *Front. Phys.* **2022**, *9*, 810394. [CrossRef]
58. Bellaiche, L.; Vanderbilt, D. Virtual crystal approximation revisited: Application to dielectric and piezoelectric properties of perovskites. *Phys. Rev. B* **2000**, *61*, 7877–7882. [CrossRef]
59. Eckhardt, C.; Hummer, K.; Kresse, G. Indirect-to-direct gap transition in strained and unstrained Sn_xGe_{1-x} alloys. *Phys. Rev. B* **2014**, *89*, 165201. [CrossRef]
60. Hohenberg, P.; Kohn, W. Inhomogeneous Electron Gas. *Phys. Rev.* **1964**, *136*, B864–B871. doi: 10.1103/PhysRev.136.B864. [CrossRef]
61. Kohn, W.; Sham, L.J. Self-Consistent Equations Including Exchange and Correlation Effects. *Phys. Rev.* **1965**, *140*, A1133–A1138. [CrossRef]
62. Kresse, G.; Hafner, J. Ab initio molecular dynamics for liquid metals. *Phys. Rev. B* **1993**, *47*, 558–561. [CrossRef]
63. Kresse, G.; Furthmüller, J. Efficiency of ab-initio total energy calculations for metals and semiconductors using a plane-wave basis set. *Comput. Mater. Sci.* **1996**, *6*, 15–50. [CrossRef]
64. Perdew, J.P.; Burke, K.; Ernzerhof, M. Generalized Gradient Approximation Made Simple. *Phys. Rev. Lett.* **1996**, *77*, 3865–3868. [CrossRef]
65. Perdew, J.P.; Ruzsinszky, A.; Csonka, G.I.; Vydrov, O.A.; Scuseria, G.E.; Constantin, L.A.; Zhou, X.; Burke, K. Restoring the Density-Gradient Expansion for Exchange in Solids and Surfaces. *Phys. Rev. Lett.* **2008**, *100*, 136406. [CrossRef]
66. Anisimov, V.I.; Zaanen, J.; Andersen, O.K. Band theory and Mott insulators: Hubbard U instead of Stoner I. *Phys. Rev. B* **1991**, *44*, 943–954. [CrossRef]
67. Togo, A.; Tanaka, I. First principles phonon calculations in materials science. *Scr. Mater.* **2015**, *108*, 1–5. [CrossRef]
68. Baroni, S.; de Gironcoli, S.; Dal Corso, A.; Giannozzi, P. Phonons and related crystal properties from density-functional perturbation theory. *Rev. Mod. Phys.* **2001**, *73*, 515–562. [CrossRef]
69. Blaha, P.; Schwarz, K.; Madsen, G.; Kvasnicka, D.; Luitz, J. wien2k. In *An Augmented Plane Wave+ Local Orbitals Program for Calculating Crystal Properties*; Vienna University of Technology: Vienna, Austria, 2001.
70. Krieger, G.; Martinelli, L.; Zeng, S.; Chow, L.; Kummer, K.; Arpaia, R.; Sala, M.M.; Brookes, N.; Ariando, A.; Viart, N.; et al. Charge and spin order dichotomy in $NdNiO_2$ driven by $SrTiO_3$ capping layer. *arXiv* **2021**, arXiv:2112.03341.
71. Rossi, M.; Osada, M.; Choi, J.; Agrestini, S.; Jost, D.; Lee, Y.; Lu, H.; Wang, B.Y.; Lee, K.; Nag, A.; et al. A Broken Translational Symmetry State in an Infinite-Layer Nickelate. *arXiv* **2021**, arXiv:2112.02484.

72. Ren, X.; Gao, Q.; Zhao, Y.; Luo, H.; Zhou, X.; Zhu, Z. Superconductivity in infinite-layer $Pr_{0.8}Sr_{0.2}NiO_2$ films on different substrates. *arXiv* **2021**, arXiv:2109.05761.
73. Wang, N.; Yang, M.; Chen, K.; Yang, Z.; Zhang, H.; Zhu, Z.; Uwatoko, Y.; Dong, X.; Jin, K.; Sun, J.; et al. Pressure-induced monotonic enhancement of Tc to over 30 K in the superconducting Pr0.82Sr0.18NiO2 thin films. *arXiv* **2021**, arXiv:2109.12811.

Review

Superconductivity and Charge Ordering in BEDT-TTF Based Organic Conductors with β''-Type Molecular Arrangement

Yoshihiko Ihara [1,*,†] and Shusaku Imajo [2,†]

1 Department of Physics, Faculty of Science, Hokkaido University, Sapporo 060-0810, Japan
2 Institute for Solid State Physics, The University of Tokyo, Kashiwa 277-8581, Japan; imajo@issp.u-tokyo.ac.jp
* Correspondence: yihara@phys.sci.hokudai.ac.jp
† These authors contributed equally to this work.

Abstract: Exotic superconductivity that appears near the charge ordering instability has attracted significant interest since the beginning of superconducting study. The discovery of possible coexistence of charge ordering and superconductivity in cuprates and kagome metals has further fascinated researchers in recent years. In this review, we focus on the BEDT-TTF-based organic superconductor with β''-type molecular packing sequence, which shows the charge ordering transition in the very vicinity of superconducting transition, and summarize the experimental results reported up to the present. At the charge ordering temperature, ultrasonic measurement detects the softening of the crystal lattice, and ^{13}C-NMR measurement shows an increase in nuclear spin-lattice relaxation rate divided by temperature $1/T_1T$. These results suggest that low-energy dynamics are activated near the charge ordering transition, leading us to invoke the charge-fluctuation mediated superconducting pairing mechanism.

Keywords: superconductivity; charge order; molecular conductor

1. Introduction

Metallic conductivity in a material is introduced by doping carriers to the conduction band. When the conduction band is empty, a material simply shows an insulating behavior and is referred to as the band insulator. Even with the carriers in the conduction band when the doping level is 1/2, the onsite electron-electron correlations disturb the itinerancy of carriers to cause another insulating state, known as the Mott–Hubbard insulator [1,2]. Prominent many-body effects near the half filling result in fascinating physics, such as metal-insulator transition and unconventional superconductivity. Moreover, at the doping level of 1/4, long-range Coulomb interactions enforce the carriers to localize at every second site, leading again to an insulating state with charge ordering [3,4]. The electrons near this charge ordered state would also host exotic ground states through the long-range electron-electron correlation effect. The effect of charge ordering on superconductivity has been intensively studied since the discovery of charge ordering in copper-oxide superconductors [5,6]. Recent discovery of superconductivity in the charge ordered state of kagome metal CsV$_3$Sb$_5$ further promotes the discussion on the novel superconducting mechanism mediated by charge fluctuations [7–9]. The increased charge fluctuations near the valence criticality in CeCu$_2$Si$_2$ were also proposed to support superconductivity [10]. Above all, organic conductors that show both superconductivity and charge ordering have been the most intensively studied for decades [11–19]. Although coexistence between superconductivity and charge instability has been found in a variety of materials with different electronic orbitals, no universal understanding on its mechanism has been given. In this review article, we focus on a BEDT-TTF-based organic conductor, which shows the superconducting and charge ordering transitions at almost the same temperature, and summarize the experimental studies reported thus far.

The BEDT-TTF based organic conductors with the chemical formula (BEDT-TTF)$_2X$ (X = monovalent anion) ideally possess 1/4 carriers per single BEDT-TTF site. In the celebrated κ-type salts, however, the conduction band is constructed from effective (BEDT-TTF)$_2$ molecular orbital formed by the strongly dimerized two BEDT-TTF molecules, and thus the half filling per (BEDT-TTF)$_2$ dimer results in Mott physics [20]. The 1/4 filling per BEDT-TTF site is restored by arranging the BEDT-TTF molecules in a way that prevents dimerization, for instance, θ-type and β-type molecular arrangements [18,19,21,22]. The charge ordered state is found in these series of organic conductors, and, most importantly, superconductivity appears near the charge criticality. In this respect, our interest is directed to the β''-type molecular arrangement, because the charge ordering temperature $T_{CO} \simeq 8.5$ K was found in the proximity to the superconducting transition temperature $T_c \simeq 7$ K in β''-(BEDT-TTF)$_4$[(H$_3$O)Ga(C$_2$O$_4$)$_3$]·PhNO$_2$ [23]. Here, we summarize the electronic properties of β''-(BEDT-TTF)$_4$[(H$_3$O)M(C$_2$O$_4$)$_3$]·G salts with several metallic ions M and guest molecules G and review the detailed experiments performed for a salt with the specific combination of M = Ga and G = PhNO$_2$.

This review article is constructed in the following way. In Section 2, we summarize the crystal structure and superconducting transition temperatures reported for a series of β''-(BEDT-TTF)$_4$[(H$_3$O)M(C$_2$O$_4$)$_3$]·G salts. (Hereafter, these chemical formulae are abbreviated as β''-M/G.) Section 3 addresses the chemical pressure effects introduced by the guest molecules. The quantum oscillation, elastic constants, ^{13}C-NMR, and EPR measurements conducted to investigate the charge ordered state in β''-Ga/PhNO$_2$ are summarized in Sections 4–7, respectively. In Section 8, we present the superconducting properties measured by the temperature, field, and field-orientation dependence of heat capacity. Finally, in Section 9 a summary of our current understanding on the basis of reported results and perspective will be given.

2. Crystal Structure and Fermi Surface

β''-(BEDT-TTF)$_4$[(H$_3$O)Ga(C$_2$O$_4$)$_3$]PhNO$_2$, which we focus on in this review, belongs to the 4:1 β''-type BEDT-TTF-tris(oxalato)metallate salts known as the Day series [24,25]. This 4:1 β''-type family, having the formula (BEDT-TTF)$_4 AM$(C$_2$O$_4$)$_3 G$, is synthesized by means of electrochemical oxidation of the organic BEDT-TTF donor with 18-crown-6 ether and counter molecule AM(C$_2$O$_4$)$_3$ in solvents G under galvanostatic current. Martin and co-workers successfully obtained pseudo-κ [26–28], α-β'' [26,29], and 2:1-β'' with 18-crown-6 [26,30–32] phases in addition to the 4:1 β''-phase by optimizing the synthesis conditions for each phase. For the 4:1 β''-phase, independent studies by Akutsu, Coronado, Martin, and Prokhorova and their co-workers report an extensive variety of M(=Fe^{3+}, Cr^{3+}, Ga^{3+}, Rh^{3+}, Ir^{3+}, Ru^{3+}, Al^{3+}, Co^{3+}) and G(=C$_5$H$_5$N, DMF, PhCN, PhNO$_2$, PhBr, PhCl, PhF, PhI, CH$_2$Cl$_2$) [23,25–28,33–38]. When M = Ga and G = PhNO$_2$, the 4:1 β''-salt is often obtained as the minor product with the major product of the semiconducting pseudo-κ phase crystals. β''-Ga/PhNO$_2$ salt crystallizes as black distorted hexagonal rods, whereas the pseudo-κ phase crystallizes as dark-brown diamond-shaped plates.

The 4:1 β''-salts crystallize in the monoclinic space group $C2/c$. As shown in Figure 1, this series has a two-dimensional layered structure composed of the BEDT-TTF layers and counter layers. The two crystallographically independent BEDT-TTF molecules, A and B, are stacked in the β''-type arrangement with weak tetramerization without strong dimerization in the a-b plane. In the counter anion layers, M^{3+} and A^+ form a hexagonal frame bridged by (C$_2$O$_4$)$^{2-}$, and G occupies the cavity. The weak tetramerization of the quarter-filled BEDT-TTF system leads to the semimetallic band structure with the compensated Fermi surface of the electron and hole pockets, as calculated by the extended-Hückel tight-binding method [39] with a unit cell transformation from the C-centered lattice to the primitive lattice [40,41], shown in Figure 1d,e. The small Fermi pockets indicate a small number of itinerant carriers, which weakens the Coulomb screening effect, and therefore, the long-range Coulomb repulsion effectively induces the instability of charge distribution.

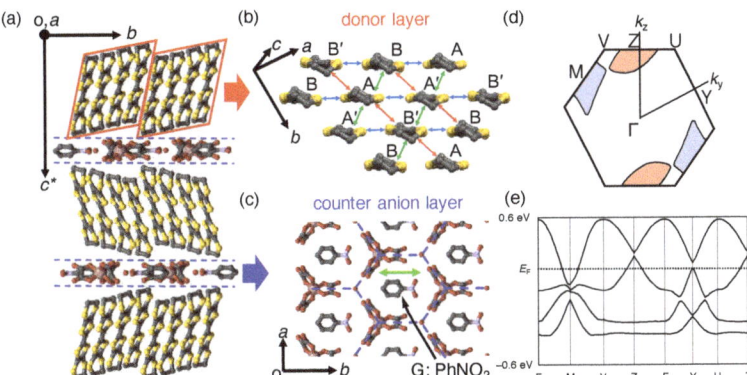

Figure 1. (**a**) Interlayer packing structure of β''-(BEDT-TTF)$_4 AM(C_2O_4)_3 G$. The red rhombi indicate tetramers of the BEDT-TTF molecules, and the blue dashed lines represent the counter anion layers. (**b**) Arrangement of the BEDT-TTF molecules in the β''-type packing motif. In this 4:1 β''-phase, there are the two crystallographically independent BEDT-TTF molecules, A and B. The arrows signify transfer integrals between them along the diagonal (red), stacking (green), and horizontal (blue) directions. (**c**) Honeycomb cavity composed of the monovalent cation A^+ and tris(oxalato)metallate $M(C_2O_4)_3{}^{2-}$ in the counter anion layer. The guest molecule G occupies this hexagonal vacancy. (**d**,**e**) Fermi surface (**d**) and band structure (**e**) of β''-(BEDT-TTF)$_4$[(H$_3$O)Ga(C$_2$O$_4$)$_3$]PhNO$_2$ (Ga/PhNO$_2$) derived from the band calculation using the extended-Hückel tight-binding method [39–41].

3. Chemical Pressure Effect

Because electronic states depend on the transfer integral between the BEDT-TTF molecules, each M/G should exhibit different low-temperature behavior through the chemical pressure effect. Figure 2a shows the temperature dependence of the out-of-plane resistance $R(T)$ reduced by $R(300\text{ K})$ for several M/G [41]. In the inset, the low-temperature region is enlarged to show the superconducting transition clearly. Whereas $R(T)/R(300\text{ K})$ above 100 K is moderate in all salts, the low-temperature $R(T)$ behavior strongly depends on the chemical pressures. For the higher-T_c salts, such as Fe/PhCN and Ga/PhNO$_2$, an abrupt increase in the resistance is observed at approximately 10 K, whereas the lower-T_c salts and non-superconducting salts do not show such semiconducting behavior. A simple deduction is that the factor producing this semiconducting nature raises T_c. Namely, the origin of the change in the electronic state around 10 K should be connected to the origin of superconductivity. To organize these results, the chemical substitution effect is summarized in view of the effective pressure applied to the electronic state. Assuming that the size of the hexagonal cavity in the counter layer governs the arrangement of the BEDT-TTF molecules and the distances between them, the size of G occupying the vacancy must be important. Because the guest molecules G orient toward the b axis as shown in Figure 1, the b-axis length should dominantly control the chemical pressure. Indeed, the longer G, such as PhCN and PhNO$_2$, gives the longer b-axis length compared to the shorter G, such as C$_5$H$_5$N and CH$_2$Cl$_2$ (Figure 2c). To ascertain the b-axis length dependence of the electronic state, we plotted T_c as a function of b-axis length in Figure 2c. The values of T_c for the series of salts reported in Refs. [23,28,34–36,38,41–43] were determined by the resistivity data presented in these studies. The large positive correlation coefficient $R \sim 0.84$ indicates that the electronic state is certainly under the influence of the b-axis length.

To understand the low-temperature electronic state in more detail, the magnetoresistance measured at 1.5 K [41] is displayed in Figure 3a. For some salts, superconductivity is observed and immediately suppressed below a few teslas. At higher fields (>10 T), Shubnikov–de Haas (SdH) oscillations are observed. Note that Fe/PhCN and Ga/PhNO$_2$ do not exhibit large SdH signals as compared to other salts; however, small oscillations certainly appear in the higher-field region as enlarged in the inset. The Fourier transform spectra of the SdH oscillations are shown in Figure 3b. As expected for a compensated metal, Fe/PhCN and Ga/PhNO$_2$ show only one signal at the SdH frequencies of 195 T and 220 T, respectively. These results are consistent with the band calculation [40,41], suggesting that the cross-sectional area of the Fermi pockets A_{FS} is approximately 10% of that of the Brillouin zone A_{BZ}. In the case of the other salts, some other additional peaks are observed in the low-frequency region (∼50 T), which means that a split of the Fermi surface occurs and changes the topology of the Fermi surface. Indeed, the band calculations for some salts [41] indicate that an additional tiny hole pocket appears around the Y point depending on transfer integrals. To study the chemical pressure effect on the SdH oscillations simply, we here focus on the signal of the electron pocket at the M point (marked with each circle). Referring to the reported data by Bangura, Coldea, Prokhorova, Uji, and ourselves [37,40–44], we show the ratio A_{FS}/A_{BZ} (c) and the effective electron mass m^*/m_e (d) as a function of b in Figure 3. These results also clarify the b-axis length dependence of the electronic state. As the elongation of the b-axis length with the longer G should reduce the amplitude of the transfer integrals in the molecular stacking, it is reasonable that increasing b leads to the diminution in A_{FS}/A_{BZ} and augmentation of m^*/m_e. This perspective is consistent with the variation of band structure calculated as a function of long-range electron correlations V normalized by the band width W [41]. That is to say, the electron correlations leading to the charge disproportion facilitate superconductivity.

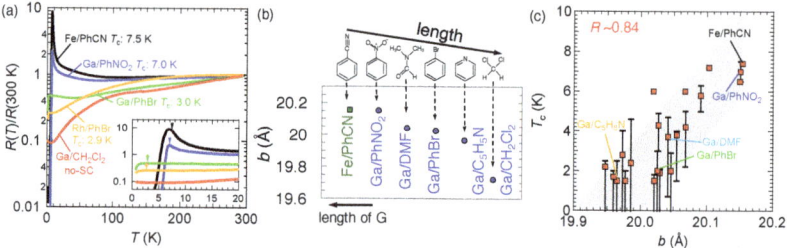

Figure 2. (a) Temperature dependence of the reduced electrical resistance $R(T)/R(300\,K)$ for some salts [41]. The inset is an expanded view below 20 K. The arrows indicate the superconducting transition temperatures T_c determined by the onset of the drop of the resistance. (b) Schematic comparison of the b-axis length of some salts depending on the length of the guest molecule G. (c) T_c vs. b of the reported M/G salts. The T_c of each salt was determined from the resistivity data reported in Refs. [23,28,34–36,38,41–43]. This plot gives a large positive correlation coefficient $R\sim 0.84$. The bars indicate the onset and zero-resistivity temperatures.

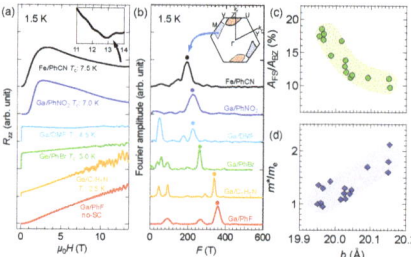

Figure 3. (a) Magnetoresistance of some salts at 1.5 K [41]. Each dataset has an offset for clarity. The inset is an enlarged plot for Fe/PhCN to highlight the quantum oscillations. (b) Fourier spectra of the SdH oscillations shown in (a). The peaks marked with the circles originate from the electron Fermi pocket around the M point shown in the inset. (c,d) Ratio of the size of the electron Fermi pocket A_{FS} and the Brillouin zone A_{BZ} (c) and the effective mass (d) as a function of the b-axis length. These results were reported in Refs. [37,40–44].

4. Charge Ordered State Viewed from High-Field Quantum Oscillations

In the last section, we elucidated that the long-range electron correlations are important for superconductivity. Considering that electron correlations of π electrons in the quarter-filled β''-type salts induce charge instability in itinerant carriers according to V/W, the charge degrees of freedom must be discussed. As introduced above, Ga/PhNO$_2$ is a good target for this discussion due to the coexistence of the charge ordering and superconductivity, and thus, we hereafter focus on Ga/PhNO$_2$.

Figure 4a shows the SdH oscillatory component in the electrical transport $\Delta R_{\rm osc}/R$ up to 60 T at various temperatures [45]. At low temperatures, the amplitude of the observed SdH oscillations dwindles above 40 T and cannot be explained by a single-component SdH oscillation. This behavior can be understood by assuming that the two oscillations interfere with each other and render the oscillation beating. Indeed, using the two-component Lifshitz–Kosevich (LK) formula, the field dependence can be reproduced, as described by the dotted curves. In Figure 4b, we present the thermal variation in the frequency F of one of the SdH signals. At higher temperatures, the F is approximately 220 T, which agrees with the value discussed above [40,41,43]. Below ~8 K, where the charge disproportion develops, the value of F decreases. This means that unbalance between the hole and electron Fermi pockets manifests in the charge ordered state, and the different cross sections of Fermi surface yield the beating of the SdH oscillations. Considering the proportional relation between F and $A_{\rm FS}$, the result indicates that the charge ordering reduces the number of itinerant carriers. Naturally, the charge ordering affects the effective electron mass. Figure 4c displays the amplitude A at 42 T as a function of temperature. The dotted curve represents a fit to the LK formula with $m^*/m_{\rm e}$ = 1.6. Above 8 K, the obtained data have larger values than the fitting curve. To make this deviation clearer, the ratio of the data and fit, $A/A(m^*/m_{\rm e} = 1.6)$, is shown in the inset. As a smaller $m^*/m_{\rm e}$ causes a larger A, the behavior indicates that the charge ordering enhances $m^*/m_{\rm e}$. As the charge ordering should connect with the localization of the electrons in the strong electron correlations, these results seem to lend support to the present scenario.

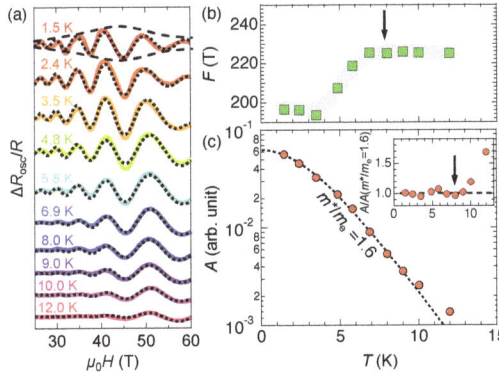

Figure 4. (a) Oscillatory component of the high-field magnetoresistance $\Delta R_{\rm osc}/R$ [45]. The dotted curves are fits to the two-component LK formula. The dashed envelope is a visual guide for the amplitude of the oscillations. (b,c) Temperature variation in the frequency F (b) and the amplitude A (c) of one of the SdH signals. The dotted curve in (c) is a calculation obtained by the LK formula with $m^*/m_{\rm e}$ = 1.6. The inset displays the ratio of A of the data and the simulation when $m^*/m_{\rm e}$ = 1.6.

5. Elastic Response to Charge Instability

As charge ordering transition can be detected by elastic properties through the coupling between an electric quadrupole moment and strain of lattice [46], the ultrasonic properties are measured by our group and the results reported in Ref. [45] are presented in Figure 5. The relative change in the longitudinal ultrasonic attenuation $\Delta\alpha$ shows the broad maximum around 9 K. The relative change in the elastic constant $\Delta C_L/C_L$ also suggests that the lattice is softened in this temperature region. The absence of the magnetic field dependence in this temperature range indicates that this anomaly is not related to superconductivity and magnetic degrees of freedom. Therefore, the lattice softening is induced by the development of the pure charge fluctuations that result in the static charge ordering at 8.5 K. This viewpoint shows a good agreement with the results of the ^{13}C-NMR [47,48] as well as the high-field SdH [45] studies discussed in Sections 4 and 6. Note that the small softening observed at 6 K (inset) is attributable to the superconducting transition due to the suppression in a magnetic field.

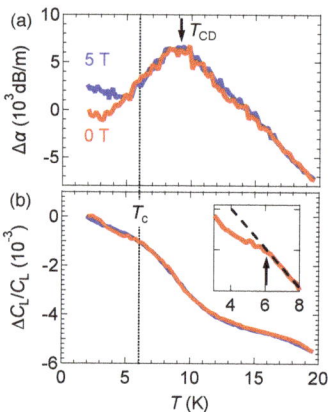

Figure 5. (**a**,**b**) Low-temperature elastic properties, (**a**) the relative change in the ultrasonic attenuation $\Delta\alpha$, and (**b**) the elastic constant $\Delta C_L/C_L$, as a function of temperature [45]. The red and blue curves are the data at 0 and 5 T, respectively. The arrow in (**a**) indicates the transition temperature of the charge ordered state T_{CO}. The inset in (**b**) is the enlarged plot around T_c. To emphasize the superconducting transition, a linear extrapolation estimated from the behavior of the normal state is shown as a dashed line in the inset.

6. Charge Ordering Observed by ^{13}C-NMR Spectroscopy

NMR spectroscopy is a powerful tool to measure the disproportionate local charges at BEDT-TTF molecules. In this section, we explore the static properties of the charge-ordered state by reviewing the highly resolved NMR spectra taken at 15 T and precisely quantify the site-charge modulation that appears near the superconducting transition [47,48].

When the external field is applied along the b axis, a single ^{13}C-NMR peak was observed at 20 K, as shown in the inset of Figure 6a. The single-peak spectrum was broadened at low temperatures, and a clearly resolved two-peak structure was observed at 1.6 K. We determined the NMR shift δ from the peak positions and display the temperature dependence of the NMR shift for each peak in Figure 6a. The peak splitting was observed below 8.5 K both at 15 T and 8 T. This peak splitting cannot be explained by the effects of a superconducting transition, because T_c is suppressed to 3 K at 15 T. Only a barely resolved kink in the NMR shift was observed at T_c. It is clear that the temperature variation of δ and the peak splitting in units of parts per million (ppm) are identical between 8 and 15 T. The field-independent peak separation confirms that electron spins are in the paramagnetic state and have polarization proportional to the external magnetic field. This indicates the absence of a spontaneous internal field, which could have been generated by some field-independent magnetic ordering. We identified that the transition at 8.5 K is the order in the charge degrees of freedom, in which the molecular site charges deviate from the formal value of $0.5e$. We assigned the broad and the sharp peak as the NMR signal from the charge-rich (R) and charge poor (P) sites, respectively, according to the discussion given later. The peak separation is related to the charge imbalance between the R and P sites and is proportional to the order parameter of the charge-ordered state.

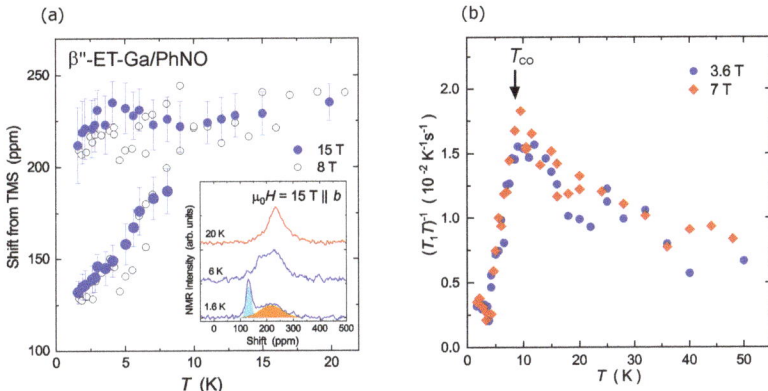

Figure 6. (a) Temperature dependence of the NMR shift at 15 T and 8 T [48]. The vertical axis is the peak shift from tetramethylsilane (TMS). The field-independent peak separation at the lowest temperature indicates the absence of any fixed internal field that could be associated with a magnetic transition. Inset shows ^{13}C-NMR spectra at various temperatures. The peak splitting starts below 8 K, and a clearly split two-peak structure was observed at 1.6 K. The ratio of the integrated area between the sharp (cyan) and broad (orange) peaks was evaluated as 0.4:1. (b) Temperature dependences of $1/T_1T$ measured at 3.6 and 7 T [47]. The peaks at T_{CO} were found in both fields, suggesting unconventional electronic state affected by long-range electron-electron correlations.

The dynamical properties of electrons associated with the charge ordering transition are investigated by measuring the nuclear spin-lattice relaxation rate $1/T_1$ at 3.6 and 7 T [47]. As shown in Figure 6b, $1/T_1T$ increases with decreasing temperature, forming a peak at T_{CO}. In general, $1/T_1T$ is proportional to the square of the density of states in the Fermi liquid state and is temperature independent. The temperature dependence of $1/T_1T$ is introduced by the enhanced magnetic fluctuations in the vicinity of magnetic criticality. For charge ordering β''-Ga/PhNO$_2$, however, weak magnetic fluctuations cannot induce a strong temperature dependence in $1/T_1T$. Charge fluctuations should be enhanced around T_{CO}, but they are not directly coupled with ^{13}C nuclear spins, because ^{13}C nuclei with a nuclear spin $I = 1/2$ do not carry an electric quadrupole moment, which can interact with charge fluctuations and cause relaxation of nuclear magnetization. The coupling between charge and magnetic fluctuations is required to increase $1/T_1T$ at T_{CO}. One possible interpretation is that the fluctuations in local spin density, which are generated by charge density fluctuations, create fluctuating hyperfine fields at the ^{13}C site. Direct observation of charge fluctuations by quadrupolar nuclear spins is desired to reveal the mechanisms of spin-charge coupling.

Below T_{CO}, the Fermi liquid behavior in $1/T_1T$ is absent until the superconducting state emerges. The non-Fermi liquid behavior is consistent with the semiconducting resistivity at the corresponding temperature range just above T_c. Because $1/T_1T$ decreases below T_{CO} following a power-law close to T^2, a clear anomaly associated with a superconducting transition was not observed at T_c. The temperature dependence of quasi particle density of states is more clearly observed from the heat capacity measurement (Section 8) [45].

The NMR intensity is proportional to the number of ^{13}C nuclei on the molecular sites. In the charge-ordered state, if the number of R sites were equal to that of P sites (PR pattern), the NMR intensity of the narrow peak would be comparable to that of the broad peak. However, as shown in the inset of Figure 6a, the intensity ratio between the peaks was 0.4:1, which indicates that there are at least twice as many R sites as P sites in the charge-ordered state (PRR pattern) [48]. The crystal structure of the β''-Ga/PhNO$_2$ salt consists of two crystallographically independent BEDT-TTF molecules, each of which forms a pair with another molecule connected by inversion symmetry. The simplest PR charge pattern can be attributed to the inversion symmetry breaking between a pair of BEDT-TTF molecules. The experimentally suggested PRR pattern addresses a more complicated charge pattern, possibly originating from the competition between the long and short range Coulomb interactions, which will be discussed later.

The large charge modulation was confirmed by the field angle dependence of NMR spectra at the lowest temperature [48]. The angle dependence of the NMR shift originates from the anisotropic dipolar Knight shift from π electrons and from the anisotropy of the chemical shift. The Knight (chemical) shift becomes maximum (minimum) when the external field is along the long axis of the π orbital, which is perpendicular to the BEDT-TTF molecular plane. To obtain the pure Knight shift contribution, the chemical shift contribution should be subtracted from the total NMR shift using a chemical shift tensor for BEDT-TTF molecules [49,50]. In Figure 7a, the Knight shift is plotted as a function of the angle between the field and the normal to the molecular plane. The amplitude of this angle dependence is proportional to the spin density in the π orbital. The contrasting behavior for the two independent sites confirms that the site charge is strongly modified from the formal value (0.5e). In particular, a very weak angle dependence for the P sites indicates that they possess a small site charge.

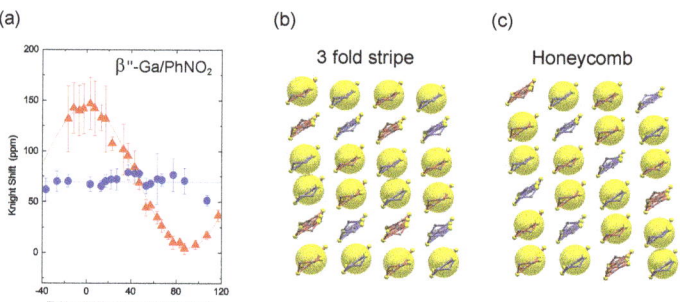

Figure 7. (a) Field direction dependence of Knight shift [48]. The field direction is defined with respect to the molecular long axis. The large amplitude for R sites indicates rich charge filling to the BEDT-TTF π orbital. (b,c) Schematic images of charge patterns in the conducting plane. The yellow balls on BEDT-TTF molecules indicate charge-rich (R) sites. The vertical P-R-R pattern can be aligned horizontally to make stripes (b), or shifted by one site to form a honeycomb structure (c).

Now we discuss a possible charge pattern in the ordered state. The NMR intensity ratio between the R and P sites suggests a PRR structure. In the two-dimensional conducting plane, we can assume PRR stripe and honeycomb structures, as shown in Figure 7b,c. The possibility of interlayer charge ordering was excluded because of the large energy cost for a charge-rich plane. To get insight into the energetic stability of the structure, we calculated the average Coulomb potential (per site) between neighboring BEDT-TTF sites, by placing point charges on molecular sites,

$$E(q) = \frac{e^2}{2\pi\epsilon_0} \langle r^{-1} \rangle \left(1 - Zq^2\right), \quad (1)$$

where $\langle r^{-1} \rangle = 1/8 \sum r_i^{-1}$ is the average of the inverse intermolecular distance over eight neighboring BEDT-TTF sites, and qe is the charge deviation of the P site from 0.5e [48]. The coefficient Z, which depends on the charge pattern, is calculated to be 0.21 and 0.41 for the PRR stripe and honeycomb structures, respectively. The larger Z value for the honeycomb structure suggests that the honeycomb structure is energetically more stable, and is thus expected to be realized. In terms of the nearest-neighbor Coulomb potential, however, the conventional PR stripe structure with equal P and R sites should be the most stable charge pattern, whereas this possibility is excluded by our NMR data. It is thus clear that other interactions should be taken into account. From theoretical studies on other BEDT-TTF salts with θ-type molecular packing, we learn that the PRR charge pattern can be stabilized in a wide parameter range in the static limit of the extended Hubbard model [51]. The exotic charge pattern appears in a θ-type structure, because the in-plane structure is closer to the triangular than to the square lattice. Considering the in-plane structure of β''-Ga/PhNO$_2$ salt to be a squeezed triangular lattice, the exotic charge pattern can be induced by the competition between the off-site and the on-site Coulomb interactions. As in β''-Ga/PhNO$_2$ salt, the charge ordering transition occurs in a metallic state, the effect of the transfer integral should certainly be taken into account [52] to explain the charge pattern and understand the superconducting pairing mechanism inside the charge ordered state.

7. EPR Measurement to Detect SC/CO Coexistence

NMR spectroscopy is one of the most powerful techniques to investigate the electronic state from a microscopic viewpoint. However, because of an insufficient spectrum resolution [inset of Figure 6a], we were not able to exclude the possibility of phase segregation. As an alternative probe, the electron paramagnetic resonance (EPR) experiment was performed by our group [53].

The EPR signal in β''-Ga/PhNO$_2$ salt originates from the π electrons in the highest occupied molecular orbital of the BEDT-TTF molecules. We succeeded in detecting the charge anomaly on the clearly resolved EPR spectrum by taking advantage of the anisotropy of g factors and the bilayer crystal structure of β''-Ga/PhNO$_2$ salt. Thus, the EPR experiment allows us to observe the phase segregation, if any, as an additional component of the EPR spectrum. The present results, which are explained by a single EPR contribution at any temperature, clearly evidence a uniform coexistence between the superconducting and charge ordering states.

We measured the temperature dependence of the EPR spectrum in the field applied parallel to the b axis (b-axis field) and to the direction 45° rotated from the b axis (45° field). In the 45° field, T_c is suppressed below 3 K by a field of approximately 300 mT. At the lowest temperature in the 45° field, a two-peak EPR spectrum was observed, as shown in Figure 8b. With increasing temperature, the peak separation becomes small, and a single peak was observed at temperatures higher than 8.5 K. We found a trace of the two-peak structure at 8 K as the wiggle around the center of the spectrum at 344.4 mT. Therefore, the peak positions were determined by the two-peak Lorentzian fit for the spectra below 8 K [solid symbols in Figure 8c] and by the single-peak Lorentzian fit above 8.5 K [open symbols in Figure 8c]. The abrupt increase in the peak separation below 8 K clearly evidences a phase transition. This anomaly agrees with the charge ordering transition at $T_{CO} = 8.5$ K determined from the ultrasonic [45] (Section 5) and ^{13}C-NMR measurements [47,48] (Section 6). The origin of the EPR peak splitting associated with charge ordering is discussed in Ref. [53].

As the charge ordering anomaly is successfully observed in the EPR spectrum, we compare the EPR spectra in the b-axis and 45° fields to unveil the relationship between the charge ordering and superconducting states. Typical EPR spectra for b-axis fields are presented in Figure 8a. In the b-axis field of 345 mT, T_c does not change because of the extremely high upper critical field ($B_{c2} > 30$ T). The effect of the superconducting transition was observed in the EPR spectrum as a reduction of the integrated intensity below 7 K [Figure 8d]. Such a decrease in intensity was not observed in the 45° field, because T_c is suppressed below 3 K. This result confirms that the electronic spins that would show superconductivity in zero field contribute to the EPR intensity when superconductivity is suppressed by the 45° field. Thus, if the superconducting part of the sample did not show the charge ordering transition, which is the case for the macroscopic phase segregation, an additional EPR peak originating from the electrons in a normal metallic state should be observed at the center of the two-peak spectrum. However, such an extra contribution was not observed at the lowest temperature of 3.6 K, as shown at the bottom of Figure 8b. The clear two-peak spectrum in the 45° field allows us to conclude that the superconducting state coexists uniformly with the charge ordered state. We note that the EPR intensity decreases gradually below T_c in the b-axis field, and finite intensity remains even at 3.6 K. This behavior contrasts with the conventional behavior expected for a homogeneous superconducting state, in which EPR signal should disappear. The EPR in the superconducting state may originate from the nearly localized electrons in the charge ordered state, for instance, the pin site in the pinball liquid state [54–56].

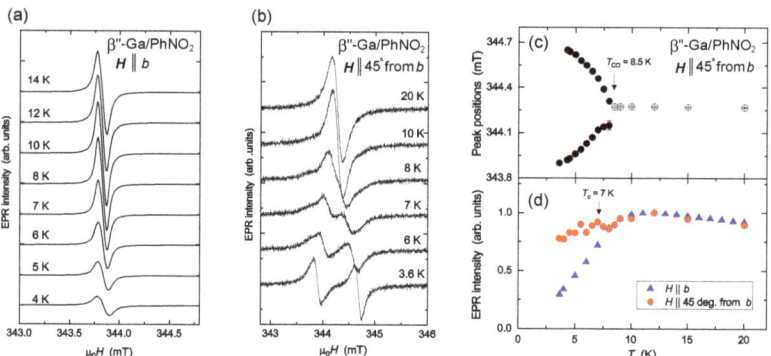

Figure 8. Temperature dependence of EPR spectra in b-axis field (**a**) and 45° field (**b**) [53]. No spectrum splitting was observed in b-axis field, whereas clear spectrum splitting was observed in 45° field below T_{CO} = 8.5 K. The EPR intensity decreases in b-axis field below $T_c \simeq 7$ K, but finite intensity was observed even at 4 K. (**c**) EPR peak position for 45° field determined by Lorentzian fit. (**d**) Temperature dependence of EPR intensity normalized at 12 K.

8. Gap Symmetry of the Charge-Fluctuation-Mediated Superconductivity

In the previous sections, we understand the properties of the charge ordering, and that superconductivity emerges from the charge ordered state. Proximity of these transition temperatures implies that superconductivity is mediated by the charge fluctuations; nevertheless, we need to know the exact nature of the superconducting state itself for the direct determination of the pairing mechanism of superconductivity. The superconducting gap function is one of the most important pieces of information for discussing the pairing mechanism. Heat capacity measurement, sensitive to low-energy excitations, is a powerful method that can scrutinize this through temperature, field, and field-angle dependence [57–60].

Figure 9a shows the low-temperature heat capacity in various fields measured by our group [45]. The zero-field data indicate that the finite intercept, which corresponds to the residual electronic heat capacity, exists. This means that the electronic state is inhomogeneous even at 0 T. Considering the cleanness of the electronic state confirmed by the SdH oscillation, the coexistence with the normal state should be an intrinsic characteristic, which is consistent with the results of the previous NMR [47,48] and EPR studies [53] discussed in Sections 6 and 7. With increasing field, the electronic heat capacity is recovered because of the suppression of superconductivity. As this field-dependent contribution relates to superconductivity, we here plot the superconducting electronic heat capacity C_{sc}, subtracting the lattice heat capacity and residual electronic component as a function of reduced temperature T/T_c in Figure 9b. The low-temperature C_{sc} exhibits neither T^2 nor T^3 dependence (dashed lines) observed in nodal superconductors, such as the κ-type organic superconductors, but is well reproduced by the solid curve for the anisotropic full-gapped model using $\Delta_0[1 + A\cos(4\phi)]^{1/2}$, where $\Delta_0 = 2.5 k_B T_c$ [61] and $A = 0.64$ ($\Delta_{max}/\Delta_{min} \sim 2.2$). Magnetic-field dependence of C_p [45] agrees with the full-gapped model because of the observation of H-linear behavior at low fields. Thus, the superconducting state of Ga/PhNO$_2$ should be fully gapped.

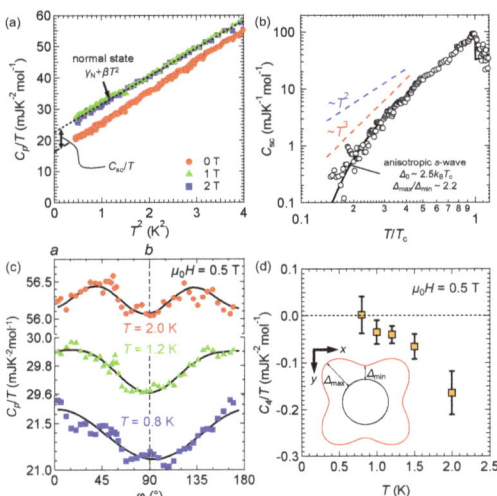

Figure 9. (a) Low-temperature heat capacity in various magnetic fields plotted as C_p/T vs. T^2 [45]. (b) Electronic heat capacity related to the superconductivity C_{sc} at 0 T. The dashed line represents T-squared relation, and the solid curve is a calculation for the anisotropic s-wave superconductivity using the gap function $\Delta_0[1 + A\cos(4\phi)]^{1/2}$. (c) In-plane angle-resolved heat capacity in a field of 0.5 T rotated from the a axis (0°) to the b axis (90°). (d) Temperature dependence of the fourfold term C_4/T. The inset is a schematic illustration of the anisotropic s-wave gap.

To deepen the understanding of the gap symmetry in more detail, in Figure 9c we present the in-plane field-angle dependence of heat capacity in a field of 0.5 T [45]. The angular dependence is simply described by the equation, $C_p/T = [C_0 + C_2\cos(2\phi) + C_4\cos(4\phi)]/T$ (solid curve). Taking account of the twofold rotational symmetry of the crystal structure [23], it is natural that the anisotropy of the Fermi velocity yields the twofold component $C_2\cos(2\phi)$. As the crystal structure does not have the fourfold symmetry, the fourfold term $C_4\cos(4\phi)$ should reflect the anisotropic part of the gap function. Hence, to shed light on the fourfold component, we present the temperature dependence of C_4/T in Figure 9d. The negative values mean that the minima of the fourfold term are located in the directions of the crystal axes. Such a fourfold periodic term is often observed in d-wave superconductivity according to the anisotropy of the gap function with four gap nodes. The correspondence between the angle-resolved heat capacity and the anisotropy of the gap function has been theoretically and experimentally verified [57–59]. Based on the zero-energy Doppler effect, the gap-node positions can be determined by the sign of C_4/T in a low-energy limit. However, in the present case, the fourfold component disappears at low temperatures. This disappearance is the feature of the anisotropic s-wave gap because the finite gap minima prohibit the quasiparticle excitations by the zero-energy Doppler effect [57]. This result also suggests that the present superconductivity is fully gapped, albeit anisotropically.

The full-gap superconductivity is distinct from the well-known dimer-Mott type organic superconductivity exhibiting d-wave symmetry with line nodes [60]. For the dimer-Mott system, the superconductivity is mediated by antiferromagnetic spin fluctuations, which originate from the strong on-site Coulomb repulsion U. Naturally, the on-site pairing is disadvantageous when utilizing U as an attractive force for the Cooper pairing, and the gap function must have nodes that render the sign of the gap function reverse. Similar to the dimer-Mott system, even for the quarter-filled system including the β''-phase, d-wave symmetry is favored by spin fluctuations when U is sufficiently stronger than V [62]. However, with increasing V, the pairing around (π,π) in the momentum space is suppressed [62], and the on-site pairing interaction becomes more advantageous to reduce the Coulomb repulsion between the nearest neighbor sites and the second nearest neighbor sites. Note the emergent gap symmetry strongly depends on the lattice geometry [62–64]; however, the charge fluctuations should facilitate the on-site pairing. Although there is no specific theoretical prediction for Ga/PhNO$_2$, the anisotropic s-wave gap function does not contradict the scenario that superconductivity in Ga/PhNO$_2$ arises from the charge fluctuations. To verify this picture, further detailed theoretical research is desirable in the future.

9. Summary

We have reviewed the experimental studies to explore the electronic properties of the organic superconductor with β''-type molecular arrangement. Among various combinations between metallic ions M and guest molecules G, Ga/PhNO$_2$ was found to show a remarkable electronic state, in which high superconducting transition sets in at a temperature very close to the charge ordering transition (Section 2). At the charge ordering transition, ^{13}C-NMR experiments revealed that the local site charge at each BEDT-TTF molecule deviates from their average value of $0.5e$ (Section 6). Associated with this charge order, the Fermi surface is deformed, as evidenced from the magnetoresistance experiment (Section 4). This is the smoking gun evidence for the charge ordering on the metallic background, which can maintain high conductivity even with a partial carrier localization. It is noteworthy that the EPR measurement suggests a uniform electronic state, excluding a possibility of phase segregation between the charge ordered insulator and metallic parts of the sample (Section 7).

At the charge ordering temperature, softening of crystal lattice was observed from the ultrasonic experiment (Section 5). ^{13}C-NMR measurement also detected an increase in the low-energy fluctuations in the temperature dependence of $1/T_1T$ (Section 6). As the peak in the temperature dependence of ultrasonic attenuation and $1/T_1T$ appears at a temperature very close to the superconducting transition temperature, one would naturally speculate that the low-energy fluctuations introduced near the charge criticality assists the superconducting pairing interaction.

The superconducting state was investigated by heat capacity measurements (Section 8). The temperature dependence and field direction dependence of C_p suggest an anisotropic s-wave superconducting state. This superconducting gap symmetry is preferable for the charge-fluctuation-induced superconductivity. However, to identify the novel superconducting mechanism in β''-Ga/PhNO$_2$ and to uncover the effect of charge fluctuations on superconductivity, further studies from both theory and experiments are required.

Author Contributions: Y.I. and S.I. have contributed equally to this article. All authors have read and agreed to the published version of the manuscript.

Funding: This work was partially supported by the Suhara Memorial Foundation and by the Japan Society for the Promotion of Science KAKENHI Grant Nos. 23740249, 19H01832, 20K14406.

Data Availability Statement: The authors declare that the data supporting the findings of this study are available within the paper. Further information can be provided by Y.I. or S.I.

Acknowledgments: We would like to acknowledge H. Akutsu, A. Akutsu-Sato, A. L. Morritt, L. Martin, Y. Nakazawa, R. Kurihara, T. Yajima, Y. Kohama, M. Tokunaga, K. Kindo, H. Seki, A. Kawamoto, M. Jeong, H. Mayaffre, C. Berthier, M. Horvatić, K. Moribe, and S. Fukuoka for collaborations.

Conflicts of Interest: The authors declare no conflict of interest.

References

1. Mott, N. F. The Basis of the Electron Theory of Metals, with Special Reference to the Transition Metals. *Proc. Phys. Soc. A* **1949**, *62*, 416. [CrossRef]
2. Hubbard, J. Electron Correlations in Narrow Energy Bands III. An Improved Solution. *Proc. R. Soc. Math. Phys. Sci.* **1964**, *281*, 401
3. Zhang, Y.; Callaway, J. Extended Hubbard Model in Two Dimensions. *Phys. Rev. B* **1989**, *39*, 9397. [CrossRef] [PubMed]
4. Kino, H.; Fukuyama, H. Phase Diagram of Two-Dimensional Organic Conductors: (BEDT-TTF)$_2$X. *J. Phys. Soc. Jpn.* **1996**, *65*, 2158–2169. [CrossRef]

5. Wu, T.; Myaffre, H.; Krämer, S.; Hovratić, M.; Berthier, C.; Hardy, W.N.; Liang, R.; Bonn, D.A.; Julien, M.-H. Magnetic-Field-Induced Charge-Stripe Order in the High-Temperature Superconductor YBa$_2$Cu$_3$O$_y$. *Nature* **2011**, *477*, 191–194. [CrossRef]
6. Croft, T. P.; Lester, C.; Senn, M.S.; Bombardi, A.; Hayden, S.M. Charge Density Wave Fluctuations in La$_{2-x}$Sr$_x$CuO$_4$ and their Competition with Superconductivity. *Phys. Rev. B* **2014**, *89*, 224513. [CrossRef]
7. Ortiz, B.R.; Gomes, L.C.; Morey, J.R.; Winiarski, M.; Bordelon, M.; Mangum, J.S.; Oswald, I., W.H.; Rodriguez-Rivera, J.A.; Neilson, J.S.; Wilson, S.D.; et al. New Kagome Prototype Materials; Discovery of KV$_3$Sb$_5$, RbV$_3$Sb$_5$, and CsV$_3$Sb$_5$. *Phys. Rev. Mater.* **2019**, *3*, 094407. [CrossRef]
8. Ortiz, B.R.; Teicher, S.M.L.; Hu, Y.; Zuo, J.L.; Sarte, P.M.; Schueller, E.C.; Abeykoon, A.M.M.; Krogstad, M.J.; Rosenkranz, S.; Osborn, R.; et al. CsV$_3$Sb$_5$; A $\mathbb{Z}_2$ Topological Kagome Metal with a Superconducting Ground State. *Phys. Rev. Lett.* **2020**, *125*, 247002. [CrossRef]
9. Liang, Z.; Hou, X.; Zhang, F.; Ma, W.; Wu, P.; Zhang, Z.; Yu, F.; Ying, J.-J.; Jiang, K.; Shan, L.; et al. Three-Dimensional Charge Density Wave and Surface-Dependent Vortex-Core States in a Kagome Superconductor CsV$_3$Sb$_5$. *Phys. Rev. X* **2021**, *11*, 031026. [CrossRef]
10. Holmes, A. T.; Jaccard, D.; Miyake, K. Signatures of Valence Fluctuations in CeCu$_2$Si$_2$ under high pressure. *Phys. Rev. B* **2004**, *69*, 024508. [CrossRef]
11. Kobayashi, H.; Kato, R.; Kobayashi, A.; Nishio, Y.; Kajita, K.; Sasaki, W. A New Molecular Superconductor, (BEDT-TTF)$_2$(I$_3$)$_{1-x}$(AuI$_2$)$_x$ ($x < 0.02$). *Chem. Lett.* **1986**, *15*, 789.
12. Mori, T.; Inokuchi, H. Superconductivity in (BEDT-TTF)$_3$Cl$_2$·2H$_2$O. *Solid. State Commun.* **1987**, *64*, 335. [CrossRef]
13. Lubczynski, W.; Demishev, S.V.; Singleton, J.; Caulfield, J.M.; du Croo de Jongh, L.; Kepert, C.J.; Blundell, S.J.; Hayer, W.; Kurmoo, M.; Day, P. A Study of the Magnetoresistance of the Charge-Transfer Salt (BEDT-TTF)$_3$Cl$_2$ · 2H$_2$O at Hydrostatic Pressures of upt 20 kbar: Evidence for a Charge-Density-Wave Ground State and the Observation of Pressure-Induced Superconductivity. *J. Phys. Condens. Matter* **1996**, *8*, 6005. [CrossRef]
14. Mori, H.; Tanaka, S.; Mori, T.; Kobayashi, A.; Kobayashi, H. Crystal Structure and Physical Properties of M = Rb and Tl Salts of (BEDT-TTF)$_2$$MM'(SCN)_4$ [M' = Co, Zn]. *Bull. Chem. Soc. Jpn.* **1998**, *71*, 797. [CrossRef]
15. Miyagawa, K.; Kawamoto, A.; Kanoda, K. Charge Ordering in a Quasi-Two-Dimensional Organic Conductor. *Phys. Rev. B* **2000**, *62*, R7679. [CrossRef]
16. Maesato, M.; Kaga, Y.; Kondo, R.; Kagoshima, S. Control of Electronic Properties of α-(BEDT-TTF)$_2$MHg(SCN)$_4$ (M = K, NH$_4$) by the Uniaxial Strain Method. *Phys. Rev. B* **2001**, *64*, 155104. [CrossRef]
17. Takano, Y.; Hiraki, K.; Yamamoto, H. M.; Nakamura, T.; Takahashi, T.; Charge Ordering in α-(BEDT-TTF)$_2$I$_3$. *Synth. Met.* **2001**, *120*, 1081. [CrossRef]
18. Nishikawa, H.; Sato, Y.; Kikuchi, K.; Kodama, T.; Ikemoto, I.; Yamada, J.; Oshio, H.; Kondo, R.; Kagoshima, S. Charge Ordering and Pressure-Induced Superconductivity in β''-(DODHT)$_2$PF$_6$. *Phys. Rev. B* **2005**, *72*, 052510. [CrossRef]
19. Morinaka, N.; Takahashi, K.; Chiba, R.; Yoshikane, F.; Niizeki, S.; Tanaka, M.; Yakushi, K.; Koeda, M.; Hedo, M.; Fujiwara, T.; et al. Superconductivity Competitive with Checkerboard-Type Charge Ordering in the Organic Conductor β-(meso-DMBEDT-TTF)$_2$PF$_6$. *Phys. Rev. B* **2009**, *80*, 092508. [CrossRef]
20. Ishiguro, T.; Yamaji, K.; Saito, G. *Organic Superconductors*; Springer:Berlin/Heidelberg, Germany, 1998.
21. Mori, H.; Tanaka, S.; Mori, T. Systematic Study of the Electronic State in θ-type BEDT-TTF Organic Conductors by Changing the Electronic Correlation. *Phys. Rev. B* **1998**, *57*, 12023. [CrossRef]
22. Nogami, Y.; Pouget, J.-P.; Watanabe, M.; Oshima, K.; Mori, H.; Tanaka, S.; Mori, T. Structural Modulation in θ-(BEDT-TTF)$_2$CsM'(SCN)$_4$ [M' = Co, Zn]. *Synth. Met.* **1999**, *103*, 1911. [CrossRef]
23. Akutsu, H.; Akutsu-Sato, A.; Turner, S.S.; Le Pevelen, D.; Day, P. Laukhin, V.; Klehe, A.-K.; Singleton, J.; Tocher, D.; Probert, M.R.; et al. Effect of Included Guest Molecules on the Normal State Conductivity and Superconductivity of β''-(ET)$_4$[(H$_3$O)Ga(C$_2$O$_4$)$_3$]G (G = Pyridine, Nitrobenzene). *J. Am. Chem. Soc.* **2002**, *124*, 12430–12431. [CrossRef]
24. Kurmoo, M.; Graham, A.W.; Day, P.; Coles, S.J.; Hursthouse, M.B.; Caulfield, J.L.; Singleton, J.; Pratt, F.L.; Hayes, W.; Ducasse, L.; et al. Superconducting and Semiconducting Magnetic Charge Transfer Salts: (BEDT-TTF)$_4$AFe(C$_2$O$_4$)$_3$·C$_6$H$_5$CN (A=H$_2$O, K, NH$_4$). *J. Am. Chem. Soc.* **1995**, *117*, 12209–12217. [CrossRef]
25. Blundell,T.J.; Brannan,M.; Mburu-Newman, J.; Akutsu, H.; Nakazawa, Y.; Imajo, S.; Martin, L. First Molecular Superconductor with the Tris(Oxalato)Aluminate Anion, β''-(BEDT-TTF)$_4$(H$_3$O)Al(C$_2$O$_4$)$_3$·C$_6$H$_5$Br, and Isostructural Tris(Oxalato)Cobaltate and Tris(Oxalato)Ruthenate Radical Cation Salts. *Magnetochemistry* **2021**, *7*, 90. [CrossRef]
26. Martin, L. Molecular conductors of BEDT-TTF with tris(oxalato)metallate anions. *Coord. Chem. Rev.* **2018**, *376*, 277–291. [CrossRef]
27. Martin, L.; Turner, S.S.; Day, P.; Guionneau, P.; Howard, J.A.K.; Hibbs, D.E.; Light, M.E.; Hursthouse, M.B.; Uruichi, M.; Yakushi, K. Crystal Chemistry and Physical Properties of Superconducting and Semiconducting Charge Transfer Salts of the Type (BEDT-TTF)$_4$[AIMIII(C$_2$O$_4$)$_3$]PhCN (AI = H$_3$O, NH$_4$, K.; MIII = Cr, Fe, Co, Al; BEDT-TTF = Bis(ethylenedithio)tetrathiafulvalene). *Inorg. Chem.* **2001**, *40*, 1363–1371. [CrossRef] [PubMed]
28. Martin, L.; Morritt, A.L.; Lopez, J.R.; Nakazawa, Y.; Akutsu, H.; Imajo, S.; Ihara, Y.; Zhang, B.; Zhange, Y.; Guof, Y. Molecular conductors from bis(ethylenedithio)tetrathiafulvalene with tris(oxalato)rhodate. *Dalton Trans.* **2017**, *46*, 9542–9548. [CrossRef]
29. Akutsu, H.; Akutsu-Sato, A.; Turner, S.S.; Day, P.; Canadell, E.; Firth, S.; Clark, R.J.H.; Yamada, J.-I.; Nakatsuji, S. Superstructures of donor packing arrangements in a series of molecular charge transfer salts. *Chem. Commun.* **2004**, 18–19. [CrossRef]

30. Martin, L.; Morritt, A.L.; Lopez, J.R.; Akutsu, H.; Nakazawa, Y.; Imajo, S.; Ihara, Y. Ambient-pressure molecular superconductor with a superlattice containing layers of tris(oxalato)rhodate enantiomers and 18-crown-6. *Inorg. Chem.* **2017**, *56*, 717–720. [CrossRef]
31. Martin, L.; Lopez, J.R.; Akutsu, H.; Nakazawa, Y.; Imajo, S. Bulk Kosterlitz–Thouless Type Molecular Superconductor β''-(BEDT-TTF)$_2$[(H$_2$O)(NH$_4$)$_2$Cr(C$_2$O$_4$)$_3$]18-crown-6. *Inorg. Chem.* **2017**, *56*, 14045–14052. [CrossRef]
32. Morritt, A.L.; Lopez, J.R.; Blundell, T.; Canadell, E.; Akutsu, H.; Nakazawa, Y.; Imajo, S.; Martin, L. 2D Molecular Superconductor to Insulator Transition in the β''-(BEDT-TTF)$_2$[(H$_2$O)(NH$_4$)$_2$M(C$_2$O$_4$)$_3$]18-crown-6 Series (M = Rh, Cr, Ru, Ir). *Inorg. Chem.* **2019**, *58*, 10656–10664. [CrossRef]
33. Akutsu-Sato, A.; Turner, S.S.; Akutsu, H.; Yamada, J.; Nakatsuji, S.; Day, P. Suppression of superconductivity in a molecular charge transfer salt by changing guest molecule: β''-(BEDT-TTF)$_4$[(H$_3$O)Fe(C$_2$O$_4$)$_3$](C$_6$H$_5$CN)$_x$(C$_5$H$_5$N)$_{1-x}$. *J. Mater. Chem.* **2007**, *17*, 2497–2499. [CrossRef]
34. Prokhorova, T.G.; Buravov, L.I.; Yagubskii, E.B.; Zorina, L.V.; Khasanov, S.S.; Simonov, S.V.; Shibaeva, R.P.; Korobenko, A.V.; Zverev, V.N. Effect of electrocrystallization medium on quality, structural features, and conducting properties of single crystals of the (BEDT-TTF)$_4$A^I[Fe^{III}(C$_2$O$_4$)$_3$]·G family. *CrystEngComm* **2011**, *13*, 537. [CrossRef]
35. Coronado, E.; Curreli, S.; Giménez-Saiz, C.; Gómez-García, C.J. The Series of Molecular Conductors and Superconductors ET$_4$[AFe(C$_2$O$_4$)$_3$]PhX (ET = bis(ethylenedithio)tetrathiafulvalene; (C$_2$O$_4$)$^{2-}$ = oxalate; A$^+$ = H$_3$O$^+$, K$^+$; X = F, Cl, Br, and I): Influence of the Halobenzene Guest Molecules on the Crystal Structure and Superconducting Properties. *Inorg. Chem.* **2012**, *51*, 1111–1126. [PubMed]
36. Prokhorova, T.G.; Zorina, L.V.; Simonov, S.V.; Zverev, V.N.; Canadell, E.; Shibaeva, R.P.; Yagubskii, E.B. The first molecular superconductor based on BEDT-TTF radical cation salt with paramagnetic tris(oxalato)ruthenate anion. *CrystEngComm* **2013**, *15*, 7048. [CrossRef]
37. Prokhorova, T.G.; Buravov, L.I.; Yagubskii, E.B.; Zorina, L.V.; Simonov, S.V.; Zverev, V.N.; Shibaeva, R.P.; Canadell, E. Effect of Halopyridine Guest Molecules on the Structure and Superconducting Properties of β''-[Bis(ethylenedithio)tetrathiafulvalene]$_4$(H$_3$O)[Fe(C$_2$O$_4$)$_3$]·Guest Crystals. *Eur. J. Inorg. Chem.* **2015**, *34*, 5611–5620. [CrossRef]
38. Prokhorova, T.G.; Yagubskii, E.B.; Zorina, L.V.; Simonov, S.V.; Zverev, V.N.; Shibaeva, R.P.; Buravov, L.I. Specific Structural Disorder in an Anion Layer and Its Influence on Conducting Properties of New Crystals of the (BEDT-TTF)$_4$A$^+$[M^{3+}(ox)$_3$]G Family, Where G Is 2-Halopyridine; M Is Cr, Ga; A$^+$ Is [K$_{0.8}$(H$_3$O)$_{0.2}$]$^+$. *Crystals* **2018**, *8*, 92. [CrossRef]
39. Mori, T.; Kobayashi, A.; Sasaki, Y.; Kobayashi, H.; Saito, G.; Inokuchi, H. The Intermolecular Interaction of Tetrathiafulvalene and Bis(ethylenedithio)tetrathiafulvalene in Organic Metals. Calculation of Orbital Overlaps and Models of Energy-band Structures. *Bull. Chem. Soc. Jpn.* **1984** *57*, 627-633. [CrossRef]
40. Uji, S.; Iida, Y.; Sugiura, S.; Isono, T.; Sugii, K.; Kikugawa, N.; Terashima, T.; Yasuzuka, S.; Akutsu, H.; Nakazawa, Y.; et al. Fulde-Ferrell-Larkin-Ovchinnikov superconductivity in the layered organic superconductor β''-(BEDT-TTF)$_4$[(H$_3$O)Ga(C$_2$O$_4$)$_3$]C$_6$H$_5$NO$_2$. *Phys. Rev. B* **2018**, *97*, 144505. [CrossRef]
41. Imajo, S.; Akutsu, H.; Akutsu-Sato, A.; Morritt, A.L.; Martin, L.; Nakazawa, Y. Effects of electron correlations and chemical pressures on superconductivity of β''-type organic compounds. *Phys. Rev. Res.* **2019**, *1*, 033184. [CrossRef]
42. Coldea, A.I.; Bangura, A.F.; Singleton, J.; Ardavan, A.; Akutsu-Sato, A.; Akutsu, H.; Turner, S.S.; Day, P. Fermi-surface topology and the effects of intrinsic disorder in a class of charge-transfer salts containing magnetic ions: β''-(BEDT-TTF)$_4$[(H$_3$O)M(C$_2$O$_4$)$_3$]Y (M = Ga, Cr, Fe; Y = C$_5$H$_5$N). *Phys. Rev. B* **2004**, *69*, 085112. [CrossRef]
43. Bangura, A.F.; Coldea, A.I.; Singleton, J.; Ardavan, A.; Akutsu-Sato, A.; Akutsu, H.; Turner, S.S.; Day, P.; Yamamoto, T.; Yakushi, K. Robust superconducting state in the low-quasiparticle-density organic metals β''-(BEDT-TTF)$_4$[(H$_3$O)M(C$_2$O$_4$)$_3$]Y: Superconductivity due to proximity to a charge-ordered state. *Phys. Rev. B* **2005**, *72*, 014543. [CrossRef]
44. Audouard, A.; Laukhin, V.N.; Brossard, L.; Prokhorova, T.G.; Yagubskii, E.B.; Canadell, E. Combination frequencies of magnetic oscillations in β''-(BEDT-TTF)$_4$(NH$_4$)[Fe(C$_2$O$_4$)$_3$]·DMF. *Phys. Rev. B* **2004**, *69*, 144523. [CrossRef]
45. Imajo, S.; Akutsu, H.; Kurihara, R.; Yajima, T.; Kohama, Y.; Tokunaga, M.; Kindo, K.; Nakazawa, Y. Anisotropic Fully Gapped Superconductivity Possibly Mediated by Charge Fluctuations in a Nondimeric Organic Complex. *Phys. Rev. Lett.* **2020**, *125*, 177002. [CrossRef] [PubMed]
46. Goto, T.; Lüthi, B. Charge ordering, charge fluctuations and lattice effects in strongly correlated electron systems. *Adv. Phys.* **2003**, *52*, 67–118. [CrossRef]
47. Ihara, Y.; Seki, H.; Kawamoto, A. ^{13}C NMR Study of Superconductivity near Charge Instability Realized in β''-(BEDT-TTF)$_4$[(H$_3$O)Ga(C$_2$O$_4$)$_3$]C$_6$H$_5$NO$_2$. *J. Phys. Soc. Jpn.* **2013**, *82*, 83701. [CrossRef]
48. Ihara, Y.; Jeong, M.; Mayaffre, H.; Berthier, C.; Horvatić, M.; Seki, H.; Kawamoto, A. ^{13}C NMR study of the charge-ordered state near the superconducting transition in the organic superconductor β''-(BEDT-TTF)$_4$[(H$_3$O)Ga(C$_2$O$_4$)$_3$]C$_6$H$_5$NO$_2$. *Phys. Rev. B* **2014**, *90*, 121106. [CrossRef]
49. Kawai, T.; Kawamoto, A. ^{13}C-NMR Study of Charge Ordering State in the Organic Conductor, α-(BEDT-TTF)$_2$I$_3$. *J. Phys. Soc. Jpn.* **2009**, *78*, 074711. [CrossRef]
50. Klutz, T.; Hennig, I.; Haeberlen, U.; Schweitzer, D. Knight Shift Tensors and π-spin Densities in the Organic Metals α_t-(BEDT-TTF)$_2$I$_3$ and (BEDT-TTF)$_2$Cu(NCS)$_2$. *Magn. Reson.* **1991**, *2*, 441. [CrossRef]
51. Mori, T. Non-Stripe Charge Order in the θ-Phase Organic Conductors *J. Phys. Soc. Jpn.* **2003**, *72*, 1469–1475. [CrossRef]

52. Merino, J.; Ralko, A.; Fratini, S. Emergent Heavy Fermion Behavior at the Wigner-Mott Transition. *Phys. Rev. Lett.* **2013**, *111*, 126403. [CrossRef] [PubMed]
53. Ihara, Y.; Moribe, K.; Fukuoka, S.; Kawamoto, A. Microscopic coexistence of superconductivity and charge order in the organic superconductor β''-(BEDT-TTF)$_4$[(H$_3$O)Ga(C$_2$O$_4$)$_3$]C$_6$H$_5$NO$_2$. *Phys. Rev. B* **2019**, *100*, 060505. [CrossRef]
54. Kaneko, M.; Ogata, M. Mean-Field Study of Charge Order with Long Periodicity in θ-(BEDT-TTF)$_2$X. *J. Phys. Soc. Jpn.* **2006**, *75*, 014710. [CrossRef]
55. Hotta, C.; Furukawa, N.; Nakagawa, A.; Kubo, K. Phase Diagram of Spinless Fermions on an Anisotropic Triangular Lattice at Half-Filling. *J. Phys. Soc. Jpn.* **2006**, *75*, 123704. [CrossRef]
56. Merino, J.; Greco, A.; Grichko, N.; Dressel, M. Non-Fermi Liquid Behavior in Nearly Charge Ordered Layered Metals. *Phys. Rev. Lett.* **2006**, *96*, 216402. [CrossRef]
57. Miranović, P.; Ichioka, M.; Machida, K.; Nakai, N. Theory of gap-node detection by angle-resolved specific heat measurement. *J. Phys. Condens. Matter* **2005**, *17*, 7971. [CrossRef]
58. Vorontsov A.B; I. Vekhter, I. Unconventional superconductors under a rotating magnetic field. I. Density of states and specific heat. *Phys. Rev. B* **2007**, *75*, 224501. [CrossRef]
59. Sakakibara, T.; Kittaka, S.; Machida, K. Angle-resolved heat capacity of heavy fermion superconductors. *Rep. Prog. Phys.* **2016**, *79*, 094002. [CrossRef]
60. Imajo, S.; Kindo, K.; Nakazawa, Y. Symmetry change of d-wave superconductivity in κ-type organic superconductors. *Phys. Rev. B* **2021**, *103*, L060508. [CrossRef]
61. Imajo, S.; Nakazawa, Y.; Kindo, K. Superconducting Phase Diagram of the Organic Superconductor κ-(BEDT-TTF)$_2$Cu[N(CN)$_2$]Br above 30 T. *J. Phys. Soc. Jpn.* **2018**, *87*, 123704. [CrossRef]
62. Kobayashi, A.; Tanaka, Y.; Ogata, M.; Suzumura, Y. Charge-Fluctuation-Induced Superconducting State in Two-Dimensional Quarter-Filled Electron Systems. *J. Phys. Soc. Jpn.* **2004**, *73*, 1115. [CrossRef]
63. Merino, J.; McKenzie, R.H. Superconductivity Mediated by Charge Fluctuations in Layered Molecular Crystals. *Phys. Rev. Lett.* **2001**, *87*, 237002. [CrossRef] [PubMed]
64. Tanaka, Y.; Yanase, T.; Ogata, M. Superconductivity due to Charge Fluctuation in θ-Type Organic Conductors. *J. Phys. Soc. Jpn.* **2004**, *73*, 2053. [CrossRef]

Article

Grain-Size-Induced Collapse of Variable Range Hopping and Promotion of Ferromagnetism in Manganite $La_{0.5}Ca_{0.5}MnO_3$

Nikolina Novosel [1,†], David Rivas Góngora [1,†,‡], Zvonko Jagličić [2,3], Emil Tafra [4], Mario Basletić [4], Amir Hamzić [4], Teodoro Klaser [5], Željko Skoko [4], Krešimir Salamon [5], Ivna Kavre Piltaver [6], Mladen Petravić [6], Bojana Korin-Hamzić [1], Silvia Tomić [1], Boris P. Gorshunov [7], Tao Zhang [8], Tomislav Ivek [1,*] and Matija Čulo [1,*]

[1] Institut za fiziku, Bijenička cesta 46, HR-10000 Zagreb, Croatia; nnovosel@ifs.hr (N.N.); d.r.gongora@smn.uio.no (D.R.G.); hamzic.bojana@gmail.com (B.K.-H.); stomic@ifs.hr (S.T.)
[2] Institute of Mathematics, Physics and Mechanics, Jadranska 19, 1000 Ljubljana, Slovenia; zvonko.jaglicic@imfm.si
[3] Faculty of Civil and Geodetic Engineering, University of Ljubljana, Jamova c. 2, 1000 Ljubljana, Slovenia
[4] Department of Physics, Faculty of Science, University of Zagreb, Bijenička cesta 32, HR-10000 Zagreb, Croatia; etafra@phy.hr (E.T.); basletic@phy.hr (M.B.); hamzic@phy.hr (A.H.); zskoko@phy.hr (Ž.S.)
[5] Ruđer Bošković Institute, Bijenička cesta 54, 10000 Zagreb, Croatia; tklaser@irb.hr (T.K.); kresimir.salamon@irb.hr (K.S.)
[6] University of Rijeka, Faculty of Physics and Center for Micro- and Nanosciences and Technologies, Radmile Matejcic 2, 51000 Rijeka, Croatia; ivna.kavre@uniri.hr (I.K.P.); mpetravic@uniri.hr (M.P.)
[7] Laboratory of Terahertz Spectroscopy, Center for Photonics and 2D Materials, Moscow Institute of Physics and Technology, National Research University, 141701 Dolgoprudny, Russia; bpgorshunov@gmail.com
[8] School of Physics and Materials Sciences, Guangzhou University, Guangzhou 510006, China; zhangtao@gzhu.edu.cn

* Correspondence: tivek@ifs.hr (T.I.); mculo@ifs.hr (M.Č.)
† These authors contributed equally to this work.
‡ Current address: Department of Physics, University of Oslo, NO-0316 Oslo, Norway.

Citation: Novosel, N.; Rivas Góngora, D.; Jagličić, Z.; Tafra, E.; Basletić, M.; Hamzić, A.; Klaser, T.; Skoko, Ž.; Salamon, K.; Kavre Piltaver, I.; et al. Grain-Size-Induced Collapse of Variable Range Hopping and Promotion of Ferromagnetism in Manganite $La_{0.5}Ca_{0.5}MnO_3$. *Crystals* **2022**, *12*, 724. https://doi.org/10.3390/cryst12050724

Academic Editors: Raghvendra Singh Yadav and Artem Pronin

Received: 16 April 2022
Accepted: 16 May 2022
Published: 19 May 2022

Publisher's Note: MDPI stays neutral with regard to jurisdictional claims in published maps and institutional affiliations.

Copyright: © 2022 by the authors. Licensee MDPI, Basel, Switzerland. This article is an open access article distributed under the terms and conditions of the Creative Commons Attribution (CC BY) license (https://creativecommons.org/licenses/by/4.0/).

Abstract: Among transition metal oxides, manganites have attracted significant attention because of colossal magnetoresistance (CMR)—a magnetic field-induced metal–insulator transition close to the Curie temperature. CMR is closely related to the ferromagnetic (FM) metallic phase which strongly competes with the antiferromagnetic (AFM) charge ordered (CO) phase, where conducting electrons localize and create a long range order giving rise to insulator-like behavior. One of the major open questions in manganites is the exact origin of this insulating behavior. Here we report a dc resistivity and magnetization study on manganite $La_{1-x}Ca_xMnO_3$ ceramic samples with different grain size, at the very boundary between CO/AFM insulating and FM metallic phases $x = 0.5$. Clear signatures of variable range hopping (VRH) are discerned in resistivity, implying the disorder-induced (Anderson) localization of conducting electrons. A significant increase of disorder associated with the reduction in grain size, however, pushes the system in the opposite direction from the Anderson localization scenario, resulting in a drastic decrease of resistivity, collapse of the VRH, suppression of the CO/AFM phase and growth of an FM contribution. These contradictory results are interpreted within the standard core-shell model and recent theories of Anderson localization of interacting particles.

Keywords: manganites; colossal magnetoresistance; metal–insulator transition; charge order; grain size; variable range hopping; Anderson localization; core–shell model

1. Introduction

In contrast to conventional materials, valence electrons of transition metal oxides are strongly correlated, giving rise to exotic ordered states such as the Mott insulator, spin density waves, charge order, etc. [1]. Very often, a single material can host more than one of these states, as well as some conventional phases, which can either coexist or be transformed from one into the other by changing temperature, pressure, doping, or magnetic field. One of the most intriguing is the transition between an insulating and a metallic phase, the two phases so dissimilar that they may be considered as fundamental states of condensed matter. It is therefore not surprising that the transition metal oxides have attracted a lot of attention and are at the heart of condensed matter physics.

Among transition metal oxides, the family of manganese oxides, widely known as manganites, today plays a prominent role within the field of strongly correlated electron systems. Although synthesized in 1950 [2], a major interest for manganites started only in 1994 with the discovery of colossal magnetoresistance (CMR)—a transition close to the Curie point from an insulating to a metallic state caused by a magnetic field [3,4]. For general reviews, see Refs. [5–8].

Manganites can be described by a general formula $R_{1-x}A_x\text{MnO}_3$, where R stands for a trivalent rare earth or Bi^{3+} cation and A for a divalent alkaline earth or Pb^{2+} cation. Substituting R for A results in the realization of many different ordered states, i.e., in rich phase diagrams. The conventional phase diagram of the titular compound $La_{1-x}Ca_x\text{MnO}_3$ [9] is shown in Figure 1a. The left part is dominated by a ferromagnetic (FM) ground state which is metallic ($0.2 \lesssim x < 0.5$), and the right part is dominated by an antiferromagnetic (AFM) charge-ordered (CO) ground state which is insulating ($0.5 < x \lesssim 0.9$), indicating a close relationship between magnetic and transport properties. At high temperatures, $La_{1-x}Ca_x\text{MnO}_3$ is in a paramagnetic (PM) insulating state for all x. The previously mentioned CMR is tied to the FM metallic part of the phase diagram.

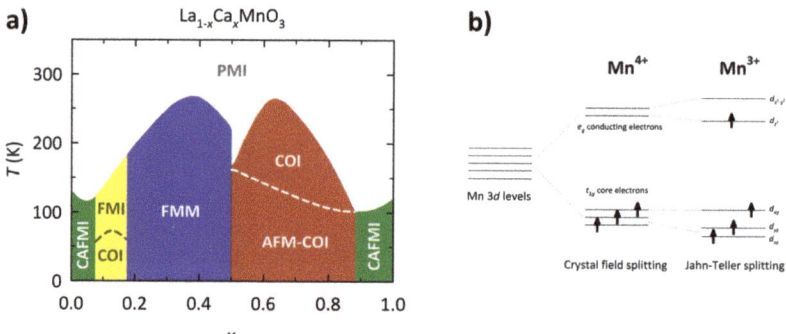

Figure 1. (a) Conventional phase diagram of $La_{1-x}Ca_x\text{MnO}_3$. PMI stands for a paramagnetic insulating phase, FMM for a ferromagnetic metallic, COI for a charge-ordered insulating, AFM for an antiferromagnetic, FMI for a ferromagnetic insulating and CAFMI for a canted antiferromagnetic insulating state. This figure is based on data from Ref. [9]. (b) Splitting of the Mn 3d levels into e_g and t_{2g} orbitals by crystal field and Jahn–Teller distortion.

$La_{1-x}Ca_x\text{MnO}_3$ should crystallize in the ideal cubic perovskite crystal structure with an Mn atom in the center of the cube, La or Ca atoms at the corners of the cube and O atoms positioned in the center of each cube face. In this way, each Mn atom is octahedrally surrounded by six O atoms. In reality, however, the perovskite structure of $La_{1-x}Ca_x\text{MnO}_3$ is distorted, first due to different size of atoms and the pores which they occupy and second due to the Jahn–Teller effect, which leads to the less symmetric orthorhombic crystal structure.

According to simple stoichiometry, pure $LaMnO_3$ ($x = 0$) contains La^{3+}, Mn^{3+} and O^{2-}, while pure $CaMnO_3$ ($x = 1$) contains Ca^{2+}, Mn^{4+} and O^{2-}. Density functional theory (DFT) studies [10,11] show that electronic properties of $LaMnO_3$ and $CaMnO_3$ are mostly determined by Mn^{3+} and Mn^{4+}, respectively, the only ions with open shells. Mn 3d orbitals are split by the oxygen octahedral field into three t_{2g} orbitals with lower energy (d_{xy}, d_{yz} and d_{xz}) and two e_g orbitals with higher energy ($d_{x^2-y^2}$ and d_{z^2}) as shown in Figure 1b. The deformation of the MnO$_6$ octahedron caused by the Jahn–Teller effect leads to an additional splitting of Mn^{3+} e_g orbitals into a higher energy $d_{x^2-y^2}$ and a lower energy d_{z^2} orbital. Mn^{4+} has three electrons which occupy only the t_{2g} orbitals, while Mn^{3+} has one more electron which occupies one of the two e_g orbitals. Due to the Hund coupling, the spins of all electrons are parallel, and each electron occupies a separate orbital, which explains why pure LaMnO$_3$ and CaMnO$_3$ are magnetic and insulating.

Substituting La^{3+} for Ca^{2+} introduces the mixed valence state Mn^{3+}/Mn^{4+}, which opens a conduction channel in La$_{1-x}$Ca$_x$MnO$_3$ via hopping of e_g electron between Mn^{3+} and Mn^{4+}. It is therefore useful to distinguish the conducting e_g electrons responsible for charge transport from the core t_{2g} electrons that give rise to a magnetic moment of Mn^{3+} and Mn^{4+} (see Figure 1b). Due to the Hund interaction, the hopping of conducting electrons depends on the angle between the magnetic moments of Mn^{3+} and Mn^{4+} and is maximal (minimal) when the moments are parallel (antiparallel). The effective hopping integral may therefore be written as $t = t_{max}\cos(\theta/2)$, where t_{max} is the maximal value and θ the angle between the moments. This, the so called double exchange mechanism [12–14], qualitatively captures the essential features of the La$_{1-x}$Ca$_x$MnO$_3$ phase diagram in Figure 1a: the metallic behavior of the FM phase ($\theta = 0$, $t = t_{max}$), the insulating behavior of the AFM phase ($\theta = \pi$, $t = 0$), and CMR (θ changes in field).

There are several more interactions that govern the physics of manganites. Namely, since an Mn^{3+} octahedron is distorted due to the Jahn–Teller effect, while Mn^{4+} is not, each hop of an e_g electron from Mn^{3+} to Mn^{4+} must be accompanied by a change of local octahedron. Such a strong electron–lattice coupling competes with electron delocalization and is believed to give rise to insulating polaron conduction in the PM state at high temperatures [15–20]. Charge ordering, i.e., a periodic arrangement of Mn^{3+} and Mn^{4+}, points towards localization of e_g electrons due to the nearest-neighbor Coulomb and/or Jahn–Teller interaction [21–27], which induces the periodic arrangement of e_g d_{z^2} orbitals as well [9,24,28,29]. Finally, there is a superexchange interaction between neighboring core t_{2g} electrons that favors AFM alignment of the Mn magnetic moments and directly competes with the FM double exchange [30–33].

The transport and magnetic properties of La$_{1-x}$Ca$_x$MnO$_3$ therefore depend on the fine interplay between the interactions that favor electron delocalization with ferromagnetism and those that favor electron localization with antiferromagnetism. The vast majority of theoretical models are focused on CMR, the FM metallic and the PM insulating state. They can be divided into models based on (i) double exchange, (ii) Anderson localization, and (iii) polarons (see Refs. [34,35]). To date, however, none of these have been successful to quantitatively capture the CMR and to fully explain the origin of the PM insulating state.

At present, it is believed that the key for understanding the manganites is phase separation on the nanoscale [35–38]—the coexistence of nanometer-size spatial regions, i.e., nanoclusters, with different electronic orders. In such a scenario, close to the boundary between the FM metallic and the PM or CO/AFM insulating phase, the system consists of FM nanoclusters embedded in a PM or CO/AFM matrix. Based on the double exchange, e_g electrons can move only within the FM nanoclusters with magnetization oriented parallel to the e_g spin. Easy alignment and/or growth of the FM nanoclusters in a magnetic field then leads to a large increase in conductivity and eventually to metallic conduction, i.e., to CMR. The CMR and metal–insulator transitions in manganites are therefore thought to rely on cluster-dynamics, rather than on the dynamics of atomic magnetic moments. Indeed, theories based on small FM clusters of randomly oriented magnetizations seem to provide a considerable improvement in our understanding of the CMR and the associated FM metallic state [39–44]. The origin of the insulating behavior in the PM and CO/AFM phases, however, still remains unresolved.

To gain insight into the origin of the insulating behavior in manganites, we recently focused on the poorly explored insulating CO/AFM part of the phase diagram $x > 0.5$ (Figure 1a). Our study on La$_{1-x}$Ca$_x$MnO$_3$ thin films indicated that dc resistivity follows the Mott three-dimensional (3D) variable-range-hopping(VRH) mechanism, which implies the crucial role of structural disorder in driving the insulating behavior through the Anderson localization of conducting e_g electrons [45]. To further explore these ideas, here we conduct a dc transport and magnetization study on polycrystalline (ceramic) La$_{1-x}$Ca$_x$MnO$_3$ samples, where the level of structural disorder can be controlled by the grain size. We focus on the very boundary between the FM metallic and CO/AFM insulating phases $x = 0.5$, where the grain size is known to have a large impact on the stability of the CO/AFM phase [46–52]. Our results show the presence of the 3D VRH in accordance with the thin film study [45] and the coexistence of both FM and CO/AFM phases. Reducing the grain size leads to the suppression of the CO/AFM state and a growth of the FM contribution. The accompanying increase in structural disorder surprisingly induces a disappearance of the VRH and pushes the system towards the metallic state, which is opposite to the Anderson localization. Such counter-intuitive behavior is discussed here in light of the standard core-shell model and recent theories of electron localization in the presence of structural disorder and electron–electron or electron–lattice interactions [53–56] that have until now not been considered in manganites.

2. Materials and Methods

Ceramic samples of half-doped manganite La$_{0.5}$Ca$_{0.5}$MnO$_3$ were prepared by the sol–gel method which is based on the esterification and polymerization reaction of ethylene glycol (EG) and ethylenediaminetetraacetic acid (EDTA). The stoichiometric amounts of La$_2$O$_3$, CaCO$_3$ and 50 % Mn(NO$_3$)$_2$ solution were used as starting materials. La$_2$O$_3$ and CaCO$_3$ were converted into metal nitrates by adding nitric acid. These metal nitrates and excessive EDTA were dissolved in distilled water to obtain a clear solution with an initial molar ratio of La:Ca:Mn = 1:1:2. The pH of the solution was adjusted to 6–7 by adding ethylenediamine, and then an appropriate amount of EG was added to the solution. Subsequently, the solution was heated with stirring to evaporate most of the solvent water. The resultant gel precursors were decomposed at about 300 °C to obtain black precursor powder which was then separated into several parts and annealed at temperatures of 600, 1100 and 1280 °C to gain samples with average grain size approximately 40, 400 and 4000 nm, respectively, labelled hereafter as S40, S400 and S4000 (see Table 1).

Table 1. Structural properties of the La$_{0.5}$Ca$_{0.5}$MnO$_3$ samples obtained from SEM imaging and Rietveld fits. In all samples, the crystallites are significantly smaller than the grains, indicating the presence of many crystallites within each grain.

Sample Label	Grain Size (nm)	Crystallite Size (nm)	a (Å)	b (Å)	c (Å)
S4000	4100 ± 1400	301 ± 10	5.4099(3)	7.6240(4)	5.4183(3)
S400	400 ± 120	28 ± 3	5.464(2)	7.787(3)	5.504(2)
S40	43 ± 13	14 ± 2	5.502(3)	7.786(4)	5.424(3)

The crystal structure and surface morphology of the samples were thoroughly investigated at room temperature by X-ray powder diffraction and scanning electron microscopy (SEM), respectively. The X-ray powder diffraction data were collected with a Bruker D8 Discover diffractometer (Bruker AXS GmbH, Karlsruhe, Germany) equipped with a LYNX-EYE XE-T detector configured in a Bragg–Brentano geometry. Data were collected in 2θ range 30–80° with a step of 0.02° and Cu source with a wavelength of 1.54060 Å with Ni filter, 2.5° Soller slit and fixed slit at 0.4 mm. The slit opening in front of the detector was 6.5 mm, and the detector opening was 1.3°, resulting in an integrating time per step of 25 s. The crystallite size was determined by the Rietveld method using X'Pert Highscore Plus software 3.0 (Malvern Panalytical, Almelo, The Netherlands). Its algorithm utilizes the following formula for crystallite size determination: $D_i = \frac{180}{\pi} \frac{\lambda}{W_i - W_{st}}$, where λ is the wavelength of the radiation used (Cu in our case), W_i is the Caglioti parameter of the investigated phase which is refined within the structure refinement, and W_{st} is the Caglioti parameter of the standard used to determine the instrumental broadening. The pseudo

Voigt function was used in both cases, and LaB$_6$ was used as the standard refined under the same conditions as the measured samples. The variance of the crystallite size is described by: $\sigma^2(D_i) = \frac{A_D^2}{4(W_i-W_{st})^3}[\sigma^2(W_i) + \sigma^2(W_{st})]$, with A_D representing the constant $180\lambda/\pi$. FWHM values as such were not directly used in this procedure, but as stated above, refined Caglioti parameter W was taken instead.

The surface morphology and grain size were examined in a Jeol JSM-7800F field emission SEM instrument by collecting the secondary electrons at a working distance of about 10 mm and with an electron beam acceleration voltage of 10 kV. The grain size for each sample was determined as an average diameter value of 100 grains on the corresponding SEM image (for details see Appendix B).

Resistivity (ρ) was measured by the standard four-contact dc technique in the temperature (T) range 4.2–300 K. The current between 1 nA and 100 µA was applied along the long axis of the samples. The electrical contacts were made by applying silver paste directly to the sample surface. The contact resistance turned out to be around 10 times smaller than the 4-point sample resistance in the whole T-range, indicating a high quality of the contacts. The resistances above 1 GΩ for the most insulating S4000 sample at $T < 34$ K were determined by two contact current measurements using the picoammeter Keithley 6487 with voltage excitation up to 10 V. A good overlap between the four-contact and the two-contact measurement implies that the quality of the contacts does not deteriorate even at the lowest T, and therefore, the two-contact measurement gives a reliable value of the sample resistivity.

Magnetization (M) measurements were conducted using commercial Quantum Design MPMS XL-5 and MPMS 3 magnetometers in magnetic fields H up to 70 kOe and in the T-range 2–300 K. Zero-field cooled (ZFC) and field cooled (FC) $M-T$ curves were measured at 0.5 K/min in the field 100 Oe. The FC curves were recorded both during cooling and warming and are standardly labeled as FCC and FCW, respectively. The magnetic hysteresis $M-H$ curves were measured at fixed temperatures always from 70 to -70 kOe and back to 70 kOe. The initial $M-H$ curve from 0 to 70 kOe was measured only at 300 K while between the two temperatures the sample was kept in the maximum field 70 kOe.

3. Results

3.1. Structural Properties

The granular structure of the ceramic La$_{0.5}$Ca$_{0.5}$MnO$_3$ samples determined by SEM is shown in Figure 2. The average grain size for the three samples S4000, S400 and S40 studied is 4100 ± 1400 nm (Figure 2a), 400 ± 120 nm (Figure 2b) and 43 ± 13 nm (Figure 2c), respectively.

Figure 2. The surface morphology of the ceramic La$_{0.5}$Ca$_{0.5}$MnO$_3$ samples. SEM images obtained from: (**a**) S4000; (**b**) S400; and (**c**) S40 with the average grain size of 4100 ± 1400, 400 ± 120 and 43 ± 13 nm, respectively. The scale bar is indicated in each image.

The dominant phase in all three samples determined by X-ray is La$_{0.5}$Ca$_{0.5}$MnO$_3$ with the space group P_{nma}. S4000 and S40 have a single phase, while S400 contains small amount of CaMnO$_3$. The X-ray scans are shown in Figure 3, and unit cell parameters as well as the crystallite sizes are listed in Table 1. Note that in all samples, each grain contains many crystallites and the smaller the grain size, the smaller the crystallite size. Therefore, going from S4000 to S40, the extrinsic structural disorder created by grain and crystallite boundaries drastically increases and is therefore expected to strongly influence the transport properties of our ceramic $La_{0.5}Ca_{0.5}MnO_3$ samples which is shown in the next section. In contrast to the extrinsic, the intrinsic structural disorder created by the La/Ca substitution is not expected to significantly change between the three samples. The same is true for any additional disorder coming from crystal defects and/or crystal impurities. For clarity from now on, we will mainly refer to the disorder created by grains but will keep in mind that crystallite boundaries, La/Ca substitution and crystal defects/impurities also contribute to the overall disorder.

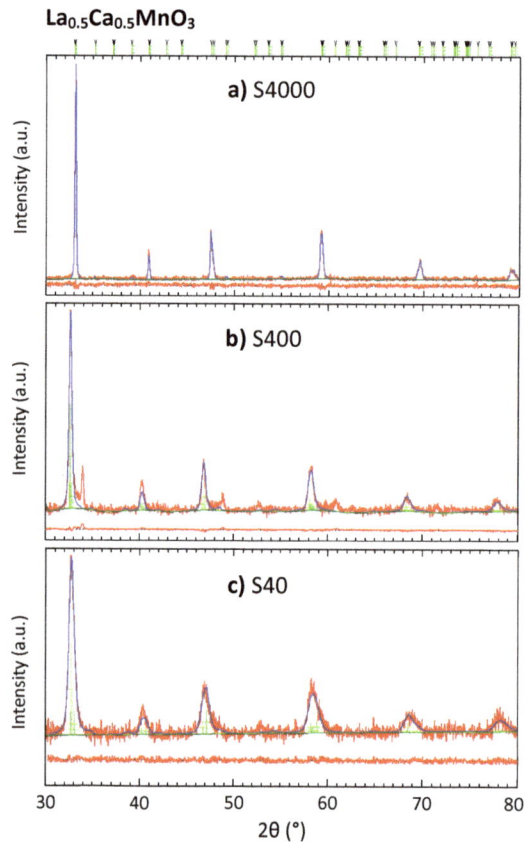

Figure 3. X-ray scans of the $La_{0.5}Ca_{0.5}MnO_3$ samples for: (**a**) S4000; (**b**) S400; and (**c**) S40. Red lines are the experimental data, blue lines are Rietveld fits, and green marks are the maximum positions. Corresponding residuals are shown below each X-ray spectrum. Several additional maximums for S400, approximately at 34°, 49° and 61°, not related to the P_{nma} structure of $La_{0.5}Ca_{0.5}MnO_3$ indicates the presence of small amounts of $CaMnO_3$.

3.2. DC Resistivity

The values of the dc resistivity at room temperature ρ_{300K} in the $La_{0.5}Ca_{0.5}MnO_3$ samples are approximately 6 mΩcm, 1.5 Ωcm and 4.4 Ωcm for S4000, S400 and S40, respectively. This huge difference in ρ_{300K} of three orders of magnitude between the samples illustrates the sensitivity of dc transport properties in $La_{0.5}Ca_{0.5}MnO_3$ to the microstructure. It seems that there is a direct correlation between the ρ_{300K} value and the level of disorder related to the grain (and crystallite) boundaries. The most ordered S4000 sample has the smallest ρ_{300K}, while the most disordered S40 sample has the largest ρ_{300K} value. In a naïve picture, such behavior can be ascribed to additional electron scattering on grain (and crystallite) boundaries which are abundant in S40 and rare in S4000 (see Table 1). However, taking into account that charge transport in $La_{0.5}Ca_{0.5}MnO_3$ close to room temperature possibly takes place via polaron hopping [45,57,58], a more complex explanation would be more appropriate which is beyond the scope of the present study.

The difference between the $La_{0.5}Ca_{0.5}MnO_3$ samples is even more pronounced in the T-dependence of resistivity shown in Figure 4a. Here the resistivities are normalized to the values at 300 K to emphasize the difference in T-evolution. As we can see, all samples show insulator-like behavior (negative $d\rho/dT$) with the ρ/ρ_{300K} ratio spanning more than 10 orders of magnitude at the lowest T. Such a large difference, once again, illustrates the sensitivity of dc transport properties to the sample microstructure. Clear temperature hysteresis of resistivity is visible only for S4000 in the T-range 100–220 K, with the resistivity in cooling being lower than the one in warming. A matching temperature hysteresis is also found in magnetization (see the next section) with indications of a slow relaxation in time. This implies that during cooling the system stays trapped in a metastable state. As we can see, the sample S4000 with the largest grains has the steepest $d\rho/dT$, i.e., it shows the most pronounced insulating behavior. On the other hand, the sample S40 with the smallest grains shows the weakest insulating behavior. This implies that with reducing the grain size, i.e., with increasing structural disorder, the system is pushed towards a metallic state. Such a counter-intuitive behavior will be discussed in Section 4.

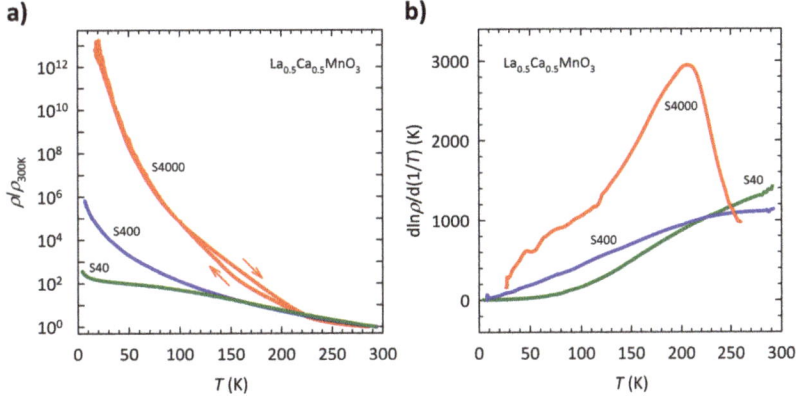

Figure 4. (a) DC resistivity and (b) its logarithmic derivative $d\ln\rho/d(1/T)$ as a function of temperature for the $La_{0.5}Ca_{0.5}MnO_3$ samples: S4000 (red line), S400 (blue line) and S40 (green line). The resistivities are normalized to the room temperature values to emphasize the difference in the T-evolution for different samples. The arrows indicate cooling and warming. The $d\ln\rho/d(1/T)$ curve for S4000 is shown only in warming for clarity.

Figure 4b shows the temperature dependence of the logarithmic resistivity derivative $d\ln\rho/d(1/T)$ for all three samples. The clear maximum in $d\ln\rho/d(1/T)$ at $T \approx 210$ K for S4000 is ascribed to the phase transition from the PM insulating to the CO insulating state (CO transition), expected from the conventional phase diagram shown in Figure 1a. Such a maximum in $d\ln\rho/d(1/T)$ is typical for $La_{1-x}Ca_xMnO_3$ compounds in the CO/AFM insulating part of the phase diagram $x > 0.5$, and as shown in Refs. [45,59], the position of the $d\ln\rho/d(1/T)$ maximum agrees well with the temperature at which the CO transition is expected according to the phase diagram. For clarity, in Figure 4b, $d\ln\rho/d(1/T)$ is shown only for warming. The cooling curve shows a similar maximum at the same T with an additional feature at a lower T which is related to the hysteretic behavior. The large width of the maximum in $d\ln\rho/d(1/T)$ for S4000 indicates that there is no long-range CO, i.e., that CO occurs only at short range. As can be seen in Figure 4b, going from S4000 to S400 the maximum in $d\ln\rho/d(1/T)$ becomes significantly broader and flatter, while for S40 it becomes indiscernible. Together with the sharp drop in resistivity, this is a strong indication that reducing the grain size suppresses the CO/AFM phase.

The logarithmic resistivity derivative $d\ln\rho/d(1/T)$ shown in Figure 4b is not constant in temperature for any of the samples, indicating the absence of conventional activated behavior across an energy gap $\rho(T) \propto \exp(\Delta/T)$. A detailed analysis showed that the insulator-like behavior in S4000 at low-T can be best fitted to the standard Mott 3D VRH mechanism $\rho(T) \propto \exp(T_0/T)^{1/4}$, where T_0 is a characteristic Mott's activation energy [60]. In general, VRH is typical for disordered systems in which a strong scattering of electrons caused by disorder leads to the localization of electronic states at the Fermi level E_F. In these so called Anderson insulators, a charge transport takes place via hopping of electrons among the localized states at E_F not only between nearest neighbors but also of variable range. This mechanism of transport can lead to insulating behavior despite the absence of an energy gap at E_F. In strongly correlated systems such as manganites, however, the situation becomes more complicated since the disorder-induced localized states may coexist with the energy gap opened by strong electron–electron interactions [53–55] or with the mobility gap opened by strong electron/lattice coupling [56], which will be addressed in Section 4. In case of manganites, the presence of VRH implies that a conducting e_g electron hops between spatially distant Mn^{3+} and Mn^{4+}.

Figure 5a shows the temperature dependence of resistivity for all samples on a $\log\rho - T^{-1/4}$ plot, suitable for the 3D VRH mechanism which should then follow a straight line. Again, only the warming curve for S4000 is shown for clarity. As we can see, $\rho(T)$ in warming for S4000 follows the 3D VRH below the CO transition in a broad T-range from $T \approx 150$ K down to $T \approx 40$ K and increases almost eight orders of magnitude. The cooling curve follows the 3D VRH in a narrower T-range ≈ 120–40 K, i.e., below the hysteretic region, where it overlaps with the warming curve. S400 and S40, though showing insulator-like behavior, do not fit well with the 3D VRH mechanism (see Appendix C).

Interestingly, the disappearance of the VRH with reducing the grain size is accompanied by the strong suppression of the CO/AFM phase (Figures 4b and 5a), implying a close relationship between them (see Section 4 for discussion). The suppression of the CO/AFM state with reducing the grain size has been well documented in the literature for the half-doped $La_{0.5}Ca_{0.5}MnO_3$ studied here [46–52], as well as for some other dopings such as $La_{0.4}Ca_{0.6}MnO_3$ [61,62] and $La_{0.25}Ca_{0.75}MnO_3$ [63] and other compositions such as $Pr_{0.5}Ca_{0.5}MnO_3$ [64,65] and $Nd_{0.5}Ca_{0.5}MnO_3$ [66]. Such behavior has been ascribed to surface effects at the grain boundaries, which destabilize the bulk long-range CO/AFM order and which become progressively more pronounced with the increase of the surface to volume ratio caused by the reduction in the grain size. As a consequence, the bulk CO/AFM state is confined only to the interior of the grains, while the grain boundaries are predominantly FM and consist of FM nanoclusters. This is the so-called core–shell model, where the core represents the CO/AFM interior of a grain, while the shell represents the FM grain boundary, schematically shown in Figure 5b. It is therefore expected that S40 with the smallest grains and a lot of grain boundaries has the largest fraction of the FM nanoclusters, while S4000 with the largest grains and few grain boundaries has the smallest fraction of the FM nanoclusters. To verify the validity of these conclusions, we performed a systematic magnetization study which is shown in the next subsection.

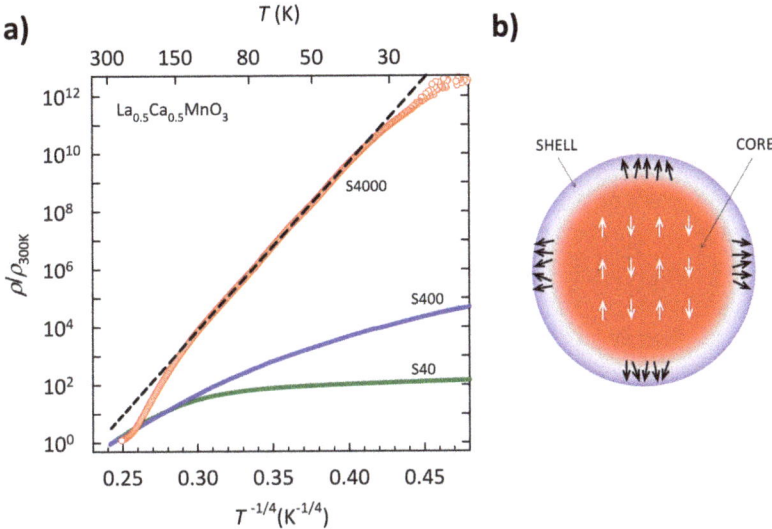

Figure 5. (a) $\log\rho - T^{-1/4}$ plot of the normalized resistivity, suitable for the Mott 3D VRH mechanism, for the $La_{0.5}Ca_{0.5}MnO_3$ samples: S4000 (red circles), S400 (blue line) and S40 (green line). Black dashed line is a fit to the 3D VRH $\rho(T) \propto \exp(T_0/T)^{1/4}$. Only the warming curve is shown for S4000 for clarity. (b) Schematic representation of the core–shell model in $La_{0.5}Ca_{0.5}MnO_3$, where the core of a grain is CO/AFM, and the shell is predominantly FM (see text).

3.3. Magnetization

Magnetization curves of the $La_{0.5}Ca_{0.5}MnO_3$ samples in the standard ZFC and FC protocols are shown in Figure 6. Sharp peak in the ZFC curve and a sudden jump in the FCC and FCW curves at $T \approx 40$ K for S4000 share a striking resemblance with the behavior of magnetization in Mn_3O_4 around its ferrimagnetic transition at $T \approx 43$–48 K [67,68]. It is known that during the sample synthesis, small amounts of Mn_3O_4 phase often appear within a manganite sample [69–71]. We therefore conclude that the behavior of magnetization for S4000 below ≈ 40 K is entirely extrinsic and related to small amounts of the Mn_3O_4 phase rather than to an intrinsic re-entrant spin glass transition, commonly reported in the literature [64,72–74]. (The absence of the re-entrant spin glass transition is additionally confirmed with the ac susceptibility measurements shown in Appendix A.) A rough estimate for the fraction of the Mn_3O_4 phase in S4000 can be obtained from the ratio of the magnetization measured in S4000 and the one measured in pure Mn_3O_4. Taking the value $M = 0.3$ emu/g for S4000 from Figure 6a, and $M = 4.5$ emu/g for Mn_3O_4 from Ref. [68], both for the FC curves in $H = 100$ Oe at the lowest T, we obtain around 7% of Mn_3O_4. Such a small fraction of Mn_3O_4 could not be detected by our X-ray measurements.

The absence of sharp features in the ZFC and FCC/FCW curves in all samples (except the extrinsic feature related to Mn_3O_4 in S4000) indicates that there are no 'clean' magnetic phase transitions which one would naïvely expect from the phase diagram in Figure 1. (The same conclusion can be drawn from the corresponding ac susceptibility, a technique which is even more sensitive for the detection of phase transitions, shown in Appendix A.) On the other hand, such behavior is consistent with the phase separation, i.e., the existence of FM and CO/AFM clusters that appear below some critical temperature [35–38]. In addition, there is a significant difference between the ZFC and FCC/FCW curves which we associate with the presence of FM nanoclusters. The FCC and FCW curves are not identical only in the case of S4000, which results in a pronounced thermal hysteresis in the same T-range 100–220 K where the hysteretic behavior of dc resistivity occurs (compare Figures 4a and 6a). The hysteretic behavior is probably related to the CO transition, visible only for S4000 as a maximum in resistivity derivative that nicely coincides with the maximum in ZFC, FCC and FCW curves at $T \approx 220$ K (compare Figures 4b and 6a). Here it is important to recall that the magnetization and dc resistivity do not probe the same electrons in $La_{0.5}Ca_{0.5}MnO_3$, since the former is related to the core t_{2g} electrons, while the latter is related to the conducting e_g electrons. The coinciding features in magnetization and dc resistivity, therefore, confirm the expected strong coupling between the t_{2g} and e_g electrons.

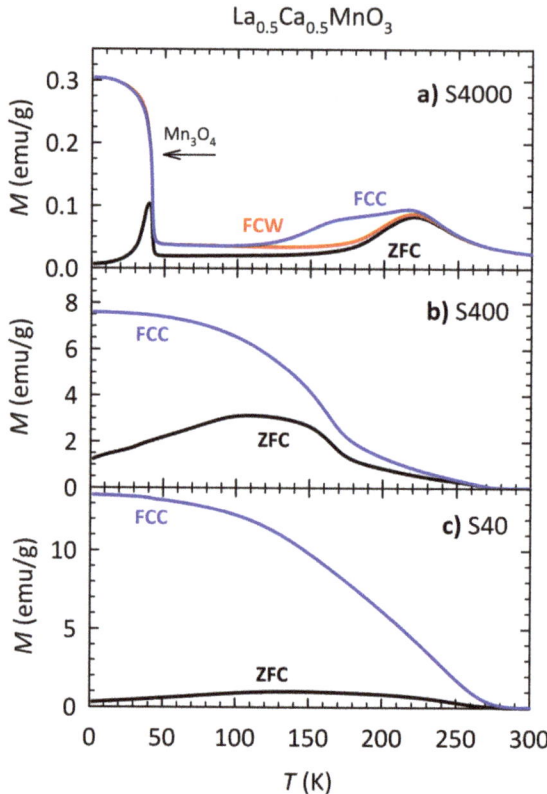

Figure 6. Standard ZFC and FC magnetization curves in $H = 100$ Oe for the $La_{0.5}Ca_{0.5}MnO_3$ samples: (a) S4000; (b) S400; and (c) S40. In the case of S4000, the FCC and FCW curves do not overlap and are indicated by the blue and red lines, respectively. In the case of S400 and S40 the FCC and FCW curves are identical (not shown) which was confirmed on other samples. The ZFC curves in all three panels are shown by the black lines. The sharp peak in ZFC and sudden jump in FCC and FCW curves at $T \approx 40$ K for S4000 are extrinsic and come from a tiny amount of Mn_3O_4 phase which often occurs during the sample synthesis [69–71].

As we can see in Figure 6, the FCC curve at low T has the highest value for S40 and the lowest for S4000, indicating that S40 has the largest and S4000 the smallest amount of FM nanoclusters in accordance with the expectation based on the core-shell model introduced in the previous section (see Figure 5). The same conclusion can be drawn from magnetic hysteresis loops shown in Figure 7. As can be seen in Figure 7a, the hysteresis loop at the lowest temperature $T = 2$ K for S40 closes at $H \approx 2$ kOe, reaching a value $M \approx 1.5$ μ_B/Mn where μ_B is the Bohr magneton. Beyond 2 kOe, the magnetization stays almost constant with a field as shown in the inset of Figure 7a. We ascribe such behavior to the FM part of the sample (the FM shells) which becomes fully saturated at 2 kOe. Taking into account that Mn^{3+} with four valence electrons contributes 4 μ_B/Mn, while Mn^{4+} with three valence electrons contributes 3 μ_B/Mn, the theoretical saturation value for $La_{0.5}Ca_{0.5}MnO_3$ (which has equal number of Mn^{3+} and Mn^{4+}) would be $M = 3.5$ μ_B/Mn in the case of all Mn spins parallel to H. Here, only the contribution of valence electrons is considered since the orbital contribution of Mn^{3+} and Mn^{4+} is quenched [75]. The significantly lower measured saturation value therefore indicates that not all magnetic moments are aligned parallel to H which is in accordance with the proposed core–shell model. The measured value $M \approx 1.5$ μ_B/Mn then implies that only around 40% of the S40 sample is FM. The rest of the sample is CO/AFM and gives only a tiny contribution to the total magnetization as visible from the small but finite slope in the $M - H$ curve beyond the saturation (inset of Figure 7a).

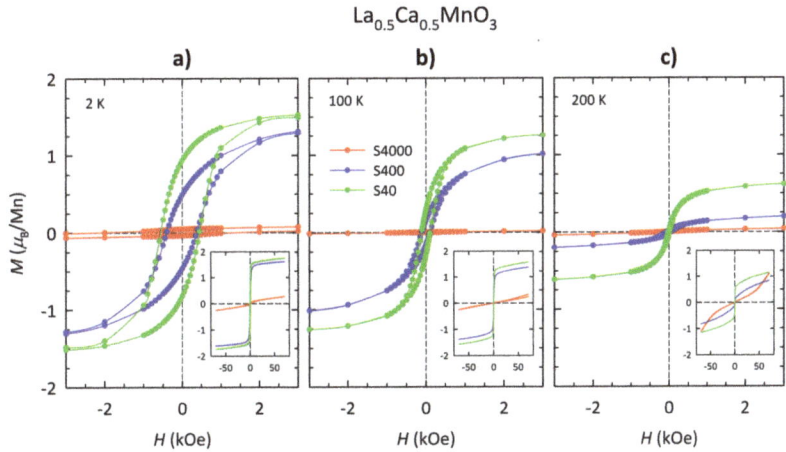

Figure 7. Magnetic hysteresis loops for the $La_{0.5}Ca_{0.5}MnO_3$ samples at: (**a**) 2 K; (**b**) 100 K; and (**c**) 200 K in the field range $-3 < H < 3$ kOe. All curves were measured in the FCC protocol. Magnetization is expressed in units of Bohr magneton μ_B per Mn atom. Red, blue and green symbols refer to the S4000, S400 and S40 sample, respectively. The full range of H is shown in insets. The low saturation values of M indicates that the FM fraction of $La_{0.5}Ca_{0.5}MnO_3$ is $< 5\%$, $\approx 35\%$ and $\approx 40\%$ for the S4000, S400 and S40 sample, respectively (see text).

Taking the measured values $M \approx 1.2$ μ_B/Mn and $M \approx 0.2$ μ_B/Mn from Figure 7a implies that around 35% of S400 and only about 5% of S4000 is FM at 2 K. The FM fraction of $La_{0.5}Ca_{0.5}MnO_3$ in S4000 is probably even lower since there is an additional contribution to the measured magnetization of the extrinsic ferrimagnetic Mn_3O_4 phase. Such a small FM fraction, together with the VRH and a clear signature of the CO transition (Figures 4b and 5a) in a sample with µm-sized grains (usually taken to represent the bulk behavior), implies that the insulating CO/AFM phase in $La_{1-x}Ca_xMnO_3$ is stable even at the very boundary with the FM metallic phase $x = 0.5$. This conclusion is in accordance with several previous reports [46,49,59].

The height and the width of hysteresis in S40 and S400 and consequently the FM fraction monotonously decrease with increasing T and become negligible on approaching room temperature (Figure 7) in accordance with the expectation based on the phase diagram. Here we recall that the hysteresis loops for all samples were measured during cooling from 300–4.2 K. Peculiar behavior is observed only in the case of S4000 within the thermal hysteretic region 220–100 K (Figure 6a) as indicated by the $M - H$ curve at 200 K that has an unexpected increase of slope at high fields (inset of Figure 7c), and the $M - H$ curve at 100 K that is not closed (inset of Figure 7b). The latter points towards slow time relaxations which imply that on cooling the system enters a metastable state with higher magnetization

(Figure 6a) and lower resistivity (Figure 4a), the origin of which is beyond the scope of the present study.

Finally, the coercive field at low T in all three samples is not completely symmetrical with respect to the origin, i.e., the hysteresis loops exhibit a small horizontal shift. Similar behavior was found also in $Pr_{0.5}Ca_{0.5}MnO_3$ ceramic samples [64] and was ascribed to the exchange bias effect related to a FM–AFM interface [64]. The presence of the exchange bias effect in our ceramic samples therefore provides additional indirect evidence for the coexistence of the AFM and FM regions, i.e., for the proposed core–shell formation depicted in Figure 5b. Direct imaging for the core–shell formation in manganites is still missing and would require studies based on local probes such as magnetic force microscopy at low temperatures.

4. Discussion

As mentioned in the introduction, the main goal of the present study is to shed more light on the origin of the insulating behavior in manganites $La_{1-x}Ca_xMnO_3$, specifically in the CO/AFM phase, which is still mostly not understood. Motivated by our previous study [45], which pointed towards the Anderson localization of conducting e_g electrons, here we explore the sensitivity of transport and magnetic properties in $La_{1-x}Ca_xMnO_3$ to the level of structural disorder controlled by the grain size. We focused on the very boundary between the FM metallic and CO/AFM insulating phases $x = 0.5$, where the grain size is known to strongly affect the stability of the CO/AFM phase [46–52]. Indeed, drastic changes in dc resistivity and magnetization were observed with varying the grain size, explained in detail in the previous sections. Here we will summarize these findings to build a picture of the insulating charge transport in the CO/AFM phase of $La_{1-x}Ca_xMnO_3$.

The absence of a clear transport gap, evident from the absence of a plateau in $d\ln\rho/d(1/T)$ at low T (Figure 4b), immediately eliminates a simple explanation of the insulating behavior based only on the double exchange, the central interaction that governs the physics of manganites. Namely, according to the double exchange, one would expect that the insulating behavior in the CO/AFM state stems from the localization of e_g electrons caused by the prohibition of tunneling from Mn^{3+} to Mn^{4+}, the magnetic moments of which are aligned antiparallel. Such a scenario, however, would necessarily result in the opening of an energy gap at E_F since an e_g electron would need extra energy to hop to overcome the strong Hund interaction in contrast to the experiment. A very similar line of reasoning also excludes localization of e_g electrons by charge ordering as a key to insulating behavior, since here an e_g electron would need extra energy to overcome the strong electron–electron and/or Jahn–Teller interaction [25–27]. Therefore, the double exchange and/or charge ordering alone cannot explain the insulating behavior of the CO/AFM phase.

Indeed, the finite density of states at E_F is supported by the presence of the Mott 3D VRH transport mechanism in S4000 (Figure 5a) which implies the existence of localized energy states at E_F. Here the localization of conducting e_g electrons is caused by structural disorder which, in contrast to the localization caused by interactions, is not accompanied by the opening of an energy gap [54]. According to such an Anderson localization scenario, an increase in the level of structural disorder induced by the reduction in grain size from 4000 to 40 nm should result in a significant increase of resistivity. However, exactly the opposite behavior was observed in our ceramic samples (Figure 4a), indicating that the insulating behavior in the CO/AFM phase cannot stem from Anderson localization alone either. This conclusion should not be surprising, however, since the sudden delocalization of e_g electrons in the FM metallic phase obviously cannot be caused solely by the level of structural disorder which is not expected to change abruptly with small changes of x around $x = 0.5$ (Figure 1a). Moreover, the progressive delocalization of e_g electrons during the metal–insulator transition induced by a high magnetic field [45] cannot be related solely to the level of structural disorder which is not expected to change during the transition.

One solution to the above problem is to postulate that the localization of conducting e_g electrons in the CO/AFM phase stems from the combined effects of the structural disorder and interactions. Indeed, by reducing the grain size not only does the VRH collapse, but also the fingerprint of the CO transition in dc resistivity disappears (see Figures 4b and 5a), suggesting that the electron–electron or electron–lattice interactions responsible for the long range CO/AFM order [25–27] also play a crucial role in the localization of conducting e_g electrons. The same conclusion can be drawn from the magnetization which shows that the disappearance of the VRH is accompanied by a significant increase of the FM contribution, i.e., a decrease of the AFM phase (Figure 6). Such behavior indicates that the grain size plays a delicate role in electron localization in $La_{0.5}Ca_{0.5}MnO_3$. On the one hand, it strengthens the localization by increasing the level of structural disorder, but on the other hand, it weakens the localization even more by destabilizing the CO/AFM phase.

There are several theoretical approaches to the problem of electron localization in the presence of both structural disorder and electron–electron interactions, the so called Mott–Anderson localization [53–55]. According to these models, the interplay between the electron–electron interactions and structural disorder is much more complex than one would expect, i.e., their effects do not necessarily reinforce each other in promoting insulating behavior. For example, for an intermediate disorder and a low interaction strength, the increase in electron–electron interactions can push the system towards the metallic state, rather than to the insulating state, by screening the disorder potential. For low disorder and intermediate interaction strength, the increase of structural disorder, instead of the insulating, can promote the metallic behavior by filling in the energy gap created by the strong electron–electron interactions. Nevertheless, for certain values of disorder and interaction strengths, the electron–electron interactions and structural disorder do reinforce each other, which would be in line with the experimental results presented here.

Even more appealing is a recent theory by Di Sante et al. [56] that focuses on the Anderson localization in the presence of electron–lattice coupling which always 'antiscreens' the disorder potential, i.e., strengthens the electron localization and therefore the insulating behavior. According to this theory, the electron–lattice coupling opens a mobility gap at E_F which separates the localized states around E_F from the delocalized states further away from E_F. (Note that there is no energy gap here since the density of states does not go to zero.) Here, the localization of conducting electrons is driven by a polaron formation, i.e., self-trapping by strong electron–lattice interactions and pinning by a disorder potential. Such a scenario could indeed be playing a role in manganites due to the strong Jahn–Teller effect believed to be responsible for polaronic conduction at high T [15–20] and for charge ordering at low T [25–27] as mentioned in the introduction. The observed VRH in S4000 would in this case be related to the states below the mobility gap which are localized by the combined effects of the structural disorder and strong electron–lattice coupling produced by the Jahn–Teller effect. The drastic decrease of resistivity with the reduction in grain size (Figure 4a), accompanied by the collapse of the VRH (Figure 5), could be ascribed to a weakening of the Jahn–Teller-induced polaron formation related to the suppression of the CO/AFM state (Figure 4b).

Interestingly, according to the same work by Di Sante et al. [56], close to the metallic state there is a regime with the so-called 'bad insulator' transport, which displays conductivity values below the Mott–Ioffe–Regel limit and an insulator-like temperature coefficient $d\rho/dT < 0$ with a finite intercept. Such peculiar behavior is often observed in strongly disordered metals [76] and here resembles the T-dependence found in the S400 and S40 ceramic samples (Figure 4a). It is reasonable to assume that with decreasing the grain size even further, $La_{0.5}Ca_{0.5}MnO_3$ would eventually end up in the metallic state. Indeed, Levy et al. [49] found the metallic behavior in $La_{0.5}Ca_{0.5}MnO_3$ ceramic samples after reducing the grain size down to 180 nm. The fact that the metallic behavior is absent in our sample S40 with significantly smaller grains illustrates the sensitivity of the transport properties of manganites to the sample preparation method that results in a significantly different level of disorder induced not only by the La/Ca substitution and grain boundaries, but also by, e.g., crystallite boundaries, which are usually ignored in the literature.

Finally, the fact that, besides the suppression of the CO/AFM phase, the collapse of the VRH is also accompanied by the growth of the FM fraction indicates the importance of the double exchange and the phase separation in manganites, which should be added to any theory of localization in manganites. The presence of the phase separation opens the possibility of a different explanation for the approach to the metallic state. If the growth of the FM fraction results in FM regions (clusters) big enough to host a Fermi surface, the metal–insulator transition becomes percolative in nature with spatially well-defined FM metallic and CO/AFM insulating regions. The drop of resistivity by more than 10 orders of magnitude between S4000 and S40 in Figure 4a without transition to the metallic state, i.e., without crossing the percolation threshold, however, cast some doubts on such a percolative nature. To resolve this issue, the studies based on local probes such as magnetic atomic force microscopy or scanning tunneling microscopy at low T are highly desirable.

5. Conclusions

In summary, our dc resistivity and magnetization study on $La_{0.5}Ca_{0.5}MnO_3$ ceramic samples clearly shows that the charge transport in the CO/AFM state at low temperatures is well described by the Mott 3D variable-range-hopping mechanism which strongly points towards the disorder-induced (Anderson) localization of conducting electrons. The drastic decrease of resistivity and collapse of the VRH with reducing the grain size, i.e., with increasing the disorder, however, drives the system in the opposite direction of the Anderson localization picture. The fact that the collapse of the VRH is accompanied by a strong suppression of the CO/AFM state implies a key role of interactions responsible for the long range order in localization of conducting electrons. In light of a recent theory of the Anderson localization in the presence of strong electron–lattice coupling, here we propose that the insulating behavior in the CO/AFM state of manganites possibly stems from polaron formation induced by the Jahn–Teller effect and enhanced by a disorder potential. The significant growth of the FM fraction with reducing the grain size, which originates from the surface effects on the grain boundaries, implies that the double exchange interaction and phase separation phenomena also play a role in the destruction of the VRH. More advanced theories of electron localization in the presence of multiple interactions would therefore be necessary to fully capture the origin of the insulating behavior in the CO/AFM state of manganites.

Author Contributions: Conceptualization, M.Č., T.I., S.T. and B.K.-H.; sample synthesis and characterization, T.Z., B.P.G.; X-ray measurements and analysis, T.K., Ž.S. and K.S.; SEM measurements and analysis, I.K.P. and M.P.; magnetization measurements and analysis, N.N. and Z.J.; dc resistivity measurements and analysis, D.R.G., T.I., M. Č., E.T., and M.B.; interpretation, M.Č., T.I., N.N., E.T., M.B., A.H., B.K.-H. and S.T.; writing—original draft preparation, M.Č.; writing—review and editing, M.Č, T.I. and N.N.; project administration, T.I. and S.T.; funding acquisition, T.I. and S.T. All authors have read and agreed to the published version of the manuscript.

Funding: This research was funded by the Croatian Science Foundation, Grants No. IP-2018-01-2730 and No. IP-2013-11-1011.

Institutional Review Board Statement: Not applicable.

Informed Consent Statement: Not applicable.

Data Availability Statement: Data is contained within the article.

Acknowledgments: We thank Ivan Balog, Mirta Herak, Goran Branković, and Jelena Vukašinović for elucidating discussions. We acknowledge support of project Cryogenic Centre at the Institute of Physics—KaCIF co-financed by the Croatian Government and the European Union through the European Regional Development Fund-Competitiveness and Cohesion Operational Programme (Grant No. KK.01.1.1.02.0012). We also acknowledge the support of project CeNIKS co-financed by the Croatian Government and the European Union through the European Regional Development Fund—Competitiveness and Cohesion Operational Programme (Grant No. KK.01.1.1.02.0013). I.K.P.

and M.P. acknowledge support from the University of Rijeka under the project number 18-144. Z.J. acknowledges also partial funding by Slovenian Research Agency (Grant No. P2-0348).

Conflicts of Interest: The authors declare no conflict of interest. The funders had no role in the design of the study; in the collection, analyses, or interpretation of data; in the writing of the manuscript, or in the decision to publish the results.

Appendix A. AC Susceptibility

Looking at the conventional phase diagram of $La_{1-x}Ca_xMnO_3$ shown in Figure 1a, we expect the phase boundary $x = 0.5$ to host both FM and CO/AFM phases in accordance with the widely accepted phase separation scenario [35–38]. Indeed, the magnetization measurements on $La_{0.5}Ca_{0.5}MnO_3$ ceramic samples shown in Figure 7 point towards the presence of two contributions, one related to the CO/AFM order placed in the interiors of grains and the other related to the FM nanoclusters placed at the grain boundaries as discussed in the main text. For completeness, here we present the corresponding magnetic ac susceptibility measurements.

Magnetic ac susceptibility is a very sensitive technique for studying magnetic properties, especially phase transitions and magnetic relaxation, and is complementary to the measurements in the static (dc) magnetic field. It measures the differential dM/dH response of the magnetization of a sample to an oscillating magnetic field $H(t) = h\cos(2\pi f t)$, where t is time, h is the field amplitude, and f is the field frequency. At low h, the magnetization oscillates with the same frequency as the field but generally with a phase shift θ, i.e., $M(t) = m\cos(2\pi f t - \theta)$. Magnetic ac susceptibility can therefore be expressed as a complex quantity $\chi' + i\chi''$, with the real component $\chi' = m\cos\theta/h$, which is in-phase with h and is related to the reversible magnetization processes, and the imaginary component $\chi'' = m\sin\theta/h$, which is out-of-phase with h and is related to the irreversible magnetization processes, i.e., the energy loss due to dissipation [77,78].

In the present study the magnetic ac susceptibility was measured using a high-resolution CryoBIND susceptometer with the driving ac field with f in the Hz–kHz range and $h = 0.9$ Oe. The measurements were performed in the T-range 4.2-320 K, during cooling and warming at the rate 1 K/min. No significant changes in χ' and χ'' were observed while changing the amplitude h of the ac field.

The measured χ' and χ'' as a function of T, normalized to the same arbitrary units by dividing with the mass of each sample, are shown in Figure A1. As we can see, χ' is significantly larger than χ'' for all three samples, indicating the expected dominance of reversible magnetization processes. Both components, χ' and χ'', increase going from S4000 to S40 in accordance with the dc magnetization shown in Figure 6. Only in the case of S4000 do the cooling and the warming curves differ, resulting in a large thermal hysteresis in χ' and χ'' in approximately the same T-range where the hysteretic behavior of dc resistivity and dc magnetization occurs. No hysteretic behavior is observed in S400 and S40.

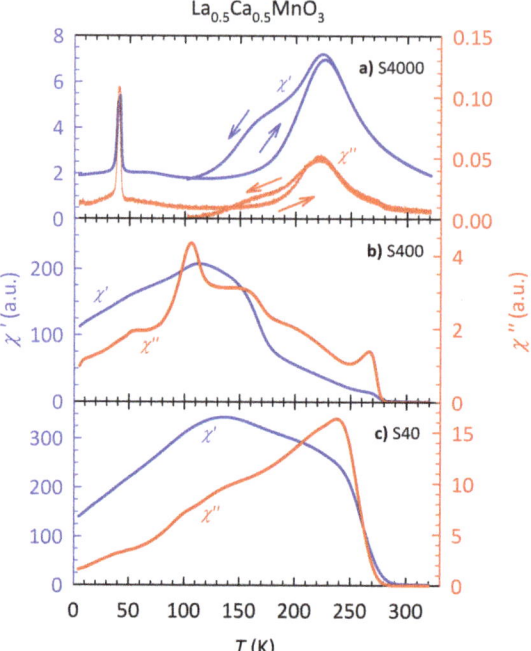

Figure A1. Magnetic ac susceptibility for the $La_{0.5}Ca_{0.5}MnO_3$ ceramic samples: (**a**) S4000; (**b**) S400; and (**c**) S40. Shown are the T-dependence of both the real χ' (blue lines, left axis), and imaginary χ'' (red lines, right axis) components, normalized to the mass of the sample. The cooling and warming curves differ only in the case of S4000.

The sharp peak in χ' and χ'' at $T \approx 40$ K for the bulk sample S4000 in Figure A1a is a clear signature of a phase transition, which is here attributed to the ferrimagnetic transition of the minority Mn_3O_4 phase, rather than to the intrinsic re-entrant spin glass transition commonly reported in the literature [64,72–74], as discussed in the main text. The absence of the spin glass transition is additionally confirmed by the frequency dependence of ac susceptibility on another $La_{0.5}Ca_{0.5}MnO_3$ sample from the same batch as S4000 shown in Figure A2. As we can see, the position of the sharp feature at $T \approx 40$ K does not change with f, which is in contrast to expectations for the spin glass transition [77–79]. The absence of additional sharp features in χ' and χ'' implies that there are no intrinsic 'clean' phase transitions, i.e., development of a 'clean' long-range order, in accordance with the phase separation into CO/AFM and FM regions.

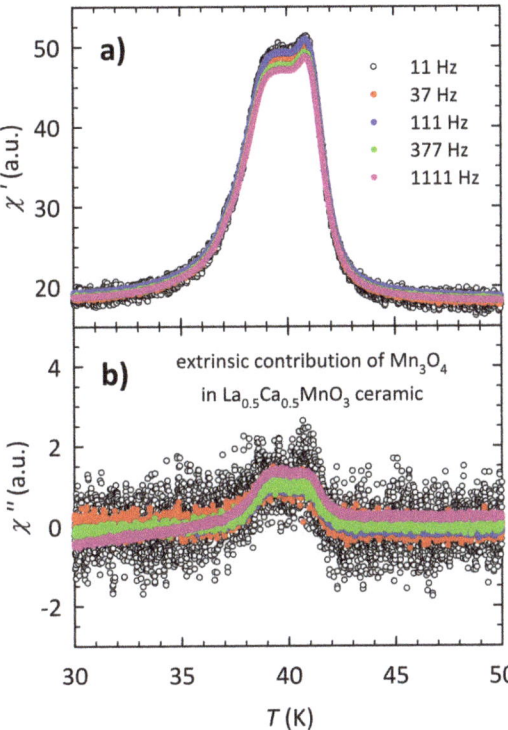

Figure A2. Temperature dependence of the (**a**) real χ'; and (**b**) imaginary χ'' part of the ac susceptibility at low T for the $La_{0.5}Ca_{0.5}MnO_3$ ceramic sample (the same batch as S4000) at different frequencies. AC susceptibility depends only weakly on frequency.

Nevertheless, χ' and χ'' exhibit non-monotonous T-dependence with several broad maximums which point towards rich magnetic behavior in $La_{0.5}Ca_{0.5}MnO_3$. The maximum in χ' at $T \approx 220$ K for S4000 in Figure A1a coincides with the similar maximum in the ZFC and FCW curves in Figure 6a, as well as with the maximum in $d\ln\rho/d(1/T)$ in Figure 4b related to the CO transition. One could therefore ascribe the maximum in χ' at $T \approx 220$ K to the development of the CO/AFM phase in the interior of grains in the S4000 sample. The finite value of χ'' with a very similar T-evolution to that of χ', however, implies that the maximum at $T \approx 220$ K is related rather to the development of the FM nanoclusters at grain boundaries, since for the AFM phase, χ'' is expected to be zero [77,78].

The increase below $T \approx 280$ K and a smooth T-dependence of χ' in S400 and S40, accompanied by a significant change in χ'' with one or more broad maximums, is also attributed to the slow development of the FM nanoclusters at the grain boundaries. The rich structure in the T-dependence of χ'' for S400 indicates the presence of various irreversible magnetization processes probably related to the the dynamics of the FM nanoclusters.

The negligible dependence of χ' and χ'' on the driving field amplitude h in all samples (not shown) indicates that all the features in the temperature plots of Figure A1 are still a part of linear response, which would not be expected if there was a conventional FM transition [77]. The dependence of χ' and χ'' on the driving field frequency is left for a future study. Especially interesting would be the frequency dependence for S4000 within the thermal hysteretic region 100–220 K, where we observed slow time evolution in the dc magnetization measurements, which points towards the presence of some form of a glassy (metastable) state.

Appendix B. Grain Size Distribution

As mentioned in the main text, the grain size of our ceramic samples was determined as an average diameter value of 100 grains on their corresponding SEM images. The grain size distributions for each sample are shown as histograms in Figure A3. The vertical black line represents the average grain size value for each sample.

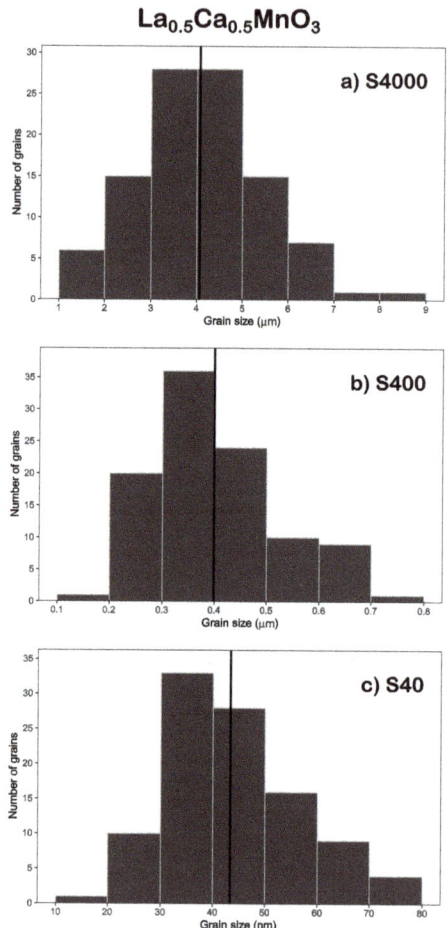

Figure A3. Grain size distribution in the $La_{0.5}Ca_{0.5}MnO_3$ ceramic samples: (**a**) S4000; (**b**) S400; and (**c**) S40. The average grain size values 4100 ± 1400, 400 ± 120 and 43 ± 13 nm for S4000, S400 and S40, respectively, are indicated by the vertical black line.

Appendix C. Fitting of Variable-Range-Hopping Mechanism

As mentioned in the main text, the resistivity of the sample S4000 follows the 3D VRH in a large T-interval 40–150 K in which it increases almost eight orders of magnitude, evident from a linear dependence on a $\log\rho$-$T^{-1/4}$ scale (Figure 5a). Although at first sight it seems that there is a linear region on the $\log\rho$-$T^{-1/4}$ plot at high T (300–140 K), for the other two samples S400 and S40, as well as at low T (140–20 K) for S40, the resistivity in that case increases by only one order of magnitude, or even less, which indicates the absence of the strong exponential T-dependence expected for the 3D VRH. To illustrate this more explicitly, we follow the procedure outlined in several papers [80–84], where one starts from the more general expression for resistivity $\rho \propto \exp(C/T)^p$, with C and p constant. Note that for $C = \Delta$ and $p = 1$ the previous equation reduces to the simple activated behavior, while for $C = T_0$ and $p = 1/4$ it recovers 3D VRH. To extract the exponent p, one defines a logarithmic resistivity derivative $W = -d(\ln\rho)/d(\ln T)$ which for the previous equation gives $p(C/T)^p$. The exponent p can then be obtained from the slope of $\ln W$ vs. $\ln T$.

The $\ln W$–$\ln T$ plots for our ceramic samples S4000, S400 and S40 are shown in Figure A4. Although the data are somewhat noisy, we can clearly see that the slope p of $\ln W$ vs. $\ln T$ agrees with the expected value $1/4$ for 3D VRH only for S4000 in the T-range 150–40 K as stated in the main text. In the case of samples S400 and S40, not only is the value of the slope of $\ln W$ vs. $\ln T$ different from $1/4$, but also the sign of the slope is incorrect in almost the whole T-range of interest. We can therefore safely conclude that the resistivity of S400 and S40 does not follow the 3D VRH mechanism.

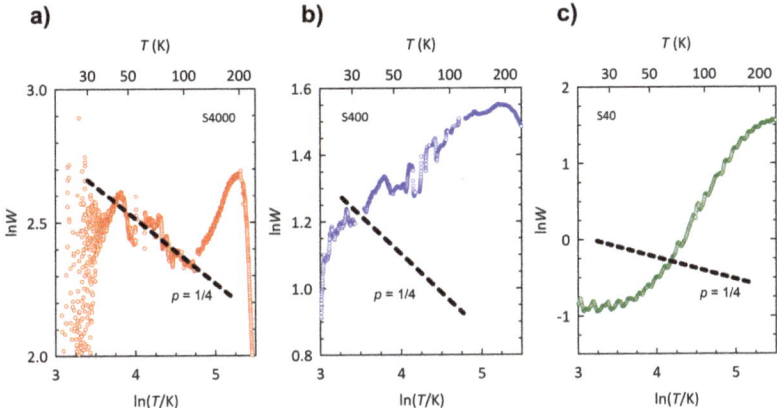

Figure A4. $\ln W$ vs. $\ln T$ for the $La_{0.5}Ca_{0.5}MnO_3$ ceramic samples: (**a**) S4000; (**b**) S400; and (**c**) S40, see text for details. Black dashed lines correspond to the slope expected for the 3D VRH ($p = 1/4$) and agree with the measured data only for S4000 in the T-range 150–40 K.

References

1. Rao, C.N.R. Transition Metal Oxides. *Annu. Rev. Phys. Chem.* **1989**, *40*, 291–326. [CrossRef]
2. Jonker, G.; Van Santen, J. Ferromagnetic compounds of manganese with perovskite structure. *Physica* **1950**, *16*, 337–349. [CrossRef]
3. Jin, S.; Tiefel, T.H.; McCormack, M.; Fastnacht, R.A.; Ramesh, R.; Chen, L.H. Thousandfold Change in Resistivity in Magnetoresistive La-Ca-Mn-O Films. *Science* **1994**, *264*, 413–415. [CrossRef] [PubMed]
4. Xiong, G.C.; Li, Q.; Ju, H.L.; Mao, S.N.; Senapati, L.; Xi, X.X.; Greene, R.L.; Venkatesan, T. Giant magnetoresistance in epitaxial $Nd_{0.7}Sr_{0.3}MnO_{3-\delta}$ thin films. *Appl. Phys. Lett.* **1995**, *66*, 1427–1429. [CrossRef]
5. Dagotto, E.; Hotta, T.; Moreo, A. Colossal magnetoresistant materials: The key role of phase separation. *Phys. Rep.* **2001**, *344*, 1–153. [CrossRef]
6. Nagaev, E. Colossal-magnetoresistance materials: Manganites and conventional ferromagnetic semiconductors. *Phys. Rep.* **2001**, *346*, 387–531. [CrossRef]
7. Salamon, M.B.; Jaime, M. The physics of manganites: Structure and transport. *Rev. Mod. Phys.* **2001**, *73*, 583–628. [CrossRef]
8. Imada, M.; Fujimori, A.; Tokura, Y. Metal–insulator transitions. *Rev. Mod. Phys.* **1998**, *70*, 1039–1263. [CrossRef]
9. Cheong, S.W.; Hwang, H.Y. Ferromagnetism vs. Charge/Orbital Ordering in Mixed-Valent Manganites. In *Colossal Magnetoresistive Oxides*; Tokura, Y., Ed.; Gordon and Breach: London, UK, 1999.
10. Kováčik, R.; Ederer, C. Effect of Hubbard U on the construction of low-energy Hamiltonians for $LaMnO_3$ via maximally localized Wannier functions. *Phys. Rev. B* **2011**, *84*, 075118. [CrossRef]
11. Nguyen, T.T.; Bach, T.C.; Pham, H.T.; Pham, T.T.; Nguyen, D.T.; Hoang, N.N. Magnetic state of the bulk, surface and nanoclusters of $CaMnO_3$: A DFT study. *Phys. Condens. Matter* **2011**, *406*, 3613–3621. [CrossRef]
12. Zener, C. Interaction between the d-Shells in the Transition Metals. II. Ferromagnetic Compounds of Manganese with Perovskite Structure. *Phys. Rev.* **1951**, *82*, 403–405. [CrossRef]
13. Anderson, P.W.; Hasegawa, H. Considerations on Double Exchange. *Phys. Rev.* **1955**, *100*, 675–681. [CrossRef]
14. Kubo, K.; Ohata, N. A Quantum Theory of Double Exchange. I. *J. Phys. Soc. Jpn.* **1972**, *33*, 21–32. [CrossRef]

15. Ohtaki, M.; Koga, H.; Tokunaga, T.; Eguchi, K.; Arai, H. Electrical Transport Properties and High-Temperature Thermoelectric Performance of $(Ca_{0.9}M_{0.1})MnO_3$ (M = Y, La, Ce, Sm, In, Sn, Sb, Pb, Bi). *J. Solid State Chem.* **1995**, *120*, 105–111. [CrossRef]
16. Jaime, M.; Salamon, M.B.; Pettit, K.; Rubinstein, M.; Treece, R.E.; Horwitz, J.S.; Chrisey, D.B. Magnetothermopower in $La_{0.67}Ca_{0.33}MnO_3$ thin films. *Appl. Phys. Lett.* **1996**, *68*, 1576–1578. [CrossRef]
17. Jaime, M.; Hardner, H.T.; Salamon, M.B.; Rubinstein, M.; Dorsey, P.; Emin, D. Hall-Effect Sign Anomaly and Small-Polaron Conduction in $(La_{1-x}Gd_x)_{0.67}Ca_{0.33}MnO_3$. *Phys. Rev. Lett.* **1997**, *78*, 951–954. [CrossRef]
18. Palstra, T.T.M.; Ramirez, A.P.; Cheong, S.W.; Zegarski, B.R.; Schiffer, P.; Zaanen, J. Transport mechanisms in doped $LaMnO_3$: Evidence for polaron formation. *Phys. Rev. B* **1997**, *56*, 5104–5107. [CrossRef]
19. Chun, S.H.; Salamon, M.B.; Han, P.D. Hall effect of $La_{2/3}(Ca,Pb)_{1/3}MnO_3$ single crystals. *J. Appl. Phys.* **1999**, *85*, 5573–5575. [CrossRef]
20. Millis, A.J. Lattice effects in magnetoresistive manganese perovskites. *Nature* **1998**, *392*, 147–150. [CrossRef]
21. Wollan, E.O.; Koehler, W.C. Neutron Diffraction Study of the Magnetic Properties of the Series of Perovskite-Type Compounds $[(1-x) La, x Ca]MnO_3$. *Phys. Rev.* **1955**, *100*, 545–563. [CrossRef]
22. Ramirez, A.P.; Schiffer, P.; Cheong, S.W.; Chen, C.H.; Bao, W.; Palstra, T.T.M.; Gammel, P.L.; Bishop, D.J.; Zegarski, B. Thermodynamic and Electron Diffraction Signatures of Charge and Spin Ordering in $La_{1-x}Ca_xMnO_3$. *Phys. Rev. Lett.* **1996**, *76*, 3188–3191. [CrossRef] [PubMed]
23. Chen, C.H.; Cheong, S.W. Commensurate to Incommensurate Charge Ordering and Its Real-Space Images in $La_{0.5}Ca_{0.5}MnO_3$. *Phys. Rev. Lett.* **1996**, *76*, 4042–4045. [CrossRef] [PubMed]
24. Mori, S.; Chen, C.H.; Cheong, S.W. Pairing of charge-ordered stripes in $(La,Ca)MnO_3$. *Nature* **1998**, *392*, 473–476. [CrossRef]
25. Zang, J.; Bishop, A.R.; Röder, H. Double degeneracy and Jahn-Teller effects in colossal-magnetoresistance perovskites. *Phys. Rev. B* **1996**, *53*, R8840–R8843. [CrossRef]
26. Mishra, S.K.; Pandit, R.; Satpathy, S. Mean-field theory of charge ordering and phase transitions in the colossal-magnetoresistive manganites. *J. Physics Condens. Matter* **1999**, *11*, 8561–8578. [CrossRef]
27. Hotta, T.; Malvezzi, A.L.; Dagotto, E. Charge-orbital ordering and phase separation in the two-orbital model for manganites: Roles of Jahn-Teller phononic and Coulombic interactions. *Phys. Rev. B* **2000**, *62*, 9432–9452. [CrossRef]
28. Zimmermann, M.v.; Hill, J.P.; Gibbs, D.; Blume, M.; Casa, D.; Keimer, B.; Murakami, Y.; Tomioka, Y.; Tokura, Y. Interplay between Charge, Orbital, and Magnetic Order in $Pr_{1-x}Ca_xMnO_3$. *Phys. Rev. Lett.* **1999**, *83*, 4872–4875. [CrossRef]
29. Hotta, T.; Takada, Y.; Koizumi, H.; Dagotto, E. Topological Scenario for Stripe Formation in Manganese Oxides. *Phys. Rev. Lett.* **2000**, *84*, 2477–2480. [CrossRef]
30. Goodenough, J.B. Theory of the Role of Covalence in the Perovskite-Type Manganites [La, $M(II)]MnO_3$. *Phys. Rev.* **1955**, *100*, 564–573. [CrossRef]
31. Bîrsan, E. The superexchange interaction influence on the magnetic ordering in manganites. *J. Magn. Magn. Mater.* **2008**, *320*, 646–650. [CrossRef]
32. Yi, H.; Yu, J.; Lee, S.I. Suppression of ferromagnetic ordering in doped manganites: Effects of the superexchange interaction. *Phys. Rev. B* **2000**, *61*, 428–431. [CrossRef]
33. Bao, W.; Axe, J.; Chen, C.; Cheong, S.W.; Schiffer, P.; Roy, M. From double exchange to superexchange in charge-ordering perovskite manganites. *Phys. B Condens. Matter* **1997**, *241–243*, 418–420. [CrossRef]
34. Dagotto, E. Open questions in CMR manganites, relevance of clustered states and analogies with other compounds including the cuprates. *New J. Phys.* **2005**, *7*, 67. [CrossRef]
35. Dagotto, E. *Nanoscale Phase Separation and Colossal Magnetoresistance*; Springer Series in Solid-State Sciences; Springer: Berlin/Heidelberg, Germany, 2003; Volume 136. [CrossRef]
36. Dagotto, E. Complexity in Strongly Correlated Electronic Systems. *Science* **2005**, *309*, 257–262. [CrossRef]
37. Nagaev, E.L. Lanthanum manganites and other giant-magnetoresistance magnetic conductors. *Physics-Uspekhi* **1996**, *39*, 781–805. [CrossRef]
38. Kagan, M.; Kugel, K.; Rakhmanov, A. Electronic phase separation: Recent progress in the old problem. *Phys. Rep.* **2021**, *916*, 1–105. [CrossRef]
39. Zhang, P.; Chern, G.W. Arrested Phase Separation in Double-Exchange Models: Large-Scale Simulation Enabled by Machine Learning. *Phys. Rev. Lett.* **2021**, *127*, 146401. [CrossRef]
40. Burgy, J.; Mayr, M.; Martin-Mayor, V.; Moreo, A.; Dagotto, E. Colossal Effects in Transition Metal Oxides Caused by Intrinsic Inhomogeneities. *Phys. Rev. Lett.* **2001**, *87*, 277202. [CrossRef]
41. Zhang, P.; Saha, P.; Chern, G.W. Machine Learning Dynamics of Phase Separation in Correlated Electron Magnets. *arXiv* **2020**, arXiv:2006.04205v1.
42. Luo, J.; Chern, G.W. Dynamics of electronically phase-separated states in the double exchange model. *Phys. Rev. B* **2021**, *103*, 115137. [CrossRef]
43. Pradhan, K.; Yunoki, S. Nanoclustering phase competition induces the resistivity hump in colossal magnetoresistive manganites. *Phys. Rev. B* **2017**, *96*, 214416. [CrossRef]
44. Burgy, J.; Moreo, A.; Dagotto, E. Relevance of Cooperative Lattice Effects and Stress Fields in Phase-Separation Theories for CMR Manganites. *Phys. Rev. Lett.* **2004**, *92*, 097202. [CrossRef] [PubMed]

45. Čulo, M.; Basletić, M.; Tafra, E.; Hamzić, A.; Tomić, S.; Fischgrabe, F.; Moshnyaga, V.; Korin-Hamzić, B. Magnetotransport properties of La$_{1-x}$Ca$_x$MnO$_3$ (0.52 ≤ x ≤ 0.75): Signature of phase coexistence. *Thin Solid Films* **2017**, *631*, 205–212. [CrossRef]
46. Freitas, R.; Ghivelder, L.; Levy, P.; Parisi, F. Magnetization studies of phase separation in La$_{0.5}$Ca$_{0.5}$MnO$_3$. *Phys. Rev. B* **2002**, *65*, 104403. [CrossRef]
47. Giri, S.K.; Nath, T.K. Suppression of Charge and Antiferromagnetic Ordering in La$_{0.5}$Ca$_{0.5}$MnO$_3$ Nanoparticles. *J. Nanosci. Nanotechnol.* **2011**, *11*, 4806–4814. [CrossRef]
48. Dhieb, S.; Krichene, A.; Boudjada, N.C.; Boujelben, W. Suppression of Metamagnetic Transitions of Martensitic Type by Particle Size Reduction in Charge-Ordered La$_{0.5}$Ca$_{0.5}$MnO$_3$. *J. Phys. Chem. C* **2020**, *124*, 17762–17771. [CrossRef]
49. Levy, P.; Parisi, F.; Polla, G.; Vega, D.; Leyva, G.; Lanza, H.; Freitas, R.; Ghivelder, L. Controlled phase separation in La$_{0.5}$Ca$_{0.5}$MnO$_3$. *Phys. Rev. B* **2000**, *62*, 6437–6441. [CrossRef]
50. Quintero, M.; Passanante, S.; Irurzun, I.; Goijman, D.; Polla, G. Grain size modification in the magnetocaloric and non-magnetocaloric transitions in La$_{0.5}$Ca$_{0.5}$MnO$_3$ probed by direct and indirect methods. *Appl. Phys. Lett.* **2014**, *105*, 152411. [CrossRef]
51. Gamzatov, A.G.; Batdalov, A.B.; Aliev, A.M.; Amirzadeh, P.; Kameli, P.; Ahmadvand, H.; Salamati, H. Influence of the granule size on the magnetocaloric properties of manganite La$_{0.5}$Ca$_{0.5}$MnO$_3$. *Phys. Solid State* **2013**, *55*, 502–507. [CrossRef]
52. Sarkar, T.; Ghosh, B.; Raychaudhuri, A.K.; Chatterji, T. Crystal structure and physical properties of half-doped manganite nanocrystals of less than 100-nm size. *Phys. Rev. B* **2008**, *77*, 235112. [CrossRef]
53. Byczuk, K.; Hofstetter, W.; Vollhardt, D. Mott-Hubbard Transition versus Anderson Localization in Correlated Electron Systems with Disorder. *Phys. Rev. Lett.* **2005**, *94*, 056404. [CrossRef] [PubMed]
54. Shinaoka, H.; Imada, M. Single-Particle Excitations under Coexisting Electron Correlation and Disorder: A Numerical Study of the Anderson–Hubbard Model. *J. Phys. Soc. Jpn.* **2009**, *78*, 094708. [CrossRef]
55. Aguiar, M.C.O.; Dobrosavljević, V.; Abrahams, E.; Kotliar, G. Critical Behavior at the Mott-Anderson Transition: A Typical-Medium Theory Perspective. *Phys. Rev. Lett.* **2009**, *102*, 156402. [CrossRef] [PubMed]
56. Di Sante, D.; Fratini, S.; Dobrosavljević, V.; Ciuchi, S. Disorder-Driven Metal-Insulator Transitions in Deformable Lattices. *Phys. Rev. Lett.* **2017**, *118*, 036602. [CrossRef] [PubMed]
57. Worledge, D.C.; Snyder, G.J.; Beasley, M.R.; Geballe, T.H.; Hiskes, R.; DiCarolis, S. Anneal-tunable Curie temperature and transport of La$_{0.67}$Ca$_{0.33}$MnO$_3$. *J. Appl. Phys.* **1996**, *80*, 5158–5161. [CrossRef]
58. Coey, J.M.D.; Viret, M.; von Molnár, S. Mixed-valence manganites. *Adv. Phys.* **1999**, *48*, 167–293. [CrossRef]
59. Zhou, H.D.; Zheng, R.K.; Li, G.; Feng, S.J.; Liu, F.; Fan, X.J.; Li, X.G. Transport properties of La$_{1-x}$Ca$_x$MnO$_3$ (0.5 ≤ x < 1). *Eur. Phys. J. B* **2002**, *26*, 467–471. [CrossRef]
60. Mott, N.F.; Davis, E.A. *Electronic Processes in Non-Crystalline Materials*, 2nd ed.; International Series of Monographs on Physics, Clarendon Press: Oxford, UK, 2012.
61. Lu, C.L.; Dong, S.; Wang, K.F.; Gao, F.; Li, P.L.; Lv, L.Y.; Liu, J.M. Charge-order breaking and ferromagnetism in La$_{0.4}$Ca$_{0.6}$MnO$_3$ nanoparticles. *Appl. Phys. Lett.* **2007**, *91*, 032502. [CrossRef]
62. Rozenberg, E.; Auslender, M.; Shames, A.I.; Mogilyansky, D.; Felner, I.; Sominskii, E.; Gedanken, A.; Mukovskii, Y.M. Nanometer size effect on magnetic order in La$_{0.4}$Ca$_{0.6}$MnO$_3$: Predominant influence of doped electron localization. *Phys. Rev. B* **2008**, *78*, 052405. [CrossRef]
63. Zhang, T.; Zhou, T.F.; Qian, T.; Li, X.G. Particle size effects on interplay between charge ordering and magnetic properties in nanosized La$_{0.25}$Ca$_{0.75}$MnO$_3$. *Phys. Rev. B* **2007**, *76*, 174415. [CrossRef]
64. Zhang, T.; Dressel, M. Grain-size effects on the charge ordering and exchange bias in Pr$_{0.5}$Ca$_{0.5}$MnO$_3$: The role of spin configuration. *Phys. Rev. B* **2009**, *80*, 014435. [CrossRef]
65. Sarkar, T.; Mukhopadhyay, P.K.; Raychaudhuri, A.K.; Banerjee, S. Structural, magnetic, and transport properties of nanoparticles of the manganite Pr$_{0.5}$Ca$_{0.5}$MnO$_3$. *J. Appl. Phys.* **2007**, *101*, 124307. [CrossRef]
66. Rao, S.S.; Tripathi, S.; Pandey, D.; Bhat, S.V. Suppression of charge order, disappearance of antiferromagnetism, and emergence of ferromagnetism in Nd$_{0.5}$Ca$_{0.5}$MnO$_3$ nanoparticles. *Phys. Rev. B* **2006**, *74*, 144416. [CrossRef]
67. Boucher, B.; Buhl, R.; Perrin, M. Magnetic Structure of Mn$_3$O$_4$ by Neutron Diffraction. *J. Appl. Phys.* **1971**, *42*, 1615–1617. [CrossRef]
68. Narayani, L.; Jagadeesha Angadi, V.; Sukhdev, A.; Challa, M.; Matteppanavar, S.; Deepthi, P.; Mohan Kumar, P.; Pasha, M. Mechanism of high temperature induced phase transformation and magnetic properties of Mn$_3$O$_4$ crystallites. *J. Magn. Magn. Mater.* **2019**, *476*, 268–273. [CrossRef]
69. Filippetti, A.; Hill, N.A. First principles study of structural, electronic and magnetic interplay in ferroelectromagnetic yttrium manganite. *J. Magn. Magn. Mater.* **2001**, *236*, 176–189. [CrossRef]
70. Tomuta, D.G.; Ramakrishnan, S.; Nieuwenhuys, G.J.; Mydosh, J.A. The magnetic susceptibility, specific heat and dielectric constant of hexagonal YMnO$_3$, LuMnO$_3$ and ScMnO$_3$. *J. Physics Condens. Matter* **2001**, *13*, 4543–4552. [CrossRef]
71. Počuča-Nešić, M.; Marinković-Stanojević, Z.; Cotič-Smole, P.; Dapčević, A.; Tasić, N.; Branković, G.; Branković, Z. Processing and properties of pure antiferromagnetic h-YMnO$_3$. *Process. Appl. Ceram.* **2019**, *13*, 427–434. [CrossRef]
72. Zhang, J.; Yu, L.; Cao, G.; Yu, Q.; Chen, M.; Cao, S. Cluster coexistence state in strongly correlation La$_{0.5}$Ca$_{0.5}$MnO$_3$ manganites. *Phys. B Condens. Matter* **2008**, *403*, 1650–1651. [CrossRef]

73. Cao, G.; Zhang, J.; Wang, S.; Yu, J.; Jing, C.; Cao, S.; Shen, X. Reentrant spin glass behavior in CE-type AFM $Pr_{0.5}Ca_{0.5}MnO_3$ manganite. *J. Magn. Magn. Mater.* **2006**, *301*, 147–154. [CrossRef]
74. Lees, M.R.; Barratt, J.; Balakrishnan, G.; Paul, D.M.; Dewhurst, C.D. Low-temperature magnetoresistance and magnetic ordering in $Pr_{1-x}Ca_xMnO_3$. *J. Physics Condens. Matter* **1996**, *8*, 2967–2979. [CrossRef]
75. Wang, K.F.; Wang, Y.; Wang, L.F.; Dong, S.; Li, D.; Zhang, Z.D.; Yu, H.; Li, Q.C.; Liu, J.M. Cluster-glass state in manganites induced by A-site cation-size disorder. *Phys. Rev. B* **2006**, *73*, 134411. [CrossRef]
76. Lee, P.A.; Ramakrishnan, T.V. Disordered electronic systems. *Rev. Mod. Phys.* **1985**, *57*, 287–337. [CrossRef]
77. Bałanda, M. AC Susceptibility Studies of Phase Transitions and Magnetic Relaxation: Conventional, Molecular and Low-Dimensional Magnets. *Acta Phys. Pol. A* **2013**, *124*, 964–976. [CrossRef]
78. Topping, C.V.; Blundell, S.J. A.C. susceptibility as a probe of low-frequency magnetic dynamics. *J. Phys. Condens. Matter* **2019**, *31*, 013001. [CrossRef]
79. Vincent, E.; Dupuis, V. Spin glasses: Experimental signatures and salient outcomes. In *Frustrated Materials and Ferroic Glasses*; Lookman, T., Ren, X., Eds.; Springer Series in Materials Science; Springer: Berlin/Heidelberg, Germany, 2017.
80. Joung, D.; Khondaker, S.I. Efros-Shklovskii variable-range hopping in reduced graphene oxide sheets of varying carbon sp^2 fraction. *Phys. Rev. B* **2012**, *86*, 235423. [CrossRef]
81. Khondaker, S.I.; Shlimak, I.S.; Nicholls, J.T.; Pepper, M.; Ritchie, D.A. Two-dimensional hopping conductivity in a δ-doped $GaAs/Al_xGa_{1-x}As$ heterostructure. *Phys. Rev. B* **1999**, *59*, 4580–4583. [CrossRef]
82. Khondaker, S.; Shlimak, I.; Nicholls, J.; Pepper, M.; Ritchie, D. Crossover phenomenon for two-dimensional hopping conductivity and density-of-states near the Fermi level. *Solid State Commun.* **1999**, *109*, 751–756. [CrossRef]
83. Chuang, C.; Puddy, R.; Lin, H.D.; Lo, S.T.; Chen, T.M.; Smith, C.; Liang, C.T. Experimental evidence for Efros–Shklovskii variable range hopping in hydrogenated graphene. *Solid State Commun.* **2012**, *152*, 905–908. [CrossRef]
84. Zabrodskii, A.G. The Coulomb gap: The view of an experimenter. *Philos. Mag. B* **2001**, *81*, 1131–1151. [CrossRef]

Article

Optical Conductivity Spectra of Charge-Crystal and Charge-Glass States in a Series of θ-Type BEDT-TTF Compounds

Kenichiro Hashimoto [1,2,*], Ryota Kobayashi [2], Satoshi Ohkura [2], Satoru Sasaki [2], Naoki Yoneyama [3], Masayuki Suda [4,5], Hiroshi M. Yamamoto [5] and Takahiko Sasaki [2]

1. Department of Advanced Materials Science, University of Tokyo, Chiba 277-8561, Japan
2. Institute for Materials Research, Tohoku University, Sendai 980-8577, Japan; kobayashiryota2046@gmail.com (R.K.); s.ohkura.fl@gmail.com (S.O.); satorusasaki0125@gmail.com (S.S.); takahiko.sasaki.d3@tohoku.ac.jp (T.S.)
3. Graduate Faculty of Interdisciplinary Research, University of Yamanashi, Kohu 400-8511, Japan; nyoneyama@yamanashi.ac.jp
4. Department of Molecular Engineering, Graduate School of Engineering, Kyoto University, Kyoto 615-8510, Japan; suda.masayuki.2c@kyoto-u.ac.jp
5. Institute for Molecular Science, Okazaki 444-8585, Japan; yhiroshi@ims.ac.jp
* Correspondence: k.hashimoto@edu.k.u-tokyo.ac.jp

Citation: Hashimoto, K.; Kobayashi, R.; Ohkura, S.; Sasaki, S.; Yoneyama, N.; Suda, M.; Yamamoto, H.M.; Sasaki, T. Optical Conductivity Spectra of Charge-Crystal and Charge-Glass States in a Series of θ-Type BEDT-TTF Compounds. *Crystals* **2022**, *12*, 831. https://doi.org/10.3390/cryst12060831

Academic Editor: Andrej Pustogow

Received: 16 May 2022
Accepted: 9 June 2022
Published: 12 June 2022

Publisher's Note: MDPI stays neutral with regard to jurisdictional claims in published maps and institutional affiliations.

Copyright: © 2022 by the authors. Licensee MDPI, Basel, Switzerland. This article is an open access article distributed under the terms and conditions of the Creative Commons Attribution (CC BY) license (https://creativecommons.org/licenses/by/4.0/).

Abstract: In the 3/4-filled band system θ-(BEDT-TTF)$_2 X$ with a two-dimensional triangular lattice, charge ordering (CO) often occurs due to strong inter-site Coulomb repulsion. However, the strong geometrical frustration of the triangular lattice can prohibit long-range CO, resulting in a charge-glass state in which the charge configurations are randomly distributed. Here, we investigate the charge-glass states of orthorhombic and monoclinic θ-type BEDT-TTF salts by measuring the electrical resistivity and optical conductivity spectra. We find a substantial difference between the charge-glass states of the orthorhombic and monoclinic systems. The charge-glass state in the orthorhombic system with an isotropic triangular lattice exhibits larger low-energy excitations than that in the monoclinic one with an anisotropic triangular lattice and becomes more metallic as the isotropy of the triangular lattice increases. These results can be understood by the different charge-glass formation mechanisms in the two systems: in the orthorhombic system, the charge-glass state originates from geometric frustration due to the equilateral triangular lattice, leading to metallic 3-fold COs, whereas in the monoclinic system, the charge-glass formation originates from geometric frustration of the isosceles triangular lattice, in which the charge-glass state is described by the superposition of insulating 2-fold stripe COs.

Keywords: strongly correlated electrons; metal-insulator transition; charge order; charge glass; charge crystal; geometrical frustration; organics; optical conductivity

1. Introduction

Charge ordering (CO), in which electrons self-organize into an alternating pattern of charge-rich and charge-poor sites owing to strong Coulomb interactions, often emerges in strongly correlated electron systems [1,2]. Inter-site Coulomb interactions play an essential role in the formation of CO, mostly leading to a long-range order. In geometrically frustrated systems, however, disordered ground states without long-range CO have been reported [3–10]. This is considered due to competition among various types of CO patterns coming from geometrical frustration [11], which prevents a specific charge configuration, similar to geometrically frustrated spin systems such as a quantum spin liquid and spin glass. Thus, geometrical frustration can suppress the tendency toward long-range CO, leading to exotic electronic states such as a charge-glass state.

Quasi-two-dimensional (quasi-2D) organic compounds with a triangular lattice, θ-$(BEDT\text{-}TTF)_2MM'(SCN)_4$ (M = Tl, Rb, Cs, M' = Zn, Co) (where BEDT-TTF denotes the donor molecule bis(ethylenedithio)tetrathiafulvalene and $MM'(SCN)_4$ represents a monovalent anion), have been extensively studied as a platform of the CO metal-insulator transition system [12–31]. In θ-$(BEDT\text{-}TTF)_2MM'(SCN)_4$, there are two crystal forms with orthorhombic ($I222$) and monoclinic ($C2$) symmetries [10,12] (see Figure 1a–d). The both crystal structures consist of an alternating stack of BEDT-TTF and anion layers, and the charge transfer between these layers leads to a quarter-filled hole band system. Figure 1b,d show the 2D molecular arrangement of each BEDT-TTF layer for the orthorhombic θ_o-type and monoclinic θ_m-type systems, respectively. The nearest-neighbor Coulomb interactions are given by V_1 and V_2 in the θ_o-type system and by V_1, V_2, and V_2' in the θ_m-type system. Note that $V_2 \approx V_2'$ for the θ_m-type system [10,12,25]. While the θ_o-type system exhibits a horizontal CO pattern (Figure 1b), the θ_m-type system shows a diagonal CO pattern (Figure 1d) [20]. The ground states of θ_o-type salts, ranging from charge ordered insulating states to a metallic state, can be tuned using the anisotropy of the nearest-neighbor Coulomb interactions on the triangular lattice, V_2/V_1, which depends on the anions [8] (see Figure 1e). Indeed, the orthorhombic θ-$(BEDT\text{-}TTF)_2TlZn(SCN)_4$ (hereafter, abbreviated as θ_o-TlZn) and θ-$(BEDT\text{-}TTF)_2RbZn(SCN)_4$ (θ_o-RbZn) with relatively anisotropic V_2/V_1 values are well-known long-range CO compounds with the transition temperature $T_{CO} \approx 240$ K [20] and 200 K [3], respectively. Such a periodic CO state can be regarded as a charge-crystal state. Importantly, this charge-crystal state can be kinetically avoided when the sample is cooled faster than a critical cooling rate, leading to a charge-glass state where the charge configurations are randomly quenched. For instance, the charge-crystal state in θ_o-RbZn can be suppressed for the critical cooling rate of $\sim$30 K/min, resulting in a charge-glass state (see Figures 1f and 2c). In θ_o-CsZn, which has a more isotropic triangular lattice, the critical cooling rate becomes much slower. As a result, the charge-glass state can be realized even upon very slow cooling (<0.1 K/min) (see Figures 1f and 2a). These experimental facts imply that geometrical charge frustration between V_1 and V_2 plays an important role for the charge-glass formation in the θ_o-type salts.

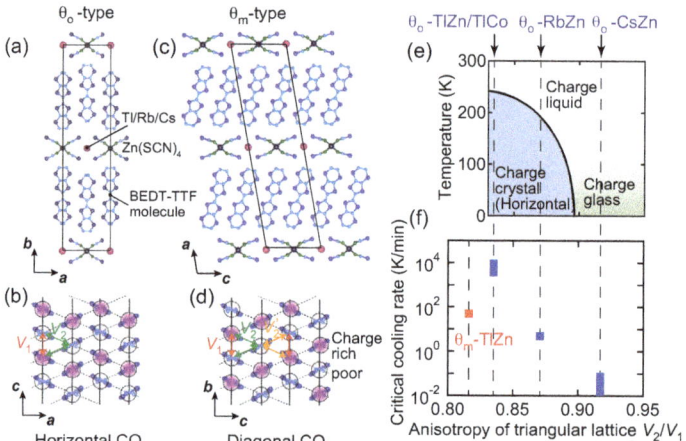

Figure 1. Crystal structure and phase diagram of θ–type (BEDT–TTF) salts. (**a**) Crystal structure of the θ_o-type salts viewed along the c-axis direction. The rectangle indicates the unit cell. (**b**) 2D conducting BEDT-TTF layers within the a-c plane for the θ_o-type salts. (**c**) Crystal structure of the θ_m-type system viewed along the b-axis direction. The parallelogram indicates the unit cell. (**d**) 2D conducting BEDT-TTF layers within the b-c plane for the θ_m-type system. (**e**) Phase diagram of the θ_o-type system a function of the anisotropy parameter, V_2/V_1. (**f**) Critical cooling rate for charge-glass formation for various θ-type salts as a function of V_2/V_1.

The monoclinic θ-(BEDT-TTF)$_2$TlZn(SCN)$_4$ (θ_m-TlZn) shows a diagonal CO at T_{CO} = 170 K [10] (see Figure 2e). In θ_m-TlZn, the long-range CO can be suppressed by rapid cooling (>~50 K/min), and the charge-glass state can be realized. Although the triangular lattice for θ_m-TlZn is more anisotropic than that for θ_o-TlZn, the critical cooling rate of θ_m-TlZn is much slower than that of θ_o-TlZn (see Figure 1f). This suggests different mechanisms of charge-glass formation between these two systems. In this study, in order to clarify the different charge-glass formation mechanisms between the orthorhombic and monoclinic θ-type salts, we measured electrical resistivity and optical conductivity spectra of θ_o-CsZn, θ_o-RbZn, and θ_m-TlZn.

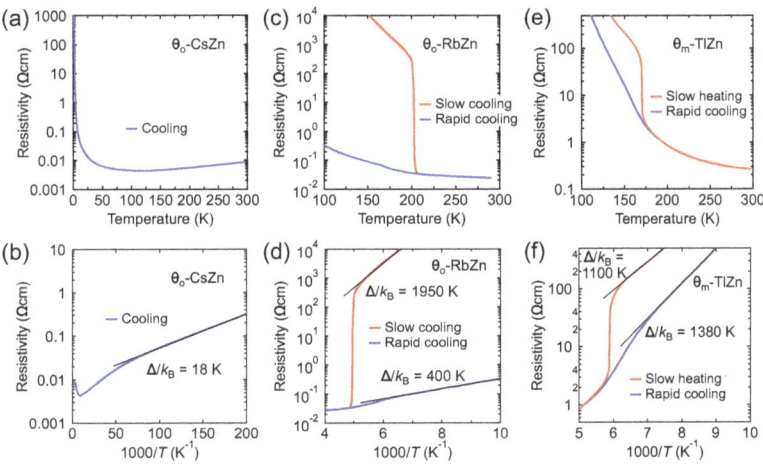

Figure 2. Resistivity and Arrhenius plot in the θ–type salts. (**a**) $\rho(T)$ curve of θ_o-CsZn measured during cooling. (**b**) Arrhenius plot of the same data in (**a**). (**c**) $\rho(T)$ curve of θ_o-RbZn measured during rapid cooling of 30 K/min (blue) and slow cooling of 0.1 K/min (red). (**d**) Arrhenius plot of the same data in (**c**). (**e**) $\rho(T)$ curve of θ_m-TlZn measured during rapid cooling of 100 K/min (blue) and slow heating after slow cooling of 0.1 K/min (red). (**f**) Arrhenius plot of the same data in (**e**). The black lines in (**b**,**d**,**f**) represent the fits to $\rho \propto \exp(\Delta/(k_B T))$.

2. Materials and Methods

Single crystals of θ_o-CsZn, θ_o-RbZn, and θ_m-TlZn were grown by the electrochemical oxidation method [17]. The typical sample size used for the resistivity and optical conductivity measurements was ~0.1 mm × 1 mm × 3 mm. The in-plane dc resistivity was measured by the 4-terminal method in the linear I-V region. The polarized optical conductivity measurements were carried out with a Fourier transform microscope spectrometer in the range of 600–8000 cm^{-1}. For θ_o-CsZn, the optical conductivity measurements in the far-infrared region (100–650 cm^{-1}) were performed using a synchrotron radiation light source at BL43IR in SPring-8. The optical conductivity was calculated through a Kramers–Kronig (KK) transformation from the optical reflectivity determined by comparison with a gold thin film evaporated on the sample surface.

3. Results

3.1. Electrical Resistivity in θ_o-CsZn, θ_o-RbZn, and θ_m-TlZn

Figure 1a,c,e show the temperature dependence of resistivity $\rho(T)$ for θ_o-CsZn, θ_o-RbZn, and θ_m-TlZn, respectively. θ_o-CsZn with the most isotropic triangular lattice shows no long-range CO and enters the charge-glass state below ~100 K. Regardless of the cooling rate of the sample, θ_o-CsZn always shows the charge-glass state. In contrast, θ_o-RbZn and θ_m-TlZn show the long-range CO transition at 200 K and 170 K, respectively, which can be suppressed by rapid cooling, resulting in the charge-glass state. Figure 1b,d,f show the Arrhenius plot of the resistivity for θ_o-CsZn, θ_o-RbZn, and θ_m-TlZn, respectively. Clear activation-type behaviors can be seen both for the charge-crystal and charge-glass states in all the salts. By fitting the data to $\rho(T) = \rho_0 \exp(\Delta/k_B T)$, we obtained the activation energies for the charge-crystal and charge-glass states as shown in Figure 2b,d,f. The activation gap for θ_o-CsZn is very small (approximately 18 K), consistent with the fact that this material is located near the phase boundary. As for θ_o-RbZn, the gap sizes of the charge-crystal and charge-glass states show a large difference. Previous X-ray diffuse scattering experiments in θ_o-CsZn have revealed a short-range 3 × 3 CO [5,13,14]. Since the 3-fold CO in θ_o-RbZn is expected to be metallic, the gap size of the charge-glass state (∼400 K) becomes much smaller than that of the charge-crystal state (∼1950 K). In contrast, in θ_m-TlZn, there is little difference in the activation gaps between the charge-crystal and charge-glass states. These differences are considered to originate from the different mechanisms of the charge-glass formation in the θ_o- and θ_m-type systems, as will be discussed later.

3.2. Optical Conductivity Spectra in θ_o-CsZn, θ_o-RbZn, and θ_m-TlZn

For a comprehensive understanding of charge-glass formation in θ-type salts, we measured the optical conductivity spectra in the series of θ-type salts. Figure 3a shows the optical conductivity spectra $\sigma_1(\omega)$ of θ_o-CsZn for $E \parallel a$ at various temperatures. At room temperature, there are two characteristic broad bands at around 1000 and 2000 cm^{-1}. In addition, the antisymmetric or antiresonance features of the vibrational modes of the BEDT-TTF molecule around 400, 900, and 1300 cm^{-1} can be seen, which become more noticeable at low temperatures [7]. Moreover, as lowering the temperature, $\sigma_1(\omega)$ in the low-energy region below ∼500 cm^{-1} is strongly enhanced. The enhancement of $\sigma_1(\omega)$ is different from a Drude response since the dc conductivity σ_{dc} in θ_o-CsZn decreases to ∼1 Ω^{-1}cm^{-1} at 4 K. The absence of a Drude peak at low temperatures is also expected from the fact that the resistivity obeys the Arrhenius law as shown in Figure 2b. Thus, the optical spectra are mainly composed of three characteristic structures with center frequencies of 100–300, 800–1000, and 2000–2500 cm^{-1} (referred to as L_{low}, L_{middle}, and L_{high}, respectively). As discussed in Ref. [7], the low-energy peak can be attributed to the short-range CO with a relatively long-period 3 × 3 CO. As for the broad bands L_{middle} and L_{high}, very similar features have been observed in other quarter-filled organic conductors close to a CO phase [32–34]. A transition between Hubbard-like bands induced by the intersite Coulomb repulsion V gives rise to a broad band in the mid-infrared region of the order of V, which corresponds to L_{high}. The other band L_{middle} is a charge-fluctuation band originating from short-range CO fluctuations.

Figure 3b shows the temperature dependence of $\sigma_1(\omega)$ of θ_o-RbZn for $E \parallel a$ measured during slow cooling (charge-crystal state) and heating after rapid cooling (charge-glass state). The optical conductivity spectra at room temperature are similar to that of θ_o-CsZn, having two broad structures at around 1000 and 2500 cm^{-1}. When the sample is slowly cooled, the CO transition occurs at 200 K, below which the optical conductivity spectra show a drastic change. The spectral weight below 3000 cm^{-1} disappears, and a clear optical gap is observed below ∼2000 cm^{-1}, which is comparable to Δ/k_B obtained from the Arrhenius plot of the resistivity in the charge-crystal state. In contrast, when the sample is quenched, the CO transition is suppressed and a charge-glass state is realized. The optical conductivity spectra in the charge-glass state share a similar shape with that above the CO transition. Thus, the optical conductivity spectra between the charge-crystal and charge-glass states show a large difference, indicating that the charge configurations are very different between these two states.

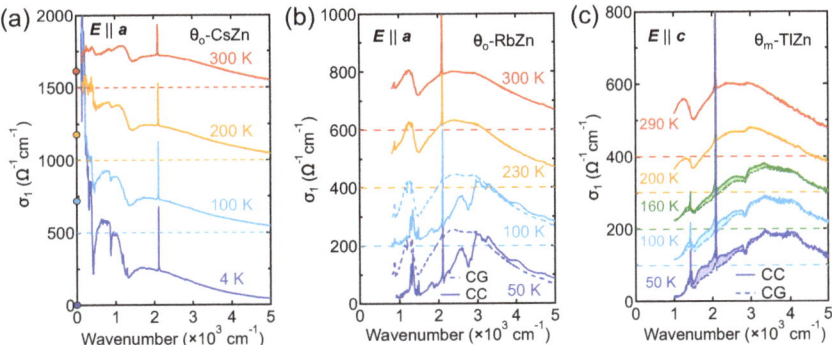

Figure 3. Optical conductivity spectra in the θ–type salts. (**a**) Optical conductivity spectra $\sigma_1(\omega)$ in θ_o-CsZn measured at several temperatures during slow cooling of 1 K/min. (**b**) Optical conductivity spectra $\sigma_1(\omega)$ in θ_o-RbZn measured at several temperatures during slow cooling of 1 K/min (solid line) and slow heating after rapid cooling of more than 50 K/min when passing through T_CO (dashed line). (**c**) Optical conductivity spectra $\sigma_1(\omega)$ in θ_m-TlZn measured at several temperatures during slow cooling of 1 K/min (solid line) and slow heating after rapid cooling of more than 50 K/min when passing through T_CO (dashed line). For clarity, the data are shifted vertically. Note that the sharp peak at approximately 2100 cm^{-1} is the CN stretching mode of SCN in the anion layer.

Figure 3c shows the temperature dependence of $\sigma_1(\omega)$ of θ_m-TlZn for $E \parallel c$ measured in the charge-crystal and charge-glass states. Although the optical conductivity spectra are slightly different in the charge-crystal and charge-glass states, the optical gaps are almost identical in the whole temperature range, which are consistent with the dc resistivity data in which the activation energies for the charge-crystal and charge-glass states are close to each other (see Figure 2f). The obtained optical gaps in the charge-crystal and charge-glass states are about 1200–1300 cm^{-1} at low temperatures, which are comparable to Δ/k_B obtained from the Arrhenius plot of the resistivity. Importantly, in the charge-crystal state, in addition to a broad peak structure around 3500 cm^{-1}, a shoulder-like feature around 2000 cm^{-1} emerges as lowering the temperature, whereas in the charge-glass state, the growth of the 2000 cm^{-1} feature seems to be frozen (see the hatched area in Figure 3). The optical spectra obtained for the charge-crystal state can be well understood by previous theoretical calculations performed for the diagonal CO phase [31], in which the first low-energy peak and the high-energy broad feature are well reproduced.

4. Discussion

We compare the optical conductivity spectra of the three salts in the charge-glass state. Figure 4 shows the optical conductivity spectra of the charge-glass states in θ_o-CsZn, θ_o-RbZn, and θ_m-TlZn. It can be clearly seen that the optical conductivity of θ_o-CsZn with the most isotropic triangular lattice shows a significant low-energy peak, and as the anisotropy of the triangular lattice increases, the spectral weight in the low-energy region shifts to a higher-energy region. This systematic evolution of the optical conductivity spectra is considered to reflect the different charge configurations in the charge-glass states in the three salts.

To discuss the difference of the optical conductivity spectra in the three salts, we consider the extended Hubbard model. The ground-state properties of θ-(BEDT-TTF)$_2X$ have been extensively studied by the extended Hubbard model on an anisotropic triangular lattice [25,27,28,30,31,35–38]. The Hamiltonian of the extended Hubbard model is given by

$$\mathcal{H}_\mathrm{EHM} = \sum_{\langle i,j\rangle\sigma}\left(-t_{ij}c_{i\sigma}^\dagger c_{j\sigma} + \mathrm{h.c.}\right) + U\sum_i n_{i\uparrow}n_{i\downarrow} + \sum_{\langle i,j\rangle}V_{ij}n_in_j, \tag{S1}$$

where $c_{i\sigma}^{\dagger}$ ($c_{i\sigma}$) is the creation (annihilation) operator for a hole at the i-th site with spin σ (↑ or ↓), n_i ($\equiv \sum_\sigma n_{i\sigma} \equiv \sum_\sigma c_{i\sigma}^{\dagger} c_{i\sigma}$) is the number operator, t_{ij} and V_{ij} are the transfer integrals and the intersite Coulomb interactions between the i-th and j-th sites, respectively, and U is the on-site Coulomb repulsion.

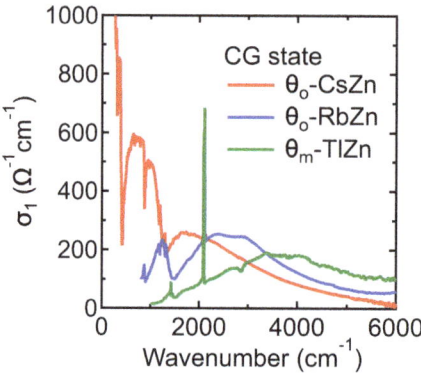

Figure 4. Comparison of optical conductivity spectra of the charge–glass states in the θ–type salts. Optical conductivity spectra of the charge-glass states in θ_o-CsZn (red), θ_o-RbZn (blue), and θ_m-TlZn (green) measured at 4 K, 50 K, and 50 K, respectively.

In Ref. [31], the polarization dependence of the optical conductivity spectra has been calculated for various charge ordering patterns, based on the extended Hubbard model. When V_1 and V_2 are close, the optical conductivity spectra for two polarization directions become isotropic, except for the difference in the magnitude of the optical conductivity spectra (which reflects the difference between the intermolecular distances in the V_1 and V_2 directions [31]), indicating the strong geometric frustration of the isotropic triangular lattice. In contrast, when V_1 is larger than V_2 (that is, in the case of diagonal CO), the optical conductivity spectra for two polarization directions become anisotropic: the optical spectra for the polarization parallel to the V_1 direction (b-axis direction in θ_m-TlZn) have only a low-energy peak, whereas the optical spectra for the polarization perpendicular to the V_1 direction (c-axis direction in θ_m-TlZn) show a step-like increase in the low-energy region, followed by a broad structure in the high-energy region. Indeed, very similar behaviors have been observed in our experimental results (see Figure 5). Such a polarization dependence in the optical spectra has also been observed in other 1D charge-ordered organic materials [39], where the polarization dependence corresponding to the stripe charge ordering pattern has been reported.

Next, we discuss the charge configurations of the charge-glass states in the θ_o-type and θ_m-type systems. When $U \gg t_{ij}$, the extended Hubbard model is compatible to the spinless fermion model (t-V model) that neglects the spin degrees of freedom. The Hamiltonian of the t-V model is given by

$$\mathcal{H}_{t\text{-}V} = \sum_{\langle i,j \rangle} \left(-t_{ij} f_i^{\dagger} f_j + \text{h.c.} + V_{ij} \tilde{n}_i \tilde{n}_j \right), \tag{S2}$$

where $f_i^{\dagger}$ (f_i) is the creation (annihilation) operator for a spinless fermion at the i-th site and $\tilde{n}_i = f_i^{\dagger} f_i$ is the number operator. It has been well established that the classical ground states of the t-V model ($t_{ij} = 0$) on an isosceles triangle lattice as shown in Figure 6b are disordered owing to geometric frustration when $V_1 \geq V_2$, whereas the vertical CO becomes a unique ground state at $V_1 < V_2$ [29,40,41]. At $V_1 > V_2$, the chain-striped states such as the horizontal and diagonal COs emerge owing to the geometric frustration between the two diagonal Coulomb interactions V_2, which is the case for θ_m-TlZn.

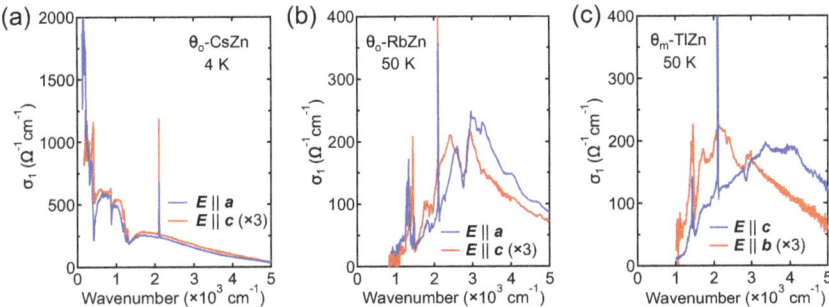

Figure 5. Polarization dependence of optical conductivity spectra in the charge–glass/crystal states of the θ–type salts. (**a**–**c**) Optical conductivity spectra of (**a**) the charge-glass state in θ_o-CsZn for $E \parallel a$ (blue) and $E \parallel c$ (red) measured at 4 K, (**b**) the charge-crystal state in θ_o-RbZn for $E \parallel a$ (blue) and $E \parallel c$ (red) measured at 50 K, and (**c**) the charge-crystal state in θ_m-TlZn for $E \parallel c$ (blue) and $E \parallel b$ (red) measured at 50 K.

When $V_1 = V_2$, the ground states include a vertical-striped state, horizontal-striped state, diagonal-striped state, and three-sublattice state, all of which are degenerate (see Figure 6a,). The three-sublattice state has been discussed in terms of a pin-ball liquid [29]. The θ_o-type compounds can be categorized in this regime. Indeed, the X-ray diffuse scattering experiments have revealed that for θ_o-CsZn, diffuse rods with $q_d = (2/3, k, 1/3)$ corresponding to a 3 × 3 CO are observed [5,13,14], while for θ_o-RbZn above T_{CO}, diffuse rods associated with a short-range 3 × 4 CO are observed at $q_d = (\pm 1/3, k, \pm 1/4)$ [4,15,16]. Such three-fold diffuse rods are different from that of θ_m-TlZn, where diffuse lines at $q_d = (1/2, l)$ corresponding to the superposition of the chain striped COs as shown in Figure 6b have been observed [10].

Based on the above calculations, we discuss the charge-glass formation mechanisms in the θ_o-type and θ_m-type systems. In the charge-glass state of the θ_o-type salts with an isotropic triangular lattice, the short-range 3-fold periodic CO patterns have been observed. Although the q vectors of the short-range COs are slightly different from the 3-fold CO shown in Figure 6a, the presence of the 3-fold COs makes the system metallic. As a result, the optical gap and Arrhenius gap in the charge-glass state become smaller than that in the charge-crystal state. On the other hand, in the charge-glass state of the θ_m-type system with an isosceles triangular lattices, the short-range 2-fold COs have been reported. In this case, the charge-glass state is described by the superposition of the insulating 2-fold stripe COs [10]. Therefore, there is no significant difference in the magnitude of charge separation between the charge-crystal and the charge-glass states. Thus, the sizes of the optical gap and Arrhenius gap become almost the same in the charge-crystal and charge-glass states. From these facts, we conclude that the charge-glass states in the θ_o-type and θ_m-type systems originate from the geometrical frustration of the equilateral and isosceles triangular lattices, respectively. Since recent thermal expansion and noise spectroscopy measurements have pointed out that the lattice degrees of freedom in addition to the electron degrees of freedom play an important role for the charge-glass formation in the θ-type BEDT-TTF compounds [42,43], elucidating the effect of lattice degrees of freedom on the charge-glass formation needs to be addressed in the future.

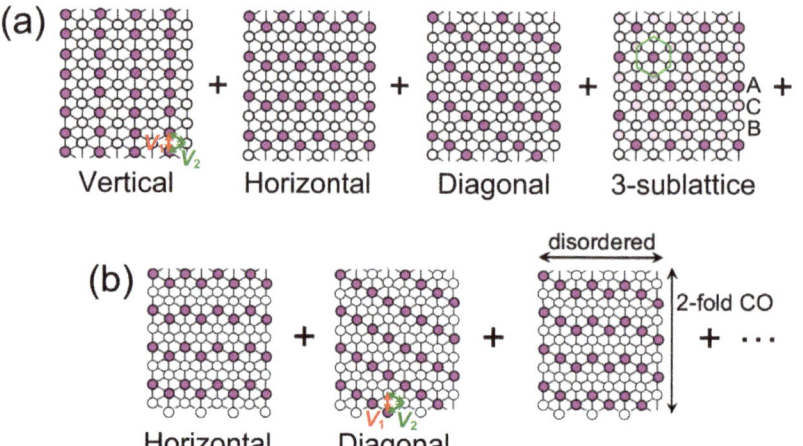

Figure 6. Schematic charge configurations on triangular lattices. (**a**) Charge configurations on the isosceles triangular lattice. Vertical, horizontal, diagonal, and three-sublattice COs are described. In the three-sublattice structure, the sublattice A is filled by one hole (pin), the sublattice B is empty, and the sublattice C is randomly occupied by the remaining holes (ball). The green hexagon stands for the unit cell. (**b**) Chain striped CO patterns on the isosceles triangular lattice, such as horizontal and diagonal COs. V_1 and V_2 ($V_1 > V_2$) are the nearest-neighbor Coulomb interactions. Since all these states are degenerate in the classical limit of the t-V model, the classical ground state can be described by the superposition of these states. The magenta and white circles represent the charge-rich and charge-poor sites, respectively.

5. Conclusions

We investigated the charge-glass states of θ_o-CsZn, θ_o-RbZn, and θ_m-TlZn by measuring the electrical resistivity and optical conductivity spectra. We find that there is a fundamental difference between the charge-glass formation mechanisms in the θ_o-type and θ_m-type systems. The charge-glass state in θ_o-CsZn exhibits large low-energy excitations, consistent with the fact that the material is located near the CO phase boundary. In θ_o-RbZn, the optical gaps between the charge-crystal and charge-glass states show a large difference, indicating that the charge configurations are very different between the two states. In contrast, the optical gap of the charge-glass state in θ_m-TlZn does not differ from that in the charge-crystal state. These results can be understood by the different charge-glass formation mechanisms in the θ_o-type and θ_m-type systems: in the θ_o-type system, the charge-glass state originates from geometric frustration due to the equilateral triangular lattice, leading to metallic 3-fold COs, whereas in the θ_m-type system, the charge-glass formation originates from geometric frustration of the isosceles triangular lattice, in which the charge-glass state can be described by the superposition of insulating 2-fold stripe COs.

Author Contributions: K.H. and T.S. conceived the project. K.H., S.O. and S.S. carried out electrical resistivity measurements. K.H., R.K., S.O., S.S. and T.S. performed optical measurements. N.Y., M.S., H.M.Y. and T.S. carried out sample preparation. K.H. wrote the manuscript. All authors have read and agreed to the published version of the manuscript.

Funding: This work was supported by Grants-in-Aid for Scientific Research (KAKENHI) (Nos. JP21H01793, JP20H05144, JP19K22123, JP19H01833, JP18KK0375, JP18H01853), and Grant-in-Aid for Scientific Research for Transformative Research Areas (A) "Condensed Conjugation" (No. JP21H05471, JP20H05869, JP20H05870) from Japan Society for the Promotion of Science (JSPS).

Institutional Review Board Statement: Not applicable.

Informed Consent Statement: Not applicable.

Data Availability Statement: The data that support the findings of this study are available from the corresponding author upon reasonable request.

Acknowledgments: We thank M. Naka, K. Yoshimi, C. Hotta, T. Thomas, and J. Müller for fruitful discussion and Y. Ikemoto and T. Moriwaki for technical assistance. Optical measurements using a synchrotron radiation light source were performed at SPring-8 with the approvals of the Japan Synchrotron Radiation Research Institute (Grant Nos. 2016A0073, 2020A0639, 2020A1065).

Conflicts of Interest: The authors declare no competing interests.

References

1. Takahashi, T.; Nogami, Y.; Yakushi, K. Charge ordering in organic conductors. *J. Phys. Soc. Jpn.* **2006**, *75*, 051008. [CrossRef]
2. Seo, H.; Merino, J.; Yoshioka, H.; Ogata, M. Theoretical aspects of charge ordering in molecular conductors. *J. Phys. Soc. Jpn.* **2006**, *75*, 051009. [CrossRef]
3. Chiba, R.; Hiraki, K.; Takahashi, T.; Yamamoto, H.M.; Nakamura, T. Charge disproportionation and dynamics in θ-(BEDT-TTF)$_2$CsZn(SCN)$_4$. *Phys. Rev. B* **2008**, *77*, 115113. [CrossRef]
4. Kagawa, F.; Sato, T.; Miyagawa, K.; Kanoda, K.; Tokura, Y.; Kobayashi, K.; Kumai, R.; Murakami, Y. Charge-cluster glass in an organic conductor. *Nat. Phys.* **2013**, *9*, 419–422. [CrossRef]
5. Sato, T.; Kagawa, F.; Kobayashi, K.; Miyagawa, K.; Kanoda, K.; Kumai, R.; Murakami, Y.; Tokura, Y. Emergence of nonequilibrium charge dynamics in a charge-cluster glass. *Phys. Rev. B* **2014**, *89*, 121102(R). [CrossRef]
6. Sato, T.; Kagawa, F.; Kobayashi, K.; Ueda, A.; Mori, H.; Miyagawa, K.; Kanoda, K.; Kumai, R.; Murakami, Y.; Tokura, Y. Systematic variations in the charge-glass-forming ability of geometrically frustrated θ-(BEDT-TTF)$_2X$ organic conductors. *J. Phys. Soc. Jpn.* **2014**, *83*, 083602. [CrossRef]
7. Hashimoto, K.; Zhan, S.C.; Kobayashi, R.; Iguchi, S.; Yoneyama, N.; Moriwaki, T.; Ikemoto, Y.; Sasaki, T. Collective excitation of a short-range charge ordering in θ-(BEDT-TTF)$_2$CsZn(SCN)$_4$. *Phys. Rev. B* **2014**, *89*, 085107. [CrossRef]
8. Oike, H.; Kagawa, F.; Ogawa, N.; Ueda, A.; Mori, H.; Kawasaki, M.; Tokura, Y. Phase-change memory function of correlated electrons in organic conductors. *Phys. Rev. B* **2015**, *91*, 041101. [CrossRef]
9. Sato, T.; Miyagawa, K.; Kanoda, K. Electronic crystal growth. *Science* **2017**, *357*, 1378–1381. [CrossRef]
10. Sasaki, S.; Hashimoto, K.; Kobayashi, R.; Itoh, K.; Iguchi, S.; Nishio, Y.; Ikemoto, Y.; Moriwaki, T.; Yoneyama, N.; Watanabe, M.; et al. Crystallization and vitrification of electrons in a glass-forming charge liquid. *Science* **2017**, *357*, 1381–1385. [CrossRef]
11. Mahmoudian, S.; Rademaker, L.; Ralko, A.; Fratini, S.; Dobrosavljević, V. Glassy dynamics in geometrically frustrated Coulomb liquids without disorder. *Phys. Rev. Lett.* **2015**, *115*, 025701. [CrossRef] [PubMed]
12. Mori, H.; Tanaka, S.; Mori, T.; Kobayashi, A.; Kobayashi, H. Crystal structure and physical properties of M = Rb and Tl salts of (BEDT-TTF)$_2$MM'(SCN)$_4$ [M' = Co, Zn]. *Bull. Chem. Soc. Jpn.* **1998**, *71*, 797–806. [CrossRef]
13. Watanabe, M.; Nogami, Y.; Oshima, K.; Mori, H.; Tanaka, S. Novel pressure-induced $2k_F$ CDW state in organic low-dimensional compound θ-(BEDT-TTF)$_2$CsCo(SCN)$_4$. *J. Phys. Soc. Jpn.* **1999**, *68*, 2654–2663. [CrossRef]
14. Nogami, Y.; Pouget, J.P.; Watanabe, M.; Oshima, K.; Mori, H.; Tanaka, S.; Mori, T. Structural modulation in θ-(BEDT-TTF)$_2$CsM'(SCN)$_4$ [M' = Co, Zn]. *Synth. Met.* **1999**, *103*, 1911. [CrossRef]
15. Watanabe, M.; Noda, Y.; Nogami, Y.; Mori, H. Investigation of X-ray diffuse scattering in θ-(BEDT-TTF)$_2$RbM'SCN$_4$. *Synth. Met.* **2003**, *135–136*, 665–666. [CrossRef]
16. Watanabe, M.; Noda, Y.; Nogami, Y.; Mori, H. Transfer integrals and the spatial pattern of charge ordering in θ-(BEDT-TTF)$_2$RbZn(SCN)$_4$ at 90 K. *J. Phys. Soc. Jpn.* **2004**, *73*, 116–122. [CrossRef]
17. Mori, H.; Tanaka, S.; Mori, T. Systematic study of the electronic state in θ-type BEDT-TTF organic conductors by changing the electronic correlation. *Phys. Rev. B* **1998**, *57*, 12023–12029. [CrossRef]
18. Tajima, H.; Kyoden, S.; Mori, H.; Tanaka, S. Estimation of charge-ordering patterns in θ-ET$_2$MM'(SCN)$_4$ (MM' = RbCo, RbZn, CsZn) by reflection spectroscopy. *Phys. Rev. B* **2000**, *62*, 9378–9385. [CrossRef]
19. Miyagawa, K.; Kawamoto, A.; Kanoda, K. Charge ordering in a quasi-two-dimensional organic conductor. *Phys. Rev. B* **2000**, *62*, R7679–R7682. [CrossRef]
20. Suzuki, K.; Yamamoto, K.; Yakushi, K. Charge-ordering transition in orthorhombic and monoclinic single-crystals of θ-(BEDT-TTF)$_2$TlZn(SCN)$_4$ studied by vibrational spectroscopy. *Phys. Rev. B* **2004**, *69*, 085114. [CrossRef]
21. Sawano, F.; Terasaki, I.; Mori, H.; Mori, T.; Watanabe, M.; Ikeda, N.; Nogami, Y.; Noda, Y. An organic thyristor. *Nature* **2005**, *437*, 522. [CrossRef] [PubMed]
22. Takahide, Y.; Konoike, T.; Enomoto, K.; Nishimura, M.; Terashima, T.; Uji, S.; Yamamoto, H.M. Current-voltage characteristics of charge-ordered organic crystals. *Phys. Rev. Lett.* **2006**, *96*, 136602. [CrossRef] [PubMed]
23. Nad, F.; Monceau, P.; Yamamoto, H.M. Effect of cooling rate on charge ordering in θ-(BEDT-TTF)$_2$RbZn(SCN)$_4$. *Phys. Rev. B* **2007**, *76*, 205101. [CrossRef]
24. Nogami, Y.; Hanasaki, N.; Watanabe, M.; Yamamoto, K.; Ito, T.; Ikeda, N.; Ohsumi, H.; Toyokawa, H.; Noda, Y.; Terasaki, I.; et al. Charge order competition leading to nonlinearity in organic thyristor family. *J. Phys. Soc. Jpn.* **2010**, *79*, 044606. [CrossRef]
25. Mori, T. Non-stripe charge order in the θ-phase organic conductors. *J. Phys. Soc. Jpn.* **2003**, *72*, 1469–1475. [CrossRef]

26. Merino, J.; Seo, H.; Ogata, M. Quantum melting of charge order due to frustration in two-dimensional quarter-filled systems. *Phys. Rev. B* **2005**, *71*, 125111. [CrossRef]
27. Kaneko, M.; Ogata, M. Mean-field study of charge order with long periodicity in θ-(BEDT-TTF)$_2$X. *J. Phys. Soc. Jpn.* **2006**, *75*, 014710. [CrossRef]
28. Watanabe, H.; Ogata, M. Novel charge order and superconductivity in two-dimensional frustrated lattice at quarter filling. *J. Phys. Soc. Jpn.* **2006**, *75*, 063702. [CrossRef]
29. Hotta, C.; Furukawa, N. Strong coupling theory of the spinless charges on triangular lattices: Possible formation of a gapless charge-ordered liquid. *Phys. Rev. B* **2006**, *74*, 193107. [CrossRef]
30. Kuroki, K. The origin of the charge ordering and its relevance to superconductivity in θ-(BEDT-TTF)$_2$X: The effect of the Fermi surface nesting and the distant electron-electron interactions. *J. Phys. Soc. Jpn.* **2006**, *75*, 114716. [CrossRef]
31. Nishimoto, S.; Shingai, M.; Ohta, Y. Coexistence of distinct charge fluctuations in θ-(BEDT-TTF)$_2$X. *Phys. Rev. B* **2008**, *78*, 035113. [CrossRef]
32. Dressel, M.; Drichko, N.; Schlueter, J.; Merino, J. Proximity of the layered organic conductors α-(BEDT-TTF)$_2$MHg(SCN)$_4$ (M = K, NH$_4$) to a charge-ordering transition. *Phys. Rev. Lett.* **2003**, *90*, 167002. [CrossRef] [PubMed]
33. Drichko, N.; Dressel, M.; Kuntscher, C.A.; Pashkin, A.; Greco, A.; Merino, J.; Schlueter, J. Electronic properties of correlated metals in the vicinity of a charge-order transition: Optical spectroscopy of α-(BEDT-TTF)$_2$MHg(SCN)$_4$ (M = NH$_4$, Rb, Tl). *Phys. Rev. B* **2006**, *74*, 235121. [CrossRef]
34. Kaiser, S.; Dressel, M.; Sun, Y.; Greco, A.; Schlueter, J.A.; Gard, G.L.; Drichko, N. Bandwidth tuning triggers interplay of charge order and superconductivity in two-dimensional organic materials. *Phys. Rev. Lett.* **2010**, *105*, 206402. [CrossRef]
35. Clay, R.T.; Mazumdar, S.; Campbell, D.K. Charge ordering in θ-(BEDT-TTF)$_2$X materials. *J. Phys. Soc. Jpn.* **2002**, *71*, 1816–1819. [CrossRef]
36. Udagawa, M.; Motome, Y. Charge ordering and coexistence of charge fluctuations in quasi-two-dimensional organic conductors θ-(BEDT-TTF)$_2$X. *Phys. Rev. Lett.* **2007**, *98*, 206405. [CrossRef]
37. Yoshimi, K.; Maebashi, H. Coulomb frustrated phase separation in quasi-two-dimensional organic conductors on the verge of charge ordering. *J. Phys. Soc. Jpn.* **2012**, *81*, 063003. [CrossRef]
38. Naka, M.; Seo, H. Long-period charge correlations in charge-frustrated molecular θ-(BEDT-TTF)$_2$X. *J. Phys. Soc. Jpn.* **2014**, *83*, 053706. [CrossRef]
39. Pustogow, A.; Treptow, K.; Rohwer, A.; Saito, Y.; Alonso, M.S.; Löhle, A.; Schlueter, J.A.; Dressel, M. Charge order in β''-phase BEDT-TTF salts. *Phys. Rev. B* **2019**, *99*, 155144. [CrossRef]
40. Wannier, H. Antiferromagnetism. The triangular Ising net. *Phys. Rev.* **1950**, *79*, 357–364. [CrossRef]
41. Houtappel, R.M.F. Order-disorder in hexagonal lattices. *Physica* **1950**, *16*, 425–455. [CrossRef]
42. Thomas, T.; Saito, Y.; Agarmani, Y.; Thyzel, T.; Hashimoto, K.; Sasaki, T.; Lang, M.; Müller, J. Involvement of structural dynamics in the charge-glass formation in molecular metals. *Phys. Rev. B* **2022**, *105*, L041114. [CrossRef]
43. Thomas, T.; Thyzel, T.; Sun, H.; Müller, J.; Hashimoto, K.; Sasaki, T.; Yamamoto, H.M. Comparison of the charge-crystal and charge-glass state in geometrically frustrated organic conductors studied by fluctuation spectroscopy. *Phys. Rev. B* **2022**, *105*, 205111. [CrossRef]

Article

A Database for Crystalline Organic Conductors and Superconductors

Owen Ganter [1,*,†], Kevin Feeny [1], Morgan Brooke-deBock [1], Stephen M. Winter [2] and Charles C. Agosta [1,*]

1. Department of Physics, Clark University, 950 Main Street, Worcester, MA 01610, USA; kfeeny@clarku.edu (K.F.); mbrookedebock@clarku.edu (M.B.-d.)
2. Department of Physics and Center for Functional Materials, Wake Forest University, 1834 Wake Forest Road, Winston-Salem, NC 27109, USA; winters@wfu.edu
* Correspondence: ganto21@wfu.edu (O.G.); cagosta@clarku.edu (C.C.A.)
† Current address: Department of Physics, Wake Forest University, 1834 Wake Forest Road, Winston-Salem, NC 27109, USA.

Abstract: We present a prototype database for quasi two-dimensional crystalline organic conductors and superconductors based on molecules related to bis(ethylenedithio)tetrathiafulvalene (BEDT-TTF, ET). The database includes crystal structures, calculated electronic structures, and experimentally measured properties such as the superconducting transition temperature and critical magnetic fields. We obtained crystal structures from the Cambridge Structural Database and created a crystal structure analysis algorithm to identify cation molecules and execute tight binding electronic structure calculations. We used manual data entry to encode experimentally measured properties reported in publications. Crystalline organic conductors and superconductors exhibit a wide variety of electronic ground states, particularly those with correlations. We hope that this database will ultimately lead to a better understanding of the fundamental mechanisms of such states.

Keywords: organic superconductor; superconductivity; charge density wave; spin density wave; spin liquid; FFLO state; materials database; data science

1. Introduction

Machine searchable databases that contain structural properties of related materials, calculated electronic structure, and measured electromagnetic properties, are providing a new way to design advanced functional materials. In addition, consolidating structural and functional information may lead to a better understanding of the microscopic mechanisms of correlated electron materials. Herein, we detail the launch of a new database of crystalline organic materials, many of which are conducting or superconducting, with the goal of motivating data-centered research to enhance the understanding of lower dimensional correlated electron materials. The database can be accessed through a website at osd.clarku.edu.

The crystalline organic materials (COM) are well suited to create this type of database first of all because they are interesting experimentally. Partially driven by their low dimensionality, this class of materials exhibits a variety of competing electronic behaviors [1–5] including metallic conductivity, Mott insulators [6–8], antiferromagnetic states [9–11], and superconductivity [12,13]. Other forms of long range charge order have also been observed, such as charge density waves (CDWs) [14–16] and spin density waves (SDWs) [17,18]. More exotic long range order, such as the quantum hall effect [18] and some of the first believable evidence for field induced inhomogeneous superconductivity (the FFLO state) [19–23] were also found in COM. There has also been discussion about the existence of tilted Dirac points [24] and spin liquids [25–27] in these organic salts.

In addition to their rich correlated electron behavior, COMs are easy to access theoretically because they form regular stoichiometric crystals based on a few common cation molecules held together with various anion complexes. It is the electron deficient cation layer that contains holes, which enables itinerant electron behavior within the layer. It is evident that the geometric arrangement of the cation molecules is a principal factor that determines the electronic ground state of the system. This so-called packing of the cation molecules can be altered either by placing different molecules into the anion layer, or by applying external pressure. By adjusting the physical parameters of the cation layer, competing correlated electron states are selectively enabled and a complex electronic phase diagram can be traversed [28–30].

There has been much theoretical progress towards understanding how the various degrees of freedom of these materials lead to the bulk electronic states that are observed [31–33]. From a theoretical perspective, the materials offer a unique window into the physics of correlated electrons, because (i) COMs span the full range of states from interaction-induced insulators to superconductors and metals, (ii) the electronic structures are relatively simple, with only one orbital per molecule typically being relevant, (iii) nearly all the materials are stoichiometric and exhibit a high degree of crystalline order, so that simple models may closely approximate experiments, and (iv) COMs are built from common molecular entities, so that variations in properties may be directly related to structural variations across vast numbers of compounds.

The theoretical study of these materials via the tight binding model, or density functional theory (DFT), is leading to a better understanding of the fundamental physics behind correlated electron systems and of quantum materials in general. Various packing sub-families are known, in which particular structural degrees of freedom, and their relationship with the underlying hopping integrals, are key inputs to predict the behavior of the system [34–37]. Highlights of the strong exchange between experiment and theory include, e.g., quantitative agreement between results of high level calculations (dynamical mean field theory) and measurements of Mott critical scaling [7] and correlation-driven crossovers in the optical response [38].

This new database was built to be used both as a way to find representative materials for targeted experiments investigating particular correlated electron states, or the proximity of competing states, and as a research tool for discovering structure–function relationships [39,40]. An aspirational use of the database would be to predict and design new materials with targeted electronic properties. To serve these functions, the database contains experimentally measured properties, crystal structures, and calculated electronic structures of quasi one and two-dimensional crystalline organic conductors and superconductors. We hope that the enhanced accessibility of information that our database provides will serve the scientific community and lead to new discoveries.

2. Website and Database

When arriving at the home page of the website, the user is presented with a list of available materials and a window in which one may specify a desired type of packing, cation molecule, and anion molecule in order to search for any matches in the database. Leaving a field blank will act as a wildcard. The user can then click on a material to navigate to the home page for that material, which shows its crystal parameters followed by an interactive set of graphics starting with rotatable views of the crystal structure, the calculated electronic structure, and the cation morphology. These views are followed by available measurements of the material properties. Individual pages for each measurement enable the user to view associated information in greater detail. When observing crystal structures, they can be viewed and filtered to include all of the atoms, only the cations, only the anions, or rectangles to represent the cations. Electronic structure diagrams can also be customized by the user to show different k-paths or results from various types of calculations.

In order to identify relevant information to include in the database, the websites of journals were automatically searched for key phrases such as "BEDT-TTF", "organic conductor", and "organic superconductor". The papers resulting from those searches were then recorded as potentially containing measurement information relevant to the database.

Each paper of interest was parsed to determine its relevance, and encode any reported measurements into the database. Candidates for crystal structures were obtained from the Cambridge Structural Database (CSD) by performing a substructure search on the set of known cations. An algorithm was then used to analyze each crystal structure and determine if it was a lower dimensional charge transfer salt of interest. Relevant crystal structures were then added to the database, and their electronic structures were computed automatically. For a smaller selection of materials DFT (WIEN2k) was also used to calculate the electronic structure to compare to the tight binding results. The organization is depicted in Figure 1. The database currently contains 110 materials, 184 crystal structures obtained from the CSD and 440 measured properties. A link is provided for each material to the corresponding CIF entry on the CSD website for the crystal structure information, and a link is provided to the paper where each measurement was found. Below we describe in more detail how the electronic calculations were made, and the methods for encoding the measurements into the database.

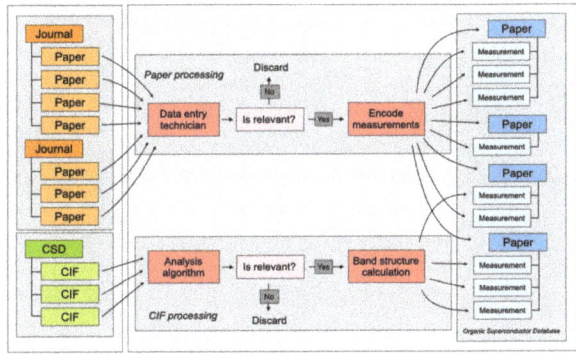

Figure 1. Diagram showing the structure and flow of information into the database.

3. Crystal Structure Analysis Algorithm

To automatically identify relevant crystal structures from the Cambridge Structural Database, we created an algorithm to assess relevance, and to perform some preliminary diagnostics. The algorithm reads a CIF file, which contains a list of the atomic coordinates within the unit cell, and the lattice information. The distance between each pair of atoms is calculated, and if that distance is below the bonding threshold distance for the given atomic species, it is assumed that a bond exists between them. In this manner, the molecules within the unit cell are identified (see Figure 2a,b). The structure of each molecule is then compared to a list of predefined structures of cation molecules such as BEDT-TTF. If the structures match, it is then known that the molecule is a cation molecule of interest. At this point it can be determined whether the material is quasi two-dimensional or not, and if so, what crystal axis is perpendicular to the layers. This is achieved by examining the overlap between cation and non-cation molecules for each Cartesian axis. If along a certain axis there are only overlapping cation molecules with no non-cation molecules, we make the assumption that this axis is in the plane of a conducting layer.

With the orientations and identities of the cations and anions automatically identified, this algorithm can automatically suggest the chemical formula and packing type of a given crystal structure. This is useful in grouping multiple structures for the same compound; however, we did not use this detected chemical formula as the compound label in the database. The name of a crystal structure is entered as denoted in the original paper to ensure that any special chemical naming conventions used by the authors are preserved in the database.

In order to display the packing geometry of the cation molecules to the user, the best fit plane of each cation is computed, and a minimum area rectangle algorithm is applied to generalize a cation molecule as a rectangle in three dimensional space. This type of generalization is common in cartoon diagrams of these materials. The resulting unit cell of cation-rectangles, shown in Figure 2c–f, is useful to inspect the packing geometry of the material. Given that the overlap of the cations is the major determinant of the electronic structure of the crystalline organics, seeing the cations as blocks to visually show the morphology of the crystal symmetry of the cations is instructive. The relative distances and angles of the rectangles can then be calculated to quantitatively analyze the packing. We also generated two-dimensional diagrams, which more simply showed the geometric orientation of the cation molecules within the conducting layers based on the angles calculated before.

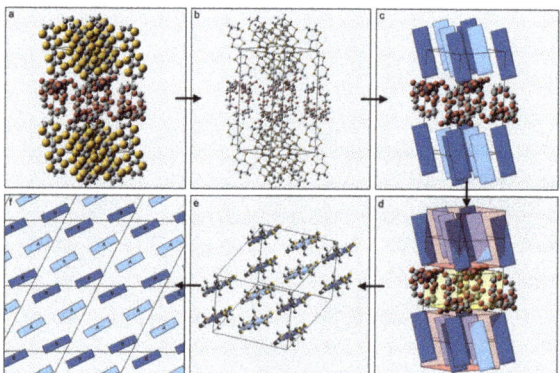

Figure 2. Views of various stages of the crystal structure analysis algorithm. The individual images show the initial unit cell consisting of atoms within a lattice (**a**), detected molecules (**b**), detected cation molecules shown as rectangles (**c**), detected anisotropy and layers (**d**), a new unit cell consisting of only the cation layer (**e**), and a two-dimensional depiction of the cation layer (**f**). In this example, the material is β''-(ET)$_4$[(H$_3$O)Cr(C$_2$O$_4$)$_3$]$_2$[(H$_3$O)$_2$]$_5$H$_2$O, as reported in [41].

4. High-Throughput Electronic Structure Calculations

High-throughput electronic structure calculations using density functional theory (DFT) have found applications in several materials databases; however, the large unit cells of these materials and low crystal symmetries make full-scale DFT calculations with plane-wave basis sets computationally expensive. This is especially problematic for cases with open-shell anions, which feature localized unpaired electrons. Unless the local correlations in the anion layer are treated explicitly (via DFT + U), anion bands may appear erroneously near the Fermi energy, yielding incorrect Fermi surfaces. More importantly however, many of our crystal structures have missing or disordered atoms, especially in the anion layer. This is particularly prevalent in anion layers that contain solvents, which are often disordered across unit cells. It is therefore necessary to reduce the computational expense and focus exclusively on the cation layers.

To calculate the electronic structure of every crystal structure in our database, we construct a two-dimensional tight binding model [42] for highest occupied molecular orbitals in the layer of cation molecules using a series of local DFT calculations. Solving the tight-binding (TB) model produces the band structure. To carry out the calculation, we used the crystal structure analysis algorithm previously described to identify all symmetrically equivalent molecules and pairs of molecules with the cation layers. We then used the method employed in [43] to estimate tight-binding hopping integrals using quantum chemistry packages (in this case ORCA [44]). Results are shown in Figure 3, for the example of α-(ET)$_2$KHg(SCN)$_4$. The method is based on calculations on pairs of molecules in which the local crystal environment is otherwise ignored, which significantly reduces computational expense. For this purpose, we used basis sets including 3-21G, 6-31G, 6-311G, and def2-SVP in conjunction with the B3LYP hybrid density functional. Localized Wannier molecular orbitals (MOs) are constructed for each molecule via maximizing the overlap with the corresponding orbital of the isolated molecules. The procedure is as follows:

1. **Obtain Isolated MOs:** For each molecular pair (labeled i, j), a calculation is first performed on the isolated molecules. From this, the MO coefficients (in the basis of Gaussian atomic orbitals) for each molecule are obtained as Φ_i^0 and Φ_j^0. These are combined as:

$$\Phi_0 = \begin{pmatrix} \Phi_i^0 & 0 \\ 0 & \Phi_j^0 \end{pmatrix} \quad (1)$$

2. **Construct Wannier Functions:** For each molecular pair, a calculation is then performed in the geometry corresponding to the crystal structure. This produces the diagonal MO energies **E**, the overlap matrix **S**, and the MO coefficients Φ. In ORCA, **S** is output in the atomic orbital basis. It is first rotated into the basis of the isolated MOs:

$$\tilde{\mathbf{S}} = \Phi_0 \, \mathbf{S} \, \Phi_0^\dagger \quad (2)$$

In this geometry, the basis of isolated MOs are no longer orthonormal. Thus, the local Wannier functions are constructed via symmetric orthornormalization, $\bar{\Phi}_0 = \tilde{\mathbf{S}}^{-1/2} \Phi_0$.

3. **Rotate Fock Matrix:** The diagonal orbital energies are then rotated into the above-defined localized MOs:

$$\mathbf{F} = \bar{\Phi}_0 \, \Phi^{-1} \, \mathbf{E} \, (\Phi^\dagger)^{-1} \, \bar{\Phi}_0^\dagger \quad (3)$$

The resulting Fock matrix has the structure:

$$\mathbf{F} = \begin{pmatrix} \mathbf{F}_{ii} & \mathbf{F}_{ij} \\ \mathbf{F}_{ji} & \mathbf{F}_{jj} \end{pmatrix} \quad (4)$$

The on-site terms $\mathbf{F}_{ii}$ and $\mathbf{F}_{jj}$ now contain both the diagonal Wannier orbital energies, and small off-diagonal "crystal field" contributions. It is advantageous to remove the latter terms via unitary transformation:

$$\bar{\mathbf{F}} = \begin{pmatrix} \mathbf{U}_i & 0 \\ 0 & \mathbf{U}_j \end{pmatrix} \mathbf{F} \begin{pmatrix} \mathbf{U}_i^\dagger & 0 \\ 0 & \mathbf{U}_j^\dagger \end{pmatrix} \quad (5)$$

where $\bar{\mathbf{F}}_{ii} = \mathbf{U}_i \, \mathbf{F}_{ii} \, \mathbf{U}_i^\dagger$ and $\bar{\mathbf{F}}_{jj} = \mathbf{U}_j \, \mathbf{F}_{jj} \, \mathbf{U}_j^\dagger$ are diagonal. The intersite hoppings can then be read from $\bar{\mathbf{F}}_{ij} = \mathbf{U}_i \, \mathbf{F}_{ij} \, \mathbf{U}_j^\dagger$.

We note, because this latter unitary transformation is different for every molecular pair, the hopping integrals obtained for different pairs represent slightly different definitions of the local Wannier functions. Nonetheless, this approximation is no more severe than the pairwise construction inherent to the method. Although this approach neglects the anion layer, the results agree well with full-scale calculations performed with Wien2k (at the GGA level) and experimental electronic structure as well (see Figure 3c); however, the former approach is much faster. With Wien2k, for example, a full DFT calculation using GGA functionals can take several days to complete (with 100 processors), while construction of a TB model with ORCA calculations takes approximately five minutes per compound (with 10 processors), even when using more expensive hybrid functionals. Such a speed-up is desirable when making high-throughput calculations for each crystal structure entry in the database.

Because the pairwise calculations are made separately, we had to adjust the signs of the resulting charge transfer integrals such that the phase of the molecular orbital on each symmetrically equivalent molecule was the same. We used the centroids of the cation molecules for the positions of the sites, disregarding the out of layer component. The filling of each model was deduced from the stoichiometry and charge of the cation molecules. Solving the tight binding Hamiltonian at each point in k-space produced the energy eigenvalues that constitute the electronic structure. In this manner, the band structure, Fermi surface, and density of states are automatically computed and may be viewed on the website. Users can interact with these data directly by selecting the k-path to use and which basis sets to display. In addition, the computed hopping integrals, and their locations in the unit cell are presented to the user in order to serve as a basis for further theoretical modeling.

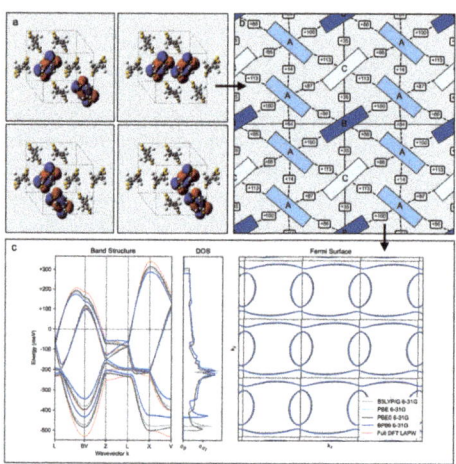

Figure 3. Views of various stages of the tight-binding electronic structure calculation. The individual images show the quantum chemistry calculation of inter-molecular charge transfer integrals (**a**), site-hopping integrals model (**b**), and electronic structure (**c**). In this example the material is α-(ET)$_2$KHg(SCN)$_4$, as reported in [39]. Note that the WIEN2k band structure in red is close to the TB band structures, especially near the Fermi level.

5. Measurements

- Locations of phase transitions:
    - Metal insulator.
    - Superconductivity, T_c, H_{c1}, H_{c2}, H_P, (where H_P is the Pauli paramagnetic limit).
    - Charge density wave.
    - Spin density wave.
    - Magnetic ordering.

  As a function of:
    - Temperature.
    - Magnetic field.
    - Pressure.
- Lattice parameters.
- Conductivity.
- London penetration depth.
- Coherence length.
- Shubnikov—de Haas and de Haas—van Alphen frequencies.
- Effective mass.
- Dingle temperature.
- Scattering time.

Each measurement entry in the database consists of three blocks of information: a block specifying the state of the system being measured, a block specifying the value of the measurement, including the error bars if available, and a block specifying the method by which the measurement was made. Our goal behind the data entry is to create a digital copy of measurement information in the precise manner in which it was specified by the authors who made the measurement. We implemented support for as many measurements as we could for this process. For example, the ability to specify a numeric value using an exact decimal, a range between two decimals, or an average decimal with a plus or minus value. Although laborious, we found that manual data entry was the most reliable way to extract measurement information from papers. We used a web application on the website for this purpose, shown in Figure 4.

Once we had a sufficient number of measurements, there were many decisions that needed to be made about how they were presented. For a deep understanding of a single material, it is necessary to see a detailed view of the actual measurements labeled by the method of measurement; for example, the superconducting transition temperature, T_c, found by resistance or specific heat, and the point on the transition curve, e.g., onset or midpoint, used to locate the transition, with citations for each measurement. For that reason, the details of the measurement method and the error bars, if given, are stored in the database. For that reason, it is also necessary to present an average value for a material when a number of materials are being compared to each other. We made the decision to discount some of the grossly outlying measurements in cases where we thought the data were not convincing. These rules for curating the data are constantly being reconsidered to present the most useful data to the community; however, the full collection of measurements will always be available so that a user of the database can analyze the published measurements with their own algorithms.

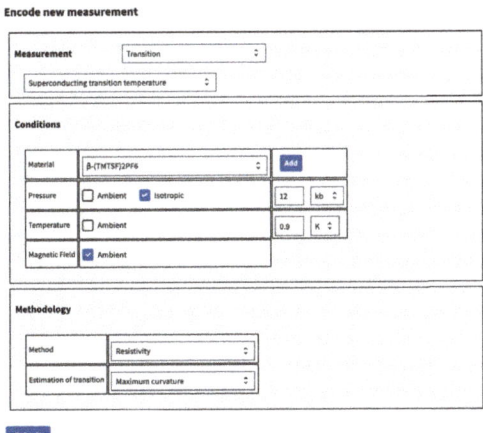

Figure 4. View of the web application used for data entry. In this example, the T_c of β-(TMTSF)$_2$PF$_6$ under 12 kbar of applied pressure is specified to be 0.9 Kelvin as reported in [12].

6. Discussion and Future Outlook

We present this database of crystalline organic conductors and superconductors as an evolving tool that will be continuously updated with new materials, features, and metrics. The goal of gathering all relevant calculation and measurement information into one central location is an arduous one, but the progress that we have made so far illustrates that it is possible. In further developing this database, we have two main goals.

Our first main goal is to populate the database with as many entries as possible. The limiting step in this process for inclusion of experimental measurements is data entry. Manual data entry of measurements from scientific articles is the only method with a high enough degree of reliability to be useful in our database. We have used members of our laboratory to perform data entry, and have trained undergraduate students as well. Currently, our database includes only a small fraction of all the relevant data that exist. In order to increase the number of measurements in our database, we will need to increase the size of our data entry team. We are considering crowdsourcing the process so that verified database users from across the world can also contribute. In addition, verified users will have the option of submitting CIF files for automatic calculation of electronic structure and tight-binding parameters. We would also like to populate the database with results from explicitly correlated theoretical methods suitable for high-throughput applications (such as density matrix embedding theory [45]). Presently, a tight-binding electronic structure is provided for each crystal structure in the database. For some of these crystal structures, missing or corrupt atoms in the anion layer prevent the use of a full LAPW DFT calculation; however, we eventually plan to include full LAPW DFT calculations for as many of the crystal structures in the database as possible. Any persons interested in becoming involved with the project can click on the orange button at the bottom of the home page to request an account. An option is also available to provide anonymous feedback.

Our second main goal regarding development of the database is to implement new features. We plan to add an interactive web interface by which users can analyze the contents of the database as a whole by correlating various calculated and measured properties. We also would like to improve the search feature of the website so that more detailed searches can be performed. There are many different avenues by which our existing work can be further developed. We are trying to create as many tools as possible to perform simple visualization and analysis of data online, such as the feature shown in Figure 5. A number of parameters can be extracted from the band structure calculations, such as the density of states at the Fermi level, and the cross sectional area of the Fermi surface. We are working on finding robust universal algorithms to calculate these and other representative values. Given this collection of the unit cell parameters, extracted electronic values, and measurement parameters, any set of data can be graphed against any other set of data, and scatter plots can be created including markers labeled with the material names. It is also possible to combine parameters with common arithmetic operations to create additional metrics. We will continue to enhance the user interface to create a more versatile and expansive analysis interface.

Following the invention and widespread availability of computers, an increasing trend towards digitization in science has taken place. Scientific databases have emerged in practically every area of study because they enable the analysis of many pieces of information, and the distribution of that information to individuals around the world. The field of data science has also grown to develop new ways of analyzing the large amount of data available. Many databases for materials science currently exist, particularly those that focus on electronic structure, crystal structure, and other measured properties [46]. Our inspiration to create this database of organic conductors and superconductors was drawn in part by the success of other databases containing density functional theory electronic structure calculations for many crystal structures [47–50]. Our goal is to gather as many different types and pieces of information related to quasi two-dimensional organic conductors and superconductors as possible. We foresee this database as having a number of different applications. Primarily, it will serve as a useful reference tool for the scientific community of organic conductors and superconductors. Database users can easily find and view crystal structures, electronic structures, and other measured properties for the materials that they are interested in. We hope to cultivate a community of scientists from across the world who are interested in using the database, contributing data to the database, and requesting new features for the database. We also look forward to analyzing the data that are stored in the database. Many techniques in the field of data science are appropriate for this purpose. In particular, certain types of data mining and machine learning have proven useful in the analysis of other materials databases [51–54]. We hope that the identification of trends between various parameters will ultimately lead to a better understanding of the fundamental mechanisms of correlated electron systems in quasi two-dimensional organic conductors and superconductors.

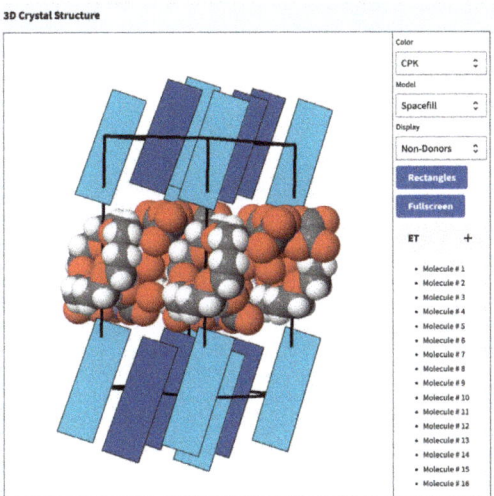

Figure 5. View of the 3D in-browser crystal structure analysis tool. The section on the right provides interactive features to the users. In this example the material is $\beta''\text{-(ET)}_4[(H_3O)Cr(C_2O_4)_3]_2[(H_3O)_2]_5H_2O$, as reported in [41].

Author Contributions: Conceptualization, O.G. and C.C.A.; Methodology, O.G., K.F., M.B.-d. and S.M.W.; Software, O.G.; Validation, K.F. and M.B.-d.; Investigation, O.G.; Resources, C.C.A. and S.M.W.; Data Curation, K.F., M.B.-d. and O.G.; Writing—Original Draft Preparation, O.G., C.C.A. and S.M.W.; Supervision, C.C.A.; Project Administration, C.C.A.; Funding Acquisition, C.C.A. and S.M.W. All authors have read and agreed to the published version of the manuscript.

Funding: This research was funded by NSF grant number DMR-1905950. S.M.W. and O.G. also acknowledge support from a pilot grant from the Center for Functional Materials, Wake Forest University.

Institutional Review Board Statement: Not applicable.

Informed Consent Statement: Not applicable.

Data Availability Statement: The data presented in this study are openly available in the described database, accessible via osd.clarku.edu.

Acknowledgments: The authors acknowledge discussions with Ben Powell. The authors thank the following people for their work in manually adding measurement entries to the database: Alireza Alipour, Abdulai Gassama, Ahad Ali Khan, Brett Laramee, Gwynnevieve Ramsey, Jade Consalvi, Meherab Hossain, and Raju Ghimire. Computations were performed on the Clark University High Performance Computing Cluster and the Wake Forest University DEAC Cluster, a centrally managed resource with support provided in part by Wake Forest University.

Conflicts of Interest: The authors declare no conflict of interest.

References

1. Jérome, D. Organic conductors: From charge density wave TTF- TCNQ to superconducting $(TMTSF)_2PF_6$. *Chem. Rev.* **2004**, *104*, 5565–5592. [CrossRef] [PubMed]
2. Kobayashi, H.; Cui, H.; Kobayashi, A. Organic metals and superconductors based on BETS (BETS = bis (ethylenedithio) tetraselenafulvalene). *Chem. Rev.* **2004**, *104*, 5265–5288. [CrossRef] [PubMed]
3. Miyagawa, K.; Kanoda, K.; Kawamoto, A. NMR studies on two-dimensional molecular conductors and superconductors: Mott transition in κ-(BEDT-TTF)$_2$X. *Chem. Rev.* **2004**, *104*, 5635–5654. [CrossRef] [PubMed]

4. Geiser, U.; Schlueter, J.A. Conducting organic radical cation salts with organic and organometallic anions. *Chem. Rev.* **2004**, *104*, 5203–5242. [CrossRef]
5. Kanoda, K.; Kato, R. Mott physics in organic conductors with triangular lattices. *Annu. Rev. Condens. Matter Phys.* **2011**, *2*, 167–188. [CrossRef]
6. Powell, B.; McKenzie, R.H. Strong electronic correlations in superconducting organic charge transfer salts. *J. Phys. Condens. Matter* **2006**, *18*, R827. [CrossRef]
7. Furukawa, T.; Miyagawa, K.; Taniguchi, H.; Kato, R.; Kanoda, K. Quantum criticality of Mott transition in organic materials. *Nat. Phys.* **2015**, *11*, 221–224. [CrossRef]
8. Gati, E.; Garst, M.; Manna, R.S.; Tutsch, U.; Wolf, B.; Bartosch, L.; Schubert, H.; Sasaki, T.; Schlueter, J.A.; Lang, M. Breakdown of Hooke's law of elasticity at the Mott critical endpoint in an organic conductor. *Sci. Adv.* **2016**, *2*, e1601646. [CrossRef]
9. Enoki, T.; Miyazaki, A. Magnetic TTF-based charge-transfer complexes. *Chem. Rev.* **2004**, *104*, 5449–5478. [CrossRef]
10. Naka, M.; Ishihara, S. Magnetoelectric effect in organic molecular solids. *Sci. Rep.* **2016**, *6*, 20781. [CrossRef]
11. Itaya, M.; Eto, Y.; Kawamoto, A.; Taniguchi, H. Antiferromagnetic Fluctuations in the Organic Superconductor κ-(BEDT-TTF)$_2$Cu(NCS)$_2$ under Pressure. *Phys. Rev. Lett.* **2009**, *102*, 227003. [CrossRef]
12. Jérome, D.; Mazaud, A.; Ribault, M.; Bechgaard, K. Superconductivity in a synthetic organic conductor (TMTSF)$_2$PF$_6$. *J. Phys. Lett.* **1980**, *41*, 95–98. [CrossRef]
13. Wosnitza, J. Superconductivity of Organic Charge-Transfer Salts. *J. Low Temp. Phys.* **2019**, *197*, 250–271. [CrossRef]
14. Latt, K.Z.; Schlueter, J.A.; Darancet, P.; Hla, S.W. Two-Dimensional Molecular Charge Density Waves in Single-Layer-Thick Islands of a Dirac Fermion System. *ACS Nano* **2020**, *14*, 8887–8893. [CrossRef]
15. Andres, D.; Kartsovnik, M.V.; Biberacher, W.; Neumaier, K.; Sheikin, I.; Müller, H.; Kushch, N.D. Field-induced charge-density-wave transitions in the organic metal α-(BEDT-TTF)$_2$KHg(SCN)$_4$ under pressure. *Low Temp. Phys.* **2011**, *37*, 762–770. [CrossRef]
16. Kondo, R.; Higa, M.; Kagoshima, S.; Hanasaki, N.; Nogami, Y.; Nishikawa, H. Interplay of charge-density waves and superconductivity in the organic conductor β-(BEDT-TTF)$_2$AuBr$_2$. *Phys. Rev. B* **2010**, *81*, 024519. [CrossRef]
17. Sasaki, T.; Lebed, A.; Fukase, T.; Toyota, N. Magnetic field response of the spin density wave in α-(BEDT-TTF)$_2$KHg(SCN)$_4$. *Synth. Met.* **1997**, *86*, 2063–2064. [CrossRef]
18. Kang, W.; Hannahs, S.T.; Chaikin, P.M. Toward a unified phase diagram in (TMTSF)$_2$X. *Phys. Rev. Lett.* **1993**, *70*, 3091–3094. [CrossRef]
19. Tanatar, M.A.; Ishiguro, T.; Tanaka, H.; Kobayashi, H. Magnetic field–temperature phase diagram of the quasi-two-dimensional organic superconductor λ-(BETS)$_2$GaCl$_4$ studied via thermal conductivity. *Phys. Rev. B* **2002**, *66*, 134503. [CrossRef]
20. Shimahara, H. Theory of the Fulde–Ferrell–Larkin–Ovchinnikov State and Application to Quasi-Low-dimensional Organic Superconductors. In *The Physics of Organic Superconductors and Conductors*; Lebed, A., Ed.; Springer: Berlin/Heidelberg, Germany, 2008; pp. 687–704. [CrossRef]
21. Cho, K.; Smith, B.E.; Coniglio, W.A.; Winter, L.E.; Agosta, C.C.; Schlueter, J.A. Upper critical field in the organic superconductor β''-(ET)$_2$SF$_5$CH$_2$CF$_2$SO$_3$: Possibility of Fulde-Ferrell-Larkin-Ovchinnikov state. *Phys. Rev. B* **2009**, *79*, 220507. [CrossRef]
22. Agosta, C. Inhomogeneous Superconductivity in Organic and Related Superconductors. *Crystals* **2018**, *8*, 285. [CrossRef]
23. Sugiura, S.; Isono, T.; Terashima, T.; Yasuzuka, S.; Schlueter, J.A.; Uji, S. Fulde-Ferrell-Larkin-Ovchinnikov and vortex phases in a layered organic superconductor. *NPJ Quantum Mater.* **2019**, *4*, 1–6. [CrossRef]
24. Hirata, M.; Ishikawa, K.; Miyagawa, K.; Tamura, M.; Berthier, C.; Basko, D.; Kobayashi, A.; Matsuno, G.; Kanoda, K. Observation of an anisotropic Dirac cone reshaping and ferrimagnetic spin polarization in an organic conductor. *Nat. Commun.* **2016**, *7*, 1–14. [CrossRef]
25. Shimizu, Y.; Miyagawa, K.; Kanoda, K.; Maesato, M.; Saito, G. Spin liquid state in an organic Mott insulator with a triangular lattice. *Phys. Rev. Lett.* **2003**, *91*, 107001. [CrossRef] [PubMed]
26. Isono, T.; Kamo, H.; Ueda, A.; Takahashi, K.; Kimata, M.; Tajima, H.; Tsuchiya, S.; Terashima, T.; Uji, S.; Mori, H. Gapless Quantum Spin Liquid in an Organic Spin-1/2 Triangular-Lattice κ-H$_3$(Cat-EDT-TTF)$_2$. *Phys. Rev. Lett.* **2014**, *112*, 177201. [CrossRef] [PubMed]
27. Miksch, B.; Pustogow, A.; Rahim, M.J.; Bardin, A.A.; Kanoda, K.; Schlueter, J.A.; Hübner, R.; Scheffler, M.; Dressel, M. Gapped magnetic ground state in quantum spin liquid candidate κ-(BEDT-TTF)$_2$Cu$_2$(CN)$_3$. *Science* **2021**, *372*, 276–279. [CrossRef] [PubMed]
28. Saito, Y.; Rösslhuber, R.; Löhle, A.; Sanz Alonso, M.; Wenzel, M.; Kawamoto, A.; Pustogow, A.; Dressel, M. Chemical tuning of molecular quantum materials κ-[(BEDT-TTF)$1-x$(BEDT-STF)x]$_2$Cu$_2$(CN)$_3$: from the Mott-insulating quantum spin liquid to metallic Fermi liquid. *J. Mater. Chem. C* **2021**, *9*, 10841–10850. [CrossRef]
29. Faltermeier, D.; Barz, J.; Dumm, M.; Dressel, M.; Drichko, N.; Petrov, B.; Semkin, V.; Vlasova, R.; Mézière, C.; Batail, P. Bandwidth-controlled Mott transition in κ-(BEDT-TTF)$_2$Cu[N(CN)$_2$]Br$_x$Cl$_{1-x}$: Optical studies of localized charge excitations. *Phys. Rev. B* **2007**, *76*, 165113. [CrossRef]
30. Kanoda, K. Mott Transition and Superconductivity in Q2D Organic Conductors. In *The Physics of Organic Superconductors and Conductors*; Lebed, A., Ed.; Springer: Berlin/Heidelberg, Germany, 2008; pp. 623–642. [CrossRef]
31. Seo, H.; Hotta, C.; Fukuyama, H. Toward systematic understanding of diversity of electronic properties in low-dimensional molecular solids. *Chem. Rev.* **2004**, *104*, 5005–5036. [CrossRef]

32. Lebed, A.G. *The Physics of Organic Superconductors and Conductors*; Springer: Berlin/Heidelberg, Germany, 2008; Volume 110. [CrossRef]
33. Powell, B.; McKenzie, R.H. Quantum frustration in organic Mott insulators: From spin liquids to unconventional superconductors. *Rep. Prog. Phys.* **2011**, *74*, 056501. [CrossRef]
34. Mori, T. Structural Genealogy of BEDT-TTF-Based Organic Conductors I. Parallel Molecules: β and β'' Phases. *Bull. Chem. Soc. Jpn.* **1998**, *71*, 2509–2526. [CrossRef]
35. Mori, T.; Mori, H.; Tanaka, S. Structural Genealogy of BEDT-TTF-Based Organic Conductors II. Inclined Molecules: θ, α, and κ Phases. *Bull. Chem. Soc. Jpn.* **1999**, *72*, 179–197. [CrossRef]
36. Mori, T. Structural Genealogy of BEDT-TTF-Based Organic Conductors III. Twisted Molecules: δ and α' Phases. *Bull. Chem. Soc. Jpn.* **1999**, *72*, 2011–2027. [CrossRef]
37. Hotta, C. Classification of Quasi-Two Dimensional Organic Conductors Based on a New Minimal Model. *J. Phys. Soc. Jpn.* **2003**, *72*, 840–853. [CrossRef]
38. Ferber, J.; Foyevtsova, K.; Jeschke, H.O.; Valentí, R. Unveiling the microscopic nature of correlated organic conductors: The case of κ-(ET)$_2$Cu[N(CN)$_2$]Br$_x$Cl$_{1-x}$. *Phys. Rev. B* **2014**, *89*, 205106. [CrossRef]
39. Mori, H.; Tanaka, S.; Oshima, M.; Saito, G.; Mori, T.; Maruyama, Y.; Inokuchi, H. Crystal and Electronic Structures of (BEDT–TTF)$_2$[MHg(SCN)$_4$](M=K and NH$_4$). *Bull. Chem. Soc. Jpn.* **1990**, *63*, 2183–2190. [CrossRef]
40. Caulfield, J.; Lubczynski, W.; Pratt, F.L.; Singleton, J.; Ko, D.Y.K.; Hayes, W.; Kurmoo, M.; Day, P. Magnetotransport studies of the organic superconductor κ-(BEDT-TTF)$_2$Cu(NCS)$_2$ under pressure: The relationship between carrier effective mass and critical temperature. *J. Phys. Condens. Matter* **1994**, *6*, 2911. [CrossRef]
41. Rashid, S.; Turner, S.S.; Day, P.; Light, M.E.; Hursthouse, M.B.; Firth, S.; Clark, R.J.H. The first molecular charge transfer salt containing proton channels. *Chem. Commun.* **2001**, 1462–1463. [CrossRef]
42. Mori, T.; Kobayashi, A.; Sasaki, Y.; Kobayashi, H.; Saito, G.; Inokuchi, H. The Intermolecular Interaction of Tetrathiafulvalene and Bis(ethylenedithio)tetrathiafulvalene in Organic Metals. Calculation of Orbital Overlaps and Models of Energy-band Structures. *Bull. Chem. Soc. Jpn.* **1984**, *57*, 627–633. [CrossRef]
43. Winter, S.M.; Riedl, K.; Valenti, R. Importance of spin-orbit coupling in layered organic salts. *Phys. Rev. B* **2017**, *95*, 060404. [CrossRef]
44. Neese, F. Software update: The ORCA program system, version 4.0. *Wiley Interdiscip. Rev. Comput. Mol. Sci.* **2018**, *8*, e1327. [CrossRef]
45. Knizia, G.; Chan, G.K.L. Density matrix embedding: A simple alternative to dynamical mean-field theory. [CrossRef]
46. Lin, L. Materials Databases Infrastructure Constructed by First Principles Calculations: A Review. *Mater. Perform. Charact.* **2015**, *4*. [CrossRef]
47. Borysov, S.S.; Geilhufe, R.M.; Balatsky, A.V. Organic materials database: An open-access online database for data mining. *PLoS ONE* **2017**, *12*, 1–14. [CrossRef]
48. Jain, A.; Ong, S.P.; Hautier, G.; Chen, W.; Richards, W.D.; Dacek, S.; Cholia, S.; Gunter, D.; Skinner, D.; Ceder, G.; et al. The Materials Project: A materials genome approach to accelerating materials innovation. *APL Mater.* **2013**, *1*, 011002. [CrossRef]
49. Topological Materials Database. Available online: https://www.topologicalquantumchemistry.org/ (accessed on 23 April 2022).
50. Rutgers DFT & DMFT Materials Database. Available online: http://hauleweb.rutgers.edu/database_w2k/ (accessed on 23 April 2022).
51. Stanev, V.; Oses, C.; Kusne, A.G.; Rodriguez, E.; Paglione, J.; Curtarolo, S.; Takeuchi, I. Machine learning modeling of superconducting critical temperature. *NPJ Comput. Mater.* **2018**, *4*, 29. [CrossRef]
52. Xie, S.R.; Stewart, G.R.; Hamlin, J.J.; Hirschfeld, P.J.; Hennig, R.G. Functional form of the superconducting critical temperature from machine learning. *Phys. Rev. B* **2019**, *100*, 1–6. [CrossRef]
53. Geilhufe, R.M.; Borysov, S.S.; Kalpakchi, D.; Balatsky, A.V. Towards novel organic high-T_c superconductors: Data mining using density of states similarity search. *Phys. Rev. Mater.* **2018**, *2*, 024802. [CrossRef]
54. Zhang, Y.; Ling, C. A strategy to apply machine learning to small datasets in materials science. *NPJ Comput. Mater.* **2018**, *4*, 25. [CrossRef]

Article

How to Recognize the Universal Aspects of Mott Criticality?

Yuting Tan [1,*], Vladimir Dobrosavljević [1,*] and Louk Rademaker [2,*]

1. Department of Physics and National High Magnetic Field Laboratory, Florida State University, Tallahassee, FL 32306, USA
2. Department of Theoretical Physics, University of Geneva, 1211 Geneva, Switzerland
* Correspondence: ytan@magnet.fsu.edu (Y.T.); vlad@magnet.fsu.edu (V.D.); louk.rademaker@gmail.com (L.R.)

Abstract: In this paper we critically discuss several examples of two-dimensional electronic systems displaying interaction-driven metal-insulator transitions of the Mott (or Wigner–Mott) type, including dilute two-dimension electron gases (2DEG) in semiconductors, Mott organic materials, as well as the recently discovered transition-metal dichalcogenide (TMD) moiré bilayers. Remarkably similar behavior is found in all these systems, which is starting to paint a robust picture of Mott criticality. Most notable, on the metallic side a resistivity maximum is observed whose temperature scale vanishes at the transition. We compare the available experimental data on these systems to three existing theoretical scenarios: spinon theory, Dynamical Mean Field Theory (DMFT) and percolation theory. We show that the DMFT and percolation pictures for Mott criticality can be distinguished by studying the origins of the resistivity maxima using an analysis of the dielectric response.

Keywords: Mott transition; quantum criticality; resistivity maxima; dielectric response; dilute 2DEGs; Mott organics; twisted transition-metal dichalcogenide bilayers; dynamical mean field theory; percolation theory; spinon theory

Citation: Tan, Y.; Dobrosavljević, V.; Rademaker, L. How to Recognize the Universal Aspects of Mott Criticality? *Crystals* **2022**, *12*, 932. https://doi.org/10.3390/cryst12070932

Academic Editor: Dmitri Donetski

Received: 4 May 2022
Accepted: 3 June 2022
Published: 30 June 2022

Publisher's Note: MDPI stays neutral with regard to jurisdictional claims in published maps and institutional affiliations.

Copyright: © 2022 by the authors. Licensee MDPI, Basel, Switzerland. This article is an open access article distributed under the terms and conditions of the Creative Commons Attribution (CC BY) license (https:// creativecommons.org/licenses/by/ 4.0/).

1. Introduction

The physics of strongly correlated matter has many faces. Still, for a majority of systems the underlying theme is the role of *"Mottness"* [1]. It is clear that if one aspect of strong correlations should be understood first, it should be the fundamental nature of the Mott metal-insulator transition [2]. Its simplest reincarnation is the transition induced by tuning the bandwidth at half-filling, a setup that produced rather spectacular advances in recent years. Several systems were identified as nearly-ideal realizations of this paradigm, allowing systematic study using a wide arsenal of experimental probes.

In this article we present an overview of three classes of two-dimensional experimental systems that exhibit bandwidth-controlled Mott criticality: dilute two-dimensional electron gases in semiconductors, "Mott-organic" compounds, and transition-metal dichalcogenide moiré systems. Thereby we aim to present the experimental facts as "bland" as possible, in Section 3, without favoring one or the other theoretical explanation. The remarkable similarities between these model systems suggests a robust universality, including characteristic behavior such as the appearance of resistivity maxima.

Possible explanations of two-dimensional Mott criticality follow in the section thereafter (Section 4), where the experimental distinguishable features of each theory takes the forefront. This is followed by a separate discussion of the largely-overlooked utility of dielectric spectroscopy in Section 5, in not only identifying phase segregation and spatial inhomogeneity, but also in revealing the thermal destruction of coherent quasiparticles associated with Landau's Fermi liquid theory.

However, first we need to address the demarcation of our topic. What makes the metal-to-insulator transition in these systems stand out from 'traditional' metal-to-insulator transitions [3,4]?

2. In Search of Mott Criticality

Condensed matter physics, or recently for sales purposes re-branded as "quantum matter physics", is the study of electric and magnetic properties of materials that surround us. The grand question that trumps all others is: how to understand which materials conduct electricity and which are insulating? Traditionally, a conducting material is called a *metal*—not to be confused with the chemical, metallurgical or astronomical meaning of that word. Our main question (metal or insulator?) has not only tremendous technological applications (in fact, all modern electronic technology depends on our ability to rapidly switch materials between metallic and insulating behavior), but also requires a thorough understanding of the problem of emergent behavior of many interacting quantum-mechanical electrons and ions.

Only at zero temperature does there exist a sharp difference between insulators and metals [4]. There are three distinct possibilities: zero conductivity $\sigma(T=0)=0$ means insulating; zero resistivity $\rho(T=0)=0$ means a superconductor; and anything in between is a metal $\sigma(T=0)=1/\rho(T=0)\neq 0$. At any nonzero temperature, an insulator typically has activated behavior $\rho(T)\sim e^{\Delta/T}$ whereas the standard Fermi liquid theory of a metal predicts a temperature-squared increase of the resistivity $\rho(T)=\rho_0+AT^2$. It has therefore become common-place to use the *derivative* of the resistivity $d\rho/dT$ as a measure of whether something is conducting ($d\rho/dT>0$) or insulating ($d\rho/dT<0$)—but this is highly misleading! As we will show later in Section 3, close to a Mott metal-insulator transition we often find non-monotonic behavior of the resistivity as a function of temperature, making the 'derivative' criterion useless. Even worse, there exist cases where the resistivity has $d\rho/dT<0$ but at zero temperature it does not diverge, signalling that this is not a true insulator (see e.g., [5,6]). Another example is the case of Mooij correlations [7,8], where the temperature-derivative of the resistivity in a metal can become negative. Consequently, since only at zero temperature the insulator/metal distinction is well-defined, we must stick with that definition. Regardless of the slope, a material is a metal if its resistivity *does not diverge* as $T \to 0$.

Many materials can be understood within the framework of *band theory* and its extensions such as Fermi liquid, Boltzmann transport, and density functional theory. This framework provides a very simple answer to the metal-or-insulator question: if the Fermi level lies in the middle of a band gap, the system is insulating; otherwise, the system behaves as a metal. This concept has the important consequence that for a crystalline material with (up to some weak disorder) well-defined unit cells, insulators can only appear when there is an even number of electrons per unit cell. Consequently, within the band theory picture, there exist only three possible routes to induce a metal-to-insulator transition: by changing the electronic density; via spontaneous symmetry breaking; or via band overlap when the filling is even. An example of the first is doping a semiconductor, which is the metal-insulator transition we induce on a daily basis inside transistors. An example of the second is the transition into antiferromagnetic ordering: when the system is at half-filling of a band (meaning one electron per unit cell), after antiferromagnetic unit cell doubling there are two electrons per unit cell, and the system can become a band insulator. The third case can be realized by for example straining a system such that the band gap changes from positive to negative.

There are, however, two main exceptions to the paradigm of band theory. On the one hand, disorder can become so large as to prevent the motion of the charge carriers—this is known as Anderson localization [9]. On the other hand, the presence of very strong electron-electron interactions can force the electrons to become "stuck" like in a traffic jam—this is known as *Mott insulation* [2]. The standard model of Mott insulation is the Hubbard model with a tight-binding Hamiltonian:

$$H = -t \sum_{\langle ij \rangle \sigma} c_{i\sigma}^\dagger c_{j\sigma} + U \sum_i n_{i\uparrow} n_{j\downarrow}, \qquad (1)$$

where t is the nearest-neighbor hopping on some lattice and U is the onsite repulsion. When $U = 0$, the system is a metal when half-filled. When $U \gg t$, it becomes energetically favorable to occupy each site with exactly one electron rather than to fill bands up to the Fermi level. The resulting Mott state can therefore not be described by band theory!

Mott insulators have been observed in a wide variety of materials, most famously transition-metal oxides, including high T_c superconducting cup rates [3]. Observing a clear transition from a standard Fermi liquid metal to a Mott insulator, however, is quite elusive. This transition can be induced either by changing the electronic density ("filling-controlled") or by changing the ratio U/t ("bandwidth-controlled"). The filling-controlled Mott transition [10] notoriously leads to a whole zoo of different instabilities, pseudogaps, and strange metal behavior, and is typically masked at low temperatures by superconductivity. The bandwidth-controlled Mott transition is, in contrast, often masked by (antiferromagnetic) spin order that hides any Mottness behind the veil of unit cell doubling.

This might, at first, suggest that *Mott criticality* is something unattainable. By "criticality" we mean that approaching the Mott transition we find vanishing energy scales, and that the resistivity curves display scaling behavior. There are, however, two clever tricks to realize Mott criticality. The first trick is *dimensionality*: a transition that is strongly first-order in $d = 3$ dimensions often becomes continuous or weakly first-order in $d = 2$ dimensions. The most striking example of this is, of course, the solidification of ^{3}He. The second, and perhaps even more important trick is *frustration*: if the lattice structure is highly frustrated (with competing magnetic interactions) one can avoid [11] antiferromagnetic ordering altogether—revealing the true Mott transition.

In this review we, therefore, focus on three classes of systems that are indeed (quasi) two-dimensional as well as frustrated: Wigner crystals in extremely dilute two-dimensional electron gases; layered Mott organic compounds; and the more recent addition of transition-metal dichalcogenide (TMD) moiré bilayers. Indeed, as we will show in Section 3, these systems all seem to exhibit remarkably similar distinct features, including clear signatures of critical resistivity scaling. Because these systems all have a fixed electron density per unit cell of $n = 1$ (at least in the insulating limit), the observed transitions are plausibly within the universality class of bandwidth-tuned Mott transitions.

A brief side-note is in order: we briefly mentioned superconductivity and disorder-induced insulators. These phases can also have a continuous transition between them, the so-called superconductor-to-insulator transition [12,13]. This, however, is an interesting topic that falls outside the scope of this review. Similarly, we also will not consider disorder-driven metal-insulator transitions [14,15], since this regime typically does not include any Mottness. More general but also somewhat older reviews of metal-insulator criticality can be found in Refs. [3,4,16–18].

3. Experiments

Given that experimental results should always be leading, the aim of this section is to introduce three material systems that are likely exhibiting a bandwidth-tuned Mott metal-insulator transition: dilute 2DEGs, organics, and moiré systems. To support the clarity of interpretation, we will stress *experimental* similarities between these systems without much room for theoretical guesswork—that is the next section's realm.

While each system has a different tuning parameter (density, pressure, or field), the *electrical resistivity* through the transition is the key observable, see Figure 1. Its behavior reveals how the transport gap Δ decreases when we approach the transition from the insulating side; as well as how the resistivity behaves on the metallic side, where Fermi liquid behavior $\rho(T) = \rho_0 + AT^2$ is typically seen at $T < T_{FL}$ with an enhanced effective mass m^*. Remarkably, in all systems one also observes distinct *resistivity maxima* at $T \sim T_{max} > T_{FL}$, signalling the breakdown of coherent transport. Crossover to the quantum critical regime is described by an additional temperature scale T_0, which is extracted from the scaling collapse of the resistivity curves as shown in Figure 2.

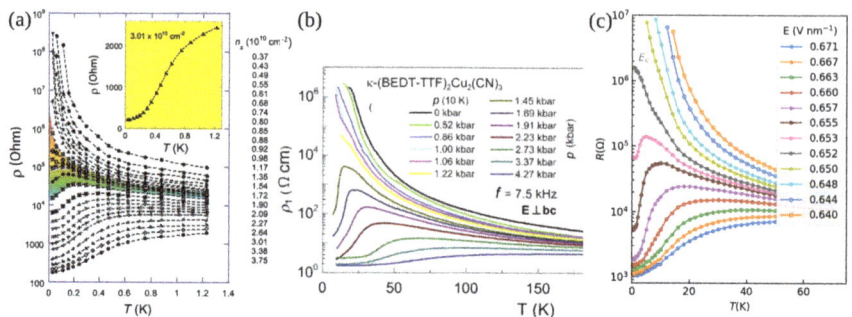

Figure 1. The key observable revealing a metal-insulator transition is the resistivity. Here we show ρ vs. T resistivity curves as a function the tuning parameter, for representative examples of the three material systems considered: (**a**) 2DEG in Si-MOSFET tuned by electronic density (reprinted with permission from Ref. [19] Copyright 2019 American Physical Society), (**b**) Mott organic material κ-(BEDT-TTF)$_2$Cu$_2$(CN)$_3$ tuned by pressure [20], and (**c**) TMD moiré bilayer MoTe$_2$/WSe$_2$ tuned by displacement field (Data imported from [21]). In all cases, one observes distinct resistivity maxima on the metallic side, at a temperature T_{max} that decreases towards the transition.

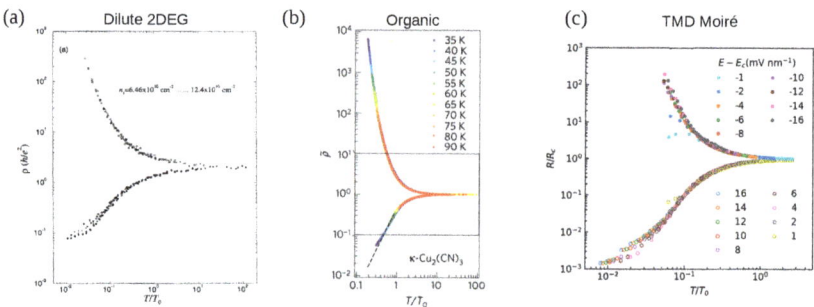

Figure 2. Critical scaling has been observed in all three experimental systems, when the resistivity is plotted versus T/T_0 where T_0 is the characteristic crossover (quantum critical) energy scale. Note that in all cases a strong "mirror" symmetry [22,23] exists between the insulating (upper) and metallic (lower) scaling branch. (**a**) In a dilute 2DEG, scaling of the bare resistivity $\rho(T)$ was achieved by simply rescaling T with $T_0 \sim |\delta|^{1.6}$ (Adapted with permission from Ref. [24] Copyright 1995 American Physical Society); (**b**) In organic compounds, the normalized resistivity $\tilde{\rho}$ is obtained by normalizing the resistivity by the critical resistivity along the Widom line. This leads to excellent scaling collapse with $T_0 \sim |\delta|^{0.60\pm0.01}$ (Adapted with permission from Ref. [25] Copyright 2015 Springer Nature); (**c**) A similar approach was followed in TMD moiré bilayer MoTe$_2$/WSe$_2$, with similar $T_0 \sim |\delta|^{0.70\pm0.05}$ (Data imported from [21]).

A practical summary of the experimental results is presented in Table 1.

Table 1. A summary of available experimental results for the three classes of systems considered. The sources (references) are given in the text below. Question-marks indicate the lack of reliable data. Fermi liquid (T^2) transport behavior has not been documented in 2DEG systems, in contrast to strong evidence for it in Mott organics and TMD moiré bilayers. Note that the characteristic energy scales Δ, $(m^*)^{-1}$, T_{FL}, T_{max}, as well as T_0 display similar continuous decrease towards the transition in all three systems, consistent with general expectations for quantum criticality. One should keep in mind that the error bars on the estimated exponent could be substantial, since the results typically depend strongly on the utilized fitting range.

System	Dilute 2DEG	Mott Organics	TMD Moiré Bilayers
Transition Type	continuous?	weakly first order (at $T < T_c \sim 0.01 T_F$)	continuous?
Δ	$\lvert n - n_c \rvert$	$\lvert P - P_c \rvert^{\nu z}$, $\nu z \approx 0.7 - 1$	$\lvert E - E_c \rvert^{\nu z}$, $\nu z \approx 0.6$
$\frac{1}{m^*}$	$\lvert n - n_c \rvert$	?	?
T_0	$\lvert n - n_c \rvert^{\nu z}$, $\nu z \approx 1.6$	$\lvert P - P_c(T) \rvert^{\nu z}$, $\nu z \approx 0.5 - 0.7$	$\lvert E - E_c \rvert^{\nu z}$, $\nu z \approx 0.7$
T_{FL}	?	$\lvert P - P_c \rvert$	$\lvert E - E_c \rvert^{\nu z}$, $\nu z \approx 0.7$
T_{max}	$\lvert n - n_c \rvert$	$\lvert P - P_c \rvert$	$\lvert E - E_c \rvert^{\nu z}$, $\nu z \approx 0.7$

3.1. Dilute 2DEG in Semiconductors

In dilute two-dimensional electron gases (2DEG) [26], the electron density can be quantified by the dimensionless parameter $r_s = 1/\sqrt{\pi n} a_B$ where n is the electron density and a_B the Bohr radius. The ratio of interaction energy versus kinetic energy scales as r_s, and therefore at large enough r_s (of the order $r_s \sim 40$ in 2D) the electrons will spontaneously crystallize into a Wigner solid. In a two-dimensional Wigner crystal, the electrons form a triangular lattice with exactly one electron per unit cell—essentially forming a frustrated Mott insulator. When the electron density n is varied, the size of the unit cell changes accordingly so that the Wigner crystal always remains fixed at one electron per unit cell. The transition from an insulating Wigner crystal to a metal can therefore be plausibly viewed as a bandwidth-tuned Mott transition. Note that this is counter-intuitive: after all, one tunes the electron density! However, what matters is the electron density *counted per unit cell* and that remains constant. This idea suggests [27–29] that the melting of a Wigner solid by increasing density should be viewed as a Wigner–Mott transition, possibly bearing many similarities to Mott transitions in narrow-band crystalline solids such as Mott organics or transition-metal oxides. If this viewpoint is correct, then the resulting metal above the transition should display resemble other strongly correlated Fermi liquids, a notion that is starting to gain acceptance on the base of recent experiments [30–32].

Experimentally, high-quality 2DEGs can be realized in metal-oxide-semiconductor field-effect devices (MOSFETs) in various semiconductors [24,30,32,33]. Through electrostatic gating the electronic density can be elegantly tuned, typically in the range of $n \sim 10^{10}$–10^{12} cm^{-2}. The peak electron mobility in ultra-clean samples can be as high as 10^4 cm^2/V s [19], which implies that down to very low temperatures the transport properties are dominated by electron-electron interactions (like Wigner crystallization) rather than extrinsic disorder effects. Lower-mobility devices have also been extensively studied (for a review see Chapter 5 of Ref. [18]), displaying different types of metal-insulator transitions displaying electron glass dynamics [34], which we will not discuss here.

Indeed, tuning the electronic density leads to insulating transport below a critical density, typically around $n_c \sim 10^{11}$ cm^{-2} [24,33], see Figure 1a. Activated behavior is often observed close to the transition [34,35], with the activation energy $\Delta \sim \lvert n - n_c \rvert$. Further on the insulating side disorder effects may become important, where Efros-Shklovskii hopping (and other effects of disorder) is also observed [24], but only at the lowest temperatures. On the metallic side, a pronounced resistivity drop (often by a factor of 10 or more) is observed [24,30] below the temperature $T_{max} \sim |n - n_c|$ which decreases as the transition is approached. Characteristic scaling of the resistivity maxima has been reported in several systems [30–32], see Figure 3, which has been interpreted as evidence for strong correlation effects. However, the expected T^2 dependence of the resistivity has not been observed, despite the reported effective mass enhancement $(m^*)^{-1} \sim |n - n_c|$ [36], characteristic of correlated Fermi liquids. Quantum critical scaling collapse of the resistivity curves has also been demonstrated [24] around the critical density, albeit excluding the lowest temperatures data, as shown in Figure 2a. This is achieved by rescaling T by a crossover scale $T_0 \sim |n - n_c|^{\nu z}$, with $\nu z \approx 1.6$. The resulting scaling function reveals surprising "mirror symmetry" [22], which was phenomenological interpreted [23] as evidence "strong-coupling quantum criticality". Similar systems to these 2DEGs include the observation of a Wigner crystal in low-density doped monolayer WSe_2 [37], where more detailed experiments still need to be performed.

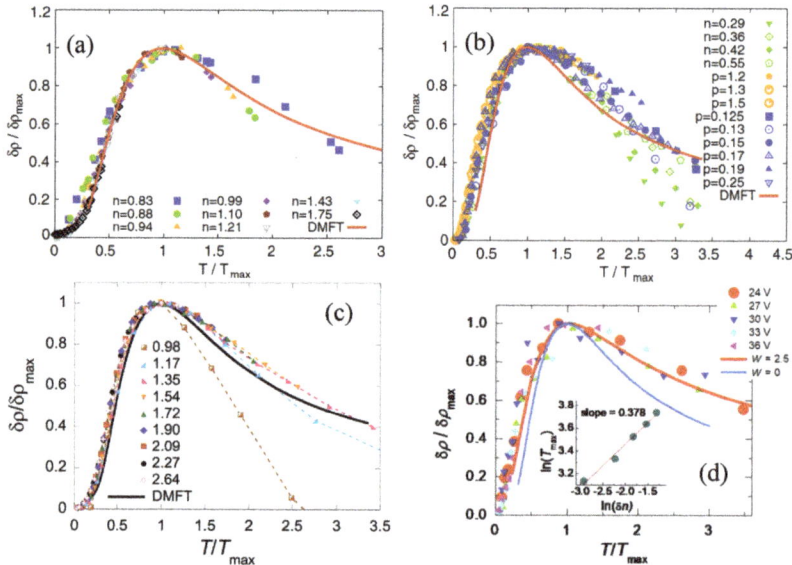

Figure 3. Characteristic scaling of the resistivity maxima has been reported in several 2DEG electron systems in semiconductors: (**a**) Si-MOSFETs (adapted with permission from Ref. [29] Copyright 2012 American Physical Society); (**b**) p-GaAs/AlGaAs quantum wells (adapted with permission from Ref. [29] Copyright 2012 American Physical Society); (**c**) SiGe/Si/SiGe quantum wells (adapted with permisssion from Ref. [30] Copyright 2020 American Physical Society); (**d**) few layered-MoS_2 (Adapted with permission from Ref. [32] Copyright 2020 American Physical Society). All data collapse to the same (theoretical) scaling function [29] obtained from the Hubbard model at half-filling, in the vicinity of the Mott point.

3.2. Organic Compounds

An organic compound [38] refers to a crystalline system where each unit cell contains an entire *molecule*, rather than just loosely bound ions. A particularly interesting class of organic compounds is based on the molecule bis-(ethylendithio)-tetrathiafulvalen (BEDT-TTF or ET), which can be fabricated with other ions into quasi-two-dimensional layered systems. Compounds based on BEDT-TTF exhibit a spectrum of interesting quantum matter phenomena, ranging from superconductivity [39] to electron glass behavior [40].

Our interest goes out especially to κ-$(ET)_2Cu[N(CN)_2]Cl$ and κ-$(ET)_2Cu_2(CN)_3$, where the molecules are organized in triangular lattice layers [38]. These materials are strongly correlated, and indeed, despite being half-filled they are insulating at ambient pressures. Due to the geometric frustration of the triangular lattice [11], no magnetic order has been observed in κ-Cu$_2$(CN)$_3$ and antiferromagnetic order only at relatively low temperatures ($T < T_N \approx 20K$) in κ-Cl. The absence of magnetic order is the strongest indication that κ-Cu$_2$(CN)$_3$ might realize a spin liquid ground state [41,42].

Upon applying pressure, a zero-temperature *first-order* phase transition brings the system into a paramagnetic metallic phase at $p_c = 122$ MPa (κ-Cu$_2$(CN)$_3$) or $p_c = 24.8$ MPa (κ-Cl), see Figure 1a [20,25]. The first order phase boundary ends in a critical point at $T_c = 20$ K or $T_c = 38$ K, respectively. It is important to emphasize that these temperatures are very small compared to the electronic energy scales. The Hubbard repulsion U and bandwidth W are both on the order of a fraction of eV [42], which implies $T_c \ll U, W$. As such, even though the observed Mott criticality appears at nonzero temperatures, much of the observed phenomena above T_c can be described *as if* the system resides in the vicinity of a quantum critical point [25].

Above T_c, a crossover pressure $P_c(T)$ can traced [25] where the measured resistivity exhibits an inflection point, see Figure 4. This defines the *"quantum Widom line"* (QWL) [43] by analogy to the standard liquid-gas crossover. Defining the critical resistivity $\rho_c(T)$ to be the resistivity along the QWL, all resistivity curves collapse onto each other when plotted as $\rho(P,T)/\rho_c(T)$ vs. $T/T_0(P)$, as shown in Figure 2b. Here the scale $T_0(P)$ reflects a critical energy scale that vanishes at the critical pressure, $T_0 \sim |P - P_c|^{0.6}$ for Cu and $T_0 \sim |p - p_c|^{0.5}$ for κ-Cl [25]. On the insulating side of the transition, the resistivity is approximately activated $\rho \sim \exp(\Delta/T)$ [42]. On the metallic side, it follows [44] the standard Fermi liquid behavior at low temperatures $\rho(T) = \rho_0 + AT^2$, up to a temperature scale T_{FL}, see Figure 5a. This destruction of the Fermi liquid seems to correspond to the appearance of a maximum in the resistivity [20].

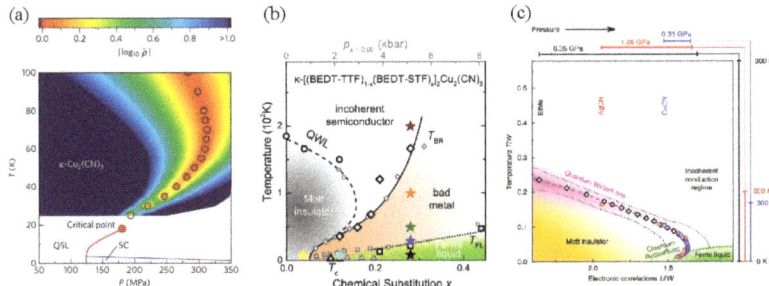

Figure 4. Finite temperature phase diagram of the Mott organic materials. (**a**) first-order phase transition line, as observed in κ-Cu$_2$(CN)$_3$ (adapted with permission from Ref. [25] Copyright 2015 Springer Nature) at $T < T_c \sim 20K$, displaying "Pomeranchuk" behavior [45], by "sloping" towards the metallic phase. The corresponding "Quantum Widom Line" [43] arises at $T > T_c$, which is identified as the center of the quantum critical region [46] with resistivity scaling [25]. (**b**) Phase diagram [44] for κ-Cu$_2$(CN)$_3$ over a broader T-range, displaying the convergence of the quantum Widom line (QWL) on the insulating side, and the "Brinkman-Rice" line ($T_{BR} = T_{max}$, which intersect at the critical end-point $T = T_c$. The Fermi-Liquid line $T_{FL} < T_{BR}$ is also shown. (**c**) The universal phase diagram for a series of spin-liquid Mott organics compounds was established [42] by rescaling the temperature T and the interaction strength U by the respective electronic bandwidth W. The parameters W and U were independently measured [42] for each material using optical conductivity.

In addition to transport measurements, and in contrast to other systems we consider, Mott organics have also been carefully investigated using optical probes. This allowed to directly identify [42] the quantum Widom line, which is back-bending towards the insulating side at higher temperatures following the closing of the Mott gap. In addition, the "Brinkman–Rice" line traced by T_{max} was identified as marking the thermal destruction of Landau quasiparticles, as seen by the vanishing of the Drude peak in the optical conductivity [44], see Figure 6.

Finally, the controversy about the presence or absence of the low-T phase coexistence region has been resolved in Mott organics, by using dielectric spectroscopy [20]. Its precise location on the phase diagram has been identified by the observation [20] of colossal dielectric response, as a smoking gun for percolative phase coexistence. In addition, the same technique was able to demonstrate the coincidence of the resistivity maxima in the (uniform) metallic phase, with the thermal destruction of Landau quasiparticles. This is seen as a dramatic drop and a change [20] of sign of the dielectric function at $T < T_{BR} = T_{max}$. These experimental results are shown in Figure 6, and discussed in more detail in Section 5.

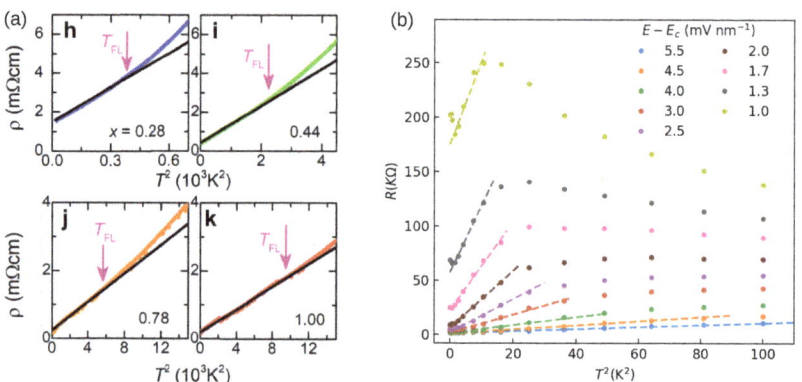

Figure 5. Fermi liquid behaviour at low temperatures, for (**a**) Mott organic material $\kappa-$[(BEDT-TTF)$_{1-x}$(BEDT-STF)$_x$]$_2$Cu$_2$(CN)$_3$ [44] and (**b**) MoTe$_2$/WSe$_2$ moiré bilayers (Data imported from [21]). Clear $\rho = \rho_0 + AT^2$ behavior is observed in both cases, up to a temperature scale T_{FL} that seems to decrease linearly towards the metal-insulator transition. The resistivity curves can be collapsed by plotting $\rho(E,T)/\rho_c(T)$ vs. T/T_0 where $T_0 \sim |E - E_c|^{0.70 \pm 0.05}$, see Figure 2c. Note that this crossover scale seems to follow both the gap size on the insulator, as well as the destruction of the Fermi liquid on the metallic side.

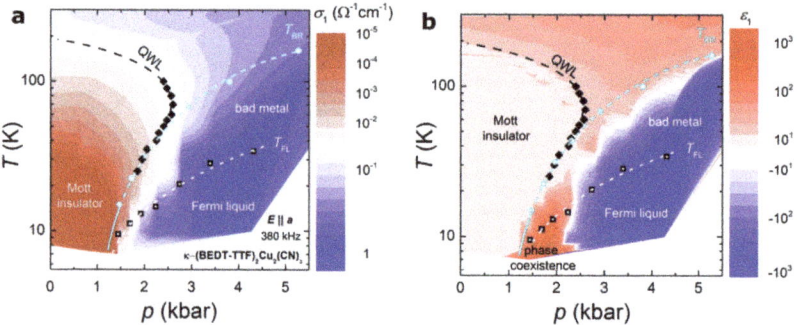

Figure 6. Transport behavior vs. dielectric response across the phase diagram of κ-Cu$_2$(CN)$_3$ [20]. (**a**) DC transport shows only very gradual change across the BR line (resistivity maxima), and cannot one see any clear indication of the phase coexistence region. (**b**) In dramatic contrast, the low-frequency dielectric function ϵ_1 assumes small positive values in the Mott insulator (pale pink), and large negative values in the quasiparticle regime (deep blue); we clearly see the boundaries of these regimes tracing the QWL and the BR line (following T_{max}), as observed in transport. Remarkably, "resilient" quasiparticles [47] persist past the Fermi Liquid line, at $T_{FL} < T < T_{BR} = T_{max}$, where bad metal behavior [48] (metallic transport above the Mott-Ioffe-Regel [49] limit is observed). At low temperature, the Mott point is buried below the phase coexistence dome, which is vividly visualized through colossal dielectric response ($\epsilon_1 \sim 10^3$–10^4).

3.3. Moiré Materials

The most recent addition to the field of strongly correlated systems are *moiré materials*. These are bilayer structures made of Van der Waals materials such as graphene and transition-metal dichalcogenides (TMDs). A lattice mismatch or relative twist angle between the layers causes a large-scale geometric "moiré" pattern. This larger unit cell (typically in the range of 5–10 nm) drastically reduces the effective electron kinetic energy such that the bandwidth is on the order of $W \sim 10$ meV. As a result, the systems become strongly correlated, with $U/W \sim 10$ or larger. Using electrostatic gating one can tune the electronic density, typically in the range of a few electrons or hole per moiré unit cell (corresponding to $n \sim 10^{12}$ cm^{-2}).

While the most famous of correlated moiré materials is without a doubt twisted bilayer graphene, more convincing evidence for Mott correlations has so far only been observed in TMD bilayers. Here, we focus on one particular system: the heterobilayer MoTe$_2$/WSe$_2$ [21]. The lattice mismatch between MoTe$_2$ and WSe$_2$ gives rise to a moiré period of $a_M \sim 5$ nm. At half-filling of the first valence band, an insulating phase appears which can be tuned into a metal by applying a vertical displacement field E. This flat valence band can be described by a spin-orbit coupled triangular lattice Hubbard model, where the displacement field E tunes the bandwidth [50].

The temperature-dependent resistivity across the transition is shown in Figure 1c. On the insulating side, the system has well-defined activated behavior of the resistivity with a gap Δ continuously vanishes as the critical displacement-field value is approached, $\Delta \sim |E - E_c|^{0.60 \pm 0.05}$. At the critical point, the resistivity is claimed to follow a powerlaw, $\rho_c \sim T^{-1.2}$, although the reliability of the low-T data may be questionable. On the metallic side, the low-T resistivity follows the Fermi liquid law $\rho(T) = \rho_0 + AT^2$ where the quadratic prefactor diverges $A \sim |E - E_c|^{-2.8 \pm 0.2}$. This clear Fermi liquid does not persist up to all temperatures, instead, a resistivity maximum appears at $T_{max} \sim |E - E_c|^{0.70 \pm 0.05}$, see Figure 5b. No magnetic order has been observed, which might be due to the geometric frustration of the triangular moiré lattice structure. Remarkably, this experiments uses the recently-developed *excitonic sensor* [21], which allows the measurements of the spin susceptibility across the transition. This reveals Curie-law behavior over a broad temperature range, thus demonstrating the presence of localized magnetic moments, as expected for a Mott system.

Finally, among other moiré systems it is worth mentioning the twisted hetero-bilayer WSe$_2$ [51]. Though it was claimed to exhibit some sort of Mott-related criticality, it is not certain whether a true insulating phase has indeed been observed given that the activated transport does not continue down to the lowest measured temperatures. In addition, insulating behavior seems to disappear above a relatively low of the order of $T^* \sim 5$–10 K, which is similar to twisted bilayer graphene, but much smaller than the estimated bandwidth of the order of $W \sim 100$–200 K. Furthermore, no clear resistivity maxima on the metallic side have been observed. Whether or not tbWSe$_2$ can be classified as a true Mott insulator is therefore quite controversial. Alternatively, the observed behavior could be a result of some sort of magnetic order, which may arise close to half filling even in weakly-coupled systems.

3.4. Universal Criticality

The resistivity curves of the three systems, as shown Figures 1 and 2, show remarkable universality, reflected in the fact that all curves can be collapsed by scaling with $T_0 \sim |\delta|^{z\nu}$ where δ is the tuning parameter and $z\nu$ the critical exponent. It is important to realize, however, that the precise scaling procedure applied was not identical in the three cases, and the resulting critical exponent also somewhat depend on the system.

So what is *different* between these systems? Let us first focus on the energy scale. The typical bandwidth W ranges from $\sim$100 s meV in Mott organics, to $\sim$10 s meV in the moiré systems, to 0.1–1 meV in the dilute 2DEGs. A finite temperature critical point is only observed in the organics, though at about $T_c \sim 1\% W$—which leaves open the possibility that a finite T critical endpoint exists in the other two systems. Indeed, most experiments are performed (so far) above the Kelvin range in moiré materials, and above 100 mK in 2DEG, which makes it hardly possible to reliably explore the T-range below few percent of the bandwidth.

Secondly, in order to achieve quantum critical scaling in Mott organics, one needs to first identify a Widom line as a demarcation of the finite-temperature crossover from insulator to metal. A similar analysis was carried out for moiré materials, although the obtained Widom line displayed no apparent "curving" as a function of temperature. This is manifestly not performed in the dilute 2DEGs. It might be interesting to see whether better collapse can be achieved through such a method.

Thirdly, with a bit of good-will, the critical exponents in organics and Moiré systems are in the same ballpark; whereas the critical exponent in the dilute 2DEGs with $z\nu = 1.60$ is significantly larger. It is also important to realize that in 2DEGs there has not been a clear observation of a Fermi liquid regime—unlike in organics and Moiré systems, see Figure 5. These ways in which 2DEGs stand out might be related to the fact that there is no underlying (Wigner) lattice on the metallic side, which could point to a perhaps nontrivial role of significant charge density fluctuations on the metallic side, an effect not present in lattice Mott systems.

4. Competing Theoretical Pictures

As we mentioned in Section 2, the observation of Mott criticality and scaling opens big questions on the theoretical front. Currently, there exist *three* main different physical pictures to address these issues.

A true Mott transition should not be hidden by some period doubling symmetry breaking. The Lieb-Schultz-Mattis theorem states that in absence of spin order, the ground state of Mott insulator must be a spin liquid [52]. This leads directly to the first theoretical picture: Mott criticality can only occur if the Mott phase is a spin liquid, where inter-site spin correlations play an important role. The theory of Senthil [53] chooses this path, by introducing an explicit spinon theory of the Mott spin liquid.

Alternatively, one focuses on local electronic processes only, ignoring inter-site spin correlations. Then the Mott transition at low temperature becomes first-order; however, it is only *weakly* first order. A first order transition line always ends at a critical point T_c, and as long as T_c is sufficiently low compared to any experimental scale, one still finds criticality and scaling. This is the picture emerging from Dynamical Mean Field Theory (DMFT) [54], a strong-coupling self-consistent approach to calculate the local electronic self-energy.

The third picture again accepts the first-order nature of a Mott transition, but this time embraces it. A first-order transition is always accompanied by a region where both phases coexist. Minor disorder or self-generated pattern formation [55,56] can smear this phase coexistence region into a continuous-looking transition exhibiting nontrivial electron dynamics. This is the 'percolation theory' picture of Mott criticality.

The goal of this review paper is to put the main theoretical predictions next to the experimental findings. As such, we will not dive into the pros and cons of each theoretical picture. A summary of the main theoretical predictions is provided in Table 2.

Table 2. A summary of predictions from competing theoretical pictures. The expected transition type differs between the three pictures, with observable differences in the behavior of the mass enhancement m^*, the Kadowaki–Woods ratio $A/(m^*)^2$, the destruction of the Fermi liquid at T_{FL}, and the appearance of a resistivity maxima at T_{max}. Details are provided in the text below.

Theory Predictions	2D Spinon Theory	DMFT	Percolation Theory
Transition Type	continuous	weakly first order (at $T < T_c \sim 0.01 T_F$)	first order
Δ	$\|g - g_c\|^{\nu z}$, $\nu z = 0.67$	$\|U - U_{c1}\|^{\nu z}$, $\nu z \approx 0.8$	remains finite
m^*	weak: $\ln \frac{1}{\|g-g_c\|}$	strong: $\|U - U_{c2}\|^{-1}$	no divergence
$A/(m^*)^2$	?	constant (KW law obeyed)	diverges: $(x_o - x_c)^{-t}$; $t = s/m$
T_{FL}	$\|g - g_c\|^{2\nu}$	$\|U - U_{c2}\|$	$T^* \sim \|x_o - x_c\|$
T_{max}	$T_{max} = \infty$	$\|U - U_{c2}\|$	$T^* \sim \|x_o - x_c\|$

4.1. Spin Liquid Picture of the Mott Point

A popular approach to describe a spin liquid state is through *spin-charge separation*. In Ref. [53], the electron is split into a charge-0 spin-1/2 fermionic *spinon f* and a charge-*e* spin-0 bosonic *chargon b*. The Mott transition, in this picture, amounts to the condensation of the chargon field, whose critical behavior falls within the 3D XY universality class. The Fermi liquid corresponds to the condensed phase of the chargon, whereas the Mott insulator corresponds to a gapped phase of the charged boson. The splitting of the electron leads to redundant degrees of freedom described by an emergent gauge field. Fluctuations of this gauge field lead to a logarithmic enhancement of the quasiparticle effective mass,

$$m^* \sim \ln \frac{1}{|g - g_c|}, \qquad (2)$$

where g is the tuning parameter and g_c is the critical value. However, as in any theory with a non-local electronic self-energy, the quasiparticle residue Z is *not* simply proportional to the inverse effective mass; instead $Z \sim |g - g_c|^\beta / \ln \frac{1}{|g-g_c|}$. Furthermore, approaching the Mott transition from the metallic side the spin susceptibility χ remains constant whereas the compressibility κ vanishes. Physically, these effects result from important inter-site spin correlations, where a gapless spin liquid can be viewed as a certain superposition of spin singlets formed by pairs of spins in the Mott insulating state. As a result, there emerges a finite gap δ to charge excitations, while the rearrangement of singlets leads to characteristic gapless spin excitations with fermionic quasiparticles. This picture is a specific realization of the famous RVB picture of Baskaran and Anderson [57], first proposed in the context of high-T_c superconductors.

Another significant consequence of describing the Mott transition as chargon condensation, is that the $T = 0$ conductivity is not continuous. The electron resistivity will display a *universal* jump from a (disorder)-dependent constant value $\rho = \rho_0$ in the Fermi liquid; to $\rho = \rho_0 + \frac{Rh}{e^2}$ (with R of order one) at the critical point; to $\rho = \infty$ in the Mott insulator. On the metallic side, the Fermi liquid is predicted to break down above $T_{FL} \sim |g - g_c|^{2\nu}$ and give rise to a *marginal* Fermi liquid state, which in turn survives up to $T_{MFL} \sim |g - g_c|^\nu$. In both cases, $\nu = 0.67$ is the 3D XY correlation length exponent. On the insulating side, the boson condensation picture implies that the charge gap vanishes as $\Delta \sim |g - g_c|^\nu$. The spinons, however, remain gapless and form a spinon Fermi surface, with low-temperature specific heat scaling as $C \sim T^{2/3}$.

Note that the original work in Ref. [53] does not directly provide a detailed description for finite temperature dependence of the resistivity, and thus no explicit prediction for a possible deviation from the Kadowaki-Woods (KW) law ($A/(m^*)^2 \approx$ constant) [58]. On the other hand, Ref. [58] presents arguments that the physical requirement for the validity of the KW law is the locality of the electronic self-energy (as in DMFT theory) [54], a condition which is not obeyed by the RVB-type spin-liquid theories such as the Senthil's spinon picture.

It should be stressed that the spinon theory makes one sharp prediction about finite temperature transport in the critical regime. Namely, the *critical resistivity curve* is predicted to assume a universal power-law form $\rho_c(T) \sim 1/T$ in $d = 3$ [59], but remain T-independent in $d = 2$, see Figure 7. This therefore leads to distinct resistivity maxima in 3D, but *not* in 2D [53], where monotonic T-dependence should be found on both sides of the transition, albeit with opposite slope. Physically, this difference reflects the proposed importance of "infrared" (IR, long distance) effects due to gauge fields, which should have strong (spatial) dimensionality dependence. Concerning quantum critical scaling, it is interesting that this theory proposes the emergence of *two* crossover temperature scales, both of which vanish at the transition.

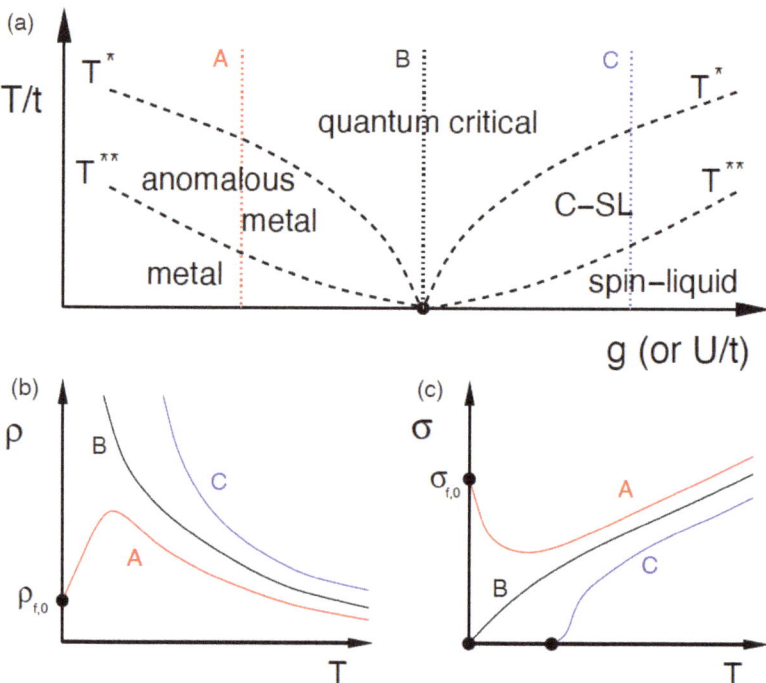

Figure 7. Predictions of the spinon theory (reprinted with permission from Ref. [59] Copyright 2009 American Physical Society). (**a**) The phase diagram features a quantum critical point at $T = 0$, and two distinct finite-T crossover scales T^* (above which the system is quantum critical) and T^{**} (below which the system is either a metal or a gapless spin liquid). (**b**) and (**c**) Resistivity and conductivity along the lines A, B and C in the phase diagram in (**a**). Critical resistivity is predicted to diverge as $\rho_c(T) \sim 1/t$ in $d = 3$, leading to resistivity maxima on the metallic side (coductivity minima). In contrast, the same theory predicts finite critical resistivity $\rho_c(T) \sim \rho^*$ in $d = 2$ [53], hence monotonic behavior on both sides of the transition and no resistivity maxima.

We finally mention that a similar spin-charge separation theory has been very recently proposed to also describe the Wigner–Mott transition in TMD bilayers, where a possible role of charge fluctuations has also been discussed for the metallic side [60,61].

4.2. Dynamical Mean Field Theory Picture of the Mott Point

Our second theory of interest is *Dynamical Mean Field Theory* (DMFT), which explicitly ignores all nonlocal (spin or charge) spatial correlations, and therefore aims to self-consistently calculate the local electronic self-energy $\Sigma(\omega)$ [54]. Physically, its real part describes the modifications of the electronic spectra, while its imaginary part encodes the frequency and temperature dependence of the electron-electron scattering rate. In this way, this theory is not limited to low-temperature excitations only, but is able to capture strong inelastic scattering at high temperatures, and therefore describe both the (coherent) Fermi liquid regime, and also the incoherent high-temperature transport, for example the famed bad metal behavior [48,62] above the MIR limit [49].

While there exist some limiting cases where an analytic solution is possible, it is mainly a numerical approach at finite temperature. DMFT is exact in the limit of large coordination, which physically corresponds to maximal magnetic frustration. Therefore, in the simplest implementation, DMFT describes Mott physics in absence of any magnetic order, nor does it include any (inter-site) spin liquid correlations. We should mention that extensions of DMFT have recently been proposed [63] that include spinon effects, based on an alternative (matrix M, N) rotor representation. This theory, which includes some dynamical effects even at the saddle-point level, suggest that coherent spinon excitations are very fragile to charge fluctuations emerging upon the closing of the Mott gap, suppressing the spin liquid correlations not only on the metallic side, but also within the critical region. We will not further discuss these most sophisticated approaches here, but will limit our attention to the predictions of the simplest single-site DMFT theory.

When applied to the single-band Hubbard model on a frustrated lattice (such as the triangular lattice), DMFT predicts that on the metallic side the quasiparticle mass diverges linearly $m^* \sim |U - U_{c2}|^{-1}$ at a critical value U_{c2} (similar to the prediction of the Brinkman–Rice (BR) theory of the Mott transition [64]). The quasiparticle weight Z is inversely proportional to (m^*). Similarly, other features of the Fermi liquid such as the Kadowaki-Woods law $A \sim (m^*)^2$ are upheld. This Fermi liquid behavior persists up to a temperature T_{FL} that vanishes linearly when approaching U_{c2}. Interesting, at $T_{max} \sim T_{FL}$ the resistivity exhibits a maximum [29]. On the insulating side, there exist no well-defined quasiparticles as the self-energy diverges, $\Sigma(\omega) \sim 1/\omega$. The electronic spectrum is split into an upper and lower Hubbard band, separated by a gap that remains nonzero at U_{c2}. The insulating state becomes unstable at a lower value of the interaction $U_{c1} < U_{c2}$, where the gap closes $\delta \sim |U - U_{c1}|^{\nu z}$, $\nu z \approx 0.8$ [65]. As a result, there emerges a low-T first-order metal-insulator transition, and an associated phase coexistence region at $U_{c1} < U < U_{c2}$ [54]. These main predictions are summarized in Figure 8.

At nonzero temperature, the first-order transition line ends at a critical point at a temperature $T_c \approx 0.015W$, significantly smaller than the bare bandwidth W. At temperatures $T \gg T_c$ the results can be viewed as effectively quantum critical [46,65]. This quantum critical regime is centered around the so-called quantum Widom line (QWL) [43], which physically represents a finite-temperature instability trajectory of the insulating phase, as shown in Figure 8a. It extends the first-order line past $T = T_c$, and can experimentally be detected from an inflection point analysis [25] of the resistivity curves. In this regime, the resistivity satisfies the scaling law $\rho(T, \delta U) = \rho_c(T) f(T/T_0(\delta U))$ with a crossover temperature scale $T_o \sim |\delta U|^{\nu z}$ where $\nu z \approx 0.6$, see Figure 8c. The crossover scale T_o is a property of the quantum critical regime and should not be confused with the low-temperature scaling of the Fermi liquid temperature T_{FL}. DMFT therefore predicts two different regimes of scaling: the quantum critical regime at $T \gg T_c$, and the metal regime dominated by scaling in T_{FL}.

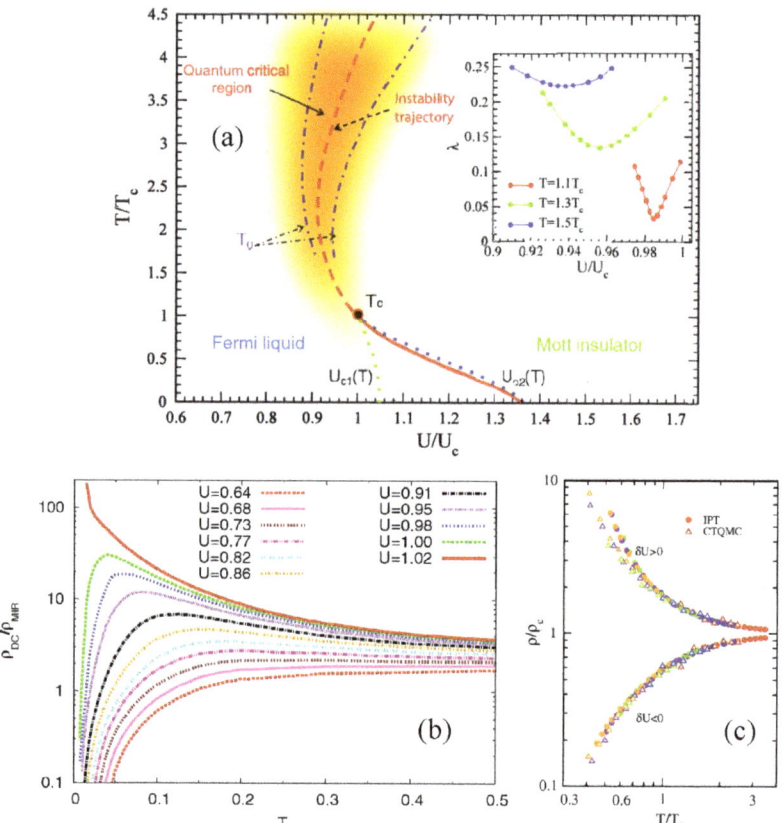

Figure 8. Predictions of DMFT theory. (**a**) Phase diagram featuring a phase coexistence region at $T < T_c$, and a Quantum Critical region centered around the Quantum Widom Line (QWL) (adapted with permission from Ref. [46] Copyright 2011 American Physical Society). (**b**) Resistivity (normalized by the Mott-Ioffe-Regel (MIR) limit) as a function of temperature T across the transition. Note the pronounced resistivity maxima on the metallic side (adapted with permission from Ref. [66] Copyright 2010 American Physical Society. (**c**) scaling collapse of the resistivity curves, displaying pronounced "mirror symmetry" of the two branches (adapted with permission from Ref. [46] Copyright 2011 American Physical Society).

4.3. Percolative Phase Coexistence Picture

In the early theories of both Mott insulators [2] and Wigner crystals, the transition from insulator to metal was often assumed to be robustly first order, at least at sufficiently low temperature. However, even the presence of weak disorder or medium-ranged interactions will create an "emulsion" (microscopic phase coexistence) with Mott/Wigner insulating "islands" in between metallic "rivers" as proposed by Spivak and Kivelson in the context of 2DEG systems in semiconductors [55,67]. If so, then tuning bandwidth and/or temperature should produce a continuous variation of the metallic fraction x. As long as it exceeds the percolation threshold ($x > x_c$), the system is conducting. At $x < x_c$ the metallic domains no longer connect across the system, and conduction stops, at least at $T = 0$. Critical behavior now arises because we dealing with a classical percolation transition.

In this picture, the $T = 0$ metal-insulator transition may occur without actually closing the insulating gap Δ. Similarly, at the percolation threshold $x = x_c$ the metallic Fermi liquid at low T is still stable, and consequently there is no strict divergence of the effective mass m^*, nor the vanishing of T_{FL} at the percolation threshold. We note, however, that such a (classical) percolation picture should apply only if the characteristic domain size is sufficiently larger than the characteristic correlation (or dephasing) length, therefore strictly speaking not at the lowest temperatures. However, finite-temperature variation of the transport properties should be adequately captured, as abundantly documented [68] in other systems featuring microscopic phase separation, such as Colossal Magneto-Resistance (CMR) manganites, for example.

There is, however, an interesting and nontrivial feature of the percolation picture pertaining to finite temperatures. Because of its localized spins, the entropy of the Mott or Wigner–Mott insulating phase should be higher than that of the of the metal (Fermi liquid). As a result, when raising the temperature in the metallic regime, the insulating volume fraction will *increase*: a manifestation of the Pomeranchuk effect [45]. As a result, the resistivity should increase up to some $T = T_{max}$, above which the metallic domains no longer connect, and resistivity will decrease again [56], leading to resistivity maxima. This qualitative picture has been advocated in the work of Spivak and Kivelson, but no concrete prediction of the precise temperature dependence of the resistivity has been made, nor how the corresponding family of curves should scale as the transition is approached.

Interestingly, the same physical picture should in fact apply not only to Wigner–Mott transitions in semiconductors, but also the to conventional Mott transition, provided that there exists a well-defined metal-insulator phase coexistence region around the Mott point. Indeed, recent work on Mott organics [20] revealed precisely such a phase coexistence region, albeit only at very low temperatures of the order of at most few percent of the (bare) Fermi energy. Here, careful theoretical modeling [20] firmly established the validity of the percolation picture, but only within a well-defined phase coexistence region. In contrast, in all the systems studied (2DEG, Mott organics, moiré), the pronounced resistivity maxima persist even much further onto the metallic side, where T_{max} can reach a substantial fraction of T_F, where phase coexistence is very unlikely. Furthermore, recent experimental work on Mott organics by Kanoda and collaborators demonstrated [69,70] the extreme fragility of such a phase coexistence region to disorder, as generally expected in 2D systems [71]. Nevertheless, it is extremely useful to have an independent experimental method to distinguish the phase coexistence region (where percolative effects are likely) from the regimes where a more uniform electron fluid/solid resides. The possibility to do so was spectacularly demonstrated in the context of Mott organics. In the next section we discuss how the dielectric response can tell which mechanism (quasiparticle destruction or percolation) is at play in a given regime.

We briefly mention that percolation effects have been also discussed in the context of spinon theory in a recent paper [72], which does require however significant disorder. On the other hand, the Spivak-Kivelson theory does not require disorder as the micro-emulsion of insulators and metals can be self-generated. This seems to be more in line with the experiments of Section 3: at least the 2DEGs [19] and the Mott organics [38] are displaying Mott criticality in the cleanest samples possible. It is therefore very plausible that most universal features observed in all critical Mott systems are not the result of disorder, but are instead the inherent manifestations of strong correlation physics.

5. Interpreting Resistivity Maxima

As we have seen from our brief theory overview above, several scenarios were proposed, with sometimes similar predictions for characteristic features seen in experiments. A notable example is the clear emergence of the resistivity maxima on the metallic side, at a temperature $T = T_{max} \gtrsim T_{FL}$, which is seen to decrease towards the transition. What is its physical content? The three theoretical pictures propose very different physical perspectives on what goes on here.

As we mentioned in Section 4.1, spinon theory [53] predicts the presence of resistivity maxima only in $d = 3$, but not in $d = 2$. However, robust resistivity maxima are clearly seen all the material systems of Section 3. An understanding of the resistivity maxima must therefore come from either the DMFT perspective on Mott physics, or from the percolative scenario. Both mechanisms provide reasonable albeit very different routes to explain the resistivity maxima. How should one distinguish them and thus identify the precise mechanism at play in a given system? Luckily, important clues were provided by recent experiments on Mott organics [20]. Here one finds two distinct regimes, both featuring similar resistivity maxima, but with very different dielectric response. One such regime is corresponds to the (spatially inhomogeneous) metal-insulator phase coexistence region, where colossal enhancement of dielectric response has been found. The other regime was found further on the metallic side, where a dramatic drop and a change of sign the dielectric constant signaled thermal destruction of coherent quasiparticles due to strong correlation effects.

In the following we show how general scaling arguments can be used within each of the two proposed scenarios, to demonstrate the general robustness of these trends, thus providing a new window of what precisely goes on near the Mott point.

5.1. Resistivity Maxima from Thermally Destroying Coherent Quasiparticles

Both experiments and theory provide evidence that a strongly correlated Fermi liquid forms on the metallic side of the Mott point, with a characteristic "Brinkman–Rice" (BR) energy scale $T_{BR} \sim 1/m^*$, which decreases towards the transition, thus characterizing the heavy quasiparticles. Inelastic electron-electron scattering increases with temperature, eventually leading to the thermal destruction of the quasiparticles around $T \sim T_{BR}$, and the associated modification of both the single particle (ARPES) spectra and the optical conductivity. At higher temperatures, transport assumes incoherent character, which can no longer be understood in terms of the quasiparticle picture or Fermi Liquid ideas alone. Its precise form generally depends on band filling and the correlation strength, but more precise predictions require a specific microscopic model and a theoretical picture.

Concrete and quantitative results, in this regime, were given by DMFT theory, which provided first insight into the origin of the resistivity maxima in certain Mott organic materials at half-filling, as well as in certain oxides. Subsequent DMFT studies stressed that the characteristic temperature scale for the resistivity maxima indeed tracks the BR scale of the quasiparticles ($T_{max} = T_{BR}$), while preserving the functional form of the resistivity curves across this coherence-incoherence crossover. This revealed the scaling behavior of the resistivity curves in the correlated metallic regime, with a universal scaling function of T/T_{max}. The predicted scaling behavior has been confirmed by a number of experiments on various systems [29,30,32], displaying even quantitative agreement with the theoretical scaling function, with no adjustable parameters.

Further optical and dielectric studies [20] in Mott organics also confirmed the predicted destruction of the Drude peak around T_{BR}, again signaling the thermal destruction of quasiparticles. They established that it dramatically affects not only DC transport, but also the dielectric response, which in this metallic regime is seen to display a dramatic drop from moderate positive values at $T > T_{drop} \sim T_{max}$ to very large but negative values at $T < T_{drop} \sim T_{max}$. These studies, combining experiments and DMFT theory, have firmly established that the dielectric response can be used to directly reveal the thermal destruction of quasiparticles around the BR temperature.

In the following, we extend the systematic studies of Ref. [29], to stress that within DMFT both DC transport and the dielectric response display the characteristic crossover behavior across T_{BR}, and the associated scaling behavior upon approaching the Mott point. To do this we calculate the dielectric function ϵ_1 as a function of temperature and interaction U, using the same setup as in our recent work [20]. For simplicity, we focus on a simple semi-circular band model at half filling, and carry out DMFT calculations using the standard CTQMC impurity solver with the the Maximum Entropy method for analytical continuation to the real axis. Just as in Ref. [29], once we get the single particle self energy from our DMFT equations, we calculate the real part of optical conductivity $\sigma_1(\omega)$ from the standard Kubo formula, and the imaginary part of optical conductivity $\sigma_2(\omega)$ using the Kramers-Kroning tranform; the (complex) dielectric function is then obtained via [73]

$$\epsilon(\omega) = 1 + 4\pi i \frac{\sigma(\omega)}{\omega}. \quad (3)$$

The results for the single-particle density of states and the optical conductivity are shown in Figure 9, and results for the DC transport and the low-frequency dielectric response are displayed in Figure 10, for the parameter range corresponding to the correlated metallic phase ($U \lesssim U_{c1}$). Here panels (a) and (b) reproduce the results of Ref. [29], showing the characteristic scaling behavior of the resistivity maxima near the Mott point. The analogous behavior for the dielectric function ϵ_1 is shown in panels (c) and (d), firmly establishing that the observed crossover behavior assumes a universal scaling form in the correlated metallic regime. The notion that the thermal destruction of quasiparticles lies behind both phenomena is seen even more clearly in Figure 11, where we show how the scale T_{max} for the resistivity maxima, and the scale T_{drop} of the dielectric response, both scale with the quasiparticles weight $Z = m/m^*$, as the transition is approached.

These results establish a way to experimentally recognize the thermal destruction of quasiparticles as a dominant mechanism behind the resistivity maxima within a correlated but uniform metallic phase. Since the correlation processes captured by DMFT are essentially *local* (i.e. "ultraviolet, UV"), these effects should not display significant dependency on spatial dimensional. Indeed, experiments have shown that similar resistivity maxima are seen within correlated metallic phases both in 2D and in 3D systems.

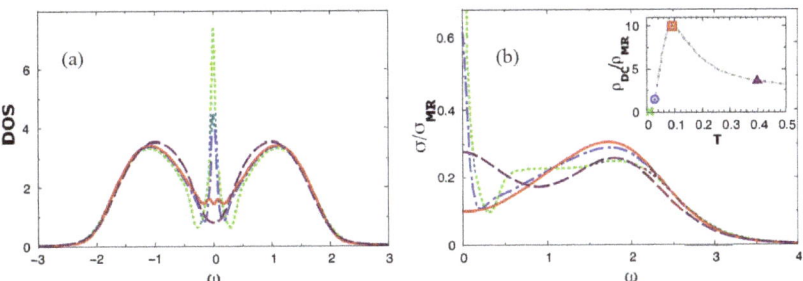

Figure 9. (a) DMFT results for the evolution of the single-particle Density of States (DOS) for several values of the temperature (reprinted with permission from Ref. [66] Copyright 2010 American Physical Society), as well as (b) that of the optical conductivity, in the strongly correlated metallic regime. Different colors correspond to the four distinctive transport regimes (inset in (b)). DOS features a distinct quasiparticle peak at low temperatures, which is thermally destroyed at temperature $T_{max} = T_{BR} \sim (m^*)^{-1}$, where the resistivity (inset of right panel) reaches a maximum. The optical conductivity displays the corresponding suppression of the low-frequency Drude peak around the same temperature.

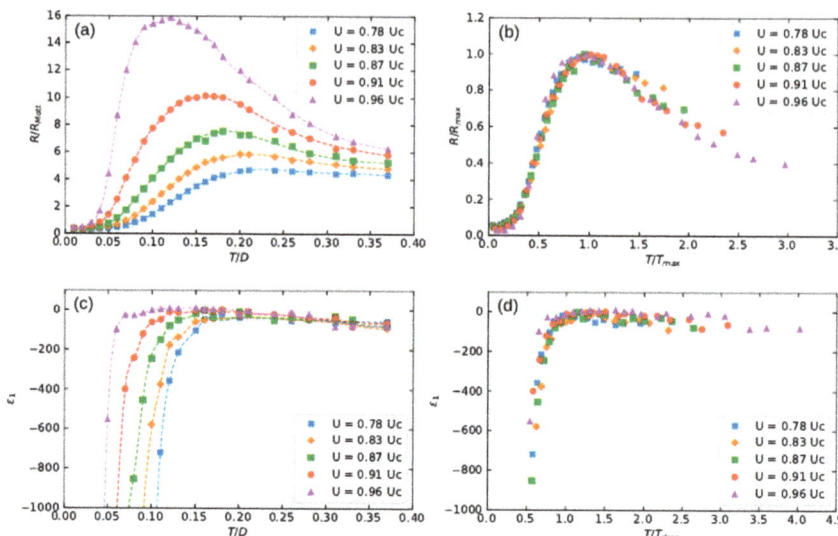

Figure 10. (**a**) DC resistivity as a function of temperature for several interaction strengths. (**b**) Scaled resistivity curves. (**c**) Real part of dielectric function ϵ_1 at $\omega/D = 0.01$, as a function of temperature for several interaction strengths. (**d**) Scaled dielectric function curves. Results are obtained for a half-filled Hubbard model solved within DMFT.

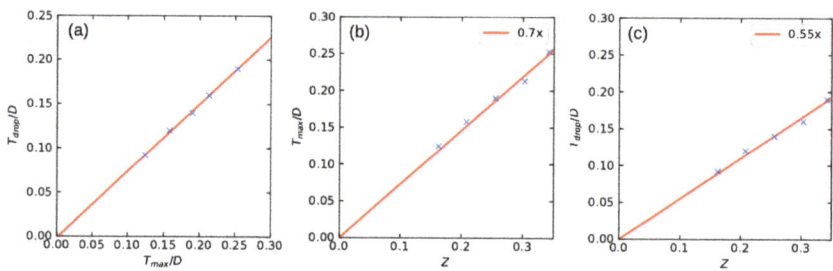

Figure 11. (**a**) T_{drop} as a function of T_{max}. (**b**) T_{max} as a function of Z. (**c**) T_{drop} as a function of Z.

Note, however, that DMFT predicts very different behavior closer to the Mott point, specifically within the phase coexistence region at $T < T_c$ and $U_{c1} < U < U_{c2}$. Here, just as around any first-order phase transition line, we expect hysteresis phenomena and inhomogeneous phase separation, where metallic and insulating domains coexist on a nano-scale. As stressed in the seminal work by Spivak and Kivelson [55], thermal effects can modify the relative volume fraction of the two coexisting phases, producing under appropriate conditions the characteristic resistivity maxima. In recent work motivated by experiments, a microscopic "hybrid-DMFT" approach was developed [20] to quantitatively describe this regime in the context of Mott organics, resulting in spectacular agreement with experiments. In the following section, however, we wish to stress that the *qualitative* aspects of this regime display a number of universal scaling features, which can be precisely understood from the perspective of percolation theory.

5.2. Percolation Scenario Due to Phase Coexistence

To focus on the universal scaling aspects of percolative processes within the metal-insulator phase coexistence region, we follow the seminal ideas of Efros and Shklovskii [74], and set up a two-component random resistor network model, with characteristic low-frequency form for the (complex) conductivity for each component:

$$\sigma_I = \sigma_I^0 \exp(-\Delta/T) - iC\omega,$$
$$\sigma_M = \frac{\sigma_M^0}{1 - i\omega\tau}. \tag{4}$$

Here we assumed activated DC transport for the insulating component, with capacitance C and a standard Drude form for the conducting component, with finite DC conductivity σ_M^0. To leading order near the percolation point, we ignore the T-dependence of σ_M^0, σ_I^0, and C, since the dominant effects come from the variation of the respective volume fractions, and the activated form of insulating transport. The temperature is expressed in the units of the activation energy Δ, which is also taken to be a constant. The corresponding expressions for the (complex) dielectric functions of the two components are given by:

$$\epsilon_I = 1 + 4\pi C + \frac{4\pi i}{\omega}\sigma_I^0 \exp(-\Delta/T),$$
$$\epsilon_M = 1 - 4\pi\tau\sigma_M^0 + \frac{4\pi i}{\omega}\sigma_M^0. \tag{5}$$

Here we ignored the capacitance of the metallic domains, which can be neglected if $\tau\sigma_M^0/C \gg 1$.

Such a random resistor network model is appropriate for any percolating two component metal-insulator system. To describe formation of the resistivity maxima, an additional physical condition has to be met, as emphasized by Spivak and Kivelson in the context of Wigner–Mott transitions, but which is in fact valid for any Mott-like system in general. As we mentioned before, this "Pomeranchuk effect" [45] requires that the first-order line (and the entire phase coexistence region) be "tilted" towards the metal, so that the higher-entropy phase emerges at higher temperatures. To schematically represent such a situation we assume that, within the phase coexistence region, the volume fraction x of the metallic component *decreases* with temperature. As an illustration, we take the following simple model:

$$x(x_o, T) = x_c + \frac{1}{2}\tanh[\frac{x_o - x^*(T)}{w(T)}], \tag{6}$$

where x_c represents the percolation threshold, $w(T) = a(T_c - T)/T_c$ defines the width of the coexistence region, and $x^*(T) = x_c + b(T/T_c)$, as illustrated in Figure 12a, for $a = 0.4$ and $b = 1$. In this model, the parameter x_o controls the metallic volume fraction at $T = 0$, which decreases at $T > 0$, and reaches the percolation threshold $x = x_c$ at $T = T^*(x_o) = T_c(x_o - x_c)/b$. Physically, the the DC resistivity will first increase with T as the metallic volume fraction decreases. Past percolation threshold, however, the metallic domains no longer connect. Transport then assumes insulating (activated) form, resulting in subsequent resistivity decrease at $T > T^*$, and the emergence of resistivity maxima around $T \sim T^*$. Similarly, the dielectric constant ϵ_1 grows (diverges) as the percolation threshold is approached from the insulating side, due to the formation of large metallic clusters with increased polarizability. On the metallic side, however, it displays a rapid decrease, dropping to large negative values within the metallic phase. As a result, dielectric response displays colossal enhancement around the percolation threshold, a phenomenon that can be viewed as a smoking gun for percolative charge dynamics.

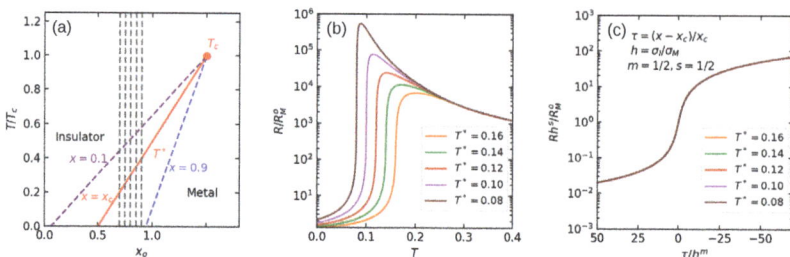

Figure 12. (a) The red line is $x = x(T^*)$. For T larger than the blue dashed line, $x = 0$. We calculate the percolation results along the grey dashed lines. (b) R/R_M^0 as a function of T for different T^*. (c) Scaled resistivity curves.

To illustrate these ideas, we use the Effective Medium Approximation (EMA) for percolation, which solves the following nonlinear equations for the complex dielectric function:

$$x\left(\frac{\epsilon_M - \epsilon}{\epsilon_M + (z/2 - 1)\epsilon}\right) + (1-x)\left(\frac{\epsilon_I - \epsilon}{\epsilon_I + (z/2 - 1)\epsilon}\right) = 0, \qquad (7)$$

and for illustration selected $\sigma_M^0/\sigma_I^0 = 100$, $\tau\sigma_M^0 = 1000$, $C = 1$, $T_c/\Delta = 0.4$, and $z = 4$ corresponding to 2D transport ($x_c = 0.5$). Precisely the anticipated behavior is observed from numerically solving the EMA equation for the corresponding DC resistivity $R = \sigma^{-1}$, as shown in Figure 12b. Here, we select several values of x_0, corresponding to $x > x_c$ (low temperature metallic regime), and plot the resistivity as a function of temperature (following dashed lines in Figure 12a. We observe distinct resistivity maxima around the temperature $T^*(x_0)$ corresponding to the percolation threshold. Note how the maxima become sharper and sharper as T^* is reduced, corresponding to the exponential (activated) decrease of the "field" $h \sim \exp\{-\Delta/T^*\}$. The expected behavior is also seen in dielectric response, as shown in Figure 13a, where we observe sharp maxima at $T \sim T^*$. Here again we see the increased "rounding" of these maxima at higher T^*, corresponding to larger $h(T^*)$. This behavior can be seen even more clearly in Figure 13b, where ϵ_1 is plotted as a function of the reduced concentration $\tau = (x(T) - x_c)/x_c$, which vanishes at $T = T^*$.

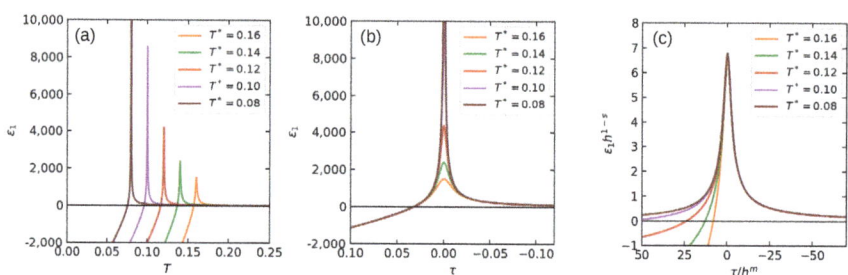

Figure 13. (a) The dielectric constant ϵ_1 as a function of T for different T^*. (b) ϵ_1 as a function of τ for different T^*. (c) Scaled dielectric function curves.

These qualitative trend can be even more rigorously described within the scaling theory for percolation, where the DC conductivity as well as the $\omega = 0$ dielectric constant are known to satisfy the following scaling relations:

$$\sigma_1(\tau, h) = \sigma_M^0 h^s F_\sigma(\tau/h^m); \quad \epsilon_1(\tau, h) = h^{s-1} F_\epsilon(\tau/h^m), \qquad (8)$$

where $\tau = (x(T) - x_c)/x_c$ measures the distance to the percolation threshold, and $h = \sigma_I/\sigma_M$ plays a role of the "symmetry breaking field", which leads to the rounding of the transition. The critical exponents s and m, as well as the crossover scaling functions F_σ and F_ϵ are universal quantities within percolation theory. To illustrate this scaling behavior within EMA, we collapse the family of resistivity curves by plotting Rh^s as a function of τ/h^m, as shown in Figure 12c, and $\epsilon_1 h^{1-s}$ as a function of s a function of τ/h^m, as shown in Figure 13c. Note how a perfect scaling collapse is observed here, but only around the peak of the dielectric response, i.e. only close to the percolation threshold. Such behavior is, in fact, not surprising, since we expect scaling phenomena to arise only within a given critical region, and not further away from the critical point.

We should stress again that all our qualitative results are rigorously valid within general percolation theory for our two-component phase coexistence model, and EMA was simply used as an illustration. EMA correctly captures the general crossover phenomena associated with percolation, but only introduces approximate values for the critical exponents $s_{EMA} = 0.5$ and $m_{EMA} = 0.5$, which are otherwise know even more accurately from numerical simulations. These details, however, are not of direct relevance for our purposes. What is important is the result that, within our "Pomeranchuk" model for phase coexistence, the percolation scenario predicts distinct resistivity maxima but also striking colossal dielectric anomalies, at the same temperature scale of $T = T^*$ which decreases towards the MIT. This behavior is in distinct contrast to the behavior we found from the DMFT picture of a correlated but uniform metallic phase, which also leads to resistivity maxima, but very different behavior of the dielectric response. This observation, which was quantitatively validated in recent experiments on Mott organics [20], thus reveals a distinct criterion to settle the long-lasting controversies between the origin of the resistivity maxima in different systems.

6. Conclusions

In this paper we discussed three different classes of physical systems which all display very similar phenomenology expected for Mott-like metal-insulator transitions. We stressed that most qualitative features are clearly seen in all these examples, including the continuous decrease of the characteristic energy scales T_{FL}, $T_{BR} = T_{max}$, T_o, Δ towards the transition, the phenomenon of quantum critical scaling seen in transport, as well as the emergence of distinct resistivity maxima on the metallic side. These observations, which are starting to portray a robust and consistent phenomenology of Mott criticality, is putting serious constraints on theory. We discussed which of these features seem compatible with various proposed theoretical pictures of the Mott point, and which ones do not.

In the final section of this paper we also presented new theoretical results, which open the possibility to precisely determine, from experiments, which mechanism dominates in which regime. We argue that the *dielectric response* offers unique insights, which so far have not been appreciated enough, as a powerful tool to distinguish between different phase coexistence and the thermal destruction of quasiparticles.

A class of issues we did not discuss in any detail in this paper is the (explicit) role of disorder around the Mott point. Given the fact that new classes of ultra-clean material are starting to emerge, with even more pronounced salient features of Mott criticality, it is becoming possible to plausibly minimize the role of disorder on experimental grounds. On the other hand, new experimental efforts are starting [69,70] to emerge in the opposite direction: to systematically add and to control the level of disorder, for example by high-energy X-ray irradiation. These fascinating research directions are guaranteed to open entirely new chapters in the study of metal-insulator quantum criticality. This will require theorists to rekindle the efforts to understand the interplay of strong correlation with disorder [75], and perhaps to develop new ideas in the process.

Author Contributions: Y.T. carried out theoretical calculations and performed the analyses. V.D. and L.R. designed the project. All authors discussed the data, interpreted the results, and wrote the paper. All authors have read and agreed to the published version of the manuscript.

Funding: This work was supported by the NSF Grant No. 1822258, and the National High Magnetic Field Laboratory through the NSF Cooperative Agreement No. 1644779 and the State of

Florida. LR acknowledges support by the Swiss National Science Foundation via Ambizione grant PZ00P2_174208.

Institutional Review Board Statement: Not applicable.

Informed Consent Statement: Not applicable.

Data Availability Statement: Data are provided: https://github.com/yutingtanphysics/ (accessed on 3 May 2022) How-to-Recognize-the-Universal-Aspects-of-Mott-Criticality.

Acknowledgments: We thank Pak Ki Henry Tsang for providing technical support in doing computer cluster calculations.

Conflicts of Interest: The authors declare no conflict of interest. The funders had no role in the design of the study; in the collection, analyses, or interpretation of data; in the writing of the manuscript, or in the decision to publish the results.

References

1. Phillips, P. Mottness. *Ann. Phys.* **2006**, *321*, 1634–1650. [CrossRef]
2. Mott, N. *Metal-Insulator Transitions*; Taylor & Francis: Abingdon, UK, 1990.
3. Imada, M.; Fujimori, A.; Tokura, Y. Metal-insulator transitions. *Rev. Mod. Phys.* **1998**, *70*, 1039–1263. [CrossRef]
4. Dobrosavljević, V.; Trivedi, N.; Valles Jr, J.M. *Conductor Insulator Quantum Phase Transitions*; Oxford University Press: Oxford, UK, 2012.
5. Cao, Y.; Fatemi, V.; Demir, A.; Fang, S.; Tomarken, S.L.; Luo, J.Y.; Sanchez-Yamagishi, J.D.; Watanabe, K.; Taniguchi, T.; Kaxiras, E.; et al. Correlated insulator behaviour at half-filling in magic-angle graphene superlattices. *Nature* **2018**, *556*, 80–84. [CrossRef] [PubMed]
6. Szentpéteri, B.; Rickhaus, P.; Vries, F.K.d.; Márffy, A.; Fülöp, B.; Tóvári, E.; Watanabe, K.; Taniguchi, T.; Kormányos, A.; Csonka, S.; et al. Tailoring the Band Structure of Twisted Double Bilayer Graphene with Pressure. *Nano Lett.* **2021**, *21*, 8777–8784. [CrossRef] [PubMed]
7. Mooij, J.H. Electrical conduction in concentrated disordered transition metal alloys. *Phys. Status Solidi A* **1973**, *17*, 521–530. [CrossRef]
8. Ciuchi, S.; Sante, D.D.; Dobrosavljević, V.; Fratini, S. The origin of Mooij correlations in disordered metals. *NPJ Quantum Mater.* **2018**, *3*, 1–6. [CrossRef]
9. Evers, F.; Mirlin, A.D. Anderson transitions. *Rev. Mod. Phys.* **2008**, *80*, 1355–1417. [CrossRef]
10. Lee, P.A.; Nagaosa, N.; Wen, X.G. Doping a Mott insulator: Physics of high-temperature superconductivity. *Rev. Mod. Phys.* **2006**, *78*, 17–85. [CrossRef]
11. Balents, L. Spin liquids in frustrated magnets. *Nature* **2010**, *464*, 199–208. [CrossRef]
12. Sondhi, S.L.; Girvin, S.M.; Carini, J.P.; Shahar, D. Continuous quantum phase transitions. *Rev. Mod. Phys.* **1997**, *69*, 315–333. [CrossRef]
13. Saito, Y.; Nojima, T.; Iwasa, Y. Highly crystalline 2D superconductors. *Nat. Rev. Mater.* **2016**, *2*, 74. [CrossRef]
14. Lee, P.A.; Ramakrishnan, T.V. Disordered electronic systems. *Rev. Mod. Phys.* **1985**, *57*, 287–337. [CrossRef]
15. Belitz, D.; Kirkpatrick, T.R. The Anderson-Mott transition. *Rev. Mod. Phys.* **1994**, *66*, 261–380. [CrossRef]
16. Abrahams, E.; Kravchenko, S.V.; Sarachik, M.P. Metallic behavior and related phenomena in two dimensions. *Rev. Mod. Phys.* **2001**, *73*, 251. [CrossRef]
17. Spivak, B.; Kravchenko, S.V.; Kivelson, S.A.; Gao, X.P.A. Colloquium: Transport in strongly correlated two dimensional electron fluids. *Rev. Mod. Phys.* **2010**, *82*, 1743–1766. [CrossRef]
18. Kravchenko, S. *Strongly Correlated Electrons in Two Dimensions*; Jenny Stanford Publishing: New York, NY, USA, 2017.
19. Melnikov, M.Y.; Shashkin, A.A.; Dolgopolov, V.T.; Zhu, A.Y.X.; Kravchenko, S.V.; Huang, S.H.; Liu, C.W. Quantum phase transition in ultrahigh mobility SiGe/Si/SiGe two-dimensional electron system. *Phys. Rev. B* **2019**, *99*, 081106. [CrossRef]
20. Pustogow, A.; Rösslhuber, R.; Tan, Y.; Uykur, E.; Böhme, A.; Wenzel, M.; Saito, Y.; Löhle, A.; Hübner, R.; Kawamoto, A.; et al. Low-temperature dielectric anomaly arising from electronic phase separation at the Mott insulator-metal transition. *NPJ Quantum Mater.* **2021**, *6*, 9. [CrossRef]
21. Li, T.; Jiang, S.; Li, L.; Zhang, Y.; Kang, K.; Zhu, J.; Watanabe, K.; Taniguchi, T.; Chowdhury, D.; Fu, L.; et al. Continuous Mott transition in semiconductor moiré superlattices. *Nature* **2021**, *597*, 350–354. [CrossRef]
22. Simonian, D.; Kravchenko, S.V.; Sarachik, M.P. Reflection symmetry at a B=0 metal-insulator transition in two dimensions. *Phys. Rev. B* **1997**, *55*, R13421–R13423. [CrossRef]
23. Dobrosavljević, V.; Abrahams, E.; Miranda, E.; Chakravarty, S. Scaling Theory of Two-Dimensional Metal-Insulator Transitions. *Phys. Rev. Lett.* **1997**, *79*, 455–458. [CrossRef]
24. Kravchenko, S.V.; Mason, W.E.; Bowker, G.E.; Furneaux, J.E.; Pudalov, V.M.; D'iorio, M. Scaling of an anomalous metal-insulator transition in a two-dimensional system in silicon at B=0. *Phys. Rev. B* **1995**, *51*, 7038–7045. [CrossRef] [PubMed]
25. Furukawa, T.; Miyagawa, K.; Taniguchi, H.; Kato, R.; Kanoda, K. Quantum criticality of Mott transition in organic materials. *Nat. Phys.* **2015**, *11*, 221–224. [CrossRef]

26. Ando, T.; Fowler, A.B.; Stern, F. Electronic properties of two-dimensional systems. *Rev. Mod. Phys.* **1982**, *54*, 437–672. [CrossRef]
27. Camjayi, A.; Haule, K.; Dobrosavljević, V.; Kotliar, G. Coulomb correlations and the Wigner–Mott transition. *Nat. Phys.* **2008**, *4*, 932–935. [CrossRef]
28. Amaricci, A.; Camjayi, A.; Haule, K.; Kotliar, G.; Tanasković, D.; Dobrosavljević, V. Extended hubbard model: Charge ordering and Wigner–Mott transition. *Phys. Rev. B* **2010**, *82*, 155102. [CrossRef]
29. Radonjić, M.M.; Tanasković, D.; Dobrosavljević, V.; Haule, K.; Kotliar, G. Wigner–Mott scaling of transport near the two-dimensional metal-insulator transition. *Phys. Rev. B* **2012**, *85*, 085133–7. [CrossRef]
30. Shashkin, A.; Melnikov, M.Y.; Dolgopolov, V.; Radonjić, M.; Dobrosavljević, V.; Huang, S.H.; Liu, C.; Zhu, A.Y.; Kravchenko, S. Manifestation of strong correlations in transport in ultraclean SiGe/Si/SiGe quantum wells. *Phys. Rev. B* **2020**, *102*, 081119. [CrossRef]
31. Shashkin, A.; Melnikov, M.Y.; Dolgopolov, V.; Radonjić, M.; Dobrosavljević, V.; Huang, S.H.; Liu, C.; Zhu, A.Y.; Kravchenko, S. Spin effect on the low-temperature resistivity maximum in a strongly interacting 2D electron system. *Sci. Rep.* **2022**, *12*, 5080. [CrossRef]
32. Moon, B.H.; Han, G.H.; Radonjić, M.M.; Ji, H.; Dobrosavljević, V. Quantum critical scaling for finite-temperature Mott-like metal-insulator crossover in few-layered MoS 2. *Phys. Rev. B* **2020**, *102*, 245424. [CrossRef]
33. Popović, D.; Fowler, A.B.; Washburn, S. Metal-Insulator Transition in Two Dimensions: Effects of Disorder and Magnetic Field. *Phys. Rev. Lett.* **1997**, *79*, 1543–1546. [CrossRef]
34. Bogdanovich, S.c.v.; Popović, D. Onset of Glassy Dynamics in a Two-Dimensional Electron System in Silicon. *Phys. Rev. Lett.* **2002**, *88*, 236401. [CrossRef] [PubMed]
35. Shashkin, A.A.; Kravchenko, S.V.; Klapwijk, T.M. Metal-Insulator Transition in a 2D Electron Gas: Equivalence of Two Approaches for Determining the Critical Point. *Phys. Rev. Lett.* **2001**, *87*, 266402. [CrossRef] [PubMed]
36. Shashkin, A.A.; Kravchenko, S.V.; Dolgopolov, V.T.; Klapwijk, T.M. Sharp increase of the effective mass near the critical density in a metallic two-dimensional electron system. *Phys. Rev. B* **2002**, *66*, 073303. [CrossRef]
37. Smolenski, T.; Dolgirev, P.E.; Kuhlenkamp, C.; Popert, A.; Shimazaki, Y.; Back, P.; Lu, X.; Kroner, M.; Watanabe, K.; Taniguchi, T.; et al. Signatures of Wigner crystal of electrons in a monolayer semiconductor. *Nature* **2021**, *595*, 53. [CrossRef] [PubMed]
38. Dressel, M.; Tomić, S. Molecular quantum materials: Electronic phases and charge dynamics in two-dimensional organic solids. *Adv. Phys.* **2020**, *69*, 1–120. [CrossRef]
39. Jerome, D. The physics of organic superconductors. *Science* **1991**, *252*, 1509–1514. [CrossRef]
40. Kagawa, F.; Sato, T.; Miyagawa, K.; Kanoda, K.; Tokura, Y.; Kobayashi, K.; Kumai, R.; Murakami, Y. Charge-cluster glass in an organic conductor. *Nat. Phys.* **2013**, *9*, 419–422. [CrossRef]
41. Kurosaki, Y.; Shimizu, Y.; Miyagawa, K.; Kanoda, K.; Saito, G. Mott Transition from a Spin Liquid to a Fermi Liquid in the Spin-Frustrated Organic Conductor $\kappa-(ET)_2Cu_2(CN)_3$. *Phys. Rev. Lett.* **2005**, *95*, 177001. [CrossRef]
42. Pustogow, A.; Bories, M.; Löhle, A.; Rösslhuber, R.; Zhukova, E.; Gorshunov, B.; Tomić, S.; Schlueter, J.A.; Hübner, T.; Hiramatsu, T.; et al. Quantum spin liquids unveil the genuine Mott state. *Nat. Mater.* **2018**, *17*, 773–777. [CrossRef]
43. Vučičević, J.; Terletska, H.; Tanasković, D.; Dobrosavljević, V. Finite-temperature crossover and the quantum Widom line near the Mott transition. *Phys. Rev. B* **2013**, *88*, 075143. [CrossRef]
44. Pustogow, A.; Saito, Y.; Löhle, A.; Sanz Alonso, M.; Kawamoto, A.; Dobrosavljević, V.; Dressel, M.; Fratini, S. Rise and fall of Landau's quasiparticles while approaching the Mott transition. *Nat. Commun.* **2021**, *12*, 1571. [CrossRef] [PubMed]
45. Pomeranchuk, I. On the thery of He^3. *Zh. Eksp. Teor. Fiz.* **1950**, *20*, 919.
46. Terletska, H.; Vucicevic, J.; Tanasković, D.; Dobrosavljević, V. Quantum Critical Transport near the Mott Transition. *Phys. Rev. Lett.* **2011**, *107*, 026401. [CrossRef]
47. Deng, X.; Mravlje, J.; Žitko, R.; Ferrero, M.; Kotliar, G.; Georges, A. How Bad Metals Turn Good: Spectroscopic Signatures of Resilient Quasiparticles. *Phys. Rev. Lett.* **2013**, *110*, 086401. [CrossRef] [PubMed]
48. Emery, V.J.; Kivelson, S.A. Superconductivity in Bad Metals. *Phys. Rev. Lett.* **1995**, *74*, 3253–3256. [CrossRef] [PubMed]
49. Hussey, N.E.; Takenaka, K.; Takagi, H. Universality of the Mott–Ioffe–Regel limit in metals. *Philos. Mag.* **2004**, *84*, 2847–2864. [CrossRef]
50. Rademaker, L. Spin-Orbit Coupling in Transition Metal Dichalcogenide Heterobilayer Flat Bands. *arXiv* **2021**, arXiv:2111.06208.
51. Ghiotto, A.; Shih, E.M.; Pereira, G.S.S.G.; Rhodes, D.A.; Kim, B.; Zang, J.; Millis, A.J.; Watanabe, K.; Taniguchi, T.; Hone, J.C.; et al. Quantum Criticality in Twisted Transition Metal Dichalcogenides. *Nature* **2021**, *597*, 345. [CrossRef]
52. Savary, L.; Balents, L. Quantum spin liquids: A review. *Rep. Prog. Phys.* **2016**, *80*, 016502. [CrossRef]
53. Senthil, T. Theory of a continuous Mott transition in two dimensions. *Phys. Rev. B* **2008**, *78*, 045109. [CrossRef]
54. Georges, A.; Kotliar, G.; Krauth, W.; Rozenberg, M.J. Dynamical mean-field theory of strongly correlated fermion systems and the limit of infinite dimensions. *Rev. Mod. Phys.* **1996**, *68*, 13–125. [CrossRef]
55. Spivak, B.; Kivelson, S.A. Phases intermediate between a two-dimensional electron liquid and Wigner crystal. *Phys. Rev. B* **2004**, *70*, 155114. [CrossRef]
56. Spivak, B.; Kivelson, S.A. Transport in two dimensional electronic micro-emulsions. *Ann. Phys.* **2006**, *321*, 2071–2115. [CrossRef]
57. Baskaran, G.; Zou, Z.; Anderson, P. The resonating valence bond state and high-Tc superconductivity—A mean field theory. *Solid State Commun.* **1987**, *63*, 973–976. [CrossRef]

58. Jacko, A.; Fjærestad, J.; Powell, B. A unified explanation of the Kadowaki–Woods ratio in strongly correlated metals. *Nat. Phys.* **2009**, *5*, 422–425. [CrossRef]
59. Podolsky, D.; Paramekanti, A.; Kim, Y.B.; Senthil, T. Mott Transition between a Spin-Liquid Insulator and a Metal in Three Dimensions. *Phys. Rev. Lett.* **2009**, *102*, 186401. [CrossRef]
60. Xu, Y.; Wu, X.C.; Ye, M.X.; Luo, Z.X.; Jian, C.M.; Xu, C. Metal-Insulator Transition with Charge Fractionalization. *arXiv* **2021**, arXiv:2106.14910.
61. Musser, S.; Senthil, T.; Chowdhury, D. Theory of a Continuous Bandwidth-tuned Wigner–Mott Transition. *arXiv* **2021**, arXiv:2111.09894.
62. Vučičević, J.; Tanasković, D.; Rozenberg, M.; Dobrosavljević, V. Bad-metal behavior reveals Mott quantum criticality in doped Hubbard models. *Phys. Rev. Lett.* **2015**, *114*, 246402.
63. Lee, T.H.; Florens, S.; Dobrosavljević, V. Fate of spinons at the Mott point. *Phys. Rev. Lett.* **2016**, *117*, 136601.
64. Brinkman, W.F.; Rice, T.M. Application of Gutzwiller's Variational Method to the Metal-Insulator Transition. *Phys. Rev. B* **1970**, *2*, 4302–4304. [CrossRef]
65. Eisenlohr, H.; Lee, S.S.B.; Vojta, M. Mott quantum criticality in the one-band Hubbard model: Dynamical mean-field theory, power-law spectra, and scaling. *Phys. Rev. B* **2019**, *100*, 155152. [CrossRef]
66. Radonjić, Miloš M.; Tanasković, D.; Dobrosavljević, V.; Haule K. Influence of disorder on incoherent transport near the Mott transition. *Phys. Rev. B* **2010**, *81*, 075118. [CrossRef]
67. Jamei, R.; Kivelson, S.; Spivak, B. Universal Aspects of Coulomb-Frustrated Phase Separation. *Phys. Rev. Lett.* **2005**, *94*, 056805. [CrossRef]
68. Dagotto, E. Complexity in Strongly Correlated Electronic Systems. *Science* **2005**, *309*, 257–262. [CrossRef]
69. Urai, M.; Furukawa, T.; Seki, Y.; Miyagawa, K.; Sasaki, T.; Taniguchi, H.; Kanoda, K. Disorder unveils Mott quantum criticality behind a first-order transition in the quasi-two-dimensional organic conductor $\kappa-(ET)_2Cu[N(CN)_2]Cl$. *Phys. Rev. B* **2019**, *99*, 245139. [CrossRef]
70. Yamamoto, R.; Furukawa, T.; Miyagawa, K.; Sasaki, T.; Kanoda, K.; Itou, T. Electronic Griffiths Phase in Disordered Mott-Transition Systems. *Phys. Rev. Lett.* **2020**, *124*, 046404. [CrossRef]
71. Aharony, A.; Imry, Y.; Ma, S.k. Lowering of Dimensionality in Phase Transitions with Random Fields. *Phys. Rev. Lett.* **1976**, *37*, 1364–1367. [CrossRef]
72. Kim, S.; Senthil, T.; Chowdhury, D. Continuous Mott transition in moiré semiconductors: Role of long-wavelength inhomogeneities. *arXiv* **2022**, arXiv:2204.10865.
73. Economou, E.N. *Green's Functions in Quantum Physics*; Springer: Berlin, Germany, 2005.
74. Efros, A.L.; Shklovskii, B.I. Critical Behaviour of Conductivity and Dielectric Constant near the Metal-Non-Metal Transition Threshold. *Phys. Stat. Sol. B* **1976**, *76*, 475. [CrossRef]
75. Miranda, E.; Dobrosavljević, V. Disorder-driven non-Fermi liquid behaviour of correlated electrons. *Rep. Prog. Phys.* **2005**, *68*, 2337. [CrossRef]

Article

Neural Network Solver for Small Quantum Clusters

Nicholas Walker [1], Samuel Kellar [1], Yi Zhang [1,2], Ka-Ming Tam [1,3,*] and Juana Moreno [1,3,*]

1. Department of Physics and Astronomy, Louisiana State University, Baton Rouge, LA 70803, USA
2. Kavli Institute for Theoretical Sciences, University of Chinese Academy of Sciences, Beijing 100190, China
3. Center for Computation & Technology, Louisiana State University, Baton Rouge, LA 70803, USA
* Correspondence: phy.kaming@gmail.com (K.-M.T.); moreno@lsu.edu (J.M.)

Abstract: Machine learning approaches have recently been applied to the study of various problems in physics. Most of these studies are focused on interpreting the data generated by conventional numerical methods or the data on an existing experimental database. An interesting question is whether it is possible to use a machine learning approach, in particular a neural network, for solving the many-body problem. In this paper, we present a neural network solver for the single impurity Anderson model, the paradigm of an interacting quantum problem in small clusters. We demonstrate that the neural-network-based solver provides quantitative accurate results for the spectral function as compared to the exact diagonalization method. This opens the possibility of utilizing the neural network approach as an impurity solver for other many-body numerical approaches, such as the dynamical mean field theory.

Keywords: metal insulator transition; anderson localization; random disorder; typical medium theory; dynamical mean field theory; coherent potential approximation; neural network; quantum impurity solver; Anderson impurity

Citation: Walker, N.; Kellar, S.; Zhang, Y.; Tam, K.-M.; Moreno, J. Neural Network Solver for Small Quantum Clusters. *Crystals* **2022**, *12*, 1269. https://doi.org/10.3390/cryst12091269

Academic Editor: Andrej Pustogow

Received: 2 August 2022
Accepted: 29 August 2022
Published: 6 September 2022

Publisher's Note: MDPI stays neutral with regard to jurisdictional claims in published maps and institutional affiliations.

Copyright: © 2022 by the authors. Licensee MDPI, Basel, Switzerland. This article is an open access article distributed under the terms and conditions of the Creative Commons Attribution (CC BY) license (https:// creativecommons.org/licenses/by/ 4.0/).

1. Introduction

A single quantum impurity is the simplest quantum many-body problem for which interaction plays a crucial role [1,2]. It was designed as a model to describe diluted magnetic impurities in a non-magnetic metallic host. In the 1960s, it was showed that the perturbation series in coupling strength diverges even with an infinitesimal anti-ferromagnetic coupling value [1]. This early unexpected result motivated the development of innovative approaches to model the strongly correlated systems [1–3].

While the physics of a single impurity problem has been rather well studied, interest in the quantum impurity problem was revived during the 1990s. This was partly due to the interest in mapping lattice models onto impurity models [4–8], since at infinite dimensions, the lattice models are equivalent to single impurity models in a mean-field represented by the density of states of the host. This approximated mapping is known as the dynamical mean-field theory. It has been further generalized to cluster impurity models which include some of the effects present in finite dimensional systems [9–11].

These mappings provide a systematic tractable approximation for the lattice models and have become a major approach in the field of strongly correlated systems [11]. Combined with density functional theory, they provide one of the best available methods for the study of the properties of materials in which strong interactions are important [12].

The mean-field where the single impurity is embedded, i.e., the density of the bath, can be rather complicated. Therefore, there is in general no analytic method for an accurate solution. Many different methods for solving the effective impurity problem have been proposed. They can be broadly divided in two categories: semi-analytic and numeric.

Between the semi-analytic methods, the most widely used one is the iterative perturbation theory. It interpolates the self-energy at both the weak and strong coupling limits and incorporates some exact constraints, such as the Luttinger theorem [13]. Another widely used semi-analytic method is the local moment approximation. It considers the perturbation on top of the strong coupling limit represented by the unrestricted Hatree Fock solution [14].

Numerical methods can also be divided into two main classes, diagonalization-based and Monte-Carlo-based. Diagonalization methods usually require discretizing the host by a finite number of so-called bath sites. The Hamiltonian which includes the bath sites and the impurity site are diagonalized exactly [15]. Another digonalization-based method is the numerical renormalization group in which the bath sites are mapped onto a one-dimensional chain of sites. The hopping amplitude decreases rapidly down the chain. The model is then diagonalized iteratively as more sites are included [16]. Density matrix renormalization group and coupled cluster theory have also been used as impurity solvers [17–20].

Hirsh and Fye were the first to propose the quantum Monte Carlo method for solving impurity problems [21]. The time axis of the simulations is divided up using the Trotter–Suzuki approximation. The interaction in each time segment is handled by the Hubbard–Stratonovich approximation [22]. The Monte Carlo method is then used to sample the Hubbard–Stratonovich fields. Continuous time quantum Monte Carlo methods have seen a lot of development over the last decade. They directly sample the partition function without using the Trotter–Suzuki approximation [23], similar to the Stochastic Series Expansion in the simulation of quantum spin models. Notably, the continuous time quantum Monte Carlo using the expansion with respect to the strong coupling limit has been proposed and complicated coupling functions beyond simple Hubbard local density-density coupling terms can now be studied [24].

On the other hand, the past few years have seen tremendous development of machine learning (ML), both algorithms and their implementation [25–27]. Many of the ML approaches in physics are designed to detect phase transitions or accelerate Monte Carlo simulations. It is a tantalizing proposal to utilize ML approaches to build a solver for quantum systems.

To build a quantum solver based on the ML approach, we need to identify the feature vector (input data) and the label (output data) for the problem. Then a large pool of data must be generated to train the model, specifically a neural network model. The Anderson impurity problem is a good test bed for the validity of such a solver. We note that similar ideas have been explored using machine learning approaches [28]. This paper is focused on using the kernel polynomial expansion and supervised ML, specifically a neural network, as the building blocks for a quantum impurity solver.

While it is relatively inexpensive to solve a single impurity problem using the above methods in modern computing facilities, current interest in the effects of disorder warrants the new requirement of solving a large set of single or few impurities problems in order to calculate the random disorder average [29–35]. Our hope is that a fast neural-network-based numerical solver in real frequency can expand the range of applicability of the recently developed typical medium theory to interacting strongly correlated systems, such as the Anderson–Hubbard model [36–39].

This paper is organized as follows. In the next section, we map the continuous Green function into a finite cluster as has been performed in many dynamical mean field theory calculations. In Section 3, we discuss the expansion of the spectral function in terms of Chebyshev polynomials. In Section 4, we discuss how to use the results from Sections 2 and 3 as the feature vectors and labels of the neural network. In Section 5, we present the spectral function calculated by the neural network approach. We conclude and discuss future work in the last section.

2. Representing the Host by a Finite Number of Bath Sites

We first identify the input and the output data of a single impurity Anderson model. The input data includes the bare density of states, the chemical potential and the Hubbard interaction on the impurity site. For a system in the thermodynamic limit, the density of states is represented by a continuous function. Since inputting a continuous function to the neural network presents a problem, we describe the continuous bath by a finite number of poles as it is performed within the exact diagonalization approach [15,40–43]. We approximate the host Green function by a cluster of bath sites,

$$G_0(i\omega_n) \approx G_0^{cl}(i\omega_n). \tag{1}$$

In the exact diagonalization method, the continuum bath is discretized and represented by a finite number of so-called bath sites, see Figure 1. Assuming that there are N_b bath sites, each bath site is characterized by a local energy (ϵ_i) and a hopping (t_i) term to the impurity site. Two additional variables, one the local Hubbard interaction (U) and the other the chemical potential (ϵ_f), are required to describe the impurity site. Therefore, there is a total of $2 + 2N_b$ variables representing the impurity problem.

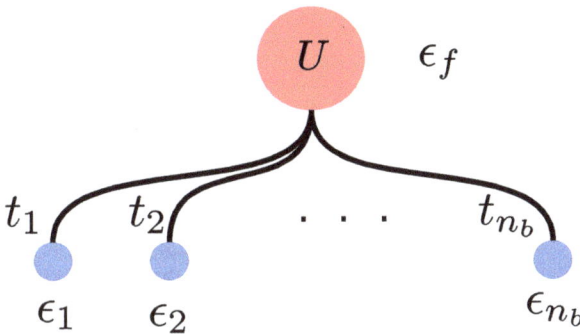

Figure 1. The cluster which represents the quantum impurity model. The red circle represents the impurity site with local interaction U and chemical potential ϵ_f. The bath sites are represented by blue circles; each of them has a local energy ϵ_i, and a hopping to the impurity site t_i.

The full Hamiltonian in the discretized form (Figure 1) is given as

$$H = \sum_{i,\sigma} t_i (c_{i,\sigma}^\dagger c_{0,\sigma} + H.c.) + \sum_{i,\sigma} \epsilon_i c_{i,\sigma}^\dagger c_{i,\sigma} + U(c_{0,\uparrow}^\dagger c_{0,\uparrow} - 1/2)(c_{0,\downarrow}^\dagger c_{0,\downarrow} - 1/2) - \epsilon_f \sum_\sigma c_{0,\sigma}^\dagger c_{0,\sigma}, \tag{2}$$

$c_{i,\sigma}^\dagger$ and $c_{i,\sigma}$ are the creation and annihilation operators for site i with spin σ, respectively. The impurity site is denoted as the 0-th site. The sum of the bath sites goes from 1 to N_b, and the spin $\sigma = \pm 1/2$.

The host Green function represented in such a finite cluster can be written exactly as,

$$G_0^{cl}(i\omega_n) = (i\omega_n + \epsilon_f - \sum_{k=1}^{N_b} \frac{t_k t_k^*}{i\omega_n - \epsilon_k})^{-1}. \tag{3}$$

Many different prescriptions for the parameterization of the host Green function have been investigated in detail for exact diagonalization solvers [44]. Conceptually, practical applications of the numerical renormalization group method also require the approximated mapping of the problem onto a finite cluster chain. Unlike the exact diagonalization method, the cluster chain can be rather large; therefore much higher accuracy can be attained in general.

These approaches do not mimic the continuum in the time dimension as it is done by continuous time Quantum Monte Carlo methods, and the mapping onto the finite cluster may represent a nuisance. Nonetheless, this is a necessity for any diagonalization-based method. In our case, the mapping presents an opportunity to naturally adapt the method to a machine learning approach in which a finite discretized set of variables is required.

Under the above approximation, the finite set of variables, $\{t_i\}, \{\epsilon_i\}, U, \epsilon_f$, can be treated as the input feature vector for the machine learning algorithm. The next question is what is the desired output or label. We will focus on the spectral function in this study. For this purpose, the next step is to represent the spectral function with a finite number of variables instead of a continuous function. The kernel polynomial method fulfills this goal [45].

3. Expanding the Impurity Green Function by Chebyshev Polynomials

In this section, we briefly discuss the kernel polynomial method for the calculation of the spectral function of a quantum interacting model. Once the host parameters, the impurity interaction and chemical potential are fixed, the ground state of the cluster is obtained using a diagonalization method for sparse matrices. We use the Lanczos approach in the present study [46,47]. Once the ground state is found, the spectral function can be calculated by applying the resolvent operator, $1/(\omega - H)$, to the ground state. A popular method is the continuous fraction expansion [47]. The challenge is that the continuous fraction tends to be under-damped and produce spurious peaks [46,47]. A more recent method is to use an orthogonal polynomial expansion. We will argue that for the application of the ML method, the polynomial expansion method tends to produce better results as we will explain later [45].

The zero temperature single particle retarded Green function corresponding to a generic many-body Hamiltonian is defined as

$$G(\omega) = \langle GS|c \frac{1}{\omega + i0^+ - H} c^\dagger |GS\rangle, \tag{4}$$

where $|GS\rangle$ is the ground state of H, and $c^\dagger$ and c are the creation and annihilation operators, respectively [45,48]. The spectral function is given as $A(\omega) = -(1/\pi) Im(G(\omega))$.

Consider the Chebyshev polynomials of the first kind defined as $T_n(x) = cos(n\, arccos(x))$. Two important properties of the Chebyshev polynomial are their orthogonality and their recurrence relation. The product of two Chebyshev polynomials integrated over $x = [-1, 1]$ and weighted by the function $w_n = \frac{1}{\pi\sqrt{1-x^2}}$ is given as

$$\int dx w_n(x) T_n(x) T_m(x) = \frac{1 + \delta_{n,0}}{2} \delta_{n,m}. \tag{5}$$

The recurrence relation is given as

$$T_n(x) = 2x T_{n-1}(x) - T_{n-2}(x). \tag{6}$$

The Chebyshev polynomials expansion method is based on the fact that the set of Chebyshev polynomials form an orthonormal basis as defined in Equation (5). Thus a function, $f(x)$, defined within the range of $x = [-1, 1]$ can be expanded as

$$f(x) = \mu_0 + 2 \sum_{n=1}^{\infty} \mu_n(x) T_n(x), \tag{7}$$

and the expansion coefficient can be obtained by the inner product of the function $f(x)$ and the Chebyshev polynomials as follow

$$\mu_n = \int_{-1}^{1} dx f(x) T_n(x) w_n(x). \tag{8}$$

Practical calculations involve truncation at a finite order. The truncation is found to be problematic when the function to expand is not smooth. In our problem, the function is the spectral function of a finite size cluster, which is a linear combination of a set of delta functions. For this reason, a direct application of the above formula will not provide a smooth function. This is analogue to the Gibbs oscillations in the Fourier expansion. The remedy is to introduce a damping factor (kernel) in each coefficient of the expansion [45,49–52]. Here we use the Jackson kernel given as

$$f(x) \approx \sum_{n=0}^{N} g_n \mu_n(x) T_n(x), \text{where} \tag{9}$$

$$g_n = \frac{(N-n+1)\cos(\frac{\pi n}{N+1}) + \sin(\frac{\pi n}{N+1})\cot(\frac{\pi}{N+1})}{N+1}. \tag{10}$$

We refer the choice of the damping factor to the review by Weiße et al. [45].
We list the steps for calculating the coefficients as follows.

1. The input bare Green function is approximated by the bare Green function of a finite size cluster. The set of parameters $\mu, \{t_i\}, \{\epsilon_i\}$ are obtained by minimizing the difference between the left-hand side and the right-hand side of Equation (1) according to some prescriptions [15,40–43].
2. The ground state ($|GS\rangle$) and the corresponding energy (E_{GS}) are obtained by the Lanczos algorithm.
3. The spectrum of the Hamiltonian is scaled within the range $[-1,1]$ as required by the Chebyshev expansion. $H \Rightarrow (H - E_{GS})/a$, where a is a real positive constant. The units of energy are also scaled in terms of a.
4. The expansion coefficients are given by the inner product between the spectral function and the Chebyshev polynomials.

$$\mu_n = <\alpha_0|\alpha_n>, \tag{11}$$

where $|\alpha_0\rangle = c^\dagger|GS>$ and $|\alpha_n\rangle = T_n(H)|\alpha_0\rangle$. With the $|\alpha_0\rangle$ and the $|\alpha_1\rangle = H|\alpha_0\rangle$ ready, all the higher order coefficients can be obtained via the recurrence relation

$$|\alpha_n\rangle = 2H|\alpha_{n-1}\rangle - |\alpha_{n-2}\rangle. \tag{12}$$

5. The spectral function is obtained by feeding the coefficients into Equation (7).

All the coefficients can be obtained by repeated use of Equation (12) which involves matrix vector multiplication. The matrix for an interacting system is usually very sparse, and the computational complexity of the matrix vector multiplication is linear with respect to the vector length, which grows as 4^{N_b+1} assuming no reduction by symmetry is applied.

Since practical calculations are limited to a finite order, N, the impurity Green function can be represented by N coefficients of the Chebyshev polynomials expansion. It is worthwhile to mention that the expansion of the Green function in terms of orthogonal polynomials is independent of the method for obtaining the coefficients of the polynomials. Instead of employing exact diagonalization, a recent study shows that one can obtain the expansion coefficients by representing the quantum states by matrix products [48].

4. Feature Vectors and the Labels for the Machine Learning Algorithm

Our strategy is to train a neural network with a large set of variables for the host, i.e., the bath sites, the impurity interaction and the impurity chemical potential. The impurity solver calculates the impurity Green function for a given bath Green function, local impurity interaction and chemical potential that is a total of $2 + 2N_b$ variables for the input.

The output is the impurity Green function which can be represented by N coefficients of the Chebyshev polynomials expansion. Using the above method, the spectral function is effectively represented in terms of N coefficients. It allows us to naturally employ the supervised learning method by identifying the $2 + 2N_b$ variables as the input feature vectors and the N variables as the output labels.

While the kernel polynomial method grows exponentially with the number of sites, the end result is represented by a finite number of coefficients which presumably does not scale exponentially with the number of sites. Once the neural network is properly trained, we can use it to predict the impurity Green function without involving a calculation which scales exponentially.

5. Results

We generated 5000 samples for training the neural networks with randomly chosen parameters which are drawn uniformly from the range listed as follows:

$$\begin{aligned} t_{i,\uparrow} &= t_{i,\downarrow} = [0, 1.5], \\ \epsilon_{i,\uparrow} &= \epsilon_{i,\downarrow} = [-5, 5], \\ U &= [0, 10], \\ \epsilon_f &= [-2.5, 2.5]. \end{aligned} \quad (13)$$

We assume that the electron bath has a symmetric density of states. That is, $t_i = t_{i+N_b/2}$ and $\epsilon_i = -\epsilon_{i+N_b/2}$ for $i = 1$ to $N_b/2$ and N_b even. This further reduces the number of variables in the feature vector to $N_b + 2$.

Before embarking on training the neural network, we would like to have some understanding of the range of Chebyshev coefficients. For this purpose, we randomly pick 32 samples and plot the coefficients in Figure 2. There are two prominent features of the coefficients: 1. There are clear oscillations and the coefficients do not decrease monotonically. 2. For all cases shown here, the coefficients essentially vanish for orders around 200 and higher. Due to these two reasons, we decide to train the neural network for the coefficients between 0 and 255.

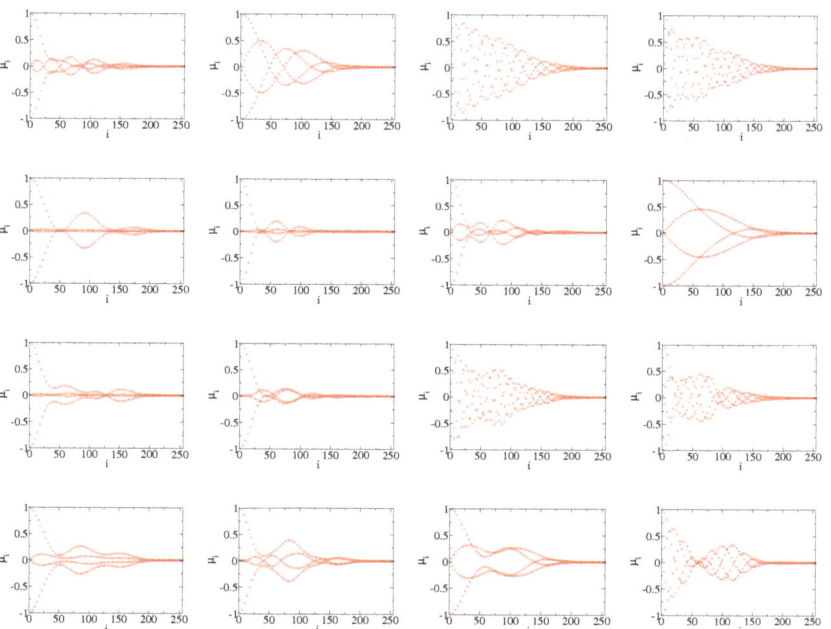

Figure 2. *Cont.*

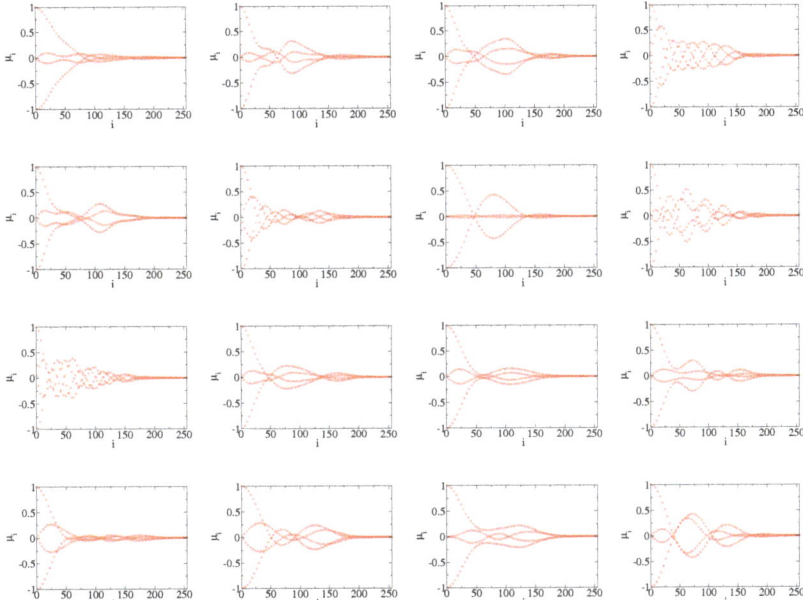

Figure 2. The coefficients of the Chebyshev polynomial expansion for 32 randomly chosen parameter sets of the finite cluster. Only the first 256 coefficients are shown, as higher order terms are vanishingly small. Only the coefficients directly calculated from the kernel polynomial method (KPM) are shown here. The coefficients obtained from the neural network match very closely with the ones from the KPM and would not be visible by laying them on the same plot, and thus they are omitted. The spectral functions and the corresponding parameters are presented in Figures 3 and 4 respectively. We demonstrate the quality of the coefficients obtained from the neural network in Figure 5. The magnitude of the the last five coefficients in each panel is smaller than 10^{-5}.

With the above approximations, the task of solving the Anderson impurity model boils down to mapping a vector containing $N_b + 2$ variables to a vector containing N coefficients. For the particular case, we study we choose $N_b = 6$ and $N = 256$. Machine learning algorithms can thus be naturally applied to this mapping.

We set up an independent dense neural network for each coefficient. The neural network has 14 layers. The input layer contains $N_{in} = N_b + 2$ units, and the output layer contains the expansion coefficient for one specific order. The 12 hidden layers have the following number of units: $2N_{in}, 2N_{in}, 4N_{in}, 4N_{in}, 8N_{in}, 8N_{in}, 8N_{in}, 8N_{in}, 4N_{in}, 4N_{in}, 2N_{in}, 2N_{in}$.

As we consider a total of 256 orders, we train 256 independent neural networks. Considering the coefficients at different orders separately may lose some information contained in the correlations between them. While it is possible to predict a few coefficients by only one neural network, we do not obtain a good prediction using a single neural network for all 256 coefficients without an elaborated fine tuning. Therefore, here instead of searching for a optimal number of coefficients to be predicted for one neural network, we consider each coefficient independently.

In Figure 3, we show the spectral functions corresponding to the same 32 samples used in Figure 2. Both the results from the direct numerical calculation based on the Lanczos method and recurrence relations and those predicted by the neural network are plotted. They basically overlap each other. There is a slight difference for the range of energies where the spectral function is nearly zero. This is perhaps due to the incomplete cancellation among the expansion terms at different orders. An improvement may be attainable if we consider the correlations of the coefficients for different orders. The input parameters of each of the 32 samples are plotted in Figure 4.

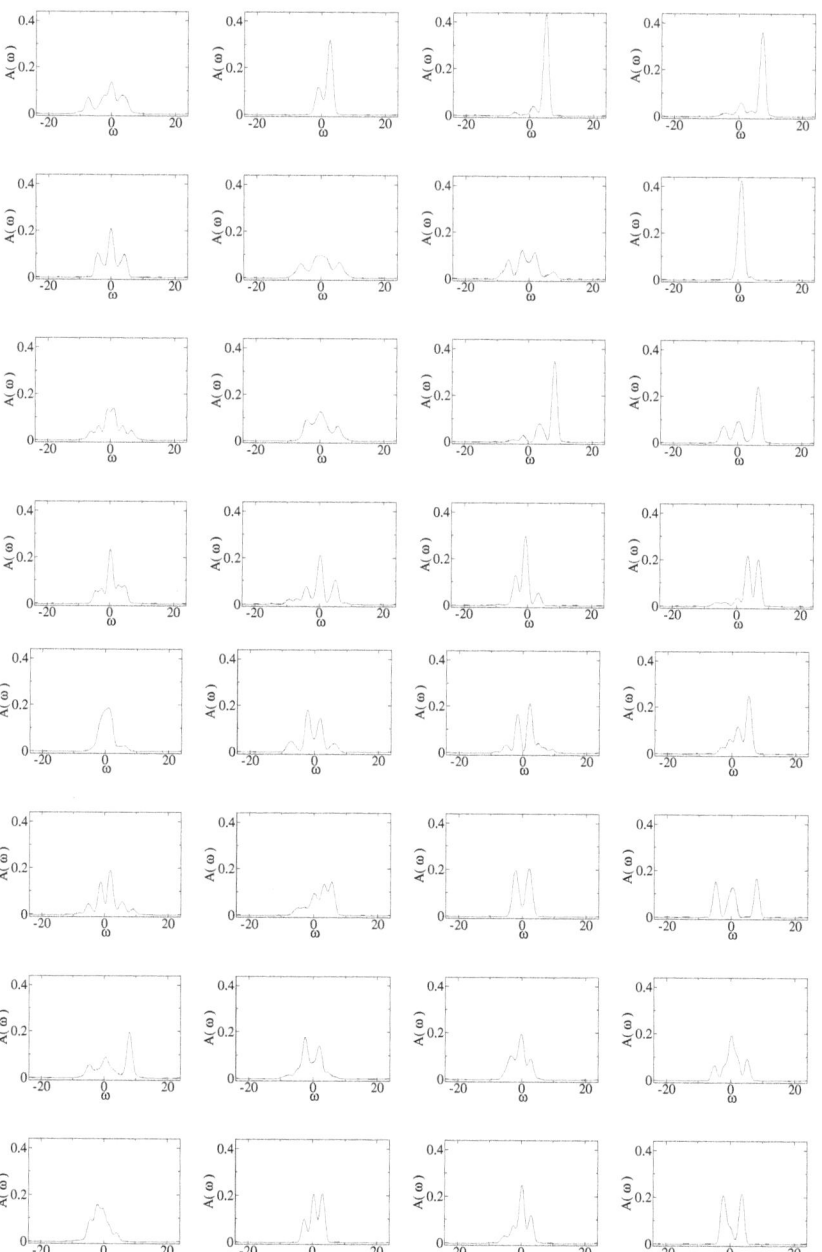

Figure 3. The spectral function, $A(\omega)$, is plotted for the same 32 randomly chosen parameter sets used in Figure 2. Both the results from the KPM and from the neural network are shown. They match each other very closely and visually overlap. A closer inspection reveals that there are slight oscillations in the spectral function when the weights are very small. This may be due to the inexact cancellations of different orders in the coefficients generated by the neural network method. In general, these oscillations are rather small and only appear when the spectral weight drops to near zero.

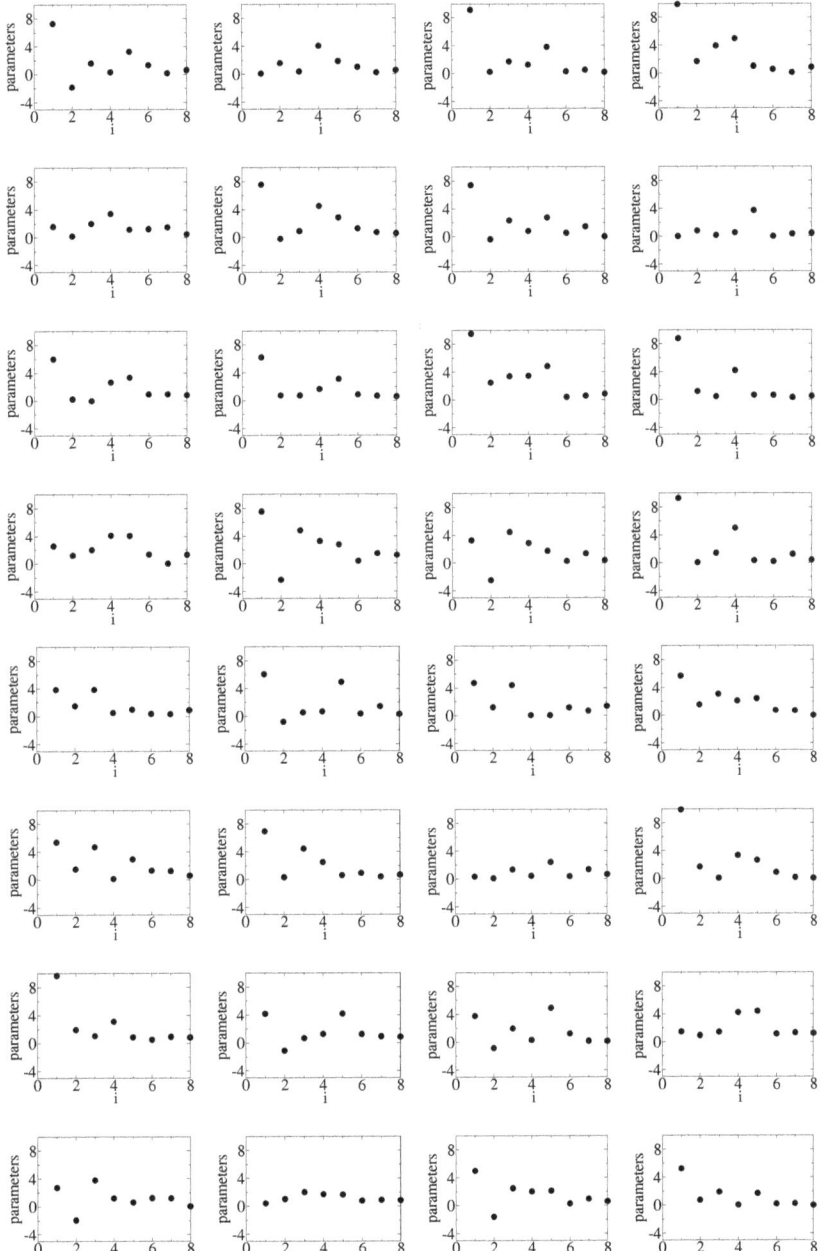

Figure 4. The input parameters of the 32 samples used in Figures 2 and 3; $i = 1$ corresponds to U; $i = 2$ corresponds to ϵ_f; $i = 3, 4, 5$ correspond to $\epsilon_1, \epsilon_2, \epsilon_3$; and $i = 6, 7, 8$ correspond to t_1, t_2, t_3.

Evidence of the capability of the neural network approach can be seen in Figure 5 where we plot the comparison of the first 32 expansion coefficients obtained by the direct numerical calculation and the neural network prediction. We find that two methods give very close results for the 1000 testing samples we consider. Please be reminded that

5000 samples are used for training. All the 1000 testing samples exhibit a linear trend. This clearly shows that a neural network is capable of providing a good prediction. There were no exceptional outliers among the 1000 testing samples we tested.

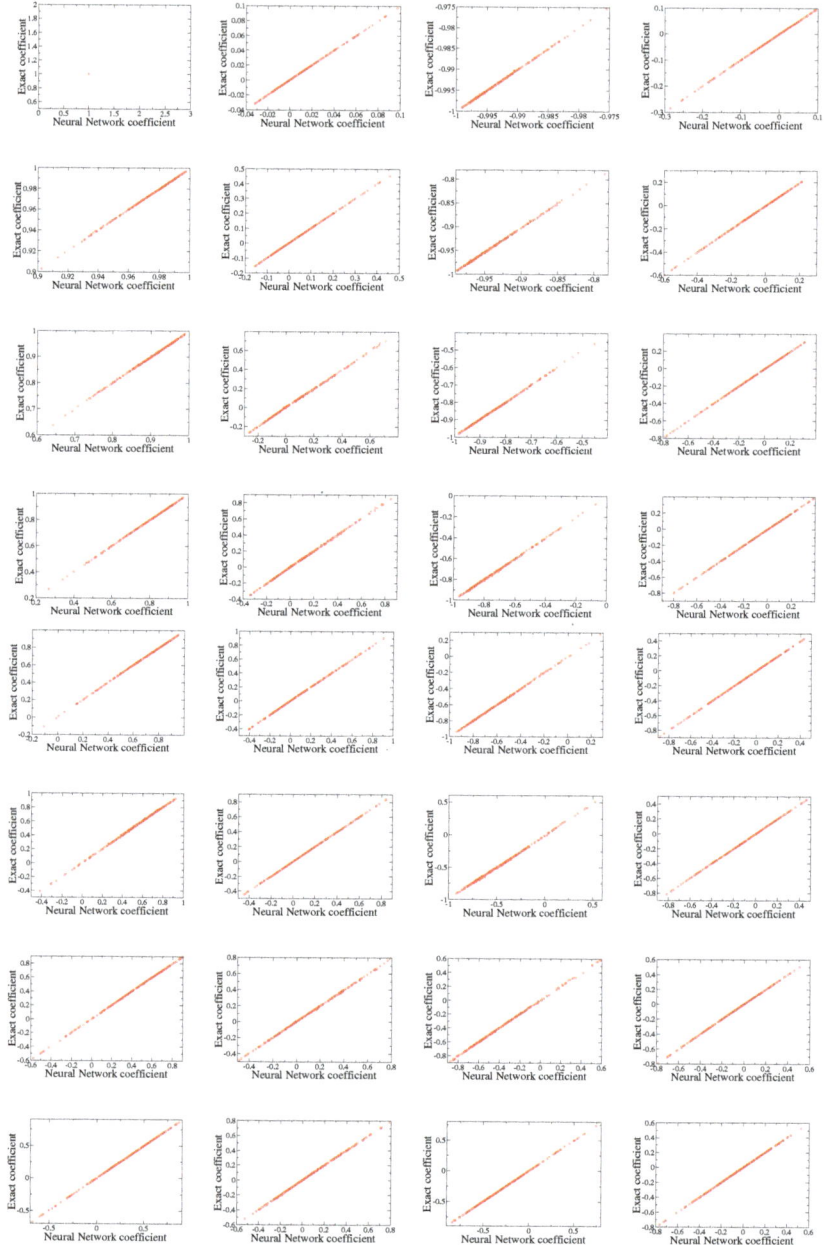

Figure 5. Comparison of the first 32 coefficients as computed by the KPM and the neural network method; 1000 samples are plotted in each figure. The figures are ordered from left to right and top to bottom from the order 0−th to the order 31−th.

6. Conclusions

We demonstrate that the supervised machine learning approach, specifically the neural network method, can be utilized as a solver for small quantum interacting clusters. This could be potentially useful for the statistical DMFT or the typical medium theory for which a large number of impurity problems have to be solved to perform the disorder averaging [29–33,36]. The main strategy is to devise a finite number of variables as the feature vector and the label for the supervised machine learning method. In line with the exact diagonalization method for the single impurity Anderson model, the feature vector is represented by the hoppings and the energies of the lattice model. The output we consider, the spectral function, is represented in terms of Chebyshev polynomials with a damping kernel. The labels are then the coefficients of the expansion. By comparing the coefficients directly calculated by the Lanczos method and the recurrence relation with the ones by the neural network, we find the agreement between these two methods is very good. Notably, among the 1000 samples we tested, there is no exceptional outlier. They all have a good agreement with the results of the numerical method.

For a simple impurity problem, the present method may not have an obvious benefit, as a rather large pool of samples have to be generated for training purposes. The situation is completely different for the study of disorder models, such as those being studied by the typical medium theory, where the present method has a clear advantage. Once the neural network is trained, the calculations are computationally inexpensive. For systems in which disorder averaging is required, this method can beat most if not all other numerical methods in term of efficiency. Moreover, the present approach is rather easy to be generalized for more complicated models, such as the few impurities models required in the dynamical cluster approximation. In addition, this method can be easily adapted to the matrix product basis proposed for the kernel polynomial expansion [48,53].

The range of parameters we choose in the present study covers the metal insulator transition of the Hubbard model within the single impurity dynamical mean field theory. An interesting question is the validity of the trained neural networks for the parameters well outside the range of the training data. We expect that the results could deteriorate without additional training data.

The size of the bath we choose in the present study is $N_b = 6$. This choice is somewhat arbitrary, for more accurate results for the single impurity problem a larger number of bath sites is preferred. The computational time required to generate the training set scales exponentially with the number of bath sites; however the computational time required to train the neural networks only scales as the third power with the number of bath sites. A larger number of bath sites should pose no problem for the present approach.

The ideas presented in this paper are rather generic. They can be generalized to the solutions by different solvers. For example, this approach can be adapted to Quantum Monte Carlo results as long as they can be represented in some kind of series expansion [54,55]. Our approach can also be adapted to predict the the coefficients of the coupled cluster theory [19,20].

Author Contributions: Conceptualization, K.-M.T.; Data curation, K.-M.T.; Formal analysis, J.M.; Funding acquisition, K.-M.T. and J.M.; Investigation, K.-M.T.; Methodology, K.-M.T., N.W., S.K. and Y.Z.; Project administration, J.M.; Supervision, J.M.; Writing—original draft, K.-M.T.; Writing—review & editing, J.M., N.W., S.K. and Y.Z. All authors have read and agreed to the published version of the manuscript.

Funding: This work is funded by the NSF Materials Theory grant DMR-1728457. Additional support was provided by NSF-EP- SCoR award OIA-1541079 (N.W. and K.-M.T.) and by the U.S. Department of Energy, Office of Science, Office of Basic Energy Sciences under Award No. DE-SC0017861 (Y.Z., J.M.).

Institutional Review Board Statement: Not applicable.

Informed Consent Statement: Not applicable.

Data Availability Statement: Correspondence and requests for data should be addressed to K.-M.T.

Acknowledgments: This work used the high performance computational resources provided by the Louisiana Optical Network Initiative (http://www.loni.org) and HPC@LSU computing, accessed from 1 January 2019 to 1 January 2022.

Conflicts of Interest: The authors declare no conflict of interest.

References

1. Kondo, J. Resistance Minimum in Dilute Magnetic Alloys. *Prog. Theor. Phys.* **1964**, *32*, 37–49. [CrossRef]
2. Anderson, P.W. A poor man's derivation of scaling laws for the Kondo problem. *J. Phys. C Solid State Phys.* **1970**, *3*, 2436–2441. [CrossRef]
3. Wilson, K.G. The renormalization group: Critical phenomena and the Kondo problem. *Rev. Mod. Phys.* **1975**, *47*, 773–840. [CrossRef]
4. Müller-Hartmann, E. The Hubbard model at high dimensions: Some exact results and weak coupling theory. *Z. Phys. B* **1989**, *76*, 211–217. [CrossRef]
5. Müller-Hartmann, E. Correlated fermions on a lattice in high dimensions. *Z. Phys. B* **1989**, *74*, 507–512. [CrossRef]
6. Metzner, W.; Vollhardt, D. Correlated Lattice Fermions in $d = \infty$ Dimensions. *Phys. Rev. Lett.* **1989**, *62*, 324–327. [CrossRef] [PubMed]
7. Bray, A.J.; Moore, M.A. Replica theory of quantum spin glasses. *J. Phys. C Solid State Phys.* **1980**, *13*, L655–L660. [CrossRef]
8. Georges, A.; Kotliar, G.; Krauth, W.; Rozenberg, M.J. Dynamical mean-field theory of strongly correlated fermion systems and the limit of infinite dimensions. *Rev. Mod. Phys.* **1996**, *68*, 13–125. [CrossRef]
9. Hettler, M.H.; Mukherjee, M.; Jarrell, M.; Krishnamurthy, H.R. Dynamical cluster approximation: Nonlocal dynamics of correlated electron systems. *Phys. Rev. B* **2000**, *61*, 12739–12756. [CrossRef]
10. Biroli, G.; Kotliar, G. Cluster methods for strongly correlated electron systems. *Phys. Rev. B* **2002**, *65*, 155112. [CrossRef]
11. Maier, T.; Jarrell, M.; Pruschke, T.; Hettler, M.H. Quantum cluster theories. *Rev. Mod. Phys.* **2005**, *77*, 1027–1080. [CrossRef]
12. Kotliar, G.; Savrasov, S.Y.; Haule, K.; Oudovenko, V.S.; Parcollet, O.; Marianetti, C.A. Electronic structure calculations with dynamical mean-field theory. *Rev. Mod. Phys.* **2006**, *78*, 865–951. [CrossRef]
13. Kajueter, H.; Kotliar, G. New Iterative Perturbation Scheme for Lattice Models with Arbitrary Filling. *Phys. Rev. Lett.* **1996**, *77*, 131–134. [CrossRef]
14. Logan, D.E.; Glossop, M.T. A local moment approach to magnetic impurities in gapless Fermi systems. *J. Phys. Condens. Matter* **2000**, *12*, 985–1028. [CrossRef]
15. Caffarel, M.; Krauth, W. Exact diagonalization approach to correlated fermions in infinite dimensions: Mott transition and superconductivity. *Phys. Rev. Lett.* **1994**, *72*, 1545–1548. [CrossRef]
16. Krishna-Murthy, H.R.; Wilkins, J.W.; Wilson, K.G. Renormalization-group approach to the Anderson model of dilute magnetic alloys. I. Static properties for the symmetric case. *Phys. Rev. B* **1980**, *21*, 1003–1043. [CrossRef]
17. Núñez Fernández, Y.; Hallberg, K. Solving the Multi-site and Multi-orbital Dynamical Mean Field Theory Using Density Matrix Renormalization. *Front. Phys.* **2018**, *6*, 13. [CrossRef]
18. Ganahl, M.; Aichhorn, M.; Evertz, H.G.; Thunström, P.; Held, K.; Verstraete, F. Efficient DMFT impurity solver using real-time dynamics with matrix product states. *Phys. Rev. B* **2015**, *92*, 155132. [CrossRef]
19. Zhu, T.; Jiménez-Hoyos, C.A.; McClain, J.; Berkelbach, T.C.; Chan, G.K.L. Coupled-cluster impurity solvers for dynamical mean-field theory. *Phys. Rev. B* **2019**, *100*, 115154. [CrossRef]
20. Shee, A.; Zgid, D. Coupled Cluster as an impurity solver for Green's function embedding methods. *arXiv* **2019**, arXiv:1906.04079.
21. Hirsch, J.E.; Fye, R.M. Monte Carlo Method for Magnetic Impurities in Metals. *Phys. Rev. Lett.* **1986**, *56*, 2521–2524. [CrossRef] [PubMed]
22. Hirsch, J.E. Discrete Hubbard-Stratonovich transformation for fermion lattice models. *Phys. Rev. B* **1983**, *28*, 4059–4061. [CrossRef]
23. Rubtsov, A.N.; Savkin, V.V.; Lichtenstein, A.I. Continuous-time quantum Monte Carlo method for fermions. *Phys. Rev. B* **2005**, *72*, 35122. [CrossRef]
24. Werner, P.; Millis, A.J. Hybridization expansion impurity solver: General formulation and application to Kondo lattice and two-orbital models. *Phys. Rev. B* **2006**, *74*, 155107. [CrossRef]
25. Carrasquilla, J.; Melko, R.G. Machine learning phases of matter. *Nat. Phys.* **2017**, *13*, 431–434. [CrossRef]
26. Huang, L.; Wang, L. Accelerated Monte Carlo simulations with restricted Boltzmann machines. *Phys. Rev. B* **2017**, *95*, 35105. [CrossRef]
27. Wang, L. Discovering phase transitions with unsupervised learning. *Phys. Rev. B* **2016**, *94*, 195105. [CrossRef]
28. Arsenault, L.F.; Lopez-Bezanilla, A.; von Lilienfeld, O.A.; Millis, A.J. Machine learning for many-body physics: The case of the Anderson impurity model. *Phys. Rev. B* **2014**, *90*, 155136. [CrossRef]
29. Dobrosavljević, V.; Pastor, A.A.; Nikolić, B.K. Typical medium theory of Anderson localization: A local order parameter approach to strong-disorder effects. *EPL* **2003**, *62*, 76–82. [CrossRef]
30. Ekuma, C.E.; Terletska, H.; Tam, K.M.; Meng, Z.Y.; Moreno, J.; Jarrell, M. Typical medium dynamical cluster approximation for the study of Anderson localization in three dimensions. *Phys. Rev. B* **2014**, *89*, 81107. [CrossRef]

31. Terletska, H.; Zhang, Y.; Chioncel, L.; Vollhardt, D.; Jarrell, M. Typical-medium multiple-scattering theory for disordered systems with Anderson localization. *Phys. Rev. B* **2017**, *95*, 134204. [CrossRef]
32. Zhang, Y.; Terletska, H.; Moore, C.; Ekuma, C.; Tam, K.M.; Berlijn, T.; Ku, W.; Moreno, J.; Jarrell, M. Study of multiband disordered systems using the typical medium dynamical cluster approximation. *Phys. Rev. B* **2015**, *92*, 205111. [CrossRef]
33. Dobrosavljević, V.; Kotliar, G. Dynamical mean–field studies of metal–insulator transitions. *Philos. Trans. R. Soc. London. Ser. A Math. Phys. Eng. Sci.* **1998**, *356*, 57–74. [CrossRef]
34. Terletska, H.; Zhang, Y.; Tam, K.M.; Berlijn, T.; Chioncel, L.; Vidhyadhiraja, N.; Jarrell, M. Systematic quantum cluster typical medium method for the study of localization in strongly disordered electronic systems. *Appl. Sci.* **2018**, *8*, 2401. [CrossRef]
35. Zhang, Y.; Nelson, R.; Siddiqui, E.; Tam, K.M.; Yu, U.; Berlijn, T.; Ku, W.; Vidhyadhiraja, N.S.; Moreno, J.; Jarrell, M. Generalized multiband typical medium dynamical cluster approximation: Application to (Ga,Mn)N. *Phys. Rev. B* **2016**, *94*, 224208. [CrossRef]
36. Ekuma, C.E.; Yang, S.X.; Terletska, H.; Tam, K.M.; Vidhyadhiraja, N.S.; Moreno, J.; Jarrell, M. Metal-insulator transition in a weakly interacting disordered electron system. *Phys. Rev. B* **2015**, *92*, 201114. [CrossRef]
37. Ulmke, M.; Janiš, V.; Vollhardt, D. Anderson-Hubbard model in infinite dimensions. *Phys. Rev. B* **1995**, *51*, 10411–10426. [CrossRef] [PubMed]
38. Semmler, D.; Byczuk, K.; Hofstetter, W. Anderson-Hubbard model with box disorder: Statistical dynamical mean-field theory investigation. *Phys. Rev. B* **2011**, *84*, 115113. [CrossRef]
39. Byczuk, K.; Hofstetter, W.; Vollhardt, D. Mott-Hubbard Transition versus Anderson Localization in Correlated Electron Systems with Disorder. *Phys. Rev. Lett.* **2005**, *94*, 056404. [CrossRef]
40. De Vega, I.; Schollwöck, U.; Wolf, F.A. How to discretize a quantum bath for real-time evolution. *Phys. Rev. B* **2015**, *92*, 155126. [CrossRef]
41. Liebsch, A.; Ishida, H. Temperature and bath size in exact diagonalization dynamical mean field theory. *J. Phys. Soc. Jpn.* **2011**, *24*, 53201. [CrossRef] [PubMed]
42. Medvedeva, D.; Iskakov, S.; Krien, F.; Mazurenko, V.V.; Lichtenstein, A.I. Exact diagonalization solver for extended dynamical mean-field theory. *Phys. Rev. B* **2017**, *96*, 235149. [CrossRef]
43. Nagai, Y.; Shinaoka, H. Smooth Self-energy in the Exact-diagonalization-based Dynamical Mean-field Theory: Intermediate-representation Filtering Approach. *J. Phys. Soc. Jpn.* **2019**, *88*, 064004. [CrossRef]
44. Sénéchal, D. Bath optimization in the cellular dynamical mean-field theory. *Phys. Rev. B* **2010**, *81*, 235125. [CrossRef]
45. Weiße, A.; Wellein, G.; Alvermann, A.; Fehske, H. The kernel polynomial method. *Rev. Mod. Phys.* **2006**, *78*, 275–306. [CrossRef]
46. Lin, H.Q. Exact diagonalization of quantum-spin models. *Phys. Rev. B* **1990**, *42*, 6561–6567. [CrossRef] [PubMed]
47. Lin, H.; Gubernatis, J. Exact Diagonalization Methods for Quantum Systems. *Comput. Phys.* **1993**, *7*, 400–407. [CrossRef]
48. Wolf, F.A.; McCulloch, I.P.; Parcollet, O.; Schollwöck, U. Chebyshev matrix product state impurity solver for dynamical mean-field theory. *Phys. Rev. B* **2014**, *90*, 115124. [CrossRef]
49. Silver, R.; Röder, H. Densities of states of mega-dimensional Hamiltonian matrices. *Int. Mod. Phys. C* **1994**, *5*, 735–753. [CrossRef]
50. Silver, R.; Röder, H.; Voter, A.; Kress, J. Kernel Polynomial Approximations for Densities of States and Spectral Functions. *J. Comput. Phys.* **1996**, *124*, 115–130. [CrossRef]
51. Silver, R.N.; Röder, H. Calculation of densities of states and spectral functions by Chebyshev recursion and maximum entropy. *Phys. Rev. E* **1997**, *56*, 4822–4829. [CrossRef]
52. Alvermann, A.; Fehske, H. Chebyshev approach to quantum systems coupled to a bath. *Phys. Rev. B* **2008**, *77*, 45125. [CrossRef]
53. Wolf, F.A.; Justiniano, J.A.; McCulloch, I.P.; Schollwöck, U. Spectral functions and time evolution from the Chebyshev recursion. *Phys. Rev. B* **2015**, *91*, 115144. [CrossRef]
54. Huang, L. Kernel polynomial representation for imaginary-time Green's functions in continuous-time quantum Monte Carlo impurity solver. *Chin. Phys. B* **2016**, *25*, 117101. [CrossRef]
55. Boehnke, L.; Hafermann, H.; Ferrero, M.; Lechermann, F.; Parcollet, O. Orthogonal polynomial representation of imaginary-time Green's functions. *Phys. Rev. B* **2011**, *84*, 75145. [CrossRef]

Review

Ingredients for Generalized Models of κ-Phase Organic Charge-Transfer Salts: A Review

Kira Riedl [1,*], **Elena Gati** [2] **and Roser Valentí** [1]

1. Institut für Theoretische Physik, Goethe-Universität Frankfurt, 60438 Frankfurt am Main, Germany
2. Max Planck Institute for Chemical Physics of Solids, 01187 Dresden, Germany
* Correspondence: riedl@itp.uni-frankfurt.de

Abstract: The families of organic charge-transfer salts κ-(BEDT-TTF)$_2 X$ and κ-(BETS)$_2 X$, where BEDT-TTF and BETS stand for the organic donor molecules $C_{10}H_8S_8$ and $C_{10}H_8S_4Se_4$, respectively, and X for an inorganic electron acceptor, have been proven to serve as a powerful playground for the investigation of the physics of frustrated Mott insulators. These materials have been ascribed a model character, since the dimerization of the organic molecules allows to map these materials onto a single band Hubbard model, in which the dimers reside on an anisotropic triangular lattice. By changing the inorganic unit X or applying physical pressure, the correlation strength and anisotropy of the triangular lattice can be varied. This has led to the discovery of a variety of exotic phenomena, including quantum-spin liquid states, a plethora of long-range magnetic orders in proximity to a Mott metal-insulator transition, and unconventional superconductivity. While many of these phenomena can be described within this effective one-band Hubbard model on a triangular lattice, it has become evident in recent years that this simplified description is insufficient to capture all observed magnetic and electronic properties. The ingredients for generalized models that are relevant include, but are not limited to, spin-orbit coupling, intra-dimer charge and spin degrees of freedom, electron-lattice coupling, as well as disorder effects. Here, we review selected theoretical and experimental discoveries that clearly demonstrate the relevance thereof. At the same time, we outline that these aspects are not only relevant to this class of organic charge-transfer salts, but are also receiving increasing attention in other classes of inorganic strongly correlated electron systems. This reinforces the model character that the κ-phase organic charge-transfer salts have for understanding and discovering novel phenomena in strongly correlated electron systems from a theoretical and experimental point of view.

Keywords: organic charge-transfer salts; magnetic exchange beyond Heisenberg; intra-dimer charge and spin degrees of freedom; electron-lattice coupling; disorder

1. Introduction

1.1. Crystal and Electronic Structure

Organic charge-transfer salts have received, in the last decades, a lot of attention for their variability as materials and their model role in understanding most of the challenging phenomena in correlated systems [1,2], such as the Mott-metal insulator transition [3–5], Mott criticality [6–9], quantum-spin liquid phases [10–13], unconventional superconductivity [14,15] or multiferroicity [16], among others. One of the most studied families is the κ phase. The κ phase in organic charge-transfer salts consists of alternation in charge donating organic BEDT-TTF (BEDT-TTF = bisethyenedithio-tetrathiafulvalene) or BETS (BETS = bisethyenedithio-tetraselenafulvalene) layers and acceptor inorganic anion layers, as depicted in Figure 1 for the example of κ-(BEDT-TTF)$_2$Cu$_2$(CN)$_3$ crystallizing in the space group $P2_1/c$.

The choice of different inorganic layers X in κ-(BEDT-TTF)$_2 X$ and κ-(BETS)$_2 X$ influences the orbital overlap between the organic molecules and allows for distinct magnetic and electronic properties in these systems. In the charge-transfer process, each molecule donates a charge of half an electron. The κ-type arrangement of organic molecules exhibits a strong dimerization of the molecules forming a triangular lattice of dimers with one hole per dimer, as illustrated in Figures 1c and 2.

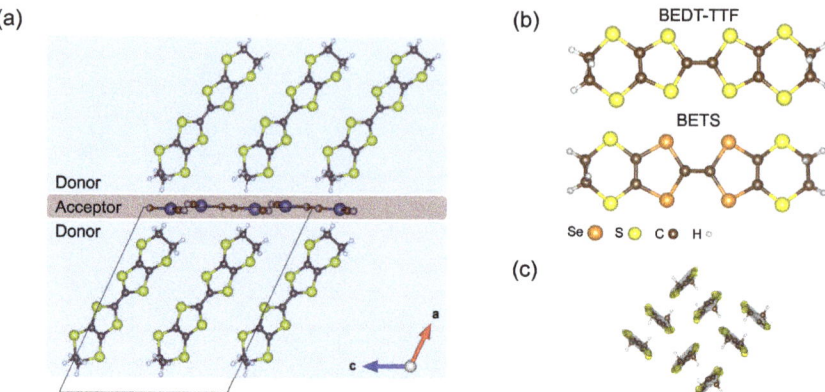

Figure 1. (a) Layered structure of organic electron donor and inorganic electron acceptor layer, shown for the example κ-(BEDT-TTF)$_2$Cu$_2$(CN)$_3$; (b) structure of BEDT-TTF and BETS molecules; and (c) κ packing motif of the organic layer of κ phase charge-transfer salts from the top view.

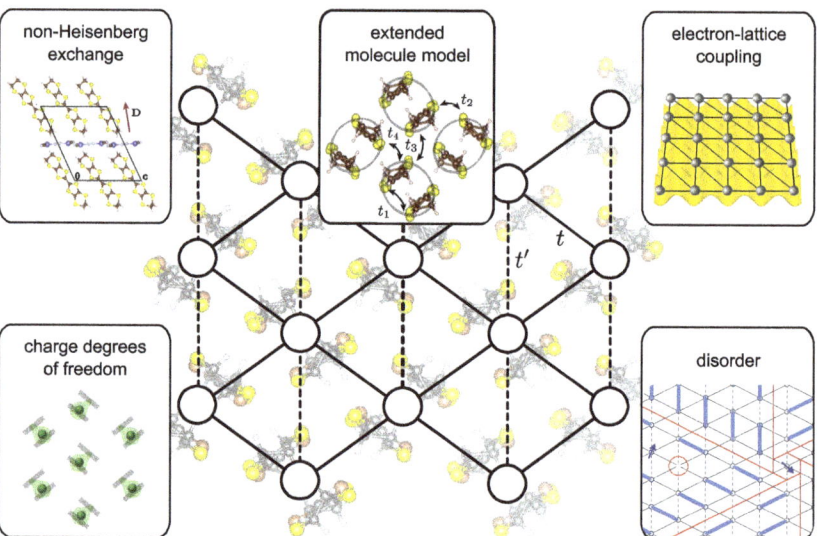

Figure 2. Center: Illustration of the mapping of the organic layer of κ phase charge-transfer salts to an anisotropic triangular lattice. The various ingredients for generalized models beyond the strongly dimerized one-band Hubbard picture discussed in this work are illustrated in the boxes.

Alternatively, one can consider the molecules within a dimer as the building blocks (see inset in Figure 2) leading to an extended molecule model with half a hole per molecule. By symmetry, there are four distinct hopping parameters between the highest occupied molecular orbitals (HOMO) $|g_i\rangle$ of the molecules. The hoppings are conventionally abbreviated as $t_{1...4}$, as shown in Figure 2. Here, t_1 is the intra-dimer hopping and one can make use of geometrical expressions $t = (t_2 + t_4)/2$ and $t' = t_3/2$ [17,18] to relate the $t_{1...4}$ hoppings of the extended molecule model to the t, t' hoppings of the dimer model shown in Figure 2. The corresponding electronic structure of the dimer model consists, then, of an anti-bonding dimer orbital [$|a\rangle = \frac{1}{\sqrt{2}}(|g_1\rangle + |g_2\rangle)$], occupied by one electron (or one hole), and of a bonding dimer orbital [$|b\rangle = \frac{1}{\sqrt{2}}(|g_1\rangle - |g_2\rangle)$] occupied by two electrons.

1.2. Phase Diagram

Under the assumption of very strong dimerization, it is well-established, as mentioned above, that the κ-phase organic charge-transfer salts are model systems to realize the half-filled one-band Hubbard model [1,2,11,18–20]:

$$\mathcal{H} = - \sum_{<i,j>,\sigma} (t_{ij} c^\dagger_{i,\sigma} c_{j,\sigma} + h.c.) + \sum_i U n_{i,\uparrow} n_{i,\downarrow} \quad (1)$$

with hopping $t_{ij} = t, t'$ (see Figure 2). One key parameter of this model is the ratio of the strength of Coulomb repulsion U to the kinetic energy $W \sim t, t'$. The κ-phase charge-transfer salts all lie in a range close to the Mott transition, i.e., $U/W \sim 1$.

The essential features of the Hubbard model can be identified in experimental *temperature – pressure* ($T - p$) phase diagrams of κ-phase organic charge-transfer salts. The application of hydrostatic pressure causes an increase in the orbital overlap and, therefore, an increase in W (or decrease in U/W). As a result, a Mott insulator is expected to undergo an insulator-to-metal transition. This is indeed observed for various Mott-insulating κ-phase charge-transfer salts, including κ-(BEDT-TTF)$_2$Cu[N(CN)$_2$]Cl [1,3,4,9,21,22] ('κ-Cl', Figure 3a) and κ-(BEDT-TTF)$_2$Cu$_2$(CN)$_3$ [5,7,23] ('κ-CuCN', Figure 3b) and κ-(BETS)$_2$Mn[N(CN)$_2$]$_3$ ('κ-Mn') [24–26]. In all cases, very moderate pressures in the order of kbar, or even less, induce a Mott metal-insulator transition (MIT). This finding demonstrates, on the one hand, the proximity of various κ-phase charge-transfer salts to the Mott MIT and, on the other hand, it also shows the tunability of these materials in laboratory settings through physical and chemical pressures [1,2].

At low temperatures, the pressure-induced Mott MIT in κ-Cl and κ-CuCN is found to be a first-order transition. Upon increasing temperature, the first-order line ends in a second-order critical endpoint at (p_{cr}, T_{cr}). Above the endpoint, only a crossover but no phase transition exists, similar to the liquid–gas transition. In the purely electronic model of the Mott transition, the Mott critical endpoint is, thus, expected to fall into the Ising universality class [27,28]. This notion of Mott criticality is further corroborated by DMFT (dynamical mean field theory) calculations [29], which are also able to predict the order of the Mott MIT. In Section 4, we will discuss the limitations of this purely electronic model when the coupling to lattice degrees of freedom becomes relevant.

Whereas the behavior very close to the first-order critical endpoint at T_{cr} is one of a classical phase transition, an intriguing observation in κ-phase charge-transfer salts is a quantum-critical scaling of measured transport in the crossover region $T \gg T_{cr}$. Such a quantum-critical scaling was predicted based on DMFT calculations of the Hubbard model [30,31]. While one would typically predict quantum-critical scaling to occur at lowest temperatures $T \to 0$, it is important to note that the dominant energy scales [20] of the Mott transition (U and W) are of the order of 1000 K $\gg T_{cr}$. Thus, for intermediate temperatures, the system effectively behaves as if the Mott critical endpoint were located at zero. In fact, frustration suppresses magnetic ordering tendencies, so that frustrated organic charge-transfer salts are suggested to show the properties of genuine Mott systems to lower temperatures [20,32]. Such an importance of magnetic frustration for investigating paramagnetic Mott metal-insulating transitions has been recently discussed as well, in the context of inorganic V$_2$O$_3$ [33].

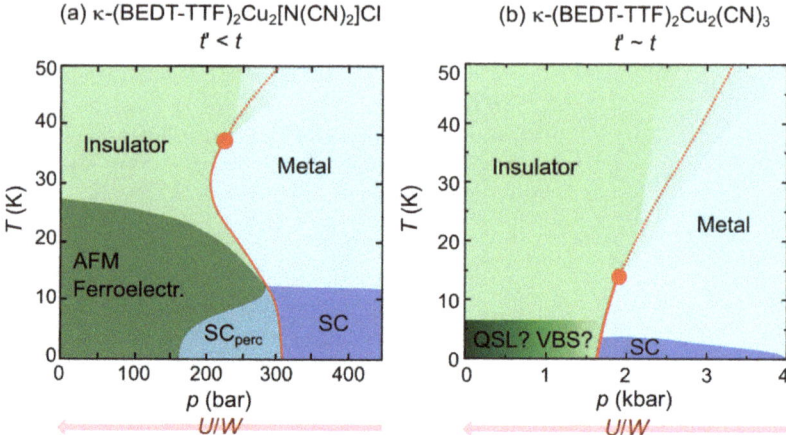

Figure 3. Experimental temperature–pressure phase diagrams of κ-phase charge-transfer salts: (**a**) κ-(BEDT-TTF)$_2$Cu[N(CN)$_2$]Cl ('κ-Cl') [1,3,4,9,16,21,22] and (**b**) κ-(BEDT-TTF)$_2$Cu$_2$(CN)$_3$ ('κ-CuCN') [7,10,23,34]. At ambient pressure and low-enough temperatures, both compounds exhibit a Mott insulating ground state (i.e., a moderate to large correlation strength, U/W). Upon applying pressure, U/W is reduced and eventually a first-order Mott metal-insulator transition is induced (red line) which ends in a second-order critical endpoint (red circle). On both sides of the Mott transition, intriguing electronic orders emerge at lowest temperatures. On the Mott insulating side, κ-Cl orders antiferromagnetically [35], which is well-understood given the anisotropy of its triangular lattice ($t' < t$). The antiferromagnetic (AFM) order is believed to be accompanied by the emergence of long-range ferroelectric order, making this compound multiferroic [16]. In contrast, κ-CuCN is characterized by an almost isotropic triangular lattice with $t' \sim t$. For a long time, it has, therefore, been considered as a candidate for quantum spin-liquid (QSL) behavior [10]. However, there exist recent results which argue in favor of the formation of a valence bond solid (VBS) [12,13,34,36–38]. On the metallic side of the Mott transition, both compounds exhibit superconductivity (SC). A region of percolative superconductivity SC$_{perc}$ can also be found in the Mott insulating state close to the metal-insulator boundary.

The spin degrees of freedom in the Mott insulating state of κ-phase charge-transfer salts become important at very low temperatures $T \to 0$. κ-Cl undergoes a transition into an antiferromagnetic (AFM) ordered ground state at $T_N \sim 27$ K [35]. In the dimer model, AFM order might be expected when the triangular lattice is very anisotropic and close to the square lattice, which is indeed the case for κ-Cl ($t'/t \approx 0.45$ [18,39]). In contrast, when the triangular lattice is more isotropic, as is the case for κ-CuCN ($t'/t \approx 0.8$–0.9 [18,39–42]), various theoretical studies of the Hubbard model on the triangular lattice at half-filling suggest the emergence of a quantum spin liquid, as, for instance, in Refs. [43–45]. In fact, κ-CuCN has been considered for a long time as a prime candidate for the realization of a quantum spin liquid [10,46] that potentially hosts gapless excitations [47]. However, thermal expansion measurements [48,49] indicate the presence of a phase transition around 6 K, often referred to as the "6 K anomaly", which might prevent the formation of a possible spin-liquid phase. In recent years, the interpretation of magnetic torque [8] and NMR [50] data suggested that a valence-bond solid rather than a quantum spin liquid might be formed in κ-CuCN [12] below $T^* \sim 6$ K. This picture was further supported by recent ESR studies confirming the presence of a spin gap opening [34] around the 6 K anomaly. Still, open questions regarding the magnetic ground state of κ-CuCN remain; see Ref. [13] for a recent review of the large set of experimentally available data on this system.

κ-Mn also shows magnetic order for $T < 30$ K in the Mott insulating state at the lowest temperature [24,26,51]. However, the magnetic order that is realized in κ-Mn is not of the Néel type, as observed in κ-Cl. In fact, it was shown that κ-Mn realizes a spin-vortex crystal order that can only be understood when taking ring-exchange interactions into account [51], as we will explain in detail in Section 2. In addition, in contrast to κ-Cl, where the magnetic ordering occurs within the Mott insulating state, the onset of magnetic order in κ-Mn coincides with a first-order metal-to-Mott insulator transition at $T_{MI} = T_N$ [24]. The ordered, low-entropy state in κ-Mn dictates a negative slope of dT_{MI}/dp in the temperature–pressure phase diagram; see also Refs. [13,38] for discussions of dT_{MI}/dp in κ-Cl and κ-CuCN.

On the metallic side of the Mott MIT, superconductivity is often found at low temperatures [52,53], as seen in the temperature–pressure phase diagrams of κ-Cl, κ-CuCN and κ-Mn, with moderate transition temperatures $T_c \sim 10$ K (see Figure 3). Note that also chemical modifications, e.g., replacing Cl completely by Br in κ-Cl, can be used to obtain superconducting metals at ambient pressure. The proximity of the superconducting phase to a (magnetic) Mott insulating phase suggests a close connection between the κ-phase organic charge-transfer salts and the high-temperature cuprate superconductors [54], where the origin of superconductivity is attributed to the presence of spin fluctuations. Actually, a few theoretical treatments of the dimer model (see, e.g., Refs. [15,55–61]) follow the scenario of superconductivity mediated by spin fluctuations.

1.3. Outline of This Review

Whereas the one-band Hubbard model on the triangular lattice clearly captures essential features of the phase diagram of the κ-phase organic charge-transfer salts, it has become evident in recent years that generalized models for κ-phase organic charge-transfer salts must contain additional contributions to accurately account for all salient features of their phase diagrams. The present review summarizes theoretical and experimental works, based on which the importance of specific, additional interactions were suggested. The paper is structured as follows (see also Figure 2 for a sketch of the synopsis). In Section 2, we discuss the role of spin-orbit coupling and higher order four-spin ring exchange couplings for the magnetic properties of the Mott insulating κ-phase charge-transfer salts. We then proceed, in Section 3, with a summary of new effects in the extended molecule model such as ferroelectricity in the Mott insulating state or mixed superconducting order parameters. In the following Section 4, we present evidence for the importance of electron-lattice coupling in these correlated electron systems. The last aspect that we will cover in Section 5 is the role of disorder for the properties of the κ-phase charge-transfer salts close to the Mott transition. In Section 6, we present a conclusion and outlook of the review and discuss specific important open questions and their relevance for the broader field of correlated electron systems.

2. Magnetic Exchange beyond Heisenberg

Deep in the Mott insulating phase, the charge degrees of freedom do not influence the low energy properties of the system and a description in terms of a purely spin-1/2 magnetic model on the triangular lattice is a suitable starting point, where one dimer represents one magnetic site (see Figure 4). The dominant magnetic exchange terms are of the Heisenberg type:

$$\mathcal{H} = \sum_{\langle ij \rangle} J_{ij} \mathbf{S}_i \cdot \mathbf{S}_j, \qquad (2)$$

with exchange J on bonds indicated by solid lines and J' on bonds indicated by dashed lines in Figure 4a. This model can be obtained from a perturbation expansion of the Hubbard model (Equation (1)) in powers of t/U [62,63]. Depending on the ratio J'/J, it is possible to scale in between the limit of a square lattice ($J'/J = 0$), an isotropic triangular lattice ($J'/J = 1$) and one-dimensional chains ($J'/J = \infty$). In the κ-phase charge-transfer salts, this ratio is determined by the nature of the anion layer. Further Heisenberg exchanges up to fourth neighbors are shown in Figure 4a.

Beyond the Heisenberg exchange there are two main contributions, which, albeit being subdominant, may change the nature of the magnetic ground state of the κ-phase charge-transfer salts completely. One is the spin-orbit coupling, which allows for the presence of anisotropic bilinear magnetic exchange, such as the Dzyaloshinskii–Moriya vector **D** (Figure 4b), changing the symmetry properties of the magnetic model severely. A second important contribution is the four-spin ring exchange K, K', which is expected to be non-negligible for materials close to the Mott MIT. Note that in the presence of a magnetic field, additional terms with an odd number of spins, such as the scalar spin chirality, are also present. We do not discuss these terms here.

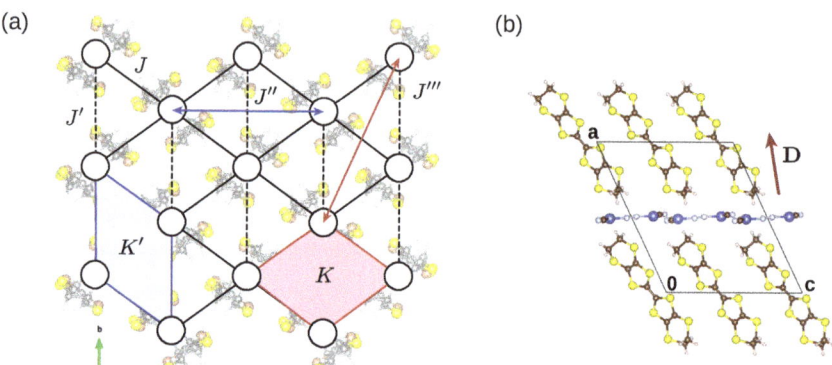

Figure 4. Magnetic exchange in κ-phase charge-transfer salts. (**a**) Heisenberg exchange terms J, J', J'', J''' and four-spin ring-exchange K and K' on the two distinct four-site plaquettes on the anisotropic triangular lattice. (**b**) The SOC-induced DM vector **D** is oriented approximately along the long side of the organic molecule, here illustrated for the example κ-(BEDT-TTF)$_2$Cu$_2$(CN)$_3$. Figures reprinted from Refs. [51,64].

2.1. Magnetic Bilinear Anisotropic Interactions

Magnetic anisotropic interactions arise due to the presence of spin-orbit coupling. Originally, this aspect was often ignored in the context of the organic charge-transfer salts due to the light nature of the Se, S, C and H atoms in the BEDT-TTF and BETS molecules. However, it was pointed out early on [65] that at least at very low temperatures the influence of anisotropic interactions may be significant to capture new magnetic phenomena. The full bilinear magnetic model for $S = 1/2$ can be expressed as follows:

$$\mathcal{H}_{(2)} = \sum_{\langle ij \rangle} J_{ij} \mathbf{S}_i \cdot \mathbf{S}_j + \mathbf{D}_{ij} \cdot (\mathbf{S}_i \times \mathbf{S}_j) + \mathbf{S}_i \cdot \mathbf{\Gamma}_{ij} \cdot \mathbf{S}_j, \qquad (3)$$

with the antisymmetric Dzyaloshinskii–Moriya (DM) vector $\mathbf{D}_{ij}$ and the symmetric pseudo-dipolar tensor $\mathbf{\Gamma}_{ij}$. Note that on the bonds indicated by the dashed lines in Figure 4a, the DM vector vanishes due to an inversion center at the bond center. To obtain an intuition about the strengths of the anisotropic terms, we consider perturbation-theory expressions in the strongly localized limit with weak SOC, i.e., $U \gg t, |\vec{\lambda}|$. Here, $\vec{\lambda}$ is a spin-dependent hopping term in the electronic Hubbard picture, arising due to the presence of SOC, $\mathcal{H}_{\text{hop}} = \sum_{ij} \mathbf{c}_i^\dagger (t_{ij} I_{2\times 2} + \frac{i}{2} \vec{\lambda}_{ij} \cdot \vec{\sigma}) \mathbf{c}_j$ [66], with the single particle operator $\mathbf{c}^\dagger = (c_{i\uparrow}^\dagger \ c_{i\downarrow}^\dagger)$ and the Pauli matrices $\vec{\sigma}$. In second-order perturbation theory, the bilinear-exchange couplings scale with $J \propto t^2/U$, $\mathbf{D} \propto (t\vec{\lambda})/U$, and $\mathbf{\Gamma} \propto (\vec{\lambda} \otimes \vec{\lambda})/U$ [64]. In κ-phase charge-transfer salts, the hopping amplitude is an order of magnitude larger than SOC, so that the Heisenberg term is expected to be dominant, with important contributions of the DM vector at very low temperatures and negligible contributions of the pseudo-dipolar tensor.

In Table 1, we list selected ab-initio results [12,51,64,67] for various κ-phase organic compounds. The nearest-neighbor Heisenberg-exchange couplings are on the order of a few hundred Kelvin, where the ratio J'/J indicates whether the material is closer to the square lattice limit ($J'/J < 1$) or whether it is approaching the limit of one-dimensional chains ($J'/J > 1$). The DM vector is, as expected, significantly smaller with $|\mathbf{D}| \sim 5\,\text{K}$

for the BEDT-TTF and $|\mathbf{D}| \sim 25$ K for the BETS compounds. The stronger anisotropic contribution in the BETS material can be directly related to the presence of the heavier Se atoms in the organic molecule (see Figure 1b). The DM vector is consistently oriented approximately along the long side of the molecule, as shown in Figure 4b for the example of κ-(BEDT-TTF)$_2$Cu$_2$(CN)$_3$. This implies that the main contribution is the component perpendicular to the triangular plane of molecular dimers. For all discussed Mott insulators in this review, this out-of-plane component has a staggered pattern, while the component along the dashed bonds in Figure 4a has a stripy pattern by symmetry [51,64].

Table 1. Representative exchange parameters in K for indicated materials κ-(BEDT-TTF)$_2$Cu[N(CN)$_2$]Cl (κ-Cl), κ-(BEDT-TTF)$_2$Ag$_2$(CN)$_3$ (κ-AgCN), κ-(BEDT-TTF)$_2$B(CN)$_4$ (κ-BCN), κ-(BETS)$_2$Mn[N(CN)$_2$]$_3$ (κ-Mn): Bilinear exchange J, J', $\mathbf{D}$ (defined in Equation (3) and Figure 4) and averaged four-spin ring exchange K, K' (defined in Equation (5) and Figure 4). The DM vectors for the $Pnma$ salts are given in the coordinate system (a, b, c), while the $P2_1/c$ values are indicated with () * and given with respect to (a, b, c^*). The ratio J'/J indicates the deviation from the isotropic triangular limit $J'/J = 1$ and K/J indicates the significance of the four-spin ring exchange. This table is not intended to be a complete representation of available results, but serves as orientation with selected values.

Material	J; J'; $\mathbf{D}$	K; K'	J'/J	K/J
κ-Cl	482; 165; $(-3.6, -3.6, -0.2)$ [64]	62; 21 [67]	0.34	0.13
κ-AgCN	250; 158; $(-2.9, -0.9, -2.9)$ * [64]	20; 13 [67]	0.63	0.08
κ-CuCN	228; 268; $(+3.3, +0.9, +1.0)$ * [64]	18; 18 [12]	1.18	0.08
κ-BCN	131; 366; $(+1.0, +4.2, -0.1)$ [64]	05; 15 [67]	2.79	0.04
κ-Mn	260; 531; $(+22.6, -1.9, +8.8)$ * [51]	16; 39 [51]	2.04	0.06

In the well-studied material κ-(BEDT-TTF)$_2$Cu[N(CN)$_2$]Cl, the presence of the DM interaction can be directly related to an observed small ferromagnetic contribution to an antiferromagnetic transition at $T_N = 27$ K by ESR [35]. This observation is consistent with a weak canting of the magnetic moments, as one would expect from the presence of a DM interaction in an AFM Néel ordered state.

Further important effects of bilinear anisotropic interactions can be detected when the spin system couples to an external magnetic field $\mathbf{H}$. To fully grasp the consequences of these anisotropic interactions, it is helpful to follow an approach first introduced in the 1990s [68]. Specific DM patterns (e.g., uniform and staggered DM patterns) allow to apply local rotations for each spin, so that the anisotropic interactions may be "gauged" away. Note that for this approach, it is assumed that $\Gamma \propto \mathbf{D} \otimes \mathbf{D}$, which is generally not perfectly fulfilled in the presence of Hund's coupling. While this condition is not fully obeyed, the general behavior of the system in the presence of an external field may still be described sufficiently with this model. A local rotation about the angle $\eta_i \Phi$, with $\eta_i = \pm 1$ for different sublattices and $\Phi = 1/2 \arctan(J/|\mathbf{D}|)$, results, then, in a bilinear isotropic Heisenberg Hamiltonian in the presence of a uniform and of a staggered magnetic field:

$$\mathcal{H}_{(2)} = \sum_{\langle ij \rangle} \tilde{J}_{ij} \tilde{\mathbf{S}}_i \cdot \tilde{\mathbf{S}}_j - \mu_B \sum_i (\mathbf{H}_{\text{eff,u}} + \eta_i \mathbf{H}_{\text{eff,s}}) \cdot \tilde{\mathbf{S}}_i. \quad (4)$$

Here, $\tilde{\mathbf{S}}$ are the rotated spin operators, $\mathbf{H}_{\text{eff,u}} \approx \mathbb{G}_u^T \cdot \mathbf{H}$ is the effective uniform magnetic field, $\mathbf{H}_{\text{eff,s}} \approx \mathbb{G}_s^T \cdot \mathbf{H} - \frac{1}{2J}((\mathbb{G}_u^T \cdot \mathbf{H}) \times \mathbf{D})$ the effective staggered magnetic field, and $\mathbb{G}_{u/s}$ the uniform/staggered contribution to the gyromagnetic tensor. In other words, applying an external uniform field $\mathbf{H}$ to an anisotropic magnetic material with a staggered DM vector is equivalent to applying an external uniform ($\mathbf{H}_{\text{eff,u}}$) and staggered field ($\mathbf{H}_{\text{eff,s}}$) to an isotropic magnetic material. Intuitively, it is then evident that in such a material the susceptibility toward a staggered order parameter is significantly enhanced, even up to large fields, in spite of the DM vector itself being comparatively small in magnitude.

This aspect was discussed in the context of analyzing the response of the QSL/VBS candidate κ-(BEDT-TTF)$_2$Cu$_2$(CN)$_3$ in the presence of an external magnetic field [64] in μSR experiments [69]. However, while this ingredient is indeed important, a consistent interpretation of the numerous available experimental observations seems to be only possible if disorder-induced orphan spins are also considered [12,34,36,37]. Hence, we will discuss the full description of this material in detail in Section 5.

2.2. Four-Spin Ring Exchange

As mentioned above, magnetic materials may be described with a pure spin model if they are in the strongly localized limit, i.e., $t/U \ll 1$. In this limit, it is possible to extract, for instance, the bilinear Heisenberg model via second-order perturbation theory from the Hubbard model, where the Heisenberg exchange scales roughly with t^2/U. Since the organic compounds considered in this review are close to the Mott MIT, these materials should, rather, be placed in the $t \lesssim U$ regime. When constructing a spin model for this class of materials, it may, therefore, be necessary to consider higher order contributions. The most dominant higher order spin contribution is an isotropic four-spin ring exchange [70]:

$$\mathcal{H}_{(4)} = \frac{1}{S^2} \sum_{\langle ijkl \rangle} K_{ijkl} \left[(\mathbf{S}_i \cdot \mathbf{S}_j)(\mathbf{S}_k \cdot \mathbf{S}_l) + (\mathbf{S}_i \cdot \mathbf{S}_l)(\mathbf{S}_j \cdot \mathbf{S}_k) - (\mathbf{S}_i \cdot \mathbf{S}_k)(\mathbf{S}_j \cdot \mathbf{S}_l) \right], \tag{5}$$

defined on a plaquette spanned by the sites $\langle ijkl \rangle$. Note that we do not consider here four-spin terms arising from spin-orbit coupling effects. In Figure 4a, the two distinct four-site plaquettes on the anisotropic triangular lattice are illustrated and labelled with K and K'. From perturbation theory, the four-site ring exchange scales with t^4/U^3, increasing in magnitude as t approaches U. For certain ratios of J'/J and K/J, ring exchange can suppress magnetic order [71,72] or induce new types of orders [73]. In Table 1, we list selected ab-initio results [12,51,64,67] for representative κ-phase salts. For all listed materials, the four-spin ring exchange is ~5–10% of the nearest-neighbor Heisenberg exchange. This is a comparatively strong contribution consistent with the proximity of the organic materials to the Mott MIT ($t \lesssim U$). Note that it is not required by symmetry that the three terms in Equation (5) have precisely identical prefactors. Since they are very similar for the considered materials, we discuss here averaged values for simplicity.

The corresponding classical phase diagram on the triangular lattice [51] is shown in Figure 5 using constraint ratios of the isotropic couplings suggested by perturbation theory [72] and setting anisotropic couplings to zero. This phase diagram contains the well-known phases of the bilinear models on the anisotropic triangular lattice, the two-sublattice Néel order ("2SL") and a spiral order with ordering wave vector $Q = (q,q)$, including the so-called "120° order" at the limit of an isotropic triangular lattice with $J'/J = 1$. In addition, a four-spin ring exchange induces novel phases, for instance four-sublattice orders, such as the non-coplanar chiral (NCC) order and the coplanar spin-vortex crystal (SVC) order. In the NCC phase, the magnetic moments are arranged such that, if they would be arranged at the corner of a tetrahedron, they would point toward or away from the center of the tetrahedron. The sign of the magnetic moments on a four-site plaquette can be summarized by a sign of the plaquette, as indicated in the NCC phase in Figure 5. The SVC phase is also a four-sublattice phase with vortex orientations, but in this case, the magnetic moments are constrained to a single plane.

Based on the ab-initio values in Table 1, three representative organic charge-transfer salts can be placed in the phase diagram in Figure 5: κ-(BEDT-TTF)$_2$Cu[N(CN)$_2$]Cl (κ-Cl), κ-(BEDT-TTF)$_2$Cu$_2$(CN)$_3$ (κ-CuCN), and κ-(BETS)$_2$Mn[N(CN)$_2$]$_3$ (κ-Mn). Within the square lattice regime ($J'/J \ll 1$), the four-spin ring exchange does not change the magnetic ground state. Consequently, while κ-Cl is found to have a significant ring-exchange contribution, its ground state is the two-sublattice AFM Néel state, consistent with experiments and as determined with the purely bilinear spin model.

In contrast, in the regime of an isotropic triangular lattice ($J'/J \approx 1$), the four-spin ring exchange starts to play an important role for the magnetic ground state, so that κ-CuCN as significant ring exchange is placed in the classical phase diagram close to phase boundaries between the multi-Q spiral phase and the NCC phase. Quantum fluctuations may suppress the long-range order, entering a QSL/VBS phase, as suggested by semi-classical [72] and DMRG [71] calculations. If disorder would be absent in κ-CuCN (see Section 5), the magnetic interactions including four-spin ring exchange would point to a QSL/VBS ground state.

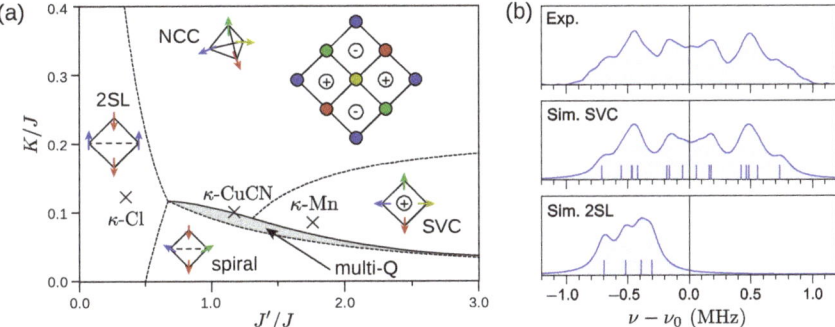

Figure 5. (a) Classical phase diagram for the four-spin ring-exchange on the anisotropic triangular lattice. Indicated are the locations for κ-(BEDT-TTF)$_2$Cu[N(CN)$_2$]Cl (κ-Cl), κ-(BEDT-TTF)$_2$Cu$_2$(CN)$_3$ (κ-CuCN), and κ-(BETS)$_2$Mn[N(CN)$_2$]$_3$ (κ-Mn), based on ab-initio calculations in Refs. [12,51,64,67] (see Table 1). The depicted phases include the two-sublattice (2SL) Néel order, non-coplanar chiral (NCC) order, and spin-vortex crystal (SVC) order. For this phase diagram, the constraints $J'/J = J'''/J'' = K'/K$, $K/J'' = 2$, and $|\mathbf{D}| = |\mathbf{\Gamma}| = 0$ were enforced. (b) ^{13}C NMR spectra measured in Ref. [74] ("Exp.") and simulated for hypothetical SVC and 2SL phases [51]. Figures adapted from Ref. [51].

A third consequence of a nonzero four-spin ring exchange is realized in κ-Mn [51], where the higher order magnetic exchange does not suppress magnetic order, but instead selects an unconventional four-sublattice magnetic order, the SVC phase, which could not be accessed without a finite K. The SVC and NCC phases introduced by a significant ring exchange (see Figure 5a) are characterized by a vector chiral order parameter $|\mathbf{v}_p|$, defined on a four-site square plaquette built by the solid bonds in Figure 4a. The vector chirality $\mathbf{v}_p = \mathbf{S}_i \times \mathbf{S}_j + \mathbf{S}_j \times \mathbf{S}_k + \mathbf{S}_k \times \mathbf{S}_l + \mathbf{S}_l \times \mathbf{S}_i$ is finite when nearest-neighbor spins are orthogonal, but second-neighbor spins are antiparallel, as realized in the NCC and SVC phases. In the two-sublattice Néel order and spiral phases, the vector chirality vanishes: $|\mathbf{v}_p| = 0$. While it is difficult to estimate the influence of quantum fluctuations based on solely the classical phase diagram, there is a some experimental evidence for κ-Mn, which seems to require such a four-sublattice chiral phase. For κ-Mn, the interpretation of the magnetic properties from experimental evidence was initially challenged by the fact that the anion layer is composed of magnetic $S = 5/2$ manganese atoms forming a distorted triangular lattice. However, the ordering temperature $T_N = 22$ K [26,74,75] of κ-Mn can be assigned to the organic BETS spins, based on the fact that the ab-initio exchange between Mn spins is with ∼1 K two orders of magnitude smaller and that the entropy change around T_N observed in specific heat measurements is too small (8% of $R \ln 2$) for the significant participation of Mn spins [51]. For comparison, in the compounds λ-(BETS)$_2$FeCl$_4$ and κ-(BETS)$_2$FeBr$_4$, the Fe^{3+} and organic spins order simultaneously [76–78]. Such features as Jaccarino–Peter superconductivity [79] and beats in the Shubnikov–de Haas effect, as reported in the latter compounds, are not observed in κ-Mn [24–26]. One example, where the experimental evidence of κ-Mn is not compatible with a 2SL or spiral order phase is the magnetic torque as a function of magnetic field [75]. As mentioned above, the out-of-plane and in-plane components of the DM vector differ in their pattern by symmetry. As a result, the vector chirality couples linearly only to $\mathbf{D}_b$, the contribution along the dashed bonds in Figure 4a, confining the spins to lying in the a^*c plane, while the 2SL and spiral phases couple only to $\mathbf{D}_{ac}$, confining the spins to the plane perpendicular to this contribution. In magnetic torque experiments, the features associated with BETS spins vanish for magnetic fields in the a^*c plane, suggesting that the field couples to an order parameter for which this plane is a special plane of symmetry. This is only true for the NCC and SVC phases. Additionally, ^{13}C NMR experiments [74,80] show signatures which are by symmetry not compatible with two-sublattice orders, but can be fitted well with the SVC order. The experimental NMR resonance [74] (top panel in Figure 5b) is symmetrical about the Larmor frequency ν_0 with a rich fine structure. The expected resonance patterns were simulated in Ref. [51] using an ab-initio hyperfine coupling tensor, Lorentzian broadening and neglecting the Mn dipolar fields. As evident from Figure 5b, the 2SL phase can be immediately ruled out, with only four distinct resonances, which are not symmetric around ν_0. In contrast, the SVC phase shows 16 distinct, symmetric resonances, allowing for an excellent simulation of the experimental data. Taking the classical phase diagram together with the experimental evidence on magnetic properties of κ-Mn, this BETS compound is a prime example for a material with the exotic magnetic order of the spin-vortex crystal phase, induced by the four-spin ring exchange.

3. New Physics in the Extended Molecule-Based Model

In this section, we focus on experimental and theoretical observations that strongly suggest that a treatment of the κ-phase charge-transfer salts in the full molecule-based model is not only needed, but also can be the source of entirely new ordering phenomena. In particular, we provide an in-depth review on the role of intra-dimer charge degrees of freedom in generating electronically driven ferroelectric ground states and we discuss the role of the anisotropy of the three inter-dimer hoppings that exist in the molecule model for the pairing symmetry of unconventional superconductivity.

3.1. Ferroelectricity in the Mott Insulating Ground State

When changing from the effective half-filled one-band dimer model on the triangular lattice with on-site Coulomb repulsion U per dimer and hopping terms t, t' between dimers to the molecule-based model, the following energy scales are at place: the hopping parameters $t_{1...4}$ as shown in Figure 2, a Coulomb repulsion $\tilde{U}$ on the molecule (please note that U was defined as the onsite Coulomb repulsion on the dimer, while $\tilde{U}$ corresponds to the on-site Coulomb repulsion on the molecule), and additional inter-molecule Coulomb repulsion terms V, which are generally expected to be smaller than $\tilde{U}$ (typically $V/\tilde{U} \sim 0.4$) [81,82]. Including V in the considerations introduces the possibility of intra-dimer charge orders that may compete with the dimer-Mott insulating state [81–83]. As a result, a new degree of freedom, namely, intra-dimer charge fluctuations, may emerge in dimerized κ-phase charge-transfer salts [81,82].

Many of the theoretical considerations regarding the inclusion of intra-dimer charge degrees of freedom into minimal low-energy models were further motivated by the simultaneous experimental discoveries of dielectric anomalies in various Mott insulating κ-phase charge-transfer salts [16,84–86]. There is a large body of evidence suggesting that these anomalies result from ferroelectricity which might be driven by the active intra-dimer charge degrees of freedom.

Dielectric spectroscopy is a common tool that is sensitive to ferroelectric order [84]. A typical signature for a ferroelectric transition is a peak in the temperature dependence of the dielectric constant $\epsilon'(T)$. Such a peak was indeed observed for various κ-phase charge-transfer salts in the Mott insulating state, as depicted in Figure 6. First, it was discovered in κ-(BEDT-TTF)$_2$Cu$_2$(CN)$_3$ (κ-CuCN, Figure 6b) below $T \lesssim 60$ K by Abdel-Jawad et al. [85]. Here, the peak of the dielectric constant shifts with frequency, which is a characteristic for a "relaxor ferroelectric". In line with this classification, the peak temperatures as a function of frequency are well-described by a Vogel–Fulcher–Tammann equation with characteristic temperature of $T_{\text{VFT}} \sim 6$ K, which interestingly corresponds to the temperature of the famous "6 K anomaly" in this compound [34,87]. A relaxor ferroelectric does not manifest long-range order, but rather represents a cluster-like order mediated by short-range correlations.

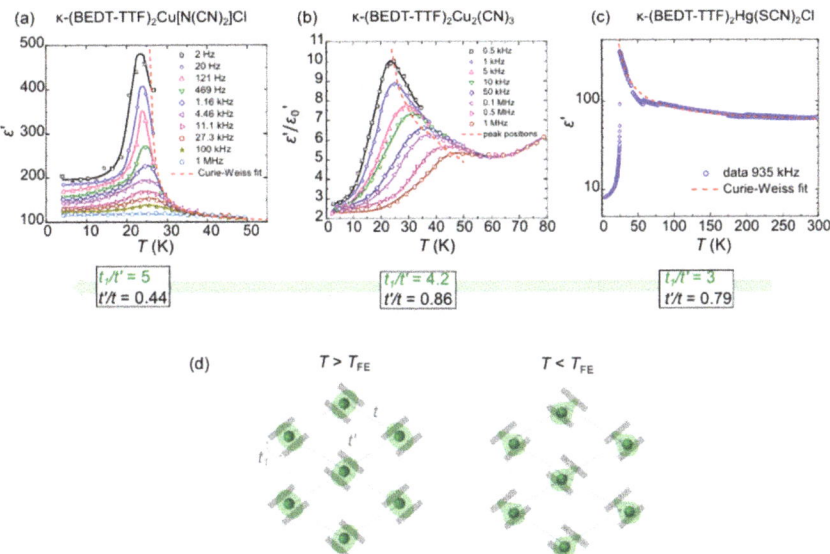

Figure 6. Role of intra-dimer charge degrees of freedom revealed by dielectric measurements: (**a**–**c**) Frequency-dependent dielectric constant ϵ' on three κ-phase organic charge-transfer salts: (**b**) κ-(BEDT-TTF)$_2$Cu[N(CN)$_2$]Cl (κ-Cl) [16,84,88], (**c**) κ-(BEDT-TTF)$_2$Cu$_2$(CN)$_3$ (κ-CuCN) [85] and (**d**) κ-(BEDT-TTF)$_2$Hg(SCN)$_2$Cl [86]. Symbols represent the measured data, solid lines are guide to the eyes. The red dashed lines represent Curie–Weiss fits of the data in (**b**,**d**), whereas they show the frequency dependence of the peak position in (**c**). These three materials differ by the degree of dimerization (t_1/t'), as well as the frustration strength (t'/t, defined in the effective dimer model). Values for t_1/t' and t'/t for each of the materials are included below the figures and taken from Refs. [18,42,86]; (**d**) Schematic view of charge distribution on the dimerized κ-phase structure. For temperatures above the ferroelectric transition temperature (i.e., $T > T_{FE}$), charge is equally distributed within a dimer. For $T < T_{FE}$, intra-dimer charge order sets in, giving rise to a macroscopic polarization.

In contrast to this, a so-called "order–disorder"-type transition, where the peak position is independent of frequency, was found in κ-(BEDT-TTF)$_2$ Cu[N(CN)$_2$]Cl (κ-Cl) [16,88], as shown in Figure 6a. This notion was further corroborated by measurements of the ferroelectric hysteresis as well as measurements of the switchability of the ferroelectric polarization. The dielectric constant in κ-Cl at high temperatures is well-described with a Curie–Weiss law with a characteristic temperature $T_{CW} \sim 27$ K. This temperature is identical to the antiferromagnetic ordering temperature T_N of this compound, making this material a realization of a multiferroic system [16]—an intriguing state of matter with cross-coupling between charge and spin degrees of freedom. Independent of the detailed behavior as a function of temperature or frequency, it can be concluded that these dielectric measurements indicate active charge degrees of freedom deep inside the Mott insulating state, which casts doubts on the stringent applicability of the dimer model.

In fact, intra-dimer charge order would be a natural explanation for the formation of ferroelectricity in the κ-phase charge-transfer salts. In general, ferroelectricity requires (i) the existence of dipoles and (ii) the breaking of inversion symmetry for the formation of a macroscopic polarization. In case of intra-dimer charge order, both conditions are naturally fulfilled in the dimerized κ-phase structure, see Figure 6d. At high temperatures above the ferroelectric transition temperature $T > T_{FE}$, the charges are localized on a dimer, corresponding to fluctuating dipoles [89]. When cooling below T_{FE}, the charges localize on one of the two organic molecules that form a dimer. As a result, a static dipole is created. At the same time, through localization on one of the molecules, inversion symmetry is broken.

Thus, overall a macroscopic polarization arises. This electronically driven mechanism for ferroelectricity, which is invoked for both salts κ-Cl and κ-CuCN [16,81,85,90,91], is different from the off-center displacement of ions in conventional ferroelectrics [92].

Clearly, the electronically driven scenario requires a proof of intra-dimer charge order, similar to the case of weakly dimerized quasi one-dimensional TMTTF-based [93] as well as α-phase BEDT-TTF salts [94,95], where charge order and ferroelectricity are well established. However, attempts to resolve a charge disproportionation within the BEDT-TTF dimer for κ-Cl and κ-CuCN have remained unsuccessful [96]. By monitoring charge-sensitive phonon vibrational modes in infrared spectroscopy, no indications for a charge-order-induced splitting of modes were observed for κ-Cl and κ-CuCN within their experimental resolution [97]. This puts an upper limit on the possible size of charge disproportionation of $\delta \sim \pm 0.005$ e. The absence of a detectable charge disproportionation has motivated proposals of alternative explanations for the observation of dielectric anomalies [5,98–100]. Prominent proposals include (i) charged domain-wall relaxations in the weak ferromagnetic state at lower temperatures for κ-Cl [98], (ii) charge defects triggered by a local inversion-symmetry breaking in the anion in κ-CuCN [99–102] and (iii) a percolative enhancement of the dielectric constant in proximity to a Mott MIT based on experiments on κ-CuCN [5,103]. Besides controversies regarding the data analysis of the dielectric data, see, e.g., Refs. [84,88,96], there is another concern that some of the alternative explanations for the dielectric anomalies involve material-specific arguments. This makes these proposals a possible explanation for the respective specific compounds, but does not answer the question of why so many κ-phase charge-transfer salts exhibit dielectric anomalies deep in the Mott insulating state. We note that the proposal that the dielectric signal even deep in the Mott insulating state arises from the spatial coexistence of insulating and correlated metallic phases close to the Mott MIT [5,103] does not necessarily explain the emergence of long-range order-disorder-type ferroelectricity.

Within the picture of intra-dimer charge order, the degree of charge disproportionation δ is expected to relate to the strength of dimerization. From ab-initio model parameters, κ-Cl and κ-CuCN both fall in the same range of rather strong dimerization with $t_1/t' \sim 4.2–5$ [18,42] (see Figure 6). Thus, a clearer case for electronically driven ferroelectricity might be made when dimerization is still dominant, but weaker than in the materials cited above. Indeed, according to DFT calculations, the related material κ-(BEDT-TTF)$_2$Hg(SCN)$_2$Cl is characterized by such a moderate degree of dimerization $t_1/t' \sim 3$ [86]. For this compound, charge order with $\delta = \pm 0.1$ e was unequivocally identified in vibrational spectroscopy [104]. Dielectric data for this compound [86] is shown in Figure 6c. The behavior of $\epsilon'(T)$ is very reminiscent of the behavior of the characteristics of well-established order-disorder type ferroelectrics [105]. Whereas in κ-(BEDT-TTF)$_2$Hg(SCN)$_2$Cl there is, thus, strong evidence for long-range charge [104] and ferroelectric order [86], there is no long-range charge order in its sister compound κ-(BEDT-TTF)$_2$Hg(SCN)$_2$Br. Nonetheless, experimental evidence from Raman spectroscopy for a "quantum dipole liquid" was presented for the Br-variant, in which electric dipoles from intra-dimer charge degrees of freedom remain fluctuating down to lowest temperatures [106]. Overall, the collection of data on both compounds κ-(BEDT-TTF)$_2$Hg(SCN)$_2$X with X = Cl, Br [86,106,107] provide strong evidence that intra-dimer charge degrees of freedom are relevant in dimerized κ-phase materials.

3.2. Superconductivity in Extended Molecule-Based Models

There are also distinct differences between the physics of the dimer model and the extended molecule-based model in the metallic regime as a result of (i) intra-dimer Coulomb interactions and (ii) the anisotropy of the inter-dimer hoppings. The relevance of the latter becomes evident in the discussion of the superconducting order parameter of the κ-phase materials. In fact, following previous work by Kuroki et al. [108], Guterding et al. [15,60] calculated the superconducting pairing symmetry of a few κ-based charge-transfer-salts superconductors using a random phase approximation (RPA) spin-fluctuation approach and hopping parameters extracted from ab-initio-based density functional theory calculations in the molecule description. It was found that the order parameter is substantially altered when considering the anisotropy in t_2 and t_4 (defined in Figure 2), which is averaged out in the dimer model. Sufficiently large anisotropy promotes a peculiar competition between square-like and diagonal hoppings, which is absent in the effective dimer model. As a result, for significant anisotropy, mixed-order parameters of type $s_\pm + d_{x^2-y^2}$ are favored over the single-component d_{xy} (in the notation of the physical Brillouin zone). Qualitatively, these results were later reinforced by solving the linearized Eliashberg equation using the two-particle self-consistent approach [61], even though there is a quantitative difference in the precise location of the crossover from d_{xy} to $s_\pm + d_{x^2-y^2}$.

The theoretically predicted mixed-order parameter in the full molecule-based model has eight nodes (see Figure 7a), giving rise to three distinct coherence peaks (Figure 7b,c). The existence of the three coherence peaks was confirmed in low-temperature scanning tunneling spectroscopy of κ-(BEDT-TTF)$_2$Cu[N(CN)$_2$]Br [60]. It is also interesting to note that calculations on the extended Hubbard model using Monte Carlo simulations found that superconductivity with mixed $s + d$ order parameter is stabilized on the verge of charge-order and dimer-Mott instabilities [109–111].

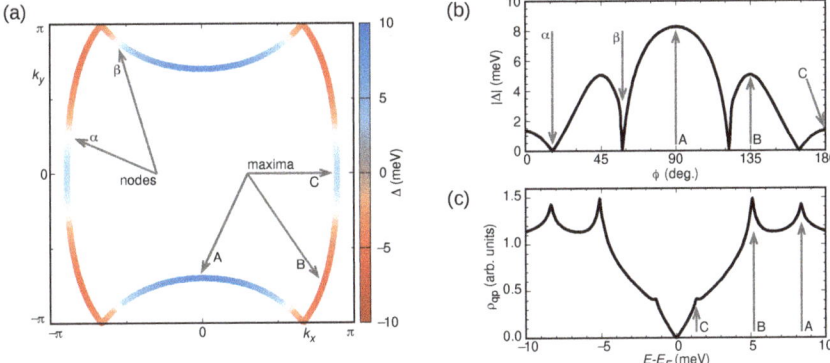

Figure 7. Theoretical results for κ-(BEDT-TTF)$_2$Cu[N(CN)$_2$]Br, calculated by an RPA spin-fluctuation approach using hopping parameters extracted from ab-initio-based density functional theory calculations in the extended molecular model: (**a**) Eight-node mixed-symmetry superconducting gap Δ, (**b**) $|\Delta|$ as a function of the angle ϕ with respect to k_x, (**c**) quasi-particle DOS ρ_{qp} in the superconducting state. The three coherence peaks A, B, C are consistent with scanning tunneling spectroscopy results [60]. Figures adapted from Ref. [60].

4. Coupling of Correlated Electrons to the Crystal Lattice

Treating the organic κ-phase charge-transfer salts or any other correlated electron system in terms of a purely electronic Hubbard model does not consider that charge and spin degrees of freedom are coupled to the crystal lattice [112,113]. That this electron-lattice coupling could have significant consequences for the behavior of the correlated electron system might already be inferred from the fact that physical pressure is often used to induce phase transitions and associated critical behaviors. For example, as outlined in Section 1.2, pressure can be used to induce the Mott MIT. Importantly, the underlying compressibility of the crystal lattice does also necessarily imply that the crystal lattice can respond to the correlated electron system in non-trivial ways. Since such a type of coupling exists in any solid-state realization of a correlated electron system, understanding the consequences of the electron–lattice coupling is nowadays central to various fields of research. A prominent example is the electronic nematic phase which is believed to be ubiquitous to the phase diagrams of many high-temperature superconductors [114–116] and which is intimately coupled to lattice instabilities [117–119]. In addition, the coupling to the lattice has received increasing levels of attention in the field of organic [34,49,87,120,121] and inorganic frustrated magnets [122–128].

In the following, we deepen the discussion of the underlying ideas by analyzing recent experimental results for the κ-phase organic salts [9,129]. The high-pressure sensitivity of these materials suggests that they are characterized by a comparatively large coupling of the electrons to the crystal lattice [130]. In this sense, the κ-phase charge-transfer salts are an ideal testground to discover and understand novel effects that arise from electron-lattice coupling, since any effect will be significantly amplified when such a coupling is naturally strong.

4.1. Critical Elasticity around the Mott Critical Endpoint

We start by a discussion of how coupling of correlated electrons to the crystal lattice impacts the pressure-induced Mott transition. For that, we review theoretical ideas and experimental results on κ-(BEDT-TTF)$_2$Cu[N(CN)$_2$]Cl (κ-Cl). The purely electronic Mott critical endpoint is characterized by a diverging response function [27,28]. When such an electronic system is embedded into a crystal lattice with finite compressibility, then the lattice responds to changes in the electronic system [112,113,131]. Sufficiently away from the critical endpoint, the crystal lattice will barely respond to the only weakly fluctuating electronic degrees of freedom. However, close to the endpoint, when the electronic system becomes critical, the internal pressure exerted by the electronic system will induce additional lattice strain. In other words, the crystal lattice is driven soft, leading to a renormalization of the system's compressibility due to a vanishing elastic modulus. These renormalization effects can successfully be captured in DMFT calculations of the compressible Hubbard model [112,113].

Importantly, the softening of the crystal lattice implies that the crystal lattice itself becomes critical when the electronic degrees of freedom become critical. This was pointed out in Ref. [131] based on considerations of an effective field theoretical description of the coupling of the Mott order parameter to lattice strain. The electronic Mott transition is, hence, expected to be preempted by an isostructural transition. Consequently, the criticality of the Mott system coupled to lattice degrees of freedom is described by mean field criticality, the universality class of isostructural solid–solid transitions, where the shear forces of the crystal lattice strongly suppress fluctuations [131,132].

Experimentally, the relevance of the theoretical considerations above was explicitly demonstrated by studies of the lattice strains as a function of pressure, i.e., strain–stress relations, around the Mott critical endpoint in κ-Cl in Ref. [9]. The experimental data is shown in Figure 8a. The key result here is the observation of pronounced non-linearities in the strain–stress relationships at temperatures higher than the critical temperature $T_{cr} \approx 37$ K. Even at a temperature of 43 K, i.e., $(T - T_{cr})/T_{cr} \sim 20\%$, the renormalization of the compressibility is found to be of the order of the compressibility itself. Thus, these results indicate a breakdown of Hooke's law induced by critical electronic degrees of freedom. The measured data do not only qualitatively agree with the physical picture described above, but also quantitatively. Following the classification as an isostructural solid–solid endpoint [131], for which the volume reflects an appropriate order parameter, the experimental data was quantitatively well described by the associated mean-field criticality [9].

Taken together, theory and experiment suggest that any Mott system that is coupled to a compressible lattice will eventually be controlled by this "critical elasticity". Importantly, this effect can be dominant in a wide range of the temperature–pressure phase diagram, as experimentally proven for κ-Cl and schematically depicted on the right side of Figure 8a [9]. Thus, the example of the Mott MIT and its criticality demonstrates nicely that electron–lattice coupling is not only a small, perturbative correction to the correlated electron problem, but an essential ingredient.

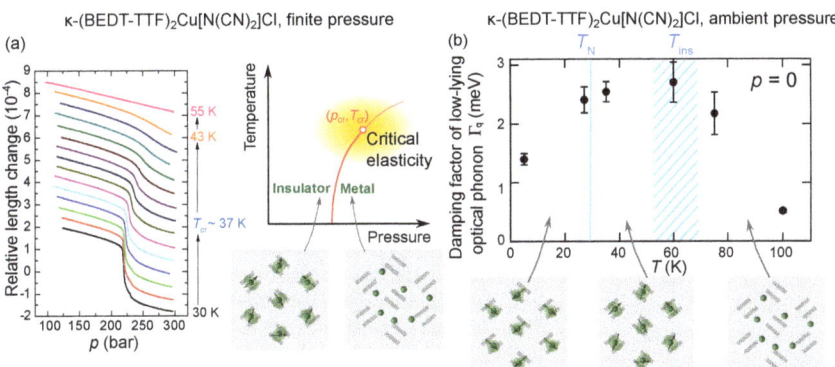

Figure 8. Coupling of correlated π-electrons with the crystal lattice in κ-(BEDT-TTF)$_2$Cu[N(CN)$_2$]Cl. (**a**) Strain–stress relationships across the pressure-induced Mott metal–insulator transition [9]: The relative length-change data on the left as a function of hydrostatic pressure indicate a wide temperature region above the critical endpoint located at $T_{cr} \sim 37$ K, in which strongly non-linear strain–stress relationships are observed. This result implies that the crystal lattice itself becomes soft as the electronic degrees of freedom become critical. This effect has been termed "critical elasticity" [131,132] and prevails in an extended region around the endpoint (right panel); (**b**) ambient-pressure study of the damping factor of a low-lying optical phonon, likely related to the BEDT-TTF breathing mode, as a function of temperature [129]. The damping factor is inversely proportional to the phonon lifetime. Three characteristic regimes were identified in the data: (i) low damping at high temperatures ($T > T_{ins}$) before the charge gap opens, (ii) high damping when the charges localize on the dimer ($T_N < T < T_{ins}$) and (iii) low damping below the onset of spin and charge order ($T < T_N$). Figures adapted from Refs. [9,129].

4.2. Phonon Anomalies Probed by Inelastic Neutron Scattering

The fact that the (de-)localization of the π electrons at the Mott MIT causes drastic anomalies in the lattice properties is a very good indication that, vice versa, anomalies in the lattice dynamics can be a very sensitive probe of underlying fluctuating electronic degrees of freedom. Thus, probing lattice dynamics can be a very important tool to detect active charge and/or spin degrees of freedom, e.g., in the Mott insulating state where the role of intra-dimer charge degrees of freedom is particularly debated (see Section 3.1, above).

Due to the large number of atoms per unit cell, there are typically many lattice and molecular vibration modes observed in, for instance, Raman or vibrational spectroscopy over a large energy range [101]. The modes of interest that are expected to be particularly sensitive to fluctuating electronic degrees of freedom are energetically low-lying phonons, in particular those that involve breathing motion within the BEDT-TTF dimer. Based on ab-initio phonon calculations, those phonons typically lie in an energy range of just a few meV, e.g., at 4.1 and 4.7 meV in κ-(BEDT-TTF)$_2$Cu$_2$(CN)$_3$ (κ-CuCN) [101]. Such low-energy phonons can be probed with high sensitivity in neutron-scattering measurements. For a long time, neutron-scattering measurements on the organics have been rare due to the typically small size of the single crystals [133]. However, technical improvements have now allowed to perform inelastic neutron-scattering measurements to probe phonon changes in κ-Cl [129] at ambient pressure and even more recently on κ-CuCN [37].

The main result of the work on κ-Cl, as shown in Figure 8b, is that the damping of a low-lying optical phonon, located around 2.6 meV, shows a strongly non-monotonic behavior as a function of temperature [129]. In particular, the phonon damping is specially high in an intermediate temperature range: When cooling κ-Cl from 100 K to below $\sim$50–60 K, the phonon damping increases rapidly, corresponding to a significantly reduced phonon lifetime. Upon further cooling below $T_N \sim 27$ K, the phonon lifetime is found to be significantly enhanced again. Interestingly, the onset of phonon damping at $\sim$50–60 K coincides with the opening of the charge gap at T_{ins} [134]. Below this temperature, the hole carriers become localized on the dimer, but the intra-dimer charge degree of freedom remains active. Only when the intra-dimer charge and the spin degrees of freedom are ordered below T_N, is a truly inelastic behavior of the phonon modes recovered (see schematics in Figure 8b). While this result suggests an intimate coupling of the lattice with charge and spin degrees of freedom, which makes the lattice unstable, it is important to note that the neutron data does not support a structurally driven damping of the phonon modes. Thus, electronic degrees of freedom (charge and spin) are the driving force behind these lattice anomalies [129].

A possible way to describe such a coupling is given by the pseudospin-phonon-coupled model, as proposed in Ref. [135], which can be extended to the coupling of phonons to other stochastic variables than only spins. According to this model, the characteristic energy of the electronic degree of freedom must be of the order of the phonon energy itself for significant phonon damping to occur. Interestingly, the intra-dimer charges that give rise to the dipole liquid in the related κ-(BEDT-TTF)$_2$Hg(SCN)$_2$Br [106] (see Section 3.1) were found to be located at $\sim 40\,\text{cm}^{-1}$, which is similar in energy to the overdamped optical phonon in κ-Cl. Thus, it seems likely that the overdamped phonons observed in inelastic neutron-scattering experiments in κ-Cl are related to fluctuating intra-dimer charge and spin degrees of freedom.

We note that the recent study of phonons in κ-CuCN [37] revealed strong similarities in the behavior of the phonon modes. This includes (i) a strongly overdamped phonon mode, when the π-electrons are localized on the dimers and (ii) a recovery of a long-lived phonon mode below the characteristic temperature of 6 K. This result clearly strengthens the notion that the enigmatic "6 K anomaly" is a result of a phase transition rather than a crossover. It was argued based on a spin-charge coupling model that the observed anomalous phonon behavior is consistent with a transition into a valence-bond solid below 6 K.

5. Role of Disorder

Disorder of various kinds is inevitably present in any real material. Even though observations of quantum oscillations in κ-phase metals in low fields and/or relatively high temperatures indicate the high crystalline quality of pristine samples [136], there is by now some body of evidence that disorder is important for understanding the properties of the κ-phase charge-transfer salts. In the present section, we discuss experimental results of the impact of controlled disorder on the properties close to the Mott MIT, as well as theoretical results on the role of disorder and resulting orphan spins in the magnetic phase.

5.1. Experimental Study of Phenomena Close to the Mott Transition under Controlled Disorder

To better grasp the effect of disorder, it is pivotal to develop means to controllably change the degree of disorder. For the κ-phase organic charge-transfer salts, two means are established with which disorder can be controlled, both reversibly and irreversibly (see Figure 9a): (i) control of the conformational degree of freedom of the ethylene endgroups (EEG) in the BEDT-TTF molecule by varying the cooling rate through the associated glass transition at T_g [137], and (ii) introduction of molecular defects, dominantly in the anion layer [138], through X-ray irradiation [139]. In the remainder of this section, we summarize recent results, based on both techniques.

First, we focus on the reversible approach of introducing disorder through controlling the EEG disorder. Those ethylene endgroups ([C$_2$H$_4$]) can adopt two different configurations, when viewed along the central C=C bond. The EEGs can either show an eclipsed or a staggered configuration, see Figure 9a. Upon cooling, the EEG tend to adopt one of the conformations. However, the ordering usually cannot be completed for kinetic reasons [130,137]. The associated relaxation becomes so slow close to T_g that equilibrium cannot be achieved, resulting in a certain amount of intrinsic structural disorder. Importantly, the amount of disorder can be controlled in a reversible manner by heating above T_g, thus melting the frozen EEG configuration, and consecutively adjusting the cooling rate through the glass transition at T_g.

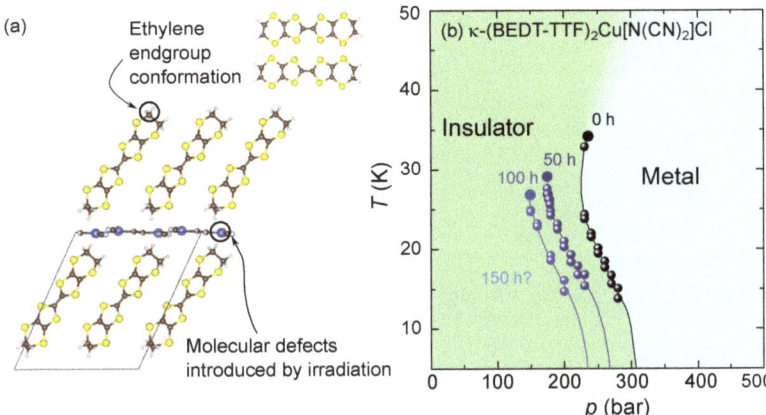

Figure 9. Experimental studies of the impact of weak disorder in κ-phase charge-transfer salts. (a) Experimentally, disorder can be deliberately introduced either by (i) creating molecular defects through X-ray irradiation [139] or (ii) by controlling the ethylene endgroup conformation via the cooling rate through the glass-forming temperature T_{glass} [137]; (b) temperature–pressure phase diagram for κ-(BEDT-TTF)$_2$Cu[N(CN)$_2$]Cl, subjected to different irradiation times [140]. Symbols represent the discontinuous Mott metal–insulator transition. (b) adapted from Ref. [140].

An important task is to quantify the amount of disorder induced by the glassy EEG freezing. This task was addressed in Ref. [141] utilizing a special heat-pulse protocol to achieve large cooling rates up to several 1000 K/min, together with modelling of the glass transition in terms of a double-well potential with realistic energy parameters [137,141,142]. It was found that EEG disorder levels of up to ∼6% might be reached in real experiments with the largest cooling speeds. Conversely, this implies for the pristine, slowly cooled crystal that intrinsic EEG disorder levels are at a maximum about 2 to 3%.

Experimentally, an increase in the residual resistivity, as well as an influence on the superconducting T_c, were observed in metallic κ-(BEDT-TTF)$_2$Cu[N(CN)$_2$]Br upon cooling with higher speed through T_g [143,144]. However, closer to the Mott transition, the effects of changing effective Hubbard parameters with EEG conformation is likely to dominate over disorder effects. This was demonstrated experimentally in Refs. [141,145], and substantiated by ab-initio calculations of the band structure of fully EEG-ordered κ-salts in Ref. [146]. In fact, it is possible to tune across the critical ratio $(U/W)_c$ for the Mott transition in deuterated κ-(BEDT-TTF)$_2$Cu[N(CN)$_2$]Br by adjusting the cooling rate through T_g. In such cooling-rate-dependent studies, important results, such as the critical slowing down of charge carriers at the Mott critical endpoint [145], were made possible.

Whereas the control of disorder through the glassy freezing of the EEG disorder is reversible, it is hard to disentangle effects that result from disorder and from changes in effective Hubbard parameters, in particular when the material is situated close to the Mott MIT. Thus, the introduction of disorder by X-ray irradiation is a promising, complementary approach to study the influence of disorder, despite its irreversibility. A detailed review of the experimental procedures and various results are given in Ref. [139].

Before turning to the results, it is important to summarize what type of disorder is created by X-ray irradiation. From studying molecular vibration modes in Cu-containing κ-phase charge-transfer salts by infrared optical spectroscopy [147], it was shown that irradiation has the strongest impact on vibrational models associated with the anion. This meets the intuitive expectation that X-rays should be mostly absorbed in the anion because they contain the heaviest atom, Cu. Thus, X-ray irradiation induces primarily molecular defects in the anion layer which creates a random lattice potential for the π carriers in the BEDT-TTF layer. This notion was corroborated by DFT calculations [138]. In addition, there is no strong indication that X-ray irradiation changes the carrier concentration in the BEDT-TTF layer [139].

By now, there are various studies which have utilized X-ray irradiation to probe the interplay of disorder and correlations close to the Mott transition in κ-phase charge-transfer salts [139,140,147–156]. In fact, the corresponding Mott–Anderson model has attracted significant attention from a theoretical perspective, leading to a large number of numerical evaluations of its properties (e.g., Refs. [157–168]). In the limit of independent electrons, the introduction of disorder is known to transform a metal into an Anderson insulator [169,170]. In the opposite limit of no disorder, but strong correlations, the latter will drive a metal–insulator transition. Naively, thus, both disorder and correlations promote insulating states. However, the interplay is much more complex, as discussed theoretically [171] and also witnessed experimentally in transport studies ($\rho(T)$) on κ-phase charge-transfer salts close to the Mott transition. Here, the impact of disorder on metallic salts on $\rho(T)$ [149,151] is found to be distinct from those of the Mott insulators κ-Cl and κ-CuCN [139,154]: For κ-(BEDT-TTF)$_2$Cu[N(CN)$_2$]Br, weak disorder increases the residual resistivity while simultaneously suppressing the superconducting critical temperature. For higher degrees of disorder, insulating behavior with $d\rho(T)/dT < 0$ is observed across the full temperature range. On the contrary, for Mott insulators, whose $\rho(T)$ shows activation-type behavior before irradiation, $\rho(T)$ decreases at any given temperature with an increasing dose of irradiation. Generally, the impact of disorder is found to be stronger for κ-phase charge-transfer salts closer to the Mott transition.

These opposing trends of the effect of disorder on the metal vs. the Mott insulator motivate a detailed study of the Mott MIT in the presence of varying degrees of disorder and pressure. Such studies were performed in Refs. [140,155] and we show the resulting phase diagram of Ref. [140] in Figure 9b for irradiation times below 150 h, corresponding to weak to moderate disorder. This study revealed a very clear tendency of the position and character of the MIT upon increasing disorder. While the MIT initially retains its discontinuous character at low temperatures, the location of the MIT is shifted to lower pressures and lower temperatures. Upon increasing irradiation further to ∼150 h, no signatures of a discontinuous MIT were detected. Thus, for low disorder, the MIT shows characteristics of the pristine Mott transition. Only above a critical disorder strength is the Mott character no longer evident.

Interestingly, a lower critical pressure p_{cr} for increased levels of disorder strengths shows equivalent features to an increased critical correlation strength $(U/W)_c$. This is consistent with theoretical predictions of a "soft Coulomb gap" [159]—disorder widens the Hubbard bands by introducing a finite spectral weight at the Fermi level. Consequently, a larger U is needed to fully open the Mott gap. This interpretation is supported by optical conductivity data on κ-(BEDT-TTF)$_2$Cu[N(CN)$_2$]Cl (κ-Cl) at ambient pressure [139] as well as from scanning tunneling microscopy studies of the normal-state density of states in κ-(BEDT-TTF)$_2$Cu[N(CN)$_2$]Br [172], even in its pristine form.

Besides the position of the metal–insulator transition in the temperature–pressure phase diagram, the behavior of κ-Cl after longer irradiation at ambient and finite pressures also reveals interesting phenomena that motivate various further studies of disordered Mott systems. Possible questions of interest include how disorder affects the nature of the transformation of metallic into insulating regions across the first-order phase transition. The thermodynamic data of Refs. [9,140] clearly indicate an increased broadening of the jump-like changes in the volume across the first-order phase transition for samples with increasing levels of irradiation-induced disorder. NMR measurements on strongly irradiated κ-Cl under hydrostatic pressure [156] revealed very slow dynamics, associated with an electronic Griffiths phase, located in proximity of the Mott phase boundary of the pristine, almost clean material. The authors of this work argued that the observed slow dynamics are incompatible with a macroscopically phase separated state and associated-domain wall motion. Further experimental studies will be useful in the future to support the intriguing physics presented above.

In addition, it was suggested that the suppression of the first-order transition stabilizes the quantum-critical behavior of the Mott transition to lower temperatures [155]. Furthermore, it was found that large irradiation times of ∼500 h suppress the antiferromagnetic order [154] that is present in pristine crystals at ambient pressure. The absence of long-range order in strongly disordered κ-Cl is particularly interesting in light of the discussion of the role of disorder in spin-liquid candidate systems. Clearly, an important step for the future is a detailed quantification of the X-ray-induced amount of disorder and its spatial distribution, which unfortunately is still lacking so far for the κ-phase charge-transfer salts. This input will be important for achieving quantitative comparisons between experiments and theories of disordered, correlated materials.

5.2. Disorder in the Magnetic Phase: Scenario of a Valence Bond Solid Host with Orphan Spins

Among the Mott insulating κ-phase charge-transfer salts, κ-(BEDT-TTF)$_2$Cu$_2$(CN)$_3$ (κ-CuCN) was recently at the center of increased interest in the context of disorder effects [12,34,36,37]. The magnetic ground state of this material is expected to have a vanishing net-magnetic moment, such as a quantum-spin-liquid (QSL) or valence-bond-solid (VBS) state. In these states, so-called "orphan spins" may emerge as a result of non-magnetic vacancies caused by defects in the anion layer, or by domain wall patterns in the VBS case, for instance, as a result of the randomized ethylene endgroup conformation discussed above (see Figure 10a).

The effects of orphan spins were argued in Ref. [12] to offer a consistent explanation for experimental observations in κ-CuCN such as magnetic torque [8] and NMR [50] measurements. This interpretation is in contrast to the originally proposed critical scenario [64,69], which was introduced based on the unconventional field-dependence of the μSR linewidth, i.e., of the magnetic susceptibility. In a critical scenario, one would expect either a uniform criticality or, as elaborated in Section 2, in the presence of a DM interaction a staggered criticality. As detailed below, both scenarios are, however, not compatible with the magnetic-torque observations.

Theoretically, the magnetic torque can be expressed as $\tau = H^2 \chi_\tau(H) f(\theta)$, where $\chi_\tau(H)$ is the torque susceptibility with a possible non-trivial field dependence and $f(\theta)$ is a function of the angle θ between magnetic field and sample (see Figure 10b). The experimental torque observations by Isono et al. [8] can be summarized as: (i) an unconventional field dependence of the torque susceptibility $\chi_\tau \propto H^{-0.8}$, (ii) a sinusoidal dependence of the torque as a function of magnetic-field angle $\tau \propto \sin(\theta - \theta_0)$, and (iii) a field-dependence of the angle shift $\theta_0(H)$. A critical scenario is not compatible with these three observations simultaneously. A uniform criticality would lead to sinusoidal-torque response for all relevant field strengths. This does not allow for a field-dependent angle-shift $\theta_0(H)$ as observed in the experiment. A staggered criticality, on the other hand, would lead to a sawtooth-shaped magnetic torque, also in contrast to the experimental observation.

A consistent explanation for all three key torque features is given by the consideration of disorder-induced orphan spins. An orphan spin can be described by a localized magnetic moment and generally consists of a broad screening cloud, which depends on the interactions of the host system. The induced magnetic moment can then be described by a uniform contribution $\sum_{i' \sim m} \langle \tilde{S}_{i'} \rangle = c_u \langle \tilde{S}_{I,m} \rangle$ and a staggered contribution $\sum_{i' \sim m} \eta_{i'} \langle \tilde{S}_{i'} \rangle = c_s \langle \tilde{S}_{I,m} \rangle$. Here, we labelled the sites surrounding the impurity site m within the screening cloud with index i' and sublattice index $\eta_{i'} = \pm 1$. In contrast to the pristine bulk case, the induced staggered magnetic moment is then parallel to the induced defect magnetization. This ensures the impurity torque has sinusoidal field-dependence, even if the staggered contribution is the dominant one, allowing for a scenario which simultaneously leads to a sinusoidal magnetic-torque shape and a field-dependent angle shift. In Figure 10c, a scenario of Ref. [12] is illustrated with the experimentally observed field exponent ($\zeta_I = 0.8$) and a dominant staggered contribution ($c_s/c_u = 10$). The result is a sinusoidal angle dependence of the total torque $\tau = \tau_B + \tau_I$. The angle shift θ_0 is field-dependent due to the relative shift in the bulk torque (with $\tau_B \propto H^2$) and the impurity torque (with $\tau_I \propto H^2 H^{-\zeta_I}$).

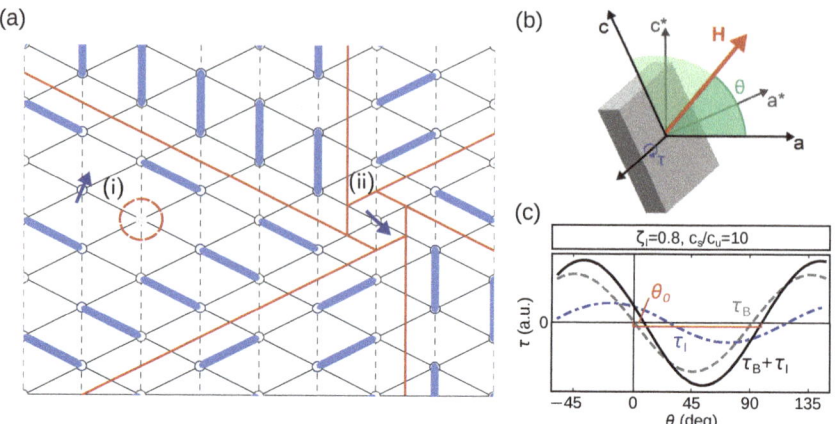

Figure 10. (**a**) Disorder scenarios in a valence-bond solid state, discussed for κ-(BEDT-TTF)$_2$Cu$_2$(CN)$_3$. Local spin 1/2, or so-called "orphan spins", may be caused by (i) vacancies in the anion layer, emphasized by the red circle or (ii) specific domain wall patterns, a possible result of randomness in the ethylene endgroup conformation. (**b**) Magnetic-torque setup, with crystal axes a and c (with $a^* \perp bc$, $c^* \perp ab$), magnetic field **H** and the angle θ between sample and field. (**c**) Theoretical magnetic torque τ dependence on field angle θ, with indication of the angle shift θ_0 and resolution of impurity (τ_I) and bulk (τ_B) contributions. Figures adapted from Ref. [12].

Considering expressions derived in the framework of finite-randomness large-spin fixed point (LSFP) [173,174], it also turns out that the critical exponent is not as sample-dependent as one might expect for a disorder phenomenon. The corresponding exponent is restricted to be $2/3 < \zeta < 1$, so that the experimentally observed $\zeta = 0.8$ falls in the middle of the possible range. These LSFP expressions also allow to predict the temperature and field dependence of the NMR linewidth ν, in good agreement with experiment [50]. They were recently employed in the analysis of a comparative study of NMR spin-relaxation-rate measurements for κ-phase compounds for which the magnetic ground state is argued not to order: κ-CuCN, κ-(BEDT-TTF)$_2$Ag$_2$(CN)$_3$ and κ-(BEDT-TTF)$_2$Hg(SCN)$_2$Cl [36]. Moreover, in κ-CuCN both, the measured NMR linewidth and magnetic torque become temperature independent around $\sim$1 K, suggesting a common impurity origin for NMR linewidth and magnetic torque. The VBS scenario was supported further recently by ESR measurements observing the opening of a spin gap below the T^* anomaly [34]. The nature of this transition was investigated in inelastic neutron scattering (INS) measurements, where the observed crossover was interpreted in favor of a VBS instead of a QSL state [37].

In addition to offering a consistent explanation of the magnetic torque, NMR, ESR, and INS results, the orphan-spin scenario could potentially also solve an open issue for κ-CuCN which was subject for debate for over a decade. The linear temperature dependence of specific heat measurements [46] was interpreted as the signature of a spinon Fermi surface, so that gapless spinon excitations of the QSL state would be present. Seemingly in contrast to this observation, thermal conductivity measurements [175] revealed $\kappa/T = 0$ for $T \to 0$, which is not compatible with the presence of low-lying fermionic excitations. Considering the presence of orphan spins, both of these observations are reasonable. In this scenario, low-lying excitations in the form of local domain walls may lead to a linear temperature dependence of specific heat. Since these excitations are local, they would not be observed in transport measurements such as thermal conductivity.

6. Conclusions and Outlook

Organic charge-transfer salts, especially those of the κ-(BEDT-TTF)$_2X$ and κ-(BETS)$_2X$ family, are considered to be model systems to explore the physics of strongly correlated electron systems in proximity of the Mott metal–insulator transition. This notion has been corroborated by a successful description of a large set of properties of the systems in terms of the one-band Hubbard model in the strongly dimerized limit as well as the nice tunability of the real materials in laboratory settings. In recent years, it became clear that all salient features of these materials, however, can only be accurately captured when including further aspects in the generalized models. In this review, we covered a selection of new aspects that have proven to be relevant to the physics of κ-phase organic charge-transfer salts. In particular, we discussed theoretical and experimental evidence for the relevance of (i) magnetic interactions beyond the Heisenberg exchange, including spin-orbit and higher order ring-exchange couplings, (ii) intra-dimer degrees of freedom in creating novel states, such as electronically driven ferroelectric states and mixed superconducting order parameters, (iii) the coupling of correlated electrons to lattice degrees of freedom and (iv) disorder. Nonetheless, a few open questions remain. Some examples are:

- As discussed in the review, ab-initio extracted magnetic models for triangular κ-phase charge-transfer salts indicate the importance of four-spin interaction terms with spin-orbit coupling effects, as well as, in the presence of a magnetic field, possible products of an odd number of spin operators, such as the scalar spin chirality. What (quantum) phases are to be expected arising from these interactions that are relevant for these materials?
- How do intra-dimer charge and spin degrees of freedom conspire in the ground-state properties of frustrated Mott insulating κ-phase charge-transfer salts? In particular, what impact do the intra-dimer charge degrees of freedom have in promoting (impeding) the formation of long-range spin order? Can these effects be theoretically described with models containing both, spin and charge degrees of freedom?
- What is the role of magnetoelastic coupling in the formation of novel states of matter, such as putative spin-liquid states in κ-(BEDT-TTF)$_2$X? Can we develop accurate models to capture these effects?
- Can we quantitatively describe the impact of disorder on the properties of these charge-transfer salts close to the Mott metal–insulator transition experimentally and theoretically? To this end, how can we accurately quantify the level of disorder in real materials and determine the nature of the disorder and their spatial distribution? Which models and methods allow to theoretically describe the interaction between disorder and bulk properties properly?
- What novel phases may be realized under non-equilibrium conditions [176]?

While this review focuses on the physics of the κ-phase organic charge-transfer salts specifically, many of the discussed aspects are relevant for the entire family of strongly correlated electron systems. For example, four-spin exchange interactions are expected to be relevant in triangular-lattice based inorganic Mott insulators with $4d$ or $5d$ transition metal ions. Likewise, as was pointed out in the past [177], charge effects might be generically relevant in frustrated Mott systems even deep in the Mott insulating phase. Recent examples are, for instance, the studies of the dielectric properties of the Kitaev magnet α-RuCl$_3$ [178]. Furthermore, the coupling of the correlated electron system to the lattice is under intensive scrutiny in the study of, for instance, unconventional superconductors (high-T_c cuprates [179], Sr$_2$RuO$_4$ [180] or Fe pnictides and chalcogenides [115]), frustrated magnets [181] and multiferroics [92,182]. Last but not least, the impact of disorder is now studied for instance in ultra-clean metals [183] and high-temperature superconductors [184,185] by deliberately introducing defects through irradiation.

Many of the insights that we reviewed in this manuscript resulted from the continuous effort of advancing theoretical and experimental methods. With new techniques becoming available in the future, there is no doubt that the organic charge-transfer salts will remain key model systems for discovering, understanding and predicting physical properties arising from strong electron correlations in real materials.

Author Contributions: All authors contributed to the conceptualization and writing of the manuscript. All authors have read and agreed to the published version of the manuscript.

Funding: This work was funded by the Deutsche Forschungsgemeinschaft (DFG, German Research Foundation) for funding through TRR 288—422213477 (project A05). EG acknowledges support from the Max Planck Society.

Data Availability Statement: Not applicable.

Acknowledgments: We wish to thank all our collaborators and colleagues for many discussions on the physics of organic charge-transfer salts.

Conflicts of Interest: The authors declare no conflict of interest.

References

1. Kanoda, K. Recent progress in NMR studies on organic conductors. *Hyperfine Interact* **1997**, *104*, 235–249. [CrossRef]
2. Toyota, N.; Lang, M.; Müller, J. *Low-Dimensional Molecular Metals*; Springer: Berlin/Heidelberg, Germany, 2007.
3. Limelette, P.; Wzietek, P.; Florens, S.; Georges, A.; Costi, T.A.; Pasquier, C.; Jérome, D.; Mézière, C.; Batail, P. Mott Transition and Transport Crossovers in the Organic Compound κ-(BEDT-TTF)$_2$Cu[N(CN)$_2$]Cl. *Phys. Rev. Lett.* **2003**, *91*, 016401. [CrossRef] [PubMed]
4. Fournier, D.; Poirier, M.; Castonguay, M.; Truong, K.D. Mott Transition, Compressibility Divergence, and the P-T Phase Diagram of Layered Organic Superconductors: An Ultrasonic Investigation. *Phys. Rev. Lett.* **2003**, *90*, 127002. [CrossRef] [PubMed]
5. Pustogow, A.; Rösslhuber, R.; Tan, Y.; Uykur, E.; Böhme, A.; Wenzel, M.; Saito, Y.; Löhle, A.; Hübner, R.; Kawamoto, A.; et al. Low-temperature dielectric anomaly arising from electronic phase separation at the Mott insulator-metal transition. *npj Quantum Mater.* **2021**, *6*, 9. [CrossRef]
6. Kagawa, F.; Miyagawa, K.; Kanoda, K. Magnetic Mott criticality in a κ-type organic salt probed by NMR. *Nat. Phys.* **2009**, *5*, 880–884. [CrossRef]
7. Furukawa, T.; Miyagawa, K.; Taniguchi, H.; Kato, R.; Kanoda, K. Quantum criticality of Mott transition in organic materials. *Nat. Phys.* **2015**, *11*, 221–224. [CrossRef]
8. Isono, T.; Terashima, T.; Miyagawa, K.; Kanoda, K.; Uji, S. Quantum criticality in an organic spin-liquid insulator κ-(BEDT-TTF)$_2$Cu$_2$(CN)$_3$. *Nat. Commun.* **2016**, *7*, 13494. [CrossRef]
9. Gati, E.; Garst, M.; Manna, R.; Tutsch, U.; Wolf, B.; Bartosch, L.; Schubert, H.; Sasaki, T.; Schlueter, J.; Lang, M. Breakdown of Hooke's law of elasticity at the Mott critical endpoint in an organic conductor. *Sci. Adv.* **2016**, *2*, e160146. [CrossRef]
10. Shimizu, Y.; Miyagawa, K.; Kanoda, K.; Maesato, M.; Saito, G. Spin Liquid State in an Organic Mott Insulator with a Triangular Lattice. *Phys. Rev. Lett.* **2003**, *91*, 107001. [CrossRef]
11. Powell, B.J.; McKenzie, R.H. Quantum frustration in organic Mott insulators: From spin liquids to unconventional superconductors. *Rep. Prog. Phys.* **2011**, *74*, 056501. [CrossRef]
12. Riedl, K.; Valentí, R.; Winter, S.M. Critical spin liquid versus valence-bond glass in a triangular-lattice organic antiferromagnet. *Nat. Commun.* **2019**, *10*, 2561. [CrossRef]
13. Pustogow, A. Thirty-Year Anniversary of κ-(BEDT-TTF)$_2$Cu$_2$(CN)$_3$: Reconciling the Spin Gap in a Spin-Liquid Candidate. *Solids* **2022**, *3*, 93–110. [CrossRef]
14. Wosnitza, J. Superconductivity of organic charge-transfer salts. *J. Low Temp. Phys.* **2019**, *197*, 250–271. [CrossRef]
15. Guterding, D.; Altmeyer, M.; Jeschke, H.O.; Valentí, R. Near-degeneracy of extended $s + d_{x^2-y^2}$ and d_{xy} order parameters in quasi-two-dimensional organic superconductors. *Phys. Rev. B* **2016**, *94*, 024515. [CrossRef]
16. Lunkenheimer, P.; Müller, J.; Krohns, S.; Schrettle, F.; Loidl, A.; Hartmann, B.; Rommel, R.; de Souza, M.; Hotta, C.; Schlueter, J.; et al. Multiferroicity in an organic charge-transfer salt that is suggestive of electric-dipole-driven magnetism. *Nat. Mater.* **2012**, *11*, 755–758. [CrossRef]
17. Komatsu, T.; Matsukawa, N.; Inoue, T.; Saito, G. Realization of Superconductivity at Ambient Pressure by Band-Filling Control in κ-(BEDT-TTF)$_2$Cu$_2$(CN)$_3$. *J. Phys. Soc. Jpn.* **1996**, *65*, 1340–1354. [CrossRef]
18. Kandpal, H.C.; Opahle, I.; Zhang, Y.Z.; Jeschke, H.O.; Valentí, R. Revision of Model Parameters for κ-Type Charge Transfer Salts: An *Ab Initio* Study. *Phys. Rev. Lett.* **2009**, *103*, 067004. [CrossRef] [PubMed]
19. Hotta, C. Theories on Frustrated Electrons in Two-Dimensional Organic Solids. *Crystals* **2012**, *2*, 1155. [CrossRef]
20. Pustogow, A.; Bories, M.; Löhle, A.; Rösslhuber, R.; Zhukova, E.; Gorshunov, B.; Tomić, S.; Schlueter, J.A.; Hübner, R.; Hiramatsu, T.; et al. Quantum spin liquids unveil the genuine Mott state. *Nat. Mater.* **2018**, *17*, 773. [CrossRef]
21. Lefebvre, S.; Wzietek, P.; Brown, S.; Bourbonnais, C.; Jérome, D.; Mézière, C.; Fourmigué, M.; Batail, P. Mott Transition, Antiferromagnetism, and Unconventional Superconductivity in Layered Organic Superconductors. *Phys. Rev. Lett.* **2000**, *85*, 5420–5423. [CrossRef]
22. Kagawa, F.; Itou, T.; Miyagawa, K.; Kanoda, K. Transport criticality of the first-order Mott transition in the quasi-two-dimensional organic conductor $\kappa-(BEDT-TTF)_2Cu[N(CN)_2]Cl$. *Phys. Rev. B* **2004**, *69*, 064511. [CrossRef]
23. Kurosaki, Y.; Shimizu, Y.; Miyagawa, K.; Kanoda, K.; Saito, G. Mott Transition from a Spin Liquid to a Fermi Liquid in the Spin-Frustrated Organic Conductor κ-(ET)$_2$Cu$_2$(CN)$_3$. *Phys. Rev. Lett.* **2005**, *95*, 177001. [CrossRef]
24. Zverev, V.N.; Kartsovnik, M.V.; Biberacher, W.; Khasanov, S.S.; Shibaeva, R.P.; Ouahab, L.; Toupet, L.; Kushch, N.D.; Yagubskii, E.B.; Canadell, E. Temperature-pressure phase diagram and electronic properties of the organic metal κ-(BETS)$_2$Mn[N(CN)$_2$]$_3$. *Phys. Rev. B* **2010**, *82*, 155123. [CrossRef]

25. Zverev, V.N.; Biberacher, W.; Oberbauer, S.; Sheikin, I.; Alemany, P.; Canadell, E.; Kartsovnik, M.V. Fermi surface properties of the bifunctional organic metal κ-(BETS)$_2$Mn[N(CN)$_2$]$_3$ near the metal-insulator transition. *Phys. Rev. B* **2019**, *99*, 125136. [CrossRef]
26. Vyaselev, O.M.; Kartsovnik, M.V.; Biberacher, W.; Zorina, L.V.; Kushch, N.D.; Yagubskii, E.B. Magnetic transformations in the organic conductor κ-(BETS)$_2$Mn[N(CN)$_2$]$_3$ at the metal-insulator transition. *Phys. Rev. B* **2011**, *83*, 094425. [CrossRef]
27. Castellani, C.; Castro, C.D.; Feinberg, D.; Ranninger, J. New Model Hamiltonian for the Metal-Insulator Transition. *Phys. Rev. Lett.* **1979**, *43*, 1957–1960. [CrossRef]
28. Kotliar, G.; Lange, E.; Rozenberg, M.J. Landau Theory of the Finite Temperature Mott Transition. *Phys. Rev. Lett.* **2000**, *84*, 5180–5183. [CrossRef] [PubMed]
29. Georges, A.; Kotliar, G.; Krauth, W.; Rozenberg, M.J. Dynamical mean-field theory of strongly correlated fermion systems and the limit of infinite dimensions. *Rev. Mod. Phys.* **1996**, *68*, 13–125. [CrossRef]
30. Vučičević, J.; Terletska, H.; Tanasković, D.; Dobrosavljević, V. Finite-temperature crossover and the quantum Widom line near the Mott transition. *Phys. Rev. B* **2013**, *88*, 075143. [CrossRef]
31. Terletska, H.; Vučičević, J.; Tanasković, D.; Dobrosavljević, V. Quantum Critical Transport near the Mott Transition. *Phys. Rev. Lett.* **2011**, *107*, 026401. [CrossRef] [PubMed]
32. Ferber, J.; Foyevtsova, K.; Jeschke, H.O.; Valentí, R. Unveiling the microscopic nature of correlated organic conductors: The case of κ-(ET)$_2$Cu[N(CN)$_2$]Br$_x$Cl$_{1-x}$. *Phys. Rev. B* **2014**, *89*, 205106. [CrossRef]
33. Leiner, J.C.; Jeschke, H.O.; Valentí, R.; Zhang, S.; Savici, A.T.; Lin, J.Y.Y.; Stone, M.B.; Lumsden, M.D.; Hong, J.; Delaire, O.; et al. Frustrated Magnetism in Mott Insulating (V$_{1-x}$Cr$_x$)$_2$O$_3$. *Phys. Rev. X* **2019**, *9*, 011035. [CrossRef]
34. Miksch, B.; Pustogow, A.; Javaheri Rahim, M.; Bardin, A.A.; Kanoda, K.; Schlueter, J.A.; Hübner, R.; Scheffler, M.; Dressel, M. Gapped magnetic ground state in quantum spin liquid candidate κ-(BEDT-TTF)$_2$Cu$_2$(CN)$_3$. *Science* **2021**, *372*, 276. [CrossRef]
35. Miyagawa, K.; Kawamoto, A.; Nakazawa, Y.; Kanoda, K. Antiferromagnetic Ordering and Spin Structure in the Organic Conductor, κ-(BEDT-TTF)$_2$Cu[N(CN)$_2$]Cl. *Phys. Rev. Lett.* **1995**, *75*, 1174–1177. [CrossRef]
36. Pustogow, A.; Le, T.; Wang, H.H.; Luo, Y.; Gati, E.; Schubert, H.; Lang, M.; Brown, S.E. Impurity moments conceal low-energy relaxation of quantum spin liquids. *Phys. Rev. B* **2020**, *101*, 140401. [CrossRef]
37. Matsuura, M.; Sasaki, T.; Naka, M.; Müller, J.; Stockert, O.; Piovano, A.; Yoneyama, N.; Lang, M. Phonon renormalization effects accompanying the 6 K anomaly in the Quantum Spin Liquid Candidate κ-(BEDT-TTF)$_2$Cu$_2$(CN)$_3$. *arXiv* **2022**, arXiv:2208.05096.
38. Pustogow, A.; Kawasugi, Y.; Sakurakoji, H.; Tajima, N. Chasing the spin gap through the phase diagram of a frustrated Mott insulator. *arXiv* **2022**, arXiv:2209.07639.
39. Koretsune, T.; Hotta, C. Evaluating model parameters of the κ- and β'-type Mott insulating organic solids. *Phys. Rev. B* **2014**, *89*, 045102. [CrossRef]
40. Nakamura, K.; Yoshimoto, Y.; Kosugi, T.; Arita, R.; Imada, M. Ab initio Derivation of Low-Energy Model for κ-ET Type Organic Conductors. *J. Phys. Soc. Jpn.* **2009**, *78*, 083710. [CrossRef]
41. Nakamura, K.; Yoshimoto, Y.; Imada, M. Ab initio two-dimensional multiband low-energy models of EtMe$_3$Sb[Pd(dmit)$_2$]$_2$ and κ-(BEDT-TTF)$_2$Cu(NCS)$_2$ with comparisons to single-band models. *Phys. Rev. B* **2012**, *86*, 205117. [CrossRef]
42. Jeschke, H.O.; de Souza, M.; Valentí, R.; Manna, R.S.; Lang, M.; Schlueter, J.A. Temperature dependence of structural and electronic properties of the spin-liquid candidate κ-(BEDT-TTF)$_2$Cu$_2$(CN)$_3$. *Phys. Rev. B* **2012**, *85*, 035125. [CrossRef]
43. Tocchio, L.F.; Feldner, H.; Becca, F.; Valentí, R.; Gros, C. Spin-liquid versus spiral-order phases in the anisotropic triangular lattice. *Phys. Rev. B* **2013**, *87*, 035143. [CrossRef]
44. Starykh, O.A. Unusual ordered phases of highly frustrated magnets: A review. *Rep. Prog. Phys.* **2015**, *78*, 052502. [CrossRef]
45. Szasz, A.; Motruk, J.; Zaletel, M.P.; Moore, J.E. Chiral spin liquid phase of the triangular lattice Hubbard model: A density matrix renormalization group study. *Phys. Rev. X* **2020**, *10*, 021042. [CrossRef]
46. Yamashita, S.; Nakazawa, Y.; Oguni, M.; Oshima, Y.; Nojiri, H.; Shimizu, Y.; Miyagawa, K.; Kanoda, K. Thermodynamic properties of a spin-1/2 spin-liquid state in a κ-type organic salt. *Nat. Phys.* **2008**, *4*, 459. [CrossRef]
47. Lee, S.S.; Lee, P.A. U(1) Gauge Theory of the Hubbard Model: Spin Liquid States and Possible Application to κ-(BEDT-TTF)$_2$Cu$_2$(CN)$_3$. *Phys. Rev. Lett.* **2005**, *95*, 036403. [CrossRef]
48. Manna, R.S.; Wolf, B.; de Souza, M.; Lang, M. High-resolution thermal expansion measurements under helium-gas pressure. *Rev. Sci. Instrum.* **2012**, *83*, 085111. [CrossRef]
49. Hartmann, S.; Gati, E.; Yoshida, Y.; Saito, G.; Lang, M. Thermal Expansion Studies on the Spin-Liquid-Candidate System κ-(BEDT-TTF)$_2$Ag$_2$(CN)$_3$. *Phys. Status Solidi* **2019**, *256*, 1800640. [CrossRef]
50. Shimizu, Y.; Miyagawa, K.; Kanoda, K.; Maesato, M.; Saito, G. Emergence of inhomogeneous moments from spin liquid in the triangular-lattice Mott insulator κ-(ET)$_2$Cu$_2$(CN)$_3$. *Phys. Rev. B* **2006**, *73*, 140407. [CrossRef]
51. Riedl, K.; Gati, E.; Zielke, D.; Hartmann, S.; Vyaselev, O.M.; Kushch, N.D.; Jeschke, H.O.; Lang, M.; Valentí, R.; Kartsovnik, M.V.; et al. Spin Vortex Crystal Order in Organic Triangular Lattice Compound. *Phys. Rev. Lett.* **2021**, *127*, 147204. [CrossRef]
52. Lang, M.; Müller, J. Quasi-Twodimensional Organic Superconductors. In *Superconductivity Review*; Citeseer: University Park, PA, USA; OPA: Amsterdam, The Netherlands, 1996.
53. Wosnitza, J. Quasi-Two-Dimensional Organic Superconductors. *J. Low Temp. Phys.* **2007**, *146*, 641–667. [CrossRef]
54. McKenzie, R.H. Similarities Between Organic and Cuprate Superconductors. *Science* **1997**, *278*, 820–821. [CrossRef]
55. Schmalian, J. Pairing due to Spin Fluctuations in Layered Organic Superconductors. *Phys. Rev. Lett.* **1998**, *81*, 4232–4235. [CrossRef]

56. Kino, H.; Kontani, H. Phase Diagram of Superconductivity on the Anisotropic Triangular Lattice Hubbard Model: An Effective Model of κ-(BEDT-TTF) Salts. *J. Phys. Soc. Jpn.* **1998**, *67*, 3691–3694. [CrossRef]
57. Kyung, B.; Tremblay, A.M. Mott transition, antiferromagnetism, and d-wave superconductivity in two-dimensional organic conductors. *Phys. Rev. Lett.* **2006**, *97*, 046402. [CrossRef]
58. Hébert, C.D.; Sémon, P.; Tremblay, A.M.S. Superconducting dome in doped quasi-two-dimensional organic Mott insulators: A paradigm for strongly correlated superconductivity. *Phys. Rev. B* **2015**, *92*, 195112. [CrossRef]
59. Watanabe, T.; Yokoyama, H.; Tanaka, Y.; Inoue, J.I. Superconductivity and a Mott Transition in a Hubbard Model on an Anisotropic Triangular Lattice. *J. Phys. Soc. Jpn.* **2006**, *75*, 074707. [CrossRef]
60. Guterding, D.; Diehl, S.; Altmeyer, M.; Methfessel, T.; Tutsch, U.; Schubert, H.; Lang, M.; Müller, J.; Huth, M.; Jeschke, H.O.; et al. Evidence for Eight-Node Mixed-Symmetry Superconductivity in a Correlated Organic Metal. *Phys. Rev. Lett.* **2016**, *116*, 237001. [CrossRef] [PubMed]
61. Zantout, K.; Altmeyer, M.; Backes, S.; Valentí, R. Superconductivity in correlated BEDT-TTF molecular conductors: Critical temperatures and gap symmetries. *Phys. Rev. B* **2018**, *97*, 014530. [CrossRef]
62. Chao, K.; Spałek, J.; Oleś, A. Canonical perturbation expansion of the Hubbard model. *Phys. Rev. B* **1978**, *18*, 3453. [CrossRef]
63. Gros, C.; Joynt, R.; Rice, T.M. Antiferromagnetic correlations in almost-localized Fermi liquids. *Phys. Rev. B* **1987**, *36*, 381. [CrossRef] [PubMed]
64. Winter, S.M.; Riedl, K.; Valentí, R. Importance of spin-orbit coupling in layered organic salts. *Phys. Rev. B* **2017**, *95*, 060404. [CrossRef]
65. Balents, L. Spin liquids in frustrated magnets. *Nature* **2010**, *464*, 199–208. [CrossRef] [PubMed]
66. Bernevig, B.A. Topological insulators and topological superconductors. In *Topological Insulators and Topological Superconductors*; Princeton University Press: Princeton, NJ, USA, 2013.
67. Riedl, K. First Principles Studies of Frustrated Spin Systems: From Low-Energy Models to Experiments. Ph.D. Thesis, Goethe-Universität Frankfurt am Main, Frankfurt am Main, Germany, 2019.
68. Shekhtman, L.; Entin-Wohlman, O.; Aharony, A. Moriya's anisotropic superexchange interaction, frustration, and Dzyaloshinsky's weak ferromagnetism. *Phys. Rev. Lett.* **1992**, *69*, 836–839. [CrossRef] [PubMed]
69. Pratt, F.L.; Baker, P.J.; Blundell, S.J.; Lancaster, T.; Ohira-Kawamura, S.; Baines, C.; Shimizu, Y.; Kanoda, K.; Watanabe, I.; Saito, G. Magnetic and non-magnetic phases of a quantum spin liquid. *Nature* **2011**, *471*, 612. [CrossRef] [PubMed]
70. Motrunich, O.I. Variational study of triangular lattice spin-1/2 model with ring exchanges and spin liquid state in κ-(ET)$_2$Cu$_2$(CN)$_3$. *Phys. Rev. B* **2005**, *72*, 045105. [CrossRef]
71. Block, M.S.; Sheng, D.N.; Motrunich, O.I.; Fisher, M.P.A. Spin Bose-Metal and Valence Bond Solid Phases in a Spin-1/2 Model with Ring Exchanges on a Four-Leg Triangular Ladder. *Phys. Rev. Lett.* **2011**, *106*, 157202. [CrossRef]
72. Holt, M.; Powell, B.J.; Merino, J. Spin-liquid phase due to competing classical orders in the semiclassical theory of the Heisenberg model with ring exchange on an anisotropic triangular lattice. *Phys. Rev. B* **2014**, *89*, 174415. [CrossRef]
73. Cookmeyer, T.; Motruk, J.; Moore, J.E. Four-spin terms and the origin of the chiral spin liquid in Mott insulators on the triangular lattice. *Phys. Rev. Lett.* **2021**, *127*, 087201. [CrossRef]
74. Vyaselev, O.M.; Kato, R.; Yamamoto, H.M.; Kobayashi, M.; Zorina, L.V.; Simonov, S.V.; Kushch, N.D.; Yagubskii, E.B. Properties of Mn^{2+} and π-Electron Spin Systems Probed by 1H and 13C NMR in the Organic Conductor κ-(BETS)$_2$Mn [N (CN)$_2$]$_3$. *Crystals* **2012**, *2*, 224–235. [CrossRef]
75. Vyaselev, O.M.; Biberacher, W.; Kushch, N.D.; Kartsovnik, M.V. Interplay between the d- and π-electron systems in magnetic torque of the layered organic conductor κ-(BETS)$_2$Mn[N(CN)$_2$]$_3$. *Phys. Rev. B* **2017**, *96*, 205154. [CrossRef]
76. Mori, T.; Katsuhara, M. Estimation of πd-Interactions in Organic Conductors Including Magnetic Anions. *J. Phys. Soc. Jpn.* **2002**, *71*, 826–844. [CrossRef]
77. Konoike, T.; Uji, S.; Terashima, T.; Nishimura, M.; Yasuzuka, S.; Enomoto, K.; Fujiwara, H.; Zhang, B.; Kobayashi, H. Magnetic-field-induced superconductivity in the antiferromagnetic organic superconductor κ-(BETS)$_2$FeBr$_4$. *Phys. Rev. B* **2004**, *70*, 094514. [CrossRef]
78. Kartsovnik, M.V.; Kunz, M.; Schaidhammer, L.; Kollmannsberger, F.; Biberacher, W.; Kushch, N.D.; Miyazaki, A.; Fujiwara, H. Interplay between conducting and magnetic systems in the antiferromagnetic organic superconductor κ-(BETS)$_2$FeBr$_4$. *J. Supercond. Nov. Magn.* **2016**, *29*, 3075–3080. [CrossRef]
79. Jaccarino, V.; Peter, M. Ultra-high-field superconductivity. *Phys. Rev. Lett.* **1962**, *9*, 290. [CrossRef]
80. Vyaselev, O.M.; Kartsovnik, M.V.; Kushch, N.D.; Yagubskii, E.B. Staggered spin order of localized π-electrons in the insulating state of the organic conductor κ-(BETS)$_2$Mn[N(CN)$_2$]$_3$. *JETP Lett.* **2012**, *95*, 565–569. [CrossRef]
81. Hotta, C. Quantum electric dipoles in spin-liquid dimer Mott insulator κ-ET$_2$Cu$_2$(CN)$_3$. *Phys. Rev. B* **2010**, *82*, 241104. [CrossRef]
82. Kaneko, R.; Tocchio, L.F.; Valentí, R.; Becca, F. Charge orders in organic charge-transfer salts. *New J. Phys.* **2017**, *19*, 103033. [CrossRef]
83. Seo, H.; Hotta, C.; Fukuyama, H. Toward Systematic Understanding of Diversity of Electronic Properties in Low-Dimensional Molecular Solids. *Chem. Rev.* **2004**, *104*, 5005–5036. [CrossRef]
84. Lunkenheimer, P.; Loidl, A. Dielectric spectroscopy on organic charge-transfer salts. *J. Phys. Condens. Matter* **2015**, *27*, 373001. [CrossRef]
85. Abdel-Jawad, M.; Terasaki, I.; Sasaki, T.; Yoneyama, N.; Kobayashi, N.; Uesu, Y.; Hotta, C. Anomalous dielectric response in the dimer Mott insulator κ-(BEDT-TTF)$_2$Cu$_2$(CN)$_3$. *Phys. Rev. B* **2010**, *82*, 125119. [CrossRef]

86. Gati, E.; Fischer, J.K.H.; Lunkenheimer, P.; Zielke, D.; Köhler, S.; Kolb, F.; von Nidda, H.A.K.; Winter, S.M.; Schubert, H.; Schlueter, J.A.; et al. Evidence for Electronically Driven Ferroelectricity in a Strongly Correlated Dimerized BEDT-TTF Molecular Conductor. *Phys. Rev. Lett.* **2018**, *120*, 247601. [CrossRef]
87. Manna, R.S.; de Souza, M.; Brühl, A.; Schlueter, J.A.; Lang, M. Lattice Effects and Entropy Release at the Low-Temperature Phase Transition in the Spin-Liquid Candidate κ-(BEDT-TTF)$_2$Cu$_2$(CN)$_3$. *Phys. Rev. Lett.* **2010**, *104*, 016403. [CrossRef] [PubMed]
88. Lang, M.; Lunkenheimer, P.; Müller, J.; Loidl, A.; Hartmann, B.; Hoang, N.H.; Gati, E.; Schubert, H.; Schlueter, J.A. Multiferroicity in the Mott Insulating Charge-Transfer Salt κ-(BEDT-TTF)$_2$Cu[N(CN)$_2$]Cl. *IEEE Trans. Magn.* **2014**, *50*, 2700107. [CrossRef]
89. Itoh, K.; Itoh, H.; Naka, M.; Saito, S.; Hosako, I.; Yoneyama, N.; Ishihara, S.; Sasaki, T.; Iwai, S. Collective Excitation of an Electric Dipole on a Molecular Dimer in an Organic Dimer-Mott Insulator. *Phys. Rev. Lett.* **2013**, *110*, 106401. [CrossRef] [PubMed]
90. Gomi, H.; Imai, T.; Takahashi, A.; Aihara, M. Purely electronic terahertz polarization in dimer Mott insulators. *Phys. Rev. B* **2010**, *82*, 035101. [CrossRef]
91. Gomi, H.; Ikenaga, M.; Hiragi, Y.; Segawa, D.; Takahashi, A.; Inagaki, T.J.; Aihara, M. Ferroelectric states induced by dimer lattice disorder in dimer Mott insulators. *Phys. Rev. B* **2013**, *87*, 195126. [CrossRef]
92. van den Brink, J.; Khomskii, D.I. Multiferroicity due to charge ordering. *J. Phys. Condens. Matter* **2008**, *20*, 434217. [CrossRef]
93. Nad, F.; Monceau, P. Dielectric Response of the Charge Ordered State in Quasi-One-Dimensional Organic Conductors. *J. Phys. Soc. Jpn.* **2006**, *75*, 051005. [CrossRef]
94. Lunkenheimer, P.; Hartmann, B.; Lang, M.; Müller, J.; Schweitzer, D.; Krohns, S.; Loidl, A. Ferroelectric properties of charge-ordered α-(BEDT-TTF)$_2$I$_3$. *Phys. Rev. B* **2015**, *91*, 245132. [CrossRef]
95. Takano, Y.; Hiraki, K.; Yamamoto, H.; Nakamura, T.; Takahashi, T. Charge disproportionation in the organic conductor, α-(BEDT-TTF)$_2$I3. *J. Phys. Chem. Solids* **2001**, *62*, 393–395. [CrossRef]
96. Tomić, S.; Dressel, M. Ferroelectricity in molecular solids: A review of electrodynamic properties. *Rep. Prog. Phys.* **2015**, *78*, 096501. [CrossRef] [PubMed]
97. Sedlmeier, K.; Elsässer, S.; Neubauer, D.; Beyer, R.; Wu, D.; Ivek, T.; Tomić, S.; Schlueter, J.A.; Dressel, M. Absence of charge order in the dimerized κ-phase BEDT-TTF salts. *Phys. Rev. B* **2012**, *86*, 245103. [CrossRef]
98. Tomić, S.; Pinterić, M.; Ivek, T.; Sedlmeier, K.; Beyer, R.; Wu, D.; Schlueter, J.A.; Schweitzer, D.; Dressel, M. Magnetic ordering and charge dynamics in κ-(BEDT-TTF)$_2$Cu[N(CN)$_2$]Cl. *J. Phys. Condens. Matter* **2013**, *25*, 436004. [CrossRef]
99. Pinterić, M.; Čulo, M.; Milat, O.; Basletić, M.; Korin-Hamzić, B.; Tafra, E.; Hamzić, A.; Ivek, T.; Peterseim, T.; Miyagawa, K.; et al. Anisotropic charge dynamics in the quantum spin-liquid candidate κ-(BEDT-TTF)$_2$Cu$_2$(CN)$_3$. *Phys. Rev. B* **2014**, *90*, 195139. [CrossRef]
100. Pinterić, M.; Lazić, P.; Pustogow, A.; Ivek, T.; Kuveždić, M.; Milat, O.; Gumhalter, B.; Basletić, M.; Čulo, M.; Korin-Hamzić, B.; et al. Anion effects on electronic structure and electrodynamic properties of the Mott insulator κ-(BEDT-TTF)$_2$Ag$_2$(CN)$_3$. *Phys. Rev. B* **2016**, *94*, 161105. [CrossRef]
101. Dressel, M.; Lazić, P.; Pustogow, A.; Zhukova, E.; Gorshunov, B.; Schlueter, J.A.; Milat, O.; Gumhalter, B.; Tomić, S. Lattice vibrations of the charge-transfer salt κ-(BEDT-TTF)$_2$Cu$_2$(CN)$_3$: Comprehensive explanation of the electrodynamic response in a spin-liquid compound. *Phys. Rev. B* **2016**, *93*, 081201. [CrossRef]
102. Lazić, P.; Pinterić, M.; Rivas Góngora, D.; Pustogow, A.; Treptow, K.; Ivek, T.; Milat, O.; Gumhalter, B.; Došlić, N.; Dressel, M.; et al. Importance of van der Waals interactions and cation-anion coupling in an organic quantum spin liquid. *Phys. Rev. B* **2018**, *97*, 245134. [CrossRef]
103. Rösslhuber, R.; Pustogow, A.; Uykur, E.; Böhme, A.; Löhle, A.; Hübner, R.; Schlueter, J.A.; Tan, Y.; Dobrosavljević, V.; Dressel, M. Phase coexistence at the first-order Mott transition revealed by pressure-dependent dielectric spectroscopy of κ-(BEDT-TTF)$_2$-Cu$_2$(CN)$_3$. *Phys. Rev. B* **2021**, *103*, 125111. [CrossRef]
104. Drichko, N.; Beyer, R.; Rose, E.; Dressel, M.; Schlueter, J.A.; Turunova, S.A.; Zhilyaeva, E.I.; Lyubovskaya, R.N. Metallic state and charge-order metal-insulator transition in the quasi-two-dimensional conductor κ-(BEDT-TTF)$_2$Hg(SCN)$_2$Cl. *Phys. Rev. B* **2014**, *89*, 075133. [CrossRef]
105. Lines, M.E.; Glass, A.M. *Principles and Applications of Ferroelectrics and Related Materials*; Related Materials; Clarendon Press: Oxford, UK, 1977.
106. Hassan, N.; Cunningham, S.; Mourigal, M.; Zhilyaeva, E.I.; Torunova, S.A.; Lyubovskaya, R.N.; Schlueter, J.A.; Drichko, N. Evidence for a quantum dipole liquid state in an organic quasi–two-dimensional material. *Science* **2018**, *360*, 1101. [CrossRef]
107. Hassan, N.M.; Thirunavukkuarasu, K.; Lu, Z.; Smirnov, D.; Zhilyaeva, E.I.; Torunova, S.; Lyubovskaya, R.N.; Drichko, N. Melting of charge order in the low-temperature state of an electronic ferroelectric-like system. *npj Quantum Mater.* **2020**, *5*, 15. [CrossRef]
108. Kuroki, K.; Kimura, T.; Arita, R.; Tanaka, Y.; Matsuda, Y. $d_{x^2-y^2}$- versus d_{xy}- like pairings in organic superconductors κ-(BEDT-TTF)$_2$X. *Phys. Rev. B* **2002**, *65*, 100516. [CrossRef]
109. Watanabe, H.; Seo, H.; Yunoki, S. Phase Competition and Superconductivity in κ-(BEDT-TTF)$_2$X: Importance of Intermolecular Coulomb Interactions. *J. Phys. Soc. Jpn.* **2017**, *86*, 033703. [CrossRef]
110. Sekine, A.; Nasu, J.; Ishihara, S. Polar charge fluctuation and superconductivity in organic conductors. *Phys. Rev. B* **2013**, *87*, 085133. [CrossRef]
111. Gomes, N.; De Silva, W.W.; Dutta, T.; Clay, R.T.; Mazumdar, S. Coulomb-enhanced superconducting pair correlations and paired-electron liquid in the frustrated quarter-filled band. *Phys. Rev. B* **2016**, *93*, 165110. [CrossRef]
112. Majumdar, P.; Krishnamurthy, H.R. Lattice Contraction Driven Insulator-Metal Transition in the $d = \infty$ Local Approximation. *Phys. Rev. Lett.* **1994**, *73*, 1525–1528. [CrossRef] [PubMed]

113. Hassan, S.R.; Georges, A.; Krishnamurthy, H.R. Sound Velocity Anomaly at the Mott Transition: Application to Organic Conductors and V_2O_3. *Phys. Rev. Lett.* **2005**, *94*, 036402. [CrossRef]
114. Kuo, H.H.; Chu, J.H.; Palmstrom, J.C.; Kivelson, S.A.; Fisher, I.R. Ubiquitous signatures of nematic quantum criticality in optimally doped Fe-based superconductors. *Science* **2016**, *352*, 958. [CrossRef]
115. Fernandes, R.; Chubukov, A.; Schmalian, J. What drives nematic order in iron-based superconductors? *Nat. Phys.* **2014**, *10*, 97. [CrossRef]
116. Achkar, A.J.; Zwiebler, M.; McMahon, C.; He, F.; Sutarto, R.; Djianto, I.; Hao, Z.; Gingras, M.J.P.; Hücker, M.; Gu, G.D.; et al. Nematicity in stripe-ordered cuprates probed via resonant X-ray scattering. *Science* **2016**, *351*, 576. [CrossRef]
117. Cano, A.; Civelli, M.; Eremin, I.; Paul, I. Interplay of magnetic and structural transitions in iron-based pnictide superconductors. *Phys. Rev. B* **2010**, *82*, 020408. [CrossRef]
118. Böhmer, A.E.; Meingast, C. Electronic nematic susceptibility of iron-based superconductors. *C. R. Phys.* **2016**, *17*, 90–112. [CrossRef]
119. Gati, E.; Xiang, L.; Bud'ko, S.L.; Canfield, P.C. Hydrostatic and Uniaxial Pressure Tuning of Iron-Based Superconductors: Insights into Superconductivity, Magnetism, Nematicity, and Collapsed Tetragonal Transitions. *Ann. Phys.* **2020**, *532*, 2000248. [CrossRef]
120. Manna, R.S.; de Souza, M.; Kato, R.; Lang, M. Lattice effects in the quasi-two-dimensional valence-bond-solid Mott insulator $EtMe_3P[Pd(dmit)_2]_2$. *Phys. Rev. B* **2014**, *89*, 045113. [CrossRef]
121. Manna, R.S.; Hartmann, S.; Gati, E.; Schlueter, J.A.; De Souza, M.; Lang, M. Low-Temperature Lattice Effects in the Spin-Liquid Candidate κ-$(BEDT-TTF)_2Cu_2(CN)_3$. *Crystals* **2018**, *8*, 87. [CrossRef]
122. Sushkov, A.B.; Jenkins, G.S.; Han, T.H.; Lee, Y.S.; Drew, H.D. Infrared phonons as a probe of spin-liquid states in herbertsmithite $ZnCu_3(OH)_6Cl_2$. *J. Phys. Condens. Matter* **2017**, *29*, 095802. [CrossRef]
123. Li, Y.; Pustogow, A.; Bories, M.; Puphal, P.; Krellner, C.; Dressel, M.; Valentí, R. Lattice dynamics in the spin-$\frac{1}{2}$ frustrated kagome compound herbertsmithite. *Phys. Rev. B* **2020**, *101*, 161115. [CrossRef]
124. Kaib, D.A.S.; Biswas, S.; Riedl, K.; Winter, S.M.; Valentí, R. Magnetoelastic coupling and effects of uniaxial strain in α-$RuCl_3$ from first principles. *Phys. Rev. B* **2021**, *103*, L140402. [CrossRef]
125. Ferrari, F.; Valentí, R.; Becca, F. Variational wave functions for the spin-Peierls transition in the Su-Schrieffer-Heeger model with quantum phonons. *Phys. Rev. B* **2020**, *102*, 125149. [CrossRef]
126. Ferrari, F.; Valentí, R.; Becca, F. Effects of spin-phonon coupling in frustrated Heisenberg models. *Phys. Rev. B* **2021**, *104*, 035126. [CrossRef]
127. Metavitsiadis, A.; Brenig, W. Phonon renormalization in the Kitaev quantum spin liquid. *Phys. Rev. B* **2020**, *101*, 035103. [CrossRef]
128. Ye, M.; Fernandes, R.M.; Perkins, N.B. Phonon dynamics in the Kitaev spin liquid. *Phys. Rev. Res.* **2020**, *2*, 033180. [CrossRef]
129. Matsuura, M.; Sasaki, T.; Iguchi, S.; Gati, E.; Müller, J.; Stockert, O.; Piovano, A.; Böhm, M.; Park, J.T.; Biswas, S.; et al. Lattice Dynamics Coupled to Charge and Spin Degrees of Freedom in the Molecular Dimer-Mott Insulator κ-$(BEDT-TTF)_2Cu[N(CN)_2]Cl$. *Phys. Rev. Lett.* **2019**, *123*, 027601. [CrossRef] [PubMed]
130. Müller, J.; Lang, M.; Steglich, F.; Schlueter, J.A.; Kini, A.M.; Sasaki, T. Evidence for structural and electronic instabilities at intermediate temperatures in κ-$(BEDT-TTF)_2X$ for X = $Cu[N(CN)_2]Cl$, $Cu[N(CN)_2]Br$ and $Cu(NCS)_2$: Implications for the phase diagram of these quasi-two-dimensional organic superconductors. *Phys. Rev. B* **2002**, *65*, 144521. [CrossRef]
131. Zacharias, M.; Bartosch, L.; Garst, M. Mott Metal-Insulator Transition on Compressible Lattices. *Phys. Rev. Lett.* **2012**, *109*, 176401. [CrossRef] [PubMed]
132. Zacharias, M.; Rosch, A.; Garst, M. Critical elasticity at zero and finite temperature. *Eur. Phys. J. Spec. Top.* **2015**, *224*, 1021–1040. [CrossRef]
133. Pintschovius, L.; Rietschel, H.; Sasaki, T.; Mori, H.; Tanaka, S.; Toyota, N.; Lang, M.; Steglich, F. Observation of superconductivity-induced phonon frequency changes in the organic superconductor kappa-$(BEDT-TTF)_2$ $Cu(NCS)_2$. *Europhys. Lett. EPL* **1997**, *37*, 627–632. [CrossRef]
134. Sasaki, T.; Ito, I.; Yoneyama, N.; Kobayashi, N.; Hanasaki, N.; Tajima, H.; Ito, T.; Iwasa, Y. Electronic correlation in the infrared optical properties of the quasi-two-dimensional κ-type BEDT-TTF dimer system. *Phys. Rev. B* **2004**, *69*, 064508. [CrossRef]
135. Yamada, Y.; Takatera, H.L.; Huber, D. Critical Dynamical Phenomena in Pseudospin-Phonon Coupled Systems. *J. Phys. Soc. Jpn.* **1974**, *36*, 641–648. [CrossRef]
136. Wosnitza, J. Fermi Surfaces of Organic Superconductors. *Int. J. Mod. Phys.* **1993**, *7*, 2707–2741. [CrossRef]
137. Müller, J.; Hartmann, B.; Rommel, R.; Brandenburg, J.; Winter, S.M.; Schlueter, J.A. Origin of the glass-like dynamics in molecular metals κ-$(BEDT-TTF)_2X$: Implications from fluctuation spectroscopy and ab initio calculations. *New J. Phys.* **2015**, *17*, 083057. [CrossRef]
138. Kang, L.; Akagi, K.; Hayashi, K.; Sasaki, T. First-principles investigation of local structure deformation induced by X-ray irradiation in κ-$(BEDT-TTF)_2Cu[N(CN)_2]Br$. *Phys. Rev. B* **2017**, *95*, 214106. [CrossRef]
139. Sasaki, T. Mott-Anderson Transition in Molecular Conductors: Influence of Randomness on Strongly Correlated Electrons in the κ-$(BEDT-TTF)_2X$ System. *Crystals* **2012**, *2*, 374–392. [CrossRef]
140. Gati, E.; Tutsch, U.; Naji, A.; Garst, M.; Köhler, S.; Schubert, H.; Sasaki, T.; Lang, M. Effects of Disorder on the Pressure-Induced Mott Transition in κ-$(BEDT-TTF)_2Cu[N(CN)_2]Cl$. *Crystals* **2018**, *8*, 38. [CrossRef]
141. Hartmann, B.; Müller, J.; Sasaki, T. Mott metal-insulator transition induced by utilizing a glasslike structural ordering in low-dimensional molecular conductors. *Phys. Rev. B* **2014**, *90*, 195150. [CrossRef]

142. Gati, E.; Winter, S.M.; Schlueter, J.A.; Schubert, H.; Müller, J.; Lang, M. Insights from experiment and ab initio calculations into the glasslike transition in the molecular conductor κ-(BEDT-TTF)$_2$Hg(SCN)$_2$Cl. *Phys. Rev. B* **2018**, *97*, 075115. [CrossRef]
143. Su, X.; Zuo, F.; Schlueter, J.A.; Kelly, M.E.; Williams, J.M. Structural disorder and its effect on the superconducting transition temperature in the organic superconductor κ-(BEDT-TTF)$_2$Cu[N(CN)$_2$]Br. *Phys. Rev. B* **1998**, *57*, R14056–R14059. [CrossRef]
144. Yoneyama, N.; Sasaki, T.; Nishizaki, T.; Kobayashi, N. Disorder Effect on the Vortex Pinning by the Cooling-Process Control in the Organic Superconductor κ-(BEDT-TTF)2Cu[N(CN)2]Br. *J. Phys. Soc. Jpn.* **2004**, *73*, 184–189. [CrossRef]
145. Hartmann, B.; Zielke, D.; Polzin, J.; Sasaki, T.; Müller, J. Critical Slowing Down of the Charge Carrier Dynamics at the Mott Metal-Insulator Transition. *Phys. Rev. Lett.* **2015**, *114*, 216403. [CrossRef]
146. Guterding, D.; Valentí, R.; Jeschke, H.O. Influence of molecular conformations on the electronic structure of organic charge transfer salts. *Phys. Rev. B* **2015**, *92*, 081109. [CrossRef]
147. Sasaki, T.; Sano, K.; Sugawara, H.; Yoneyama, N.; Kobayashi, N. Influence of randomness on the Mott transition in κ-(BEDT-TTF)2X. *Phys. Status Solidi* **2012**, *249*, 947–952. [CrossRef]
148. Analytis, J.G.; Ardavan, A.; Blundell, S.J.; Owen, R.L.; Garman, E.F.; Jeynes, C.; Powell, B.J. Effect of Irradiation-Induced Disorder on the Conductivity and Critical Temperature of the Organic Superconductor κ-(BEDT-TTF)$_2$Cu(SCN)$_2$. *Phys. Rev. Lett.* **2006**, *96*, 177002. [CrossRef]
149. Sano, K.; Sasaki, T.; Yoneyama, N.; Kobayashi, N. Electron Localization near the Mott Transition in the Organic Superconductor κ-(BEDT-TTF)$_2$Cu[N(CN)$_2$]Br. *Phys. Rev. Lett.* **2010**, *104*, 217003. [CrossRef]
150. Yoneyama, N.; Furukawa, K.; Nakamura, T.; Sasaki, T.; Kobayashi, N. Magnetic Properties of X-ray Irradiated Organic Mott Insulator κ-(BEDT-TTF)$_2$Cu[N(CN)$_2$]Cl. *J. Phys. Soc. Jpn.* **2010**, *79*, 063706. [CrossRef]
151. Sasaki, T.; Oizumi, H.; Honda, Y.; Yoneyama, N.; Kobayashi, N. Suppression of Superconductivity by Nonmagnetic Disorder in Organic Superconductor κ-(BEDT-TTF)$_2$Cu(NCS)$_2$. *J. Phys. Soc. Jpn.* **2011**, *80*, 104703. [CrossRef]
152. Antal, A.; Fehér, T.; Yoneyama, N.; Forró, L.; Sasaki, T.; Jánossy, A. Spin and Charge Transport in the X-ray Irradiated Quasi-2D Layered Compound: κ-(BEDT-TTF)$_2$Cu[N(CN)$_2$]Cl. *Crystals* **2012**, *2*, 579–589. [CrossRef]
153. Sasaki, S.; Iguchi, S.; Yoneyama, N.; Sasaki, T. X-ray Irradiation Effect on the Dielectric Charge Response in the Dimer–Mott Insulator κ-(BEDT-TTF)$_2$Cu$_2$(CN)$_3$. *J. Phys. Soc. Jpn.* **2015**, *84*, 074709. [CrossRef]
154. Furukawa, T.; Miyagawa, K.; Itou, T.; Ito, M.; Taniguchi, H.; Saito, M.; Iguchi, S.; Sasaki, T.; Kanoda, K. Quantum Spin Liquid Emerging from Antiferromagnetic Order by Introducing Disorder. *Phys. Rev. Lett.* **2015**, *115*, 077001. [CrossRef]
155. Urai, M.; Furukawa, T.; Seki, Y.; Miyagawa, K.; Sasaki, T.; Taniguchi, H.; Kanoda, K. Disorder unveils Mott quantum criticality behind a first-order transition in the quasi-two-dimensional organic conductor κ-(ET)$_2$Cu[N(CN)$_2$]Cl. *Phys. Rev. B* **2019**, *99*, 245139. [CrossRef]
156. Yamamoto, R.; Furukawa, T.; Miyagawa, K.; Sasaki, T.; Kanoda, K.; Itou, T. Electronic Griffiths Phase in Disordered Mott-Transition Systems. *Phys. Rev. Lett.* **2020**, *124*, 046404. [CrossRef]
157. Belitz, D.; Kirkpatrick, T.R. The Anderson-Mott transition. *Rev. Mod. Phys.* **1994**, *66*, 261–380. [CrossRef]
158. Dobrosavljević, V.; Kotliar, G. Mean Field Theory of the Mott-Anderson Transition. *Phys. Rev. Lett.* **1997**, *78*, 3943–3946. [CrossRef]
159. Shinaoka, H.; Imada, M. Soft Hubbard Gaps in Disordered Itinerant Models with Short-Range Interaction. *Phys. Rev. Lett.* **2009**, *102*, 016404. [CrossRef]
160. Chiesa, S.; Chakraborty, P.B.; Pickett, W.E.; Scalettar, R.T. Disorder-Induced Stabilization of the Pseudogap in Strongly Correlated Systems. *Phys. Rev. Lett.* **2008**, *101*, 086401. [CrossRef]
161. Heidarian, D.; Trivedi, N. Inhomogeneous Metallic Phase in a Disordered Mott Insulator in Two Dimensions. *Phys. Rev. Lett.* **2004**, *93*, 126401. [CrossRef]
162. Pezzoli, M.E.; Becca, F. Ground-state properties of the disordered Hubbard model in two dimensions. *Phys. Rev. B* **2010**, *81*, 075106. [CrossRef]
163. Tanasković, D.; Dobrosavljević, V.; Abrahams, E.; Kotliar, G. Disorder Screening in Strongly Correlated Systems. *Phys. Rev. Lett.* **2003**, *91*, 066603. [CrossRef]
164. Aguiar, M.C.O.; Dobrosavljević, V.; Abrahams, E.; Kotliar, G. Critical Behavior at the Mott-Anderson Transition: A Typical-Medium Theory Perspective. *Phys. Rev. Lett.* **2009**, *102*, 156402. [CrossRef]
165. Aguiar, M.C.O.; Dobrosavljević, V. Universal Quantum Criticality at the Mott-Anderson Transition. *Phys. Rev. Lett.* **2013**, *110*, 066401. [CrossRef]
166. Bragança, H.; Aguiar, M.C.O.; Vučičević, J.; Tanasković, D.; Dobrosavljević, V. Anderson localization effects near the Mott metal-insulator transition. *Phys. Rev. B* **2015**, *92*, 125143. [CrossRef]
167. Byczuk, K.; Hofstetter, W.; Vollhardt, D. Mott-Hubbard Transition versus Anderson Localization in Correlated Electron Systems with Disorder. *Phys. Rev. Lett.* **2005**, *94*, 056404. [CrossRef]
168. Lee, H.; Jeschke, H.O.; Valentí, R. Competition between disorder and Coulomb interaction in a two-dimensional plaquette Hubbard model. *Phys. Rev. B* **2016**, *93*, 224203. [CrossRef]
169. Anderson, P.W. Absence of Diffusion in Certain Random Lattices. *Phys. Rev.* **1958**, *109*, 1492–1505. [CrossRef]
170. Lee, P.A.; Ramakrishnan, T.V. Disordered electronic systems. *Rev. Mod. Phys.* **1985**, *57*, 287–337. [CrossRef]
171. Radonjić, M.c.v.M.; Tanasković, D.; Dobrosavljević, V.; Haule, K. Influence of disorder on incoherent transport near the Mott transition. *Phys. Rev. B* **2010**, *81*, 075118. [CrossRef]
172. Diehl, S.; Methfessel, T.; Tutsch, U.; Müller, J.; Lang, M.; Huth, M.; Jourdan, M.; Elmers, H.J. Disorder-induced gap in the normal density of states of the organic superconductor κ-(BEDT-TTF)$_2$Cu[N(CN)$_2$]Br. *J. Phys. Condens. Matter* **2015**, *27*, 265601. [CrossRef]

173. Westerberg, E.; Furusaki, A.; Sigrist, M.; Lee, P.A. Random Quantum Spin Chains: A Real-Space Renormalization Group Study. *Phys. Rev. Lett.* **1995**, *75*, 4302–4305. [CrossRef]
174. Westerberg, E.; Furusaki, A.; Sigrist, M.; Lee, P.A. Low-energy fixed points of random quantum spin chains. *Phys. Rev. B* **1997**, *55*, 12578–12593. [CrossRef]
175. Yamashita, M.; Nakata, N.; Kasahara, Y.; Sasaki, T.; Yoneyama, N.; Kobayashi, N.; Fujimoto, S.; Shibauchi, T.; Matsuda, Y. Thermal-transport measurements in a quantum spin-liquid state of the frustrated triangular magnet κ-(BEDT-TTF)$_2$Cu$_2$(CN)$_3$. *Nat. Phys.* **2009**, *5*, 44. [CrossRef]
176. Buzzi, M.; Nicoletti, D.; Fava, S.; Jotzu, G.; Miyagawa, K.; Kanoda, K.; Henderson, A.; Siegrist, T.; Schlueter, J.; Nam, M.S.; et al. Phase Diagram for Light-Induced Superconductivity in κ-(ET) 2- X. *Phys. Rev. Lett.* **2021**, *127*, 197002. [CrossRef]
177. Khomskii, D.I. Spin chirality and nontrivial charge dynamics in frustrated Mott insulators: Spontaneous currents and charge redistribution. *J. Phys. Condens. Matter* **2010**, *22*, 164209. [CrossRef]
178. Mi, X.; Hou, D.; Wang, X.; Liu, C.; Xiong, Z.; Li, H.; Wang, A.; Chai, Y.; Qi, Y.; Li, W.; et al. Observation of Ferroelectricity in the Kitaev Paramagnetic State of α-RuCl$_3$. *arXiv* **2022**, arXiv:2205.09530.
179. Vojta, M. Lattice symmetry breaking in cuprate superconductors: stripes, nematics, and superconductivity. *Adv. Phys.* **2009**, *58*, 699–820. [CrossRef]
180. Mackenzie, A.P.; Thomas Scaffidi, C.W.H.; Maeno, Y. Even odder after twenty-three years: The superconducting order parameter puzzle of Sr$_2$RuO$_4$. *npj Quantum Mater.* **2017**, *40*, 40. [CrossRef]
181. Wosnitza, J.; Zvyagin, S.A.; Zherlitsyn, S. Frustrated magnets in high magnetic fields—Selected examples. *Rep. Prog. Phys.* **2016**, *79*, 074504. [CrossRef]
182. Spaldin, N.A.; Ramesh, R. Advances in magnetoelectric multiferroics. *Nat. Mater.* **2019**, *18*, 203–212. [CrossRef]
183. Sunko, V.; McGuinness, P.H.; Chang, C.S.; Zhakina, E.; Khim, S.; Dreyer, C.E.; Konczykowski, M.; Borrmann, H.; Moll, P.J.W.; König, M.; et al. Controlled Introduction of Defects to Delafossite Metals by Electron Irradiation. *Phys. Rev. X* **2020**, *10*, 021018. [CrossRef]
184. Cho, K.; Kończykowski, M.; Teknowijoyo, S.; Tanatar, M.A.; Prozorov, R. Using electron irradiation to probe iron-based superconductors. *Supercond. Sci. Technol.* **2018**, *31*, 064002. [CrossRef]
185. Leroux, M.; Mishra, V.; Ruff, J.P.C.; Claus, H.; Smylie, M.P.; Opagiste, C.; Rodière, P.; Kayani, A.; Gu, G.D.; Tranquada, J.M.; et al. Disorder raises the critical temperature of a cuprate superconductor. *Proc. Natl. Acad. Sci. USA* **2019**, *116*, 10691–10697. [CrossRef]

MDPI
St. Alban-Anlage 66
4052 Basel
Switzerland
Tel. +41 61 683 77 34
Fax +41 61 302 89 18
www.mdpi.com

Crystals Editorial Office
E-mail: crystals@mdpi.com
www.mdpi.com/journal/crystals

www.ingramcontent.com/pod-product-compliance
Lightning Source LLC
LaVergne TN
LVHW070137100526
838202LV00015B/1839